Dictionary of Biology

Everyday Handbooks

Dictionary of Biology

Edwin B. Steen

Barnes & Noble, Publishers • New York
Founded 1873

Library of Congress Catalog Card Number: 70-156104
SBN 389 00333 6

Distributed
In Canada
by McGraw-Hill Company of Canada Ltd.,
Toronto
In Australia and New Zealand
    by Hicks, Smith & Sons Pty. Ltd.,
    Sydney and Wellington
In the United Kingdom, *Europe*,
and South Africa
    by Chapman & Hall Ltd., London

Published in the United States of America
by Barnes & Noble, Publishers
A Division of Amtel, Inc.

Preface

This dictionary, which comprises approximately 12,000 terms, was written to provide students of biology at all levels with a handy and reliable source for the meanings of terms in the field of biology in general and definitions pertinent to the various allied sciences. An effort was made to bring together in one volume not only commonly used terms but also the more specialized and less frequently employed terms, and especially new terms which have been introduced and have come into usage with the recent developments of modern biology.

To incorporate as many terms as possible, the definitions are frequently presented in brief analytical or functional phrases, thus deliberately avoiding the more comprehensive type of treatment appropriate to larger reference works. Taxonomic names, except for some of the major groups, are not included as entries with definitions; however, under their common names, representatives of the principal groups are listed and the groups are characterized.

The author wishes to thank the many scientists who have unknowingly contributed to this *Dictionary of Biology*, especially the writers of scientific papers and the authors of texts in the fields of zoology, botany, anatomy and physiology, and the various subsciences such as embryology, histology, ecology, genetics, endocrinology, parasitology, and others. These writers have provided the basic information from which the author has attempted to arrive at definitions that are concise and accurate.

The author is particularly grateful to all those who have assisted in the creation of this book, especially to Dr. Richard Brewer, Dr. Joseph Engemann, and Mr. Frank Hinds of Western Michigan University for their critical readings and their helpful comments. He is further indebted to Miss Margaret Kohler for her substantive suggestions and emendations and to members of the editorial staff of Barnes & Noble, Inc., for their assistance in bringing this project nearly a decade in preparation to completion.

Using This Dictionary

A dictionary may be used simply to obtain the meaning of an unknown term. However, its value can be greatly enhanced if it is also utilized as a text. Since this dictionary lists terms in all the principal subdivisions of biology, the reading of entries on a single page will provide the reader with a knowledge of new words and will usually reward him with new information in one or several fields. A further systematized study of words and their meanings will greatly broaden his general knowledge and increase his understanding and appreciation of the world about him.

Following good dictionary style, all entries are listed in alphabetical order. Boldface numerals (**1**, **2**) separate multi-sense entries; boldface letters (**a**, **b**) identify coordinate subsenses. Subject labels (*bot.*, *zool.*) are often abbreviated and are set in italic. Cross references are identified by the words *see* or *compare* set in italic. A word appearing in small capital letters (TRIPTON) signifies that that word is itself defined at the appropriate alphabetical point in the book. A word may be followed by *q.v.* (*quod vide*) if the author feels that the user would find it especially helpful to inform himself of that word's meaning.

The following are standard abbreviations that are used in this dictionary.

anat.	anatomy	*ethnol.*	ethnology
bacteriol.	bacteriology	*entomol.*	entomology
biochem.	biochemistry	*med.*	medicine
biol.	biology	*microbiol.*	microbiology
bot.	botany	*ornith.*	ornithology
chem.	chemistry	*pathol.*	pathology
ecol.	ecology	*physiol.*	physiology
embryol.	embryology	*zool.*	zoology

A

aardvark. A solitary, nocturnal, burrowing mammal (*Orycter-opus afer*) of Africa, the single representative of the order Tubulidentata; also called ant bear or earth pig.

aardwolf. A hyenalike mammal (*Proteles cristata*) of Africa.

abalone. An edible, gastropod mollusk (*Haliotis*) with a broad, oval foot by which it clings to rocks; its shell is the source of mother-of-pearl.

abaxial. 1. *zool.* Outside of or directed away from the main or central axis. **2.** *bot.* **a.** Facing away from the stem, relative to a leaf surface. **b.** Lying outside a seed axis, relative to an embryo.

abdomen. 1. In mammals, the region between the thorax and the pelvis; the belly or its cavity; in other vertebrates, the corresponding region. **2.** In arthropods, the region usually segmented posterior to the thorax.

abducens. In vertebrates, the sixth cranial nerve consisting principally of motor fibers terminating in the lateral rectus muscle of the eye.

abduct. To draw away from a median plane or axis.

abductor. A muscle which abducts a part.

aberrant. Deviating from type.

abient. Turning away from or avoiding a source of stimulation.

abiogenesis. The theory that living things can originate from nonliving material: SPONTANEOUS GENERATION.

abioseston. Nonliving components of the seston: TRIPTON.

ablastin. An antibodylike substance in rats which inhibits the multiplication of trypanosomes.

ablation. Removal of a part as by amputation or excision.

abomasum. The fourth or true stomach of a ruminant; also called rennet.

aboral. Away from or opposite the mouth.

abort. **1.** To miscarry or give birth prematurely. **2.** To check or fall short of full development. **3.** To remain rudimentary or fail to reach maturity.

abranchiate. Without gills.

abrupt. Terminating suddenly as though cut off.

abscission. **1.** Removal of a part by cutting. **2.** *bot.* Separation of a leaf from a plant by dissolution of the cementing material between cells in the abscission layer.

abscission layer. A layer of thin-walled cells at the base of leaf petioles or branches in certain woody plants, dissolution of which results in separation of the leaf, branch, or fruit from the stem.

absolute. **1.** Free from impurities or imperfections; perfect. **2.** Not related to other factors, as *absolute* growth.

absorb. **1.** To take in or imbibe, as fluids or gases through a cell membrane. **2.** To take a substance, as water or nutrients, into the body through the skin or mucous membranes, or, in plants, through root hairs. **3.** To take in and stop radiant energy, as heat and ultraviolet rays.

absorption. The act of absorbing.

abstriction. The separation and discharge of a part as in the formation of spores or conidia in various ascomycetes.

abyssal. Pertaining to the ocean depths.

abyssal realm. The portion of the marine realm below the level to which light penetrates, including all waters below a level of 600 feet.

abyssobenthos. All organisms of the abyssal zone.

acantha. A thorn or spine.

acanthaceous. Possessing or bearing spines.

acanthella. A stage following the acanthor.

acanthocephalan. A parasitic, spiny-headed worm of the phylum Acanthocephala inhabiting the intestines of vertebrates. Immature stages develop in arthropods.

acanthoid. Spinelike.

acanthor. The first larval stage of an acanthocephalan usually found in insect larvae.

acariasis. Infestation with ticks or mites.

acarid. Any arachnid of the order Acarina, which includes the ticks and mites.

acarology. The study of ticks and mites.

acarpic. Lacking fruit.

acarpotropic. Not shedding its fruit.

acaulescent. Lacking or apparently lacking a stem.

accentor. A hedge sparrow of the family Prunellidae, as *Prunella modularis* of Great Britain.

acceptor. A substance which accepts or combines with a product of a chemical reaction, especially a substance such as a cytochrome or a vitamin which accepts and transfers hydrogen atoms in cellular respiration. *Compare* donor.

accessory. Accompanying as a subordinate; aiding or contributing in a secondary way; additional.

accessory nerve. *See* spinal accessory nerve.

acclimate. To acclimatize.

acclimatization. The act or process of becoming adjusted to or tolerant of a new and different climate.

accommodation. 1. Adjustment or adaptation of an organ, structure, or an organism to a particular condition. 2. *physiol.* The automatic adjustment of the eye for viewing objects at different distances, accomplished in man by change in shape of the lens.

accresent. Increasing in size with age, as growth of the calyx after pollination.

accretion. Growth or increase in size by external addition.

accumbent. Reclining; lying against another body.

acellular. Lacking in cells; not composed of cells.

acentric. Lacking a centromere, with reference to a chromosome or chromosome fragment.

acephalous. Lacking a head.

acerose. Sharp-pointed; needlelike.

acerous. Lacking horns, tentacles, or antennae.

acervulus. A fruiting body in certain fungi in which conidia and conidiophores are produced.

acetabulum. 1. The cup-shaped socket on the lateral surface of the hipbone of tetrapods. **2.** The ventral sucker of a trematode. **3.** The genital sucker of a mite.

acetification. The production of vinegar (acetic acid, CH_3COOH) from alcohol by aerobic, acetic acid bacteria of the genus *Acetobacter*.

acetylcholine. A neurohumoral substance produced at the ends of cholinergic nerve fibers and at the motor end plates in skeletal muscle. It is an acetyl ester of choline and acts as a transmitter agent.

achaetous or **achetous.** Without setae or bristles.

achene or **akene.** A simple, one-seeded, indehiscent fruit in which the seed is attached to the pericarp at only one point, as in the sunflower, buttercup, and sycamore.

achilary. Lacking a lip.

achlamydeous. Lacking a calyx or corolla.

achromatic. 1. Lacking in color. **2.** Not staining readily, with reference to tissues. **3.** Refracting light without decomposing it into its constituent colors, with reference to lenses.

achromatic apparatus. In mitosis, the spindle fibers and cell centers, which do not stain readily. Also called achromatic figure.

acicula pl. **aciculae.** A needlelike spine or bristle.

acicular. Sharp; needlelike.

aciculum. A chitinous, rodlike structure or seta in the parapodia of polychaete annelids.

acid. 1. A sour substance, typically soluble in water, which neutralizes bases to form salts. **2.** A compound containing hydrogen atoms which are capable of being replaced by a positive element or radical. **3.** A substance which gives up protons (H^+) to another substance.

acidophilic. Staining readily with acid stains, as eosin.

acinar. Composed of acini, as an acinar gland; comprising acini, as acinar cells.

acinus pl. **acini.** A saclike secreting unit of a gland, especially one with a narrow lumen. *See* alveolus.

[1]acoelomate. Lacking a coelom.

[2]acoelomate. Any member of the Acoelomata, a group of metazoans comprising the phyla Porifera, Coelenterata, and Ctenophora characterized by lack of a coelum.

acoelous. Lacking a coelom; acoelomate.

acontium pl. **acontia.** One of numerous, threadlike, nematocyst-bearing structures borne at the bases of mesenteries in sea anemones.

acorn. The nut or fruit of the oak tree.

acorn worm. A worm of the group Enteropneusta, *q.v.*

acoustic. Auditory; pertaining to the sense of hearing or sound.

acoustic nerve. *See* vestibulocochlear nerve.

acquired. **1.** Developed as a result of the effects of the environment; not inherited. **2.** Gained from experience and training.

acquired character. *genetics.* A trait or character which is not inherited; one which has developed as a result of the effects of the environment or from use or disuse.

acraniate. Any member of the Acraniata, a group of chordates comprising the subphyla Cephalochordata and Urochordata characterized by absence of a brain and cranium; the protochordates.

acrasin. A steroid substance secreted by the amoebae of acrasiomycetes which induces the movement of cells toward an aggregation center.

acrasiomycete. Any member of the class Acrasiomycetes, comprising the cellular slime molds which possess a unicellular, amoeboid assimilative stage.

acraspedote. Lacking a velum, with reference to medusae.

acrocarpic. Having the fruiting body at the tip of the main axis, as certain mosses.

acrocentric. Having a centromere located at the end of a chromosome.

acrodont. Designating a type of tooth attachment in which sockets are lacking, the teeth being fused to the bone, as in *Sphenodon*.

acrogynous. Developing at the tip of the main shoot, with reference to the archegonia of certain bryophytes. *Compare* anacrogynous.

acromion. A process forming the tip of the spine of the scapula.

acron. The anterior, unsegmented head in embryonic or ancestral arthropods.

acropetal. Ascending; developing successively toward an apex or summit, with reference to leaves or flowers. *Compare* basipetal.

acrophilous. Inhabiting regions of high altitude.

5

acrosome. The anterior portion of a sperm head which forms a caplike structure over the nucleus.

ACTH. Adrenocorticotrophic hormone, *q.v.*

actin. A muscle protein which combines with myosin to form a contractile protein complex, actomyosin.

actiniarian. A zoantharian coelenterate of the order Actiniaria which includes the sea anemones.

actinomorphic. Regular; radially symmetrical, with reference to flowers. *Compare* zygomorphic.

actinostele. A protostele with radial arrangement of xylem and phloem.

actinotrichium pl. **actinotrichia.** A minute, horny rod, a number of which develop at the tips of fins in certain bony fishes.

actinotroch larva. In the phylum Phoronida, an elongated, ciliated trochophorelike larva.

actinula. The creeping larval stage of certain tubularian hydrozoans.

action potential. A temporary change in potential which occurs across the surface membrane of a cell, especially a muscle or nerve cell following stimulation.

active transport. The transportation or movement of substances through cell membranes against a concentration gradient with the expenditure of energy.

aculeate. Bearing spines or prickles.

aculeus. 1. A prickle or spine, as that of a rose. **2.** A sting or stinger, as that of a honeybee.

acumen. The subtriangular apex of the rostrum in certain crustaceans.

acuminate. Tapering to a point.

acute. Sharp at the end, as an acute leaf.

acyclic. 1. Not occurring in cycles. **2.** *bot.* **a.** Not arranged in circles or whorls. **b.** Designating a flower with parts inserted spirally on the receptacle.

adactyl. Lacking fingers or toes.

adaptation. 1. Modification of an organism in structure or function in adjusting to a new condition or environment. **2.** *physiol.* **a.** The adjustment of the eyes for vision in dim light (dark adaptation) or vision in bright light (light adaptation). **b.** Reduced responsiveness of a sense organ following repeated stimulation.

adaptive radiation. The development from a single ancestral group of a variety of forms adapted for various types of environments, as walking, burrowing, flying, climbing, leaping, and swimming mammals.

adaxial. 1. Facing or directed towards an axis. **2.** *bot.* Facing the stem, with reference to a leaf surface; ventral.

addax. A spiral-horned antelope (*Addax nasomaculata*) of the Sahara.

adder. 1. One of a number of venomous snakes, as the puff adder (*Bitis arietans*) of Africa and the European viper or adder (*Vipera berus*) of the family Viperidae. **2.** In the United States, one of a number of harmless snakes, as the hognose snake.

adduct. To draw toward a median plane or axis.

adductor. That which adducts, especially a muscle.

adenohypophysis. The portion of the pituitary gland which develops from Rathke's pouch. It comprises the pars tuberalis and pars distalis of the anterior lobe, and the pars intermedia.

adenoids. A mass of hypertrophied pharyngeal tonsil sometimes blocking the posterior nares.

adenosine monophosphate. A compound composed of adenine, d-ribose, and phosphoric acid, of importance in the release of energy for cellular activity. Also called AMP, adenylic acid, adenine ribotide.

adenosine triphosphate. A compound present in all cells which provides energy derived from food or sunlight for energy-expending processes, as protein synthesis, muscle contraction, impulse conduction, and glandular secretion. Upon hydrolysis, it yields adenosine diphosphate (ADP) and a simple phosphate — also called ATP.

adhesive. Sticky; tending to adhere, as *adhesive* cells, glands, and pads of various invertebrates.

adient. Turning toward or approaching a source of stimulation.

adipose. Fatty; fatlike.

adipose fin. In fishes, a fleshy fin located posterior to the dorsal fin.

adipose tissue. A form of connective tissue consisting of groups of fat cells supported by areolar tissue.

adiposis. Excessive accumulation of fat; obesity.

aditus. *anat.* An entryway.

7

adjuvant. An agent which helps, enhances, or increases the effectiveness of another substance, as a medicine or medical treatment.

adnate. United or fused to an organ or structure of a different type.

adnation. *bot.* The fusion of members of different floral circles.

adneural gland. *See* subneural gland.

adolescariae. The cercariae and metacercariae of flukes. *Compare* marita, parthenita.

adoral. Near or toward the mouth. *Compare* aboral.

ADP. Adenosine diphosphate. *See* adenosine triphosphate.

adrectal. Near or close to the rectum.

adrenal gland. In vertebrates, a gland of internal secretion located near the kidney. In mammals, it consists of an outer cortex of mesodermal origin and a medulla of ectodermal origin. The cortex secretes a number of steroid hormones (sex hormones, glucocorticoids, mineralocorticoids); the medulla secretes epinephrine and norepinephrine; also called suprarenal gland.

adrenergic. Denoting nerve fibers which produce norepinephrine at their endings.

adrenocortical. Pertaining to the adrenal cortex.

adrenocorticotrophic hormone. A hormone secreted by the anterior lobe of the hypophysis which stimulates secretion by the adrenal cortex, especially the secretion of glucocorticoids — also called ACTH.

adsorption. The ability to attract and concentrate upon surfaces molecules of gases, liquids, and dissolved solids, as the *adsorption* of gas or smoke particles by powdered charcoal.

adult. A fully grown, sexually mature individual.

adventitious. 1. Additional; accidental; foreign; casual. **2.** Occurring in unusual or abnormal positions, as *adventitious* roots or buds.

aecium pl. **aecia.** A cuplike structure in rust fungi which produces asexual spores (aeciospores).

aedeagus. The male copulatory organ of insects.

aeration. 1. Exposure to air. **2.** Impregnation of a fluid with air or oxygen. **3.** Oxygenation of the blood in the lungs.

aerator. A device for supplying air or oxygen.

aerenchyma. Parenchyma containing air spaces present in the cortex of roots and leaves of various aquatic plants.

aerial. 1. Of or pertaining to air. **2.** Living in or frequenting the air. **3.** Occurring in, produced by, or found in air.

aerial roots. Roots which develop partly or entirely above the ground, as the roots of orchids.

aerobe. An organism which requires free oxygen for maintenance of life.

aerobic. Pertaining to life in or conditions requiring free oxygen.

aerobiology. The study of organisms of their products (spores, pollen) present in the atmosphere.

aerobiosis. Life necessitating the presence of air or free oxygen.

aerophyte. A plant which thrives in air, securing its water from rain or dew, as an orchid.

aeroplankton. Collectively, the microorganisms present in air.

aesthetasc. An olfactory hair on the antennule of certain crustaceans.

aesthete. A sense organ, especially in an invertebrate, as those present in chitons and certain arthropods.

aestival. *See* estival.

aestivation. *See* estivation.

aethalium. A sessile, rounded fruiting body in certain slime molds formed by the heaping up of the plasmodium into a rounded mass.

afferent. Conveying or conducting toward, as *afferent* blood vessels.

afoliate. Lacking leaves: APHYLLOUS.

afterbirth. The placenta and associated structures which are expelled following parturition.

afterdischarge. 1. In receptors, the continuation of the discharge of impulses after stimulation has ceased. **2.** In reflexes, the continuation of the motor response after stimulation of the receptor has ceased.

afterimage. Persistence of the sensation of light after cessation of the light stimulus. In color vision, the afterimage may be positive (of the same color) or negative (of the complementary color).

afterripening. A period of dormancy exhibited by certain plants (pine, apple, pear) in which germination fails to occur under ordinary germinative conditions, the seeds requiring a period of time and specific conditions such as low temperature in order to bring about proper physiological conditions for the development of the embryo. The period varies in length from days to years.

aftershaft. An accessory feather attached at the junction of the rachis and quill.

agamete. A single cell other than a gamete which is capable of developing into a new organism.

agamic. Asexual; reproduction without union of sex cells; agamous.

agamogenesis. ASEXUAL REPRODUCTION.

agamogony. Asexual reproduction as occurs in protozoa: SCHIZOGONY.

agar. A non-nitrogenous, gumlike material obtained from certain seaweeds, especially *Gelidium corneum*, a red alga, and used extensively in the preparation of culture media in microbiology and as a stabilizer of emulsions.

agaric. A gill fungus of the order Agaricales; a mushroom or toadstool.

agastric. Lacking a digestive tract or cavity.

age. 1. The time from birth to the present of any living individual. 2. A particular period of life or development, as old *age*, the *Age* of Fishes, the Stone *Age*.

agenesis or **agenesia.** 1. A condition in which a structure fails to develop or development is incomplete; aplasia. 2. Inability to produce offspring.

ageusia. Loss or impairment of the sense of taste.

agger. *anat.* A projection or prominence.

agglomerate. 1. To group or gather together into a ball or mass, as in certain protozoans. 2. Gathered together into a dense cluster, as in the heads of certain flowers.

agglutinate. To cluster and adhere together; to collect in masses.

agglutinin. An antibody present in normal or immune serum which brings about agglutination of its antigen.

agglutinogen. A substance which when injected into an animal induces the development of a specific agglutinin.

aggregate. Clustered together to form a dense mass or head, as an *aggregate* fruit.

aglomerular. Lacking glomeruli, with reference to kidneys or kidney tubules.

aglyphic. Having solid, conical teeth, with reference to snakes.

agnathous. 1. Lacking jaws. 2. Pertaining to the Agnatha, a class of primitive, jawless vertebrates which includes the ostracoderms and cyclostomes.

agnosia. Loss of the ability to perceive or recognize persons or things.

agonisis. CERTATION.

agonist. A primary muscle which, upon contracting, moves a part.

¹agouti. A small South American rodent (*Dasyprocta*) of the family Dasyproctidae.

²agouti. A pattern of coloration common in wild animals, especially rodents, in which individual hairs are mostly black with a narrow yellow band near the tip.

agranulocyte. A nongranular leukocyte (lymphocyte or monocyte).

agrostology. The study of grasses.

agynic. 1. Lacking pistils. 2. Free from the ovary, said of pistils.

ai. The three-toed sloth (*Bradypus tridactylus*).

aianthous. Blooming continuously; having persistent flowers.

aigrette. 1. *ornith.* A plume or tuft, especially the beautiful plume borne by egrets during the breeding season. 2. *bot.* The feathery crown of a seed, as that of a dandelion.

air bladder. 1. In fishes, the swim bladder. 2. In certain seaweeds, an air-filled chamber which serves as a float.

air sac. 1. In birds, one of a number of thin-walled sacs attached to the smaller bronchial tubes, some extending into the visceral spaces and some continuing as air cavities into the larger bones. 2. In insects, a thin-walled, distensible dilatation of a tracheal tube. 3. In mammals, an alveolar sac of the lungs.

airway. *ornith.* A well-established route followed by migratory birds, especially waterfowl, in their seasonal migrations.

akaryote. 1. Lacking a nucleus. 2. A non-nucleated cell.

akene. *See* achene.

akinete. A sporelike resting stage in certain green algae in which the newly formed cell is fused to the wall of the parent cell.

11

ala pl. **alae. 1.** A wing or winglike process. **2.** A longitudinal fin or finlike structure present on certain nematodes.

alar. Pertaining to, resembling, or shaped like a wing.

alate. 1. Bearing wings or alae. **2.** *bot.* Bearing winglike processes.

albacore. The tuna or tunny, *q.v.*

albatross. A large, soaring, oceanic bird of the family Diomedeidae, as *Diomeda nigripes*, the black-footed albatross.

albino. 1. A person or animal with a congenital deficiency in natural pigment. **2.** *bot.* **a.** A plant in which chloroplasts are lacking. **b.** A plant with white instead of colored flowers.

albumen. 1. Egg white, consisting principally of albumin. **2.** Stored food in the seed of a plant. **3.** ALBUMIN.

albumin. One of a group of simple proteins present in blood serum, milk, muscle, and other animal tissues; also found in some plant tissues and fluids. Examples are ovalbumin, serum albumin, lactalbumin.

albuminous. Containing endosperm at maturity, with reference to seeds.

alcohol. An organic compound containing a functional hydroxyl (OH) group. Grain alcohol or ethanol (C_2H_5OH) is a colorless, volatile, inflammable liquid produced in alcoholic fermentation. It is also obtained from distillation of fermented grain or saccharine liquids or produced synthetically. In biology it is widely used as a solvent, preservative, fixative, and dehydrating agent. *See* methyl alcohol.

alcyonarian. Any anthozoan of the subclass Alcyonaria which includes the precious corals, sea fans, and sea feathers.

alder. A tree or shrub of the genus *Alnus*, family Corylaceae.

alderfly. An insect of the order Megaloptera (sometimes considered as a suborder of Neuroptera) with carnivorous, aquatic larvae. Most species belong to the genus *Sialis*.

aldosterone. A steroid hormone (a mineralocorticoid) secreted by the adrenal cortex.

alecithal. Possessing little or no yolk, said of ova as those of placental mammals.

alepidote. Lacking scales.

aleppo galls. Galls which form on twigs of the aleppo oak, a source of tannin.

aleurone. Protein occurring as small particles in the endosperm of seeds of various plants, especially cereals.

alewife. A clupeid fish (*Pomolobus pseudoharengus*), an anadromous fish of the Atlantic coast and also common in the Great Lakes.

alfalfa. A leguminous forage plant (*Medicago sativa*).

alga pl. **algae.** Any of a number of simple plants of the categories Monera and Protista possessing chlorophyll and capable of carrying on photosynthesis. *See* Algae.

Algae. Formerly a subdivision of the division Thallophyta comprising plants with an undifferentiated body and cells containing chlorophyll. The group has been divided into a number of divisions as follows: Cyanophyta (blue-green algae), Pyrrophyta (fire algae), Chrysophyta (golden algae and diatoms), Phaeophyta (brown algae), Rhodophyta (red algae), Euglenophyta (euglenids), Xanthophyta (yellow-green algae and chloromonads), and Chlorophyta (green algae).

algin. Sodium alginate, a purified carbohydrate obtained from brown seaweeds, widely used for its water-binding, thickening, and emulsifying properties.

alginic acid. A gelatinous polysaccharide obtained from marine algae of the *Fucus* type, used as an emulsifying agent.

aliferous. Possessing or bearing wings.

aliform. Resembling a wing.

alima. The larval stage of stomatopod crustaceans.

alimentary. 1. Pertaining to food or nutrition. 2. Nourishing or nutritious.

alimentary canal. The digestive tract from mouth to anus; the gastrointestinal tract.

alkali. A substance which neutralizes an acid, usually a metallic hydroxide; a base.

alkaloid. Any of a group of naturally occurring, nitrogen-containing compounds usually of plant origin. They resemble alkalies, are bitter in taste, and most are active physiologically, some being extremely poisonous. Many are of medical importance, as atropine, ergotamine, ephedrine, reserpine, caffeine, morphine, quinine, and strychnine.

alkalosis. A condition in which there is an increase in alkaline substances in the blood, especially sodium bicarbonate.

13

alkaptonuria. A metabolic condition characterized by the presence of alkapton (homogentisic acid) in the urine; inherited as a mendelian recessive trait.

allantois. In amniotes, an extra embryonic membrane which develops as a ventral outgrowth of the hindgut. In birds and reptiles, it forms a large sac which functions in respiration and excretion; in some mammals, it forms a part of the chorio-allantoic placenta; in man, it is a small vestigial tubelike structure.

allele. One of a pair of genes which occupy corresponding loci in homologous chromosomes. The term is also applied to traits associated with the genes.

allelomorph. An allele.

allergen. An agent capable of inducing a state of allergy, as pollen, drugs, dust, animal hair, and certain foods, especially proteins.

allergy. An antigen-antibody reaction occurring in certain individuals upon exposure to substances which under similar conditions are innocuous to other individuals. The exciting agents, called allergens, act as antigens causing the production of specific antibodies in sensitive persons. Common manifestations are hay fever, asthma, urticaria, gastrointestinal disturbances, and serum sickness.

alligator. A large reptile of the genus *Alligator* resembling a crocodile but differing in possession of a shorter and broader snout. *A. mississippiensis* inhabits North America; *A. sinensis* inhabits China.

alligator gar. The gar, *q.v.*

allochronic. Not occurring at the same time level; not contemporary, said of species.

allochthonous. 1. Brought in from an outside source, as sediment in a lake. 2. Exotic; imported or having migrated from another area, said of species.

allocryptic. Concealing; designating animals which cover themselves for concealment.

allogamy. *bot.* CROSS-FERTILIZATION. *Compare* autogamy.

allograph. A homograph.

allometry. 1. The phenomena of relative growth; heterauxesis. 2. A study of changes in shape or proportions with increase in size.

allopatric. Not occurring together or inhabiting the same area.

allopelagic. Pertaining to aquatic organisms found at various depths.

allopolyploid. A polyploid arising from hybridization of two species or genera with resulting duplication of chromosome complement.

all-or-none law. A principle in physiology to the effect that an irritable structure, when stimulated, responds to its fullest extent or not at all. It applies to skeletal muscle (a single fiber), cardiac muscle (the heart as a whole), and a single nerve fiber.

allosematic. Having protective marks or coloration resembling that of dangerous or distasteful forms. *See* mimicry.

allotropous. **1.** Having nectar readily available to insects, said of flowers. **2.** Not restricted to certain kinds of flowers, said of insects.

allotype. A paratype of the opposite sex.

alpaca. A domestic form of the llama bred for its fine wool.

alpine. **1.** Pertaining to or inhabiting the Alps. **2.** Pertaining to habitats at high altitudes, especially those above the timberline.

alternate. *bot.* Having parts not opposite or whorled, as *alternate* leaves.

alternation of generations. The alternation of a sexual with an asexual generation; heterogenesis; metagenesis.

altricial. *ornith.* Designating birds whose young are hatched in an immature and helpless condition and remain in the nest for a period of time. *See* nidicolous; *compare* precocial.

alula. **1.** In birds, the spurious or bastard wing; the feathered thumb. **2.** In insects, a calypter.

alveolar. Of, pertaining to, or composed of alveoli.

alveolate. Pitted and having the appearance of a honeycomb.

alveolus pl. **alveoli.** **1.** *anat.* A small pit, hollow, or cavity, as **a.** the socket of a tooth; **b.** an air cell of a lung; **c.** the terminal secreting portion of an alveolar gland. **2.** *bot.* A pit or depression with angular sides.

amanita. A white-spored fungus of the genus *Amanita*, as *A. muscaria*, the fly agaric or "Death Angel," and *A. phalloides*, both highly poisonous mushrooms.

amaurotic. Pertaining to or affected by blindness.

amazon. **1.** A slave-making ant of the genus *Polyergus*. **2.** A South American parrot of the genus *Amazona*.

ambergris. A concretion from the intestinal tract of the sperm whale found floating in tropical waters, used in perfumery for the fixing of odors.

ambidextrous. Using both hands equally well.

ambilateral. Pertaining to or affecting both sides.

ambrosia. **1.** Food prepared by fungus beetles (ambrosia beetles) of the genus *Xyleborus* for their young. *See* fungus garden. **2.** The common ragweed (*Ambrosia artemisii folia*), its pollen being a common allergen.

ambulacral area. One of five radially disposed sections of plates in echinoid echinoderms through which tube feet extend.

ambulacral groove. A groove on the ventral surface of each ray of a sea star which contains the podia or tube feet.

ambulatory. Capable of walking or moving from place to place.

ameba or **amoeba.** A simple unicellular or acellular organism found in fresh or marine water or living parasitically in animals. Each is a naked bit of protoplasm moving by means of pseudopodia and reproducing by simple fission. Common genera are *Amoeba* and *Entamoeba* of the class or superclass Sarcodina.

amebiasis. Infection by parasitic amebae, especially *Entamoeba histolytica*.

amebocyte. A primitive, connective tissue cell capable of independent amebalike movement present in the tissues or body fluids of various invertebrates; any ameboid cell.

ameboid. Resembling an ameba in form, structure, or movement.

amebula. An amebalike stage in the life cycle of various parasitic protozoans.

ameloblast. An enamel-forming cell in the enamel organ of a developing tooth.

ament or **amentum.** A catkin.

amentaceous. Bearing catkins.

ameristic. Not segmented.

ametabolous. Without metamorphosis, with reference to development of insects.

amictic. Designating a mode of development in rotifers in which females produce thin-shelled, diploid ova which develop by parthenogenesis into *amictic* females.

amino acid. One of a large group of organic compounds characterized by the presence of both an amino (NH_2) radical and a carboxyl (COOH) radical. They are the building blocks for the synthesis of proteins and the end products of protein digestion.

amitosis. Simple cell division in which the nucleus and cytoplasm divide by constriction without the formation of a mitotic figure. *Compare* mitosis.

amixis. Failure of interbreeding to occur between different races or species.

ammochaeta. A bristle on the head of desert ants used in cleaning sand from the anterior legs.

ammocoetes larva. The larva of the sea lamprey, which develops in freshwater streams, then migrates to the sea or certain inland lakes.

ammonia. NH_3, a colorless, pungent gas readily soluble in water with which it combines to form ammonium hydroxide (NH_4OH). It is formed in the oxidative deamination of amino acids.

ammonification. 1. Impregnation with ammonia. **2.** The conversion of complex nitrogenous compounds into simpler compounds and ultimately ammonia by hydrolysis.

ammonite. A fossil mollusk of the order Ammonoidea, class Cephalopoda, common in the Mesozoic era.

ammonotelic. Excreting ammonia as the principal product of nitrogen metabolism. *Compare* ureotelic, uricotelic.

amnion. 1. In amniotes, an extraembryonic membrane which forms a fluid-filled sac enclosing the embryo. **2.** *entomol.* A membrane covering the germ band with which it is continuous at its margins.

amniote. Any member of the Amniota, higher vertebrates whose embryos during development are enclosed within an amnion. Includes the reptiles, birds, and mammals. *Compare* anamniote.

amoe-. For words beginning with AMOE not found here, see AME-.

amorphous. Lacking definite form or shape; lacking observable differentiation in structure.

AMP. Adenosine monophosphate.

ampherotoky. Production of offspring of both sexes by parthenogenetic animals; amphitoky.

amphibian. Any member of the Amphibia, a class of cold-blooded vertebrates which includes the caecilians, frogs, toads, and salamanders. Larvae are aquatic and breathe by gills; adults breathe by lungs; about 2500 living species.

amphibious. Capable of living both in water and on land.

amphiblastula. The free-swimming larva of certain sponges with flagellated cells at one end and large, nonflagellated cells at the other.

amphicarpous. Producing two kinds of fruit.

amphichrony. Producing flowers of two different colors on the same stem.

amphicoelous. Biconcave, with reference to the centra of vertebrae.

amphicribal. Designating a type of vascular bundle in which the phloem surrounds the xylem.

amphicyte. A satellite cell.

amphid. A chemoreceptor located anteriorly in nematodes. Also called lateral organ.

amphidiploid. ALLOPOLYPLOID.

amphigastria. Reduced leaves or scales on the ventral surface of certain liverworts.

amphigenous. Growing entirely around or over a surface.

amphigony. SEXUAL REPRODUCTION.

amphimixis. True sexual reproduction in which male and female gametes, each with a different set of chromosomes, unite.

amphineuran. A chiton; a mollusk of the class Amphineura.

amphioxus. A lancelet.

amphiphloic. With phloem on both sides of the xylem.

amphipod. A crustacean of the order Amphipoda which includes the sand fleas, beach fleas, and related forms. *Gammarus* and *Hyalella* are common freshwater forms.

amphistome. A fluke of the suborder Amphistomata with acetabulum located at the posterior end of the body.

amphistylic. A method of jaw suspension found in a few primitive elasmobranchs in which the palatopterygoquadrate bar and hyomandibular cartilage are united to the chondrocranium.

amphithecium. The outer layer of cells in the sporangium of certain bryophytes.

amphitoky. AMPHEROTOKY.

amphitrichous. With one or a tuft of flagella at each pole, as in certain bacteria.

amphitropous. *bot.* Having ovary inverted and attached at the middle.

amphivasal. With xylem surrounding or on two sides of the central phloem.

amphoteric. 1. Possessing opposite traits or characters. 2. Capable of acting both as a base and an acid.

amplectant. Clasping or embracing, said of tendrils.

amplexicaul. Clasping or embracing a stem, with reference to leaves.

amplexus. Sexual embrace without true intercourse occurring, as in frogs.

ampulla. 1. *zool.* A membranous sac or vesicle or a dilatation of a tube. 2. *bot.* A bladderlike sac or vesicle, as that attached to the leaves or roots of certain aquatic plants.

amylase. Any amylolytic enzyme which catalyzes the hydrolysis of starch to sugar, as amylopsin, a pancreatic amylase; ptyalin, a salivary amylase.

amyloid. A complex protein deposited in the tissues as a result of degenerative processes, resembling starch in that it stains blue-black upon the addition of iodine.

amylopectin. The relatively insoluble component of starch consisting of a branched chain of glucose units often combined with phosphoric acid, which in solution forms a paste and reacts with iodine to give the characteristic red-violet color. *Compare* amylose.

amyloplastid. A plastid which stores starch.

amylopsin. A pancreatic enzyme which catalyzes the hydrolysis of starch to dextrins and finally to maltose.

amylose. A soluble component of starch consisting of an unbranched chain of glucose units that reacts with iodine to give the characteristic blue-black color. *Compare* amylopectin.

amyotonia. Lack of muscle tone.

anabiosis. A state of suspended animation during which organisms are capable of withstanding extreme desiccation and low temperatures as exhibited by moss-inhabiting rotifers and tardigrades.

anabolism. a. That phase of metabolism which involves synthetic reactions leading to the storage of energy; constructive metabolism. **b.** Conversion of simple substances into more complex compounds; assimilation. *See* metabolism; *compare* catabolism.

anaconda. A large South American python (*Eunectes murinus*) of the family Boidae.

anacrogynous. Not possessing terminal archegonia. *Compare* acrogynous.

anadromous. Ascending rivers to spawn, as salmon, shad, striped bass, and some trout. *Compare* catadromous.

anaerobe. An anaerobic organism.

anaerobic. Capable of growing in the absence of molecular oxygen.

anagenesis. Regeneration of tissues.

anal. Of, pertaining to, or located near the anus.

analgesic. An agent which relieves pain without altering consciousness.

analogous. a. Corresponding to or resembling another structure in function but differing in fundamental structure. **b.** Bearing a superficial resemblance, as the wing of an insect and wing of a bat. *Compare* homologous.

anamniote. Any member of the Anamniota, a group of vertebrates which do not develop an amnion. Includes the cyclostomes, fishes, and amphibians. *Compare* amniote.

anamorphic. Designating a type of development in which there are only gradual changes from one stage to the next. *See* epimorphic, metamorphosis.

anandrous. Having no stamens, said of pistillate flowers.

anaphase. The stage in mitosis or meiosis during which the chromatids or halves of chromosome pairs move from the equatorial plane to their respective poles.

anaphylaxis. A state of excessive sensitivity which develops in an animal following injection of an antigen, as a protein. Upon reinjection of the antigen, marked reactions occur and in some cases, a state of shock (anaphylactic shock) may develop with marked circulatory disturbances and possible death.

anaplasia. Reversion of cells to an embryonic type, as may occur in certain types of tumors.

anapolytic. Designating tapeworms which fail to shed the terminal proglottid when it is ripe.

anarthrous. Lacking a joint or joints.

anastomosis. 1. The union of a number of structures as blood vessels, nerves, or veins of a leaf so that a branching network is formed. **2.** The joining by surgery of two hollow structures.

anastral. Lacking an aster, with reference to the achromatic figure.

anatomy. 1. The science which deals with the structure of animals or plants. **2.** The structural organization or makeup of an organism, or a treatise dealing with the same.

anatoxin. A toxin modified by heat or chemical treatment so that its toxic properties are lost but its antigenic properties retained.

anatropous. *bot.* Inverted, with reference to a developing ovule which is turned so that the micropyle lies close to the funiculus, with chalaza at the opposite end. *Compare* atropous.

ancestrula. In bryozoans, the first zooid which, by budding, gives rise to daughter zooecia.

anchor. A curved hook in the opisthaptor of a monogenetic fluke.

anchor worm. A copepod, *Lernaea* sp., parasitic on fishes and amphibians.

anchovy. Any of a number of small, herringlike, food fishes of the family Engraulidae, as *Engraulis encrasicholus* of the Mediterranean, and *Anchoa mitchilla* of the Atlantic and Gulf coasts of the United States.

ancipital. Having two edges, said of flattened stems.

ancylostomiasis. Hookworm disease resulting from infestation with *Ancylostoma duodenale*.

andalusian fowl. The blue hybrid offspring of a cross between black and white-splashed domestic fowl.

androconia. Specialized scales on the wings of certain male lepidopterans which serve as outlets of odoriferous glands.

androdioecious. Having male (staminate) and perfect flowers on separate plants.

androecium. Collectively, the stamens of a flower.

androgamete. A male gamete or germ cell; a spermatozoan.

androgametophyte. 1. a. The male gametophyte. **b.** The gametophyte which produces microgametes. **2.** In seed plants, the germinated pollen grain.

androgen. a. A substance which promotes development and functioning of male genital organs and development of secondary sexual characteristics. **b.** A masculinizing substance. **c.** a male sex hormone, as testosterone.

androgenesis. Development of a fertilized ovum from which the egg nucleus has been experimentally removed or inactivated. *Compare* pseudogamy.

androgenous. Pertaining to the production of males or male gametes.

androgynous. **1.** *zool.* Hermaphroditic; having characteristics of both sexes. **2.** *bot.* Possessing both staminate and pistillate flowers in the same cluster.

android. Resembling a male.

andromonoecious. Possessing male (staminate) and perfect flowers on the same plant.

androphore. A supporting structure bearing raised stamens.

androsperm. A male-producing spermatozoan; one containing the Y-chromosome.

androsporangium. In seed plants, a sporangium that produces androspores; a microsporangium.

androspore. **1.** In seed plants, a spore that develops into a male gametophyte; a microspore or pollen grain. **2.** In certain algae, a form of zoospore which gives rise to dwarf male plants, as in the Oedogoniales.

androsporophyll. A microsporophyll.

androstane. A steroid, $C_{19}H_{32}$, the parent substance of androgenic hormones.

androsterone. An androgenic steroid present in the urine of men and women, considered to be a catabolic product of testosterone.

androstrobilus. A strobilus which bears microsporangia or pollen sacs.

anecdysis. A condition in which molting ceases and no further growth occurs, as in crustaceans.

anemia. A pathological condition characterized by a marked reduction in circulating red blood cells, or hemoglobin, or both.

anemochorous. Dispersed by the wind.

anemone. **1.** *bot.* A plant of the genus *Anemone* of the crowfoot family (Ranunculaceae). **2.** *zool.* A sea anemone.

anemophilous. Pollinated by the wind.

anemotaxis. A turning of an animal in response to air currents.

anemotropism. Orientation of a plant or a part of it in response to air currents.

anesthesia. Loss of sensation, as that produced by drugs.

anesthetic. 1. Pertaining to or causing anesthesia. 2. *zool.* A drug which reduces motility.

anestrus. In a female mammal, the interval of sexual inactivity between periods of heat (estrus). Also called diestrum.

aneuploidy. *genetics.* A condition in which the chromosome number is not an exact multiple of the basic number in the genome. It includes monosomic, trisomic, and tetrasomic individuals.

aneurysm. A circumscribed dilatation in the wall of an artery forming a soft, pulsating, blood-filled tumor.

angiogenesis. The development of blood vessels.

angiology. The study of blood and lymphatic vessels.

angiolysis. The atrophy or obliteration of functional blood vessels in embryonic or postnatal development.

angioma. A tumor composed principally of blood or lymphatic vessels.

angiosperm. Any flowering plant. *See* anthophyte.

angle gland. A gland at the corner of the mouth in birds.

angler. One of a number of marine fishes of the order Pediculati which bear fleshy appendages on their heads which serve as lures attracting smaller fishes as their prey, as *Lophius*, the fishing frog or goosefish.

anglewing. A small butterfly of the genus *Polygonia*, called the "comma" because of commalike markings.

angleworm. An earthworm used extensively as fish bait.

Angora hair. A hair which grows to a considerable length before being shed, as those of the head.

Angora rabbit. A long-haired domestic breed raised for its fine wool.

angular bone. A dermal bone in the lower jaw of certain vertebrates.

anhinga. The water turkey or snakebird.

anhydrous. Completely lacking in water.

ani. A cuckoolike bird of the genus *Crotophagus* ranging from the United States to Argentina.

animal. Any member of a group of living things distinguished from plants by the following general characteristics: a more compact and definite body form; absence of rigid cell walls of cellulose; power of locomotion and quick response to stimuli; inability to manufacture foods from inorganic substances. *Compare* plant, protist.

animalcule. Any minute or microscopic animal; a protozoan.

animal pole. The region of an ovum which contains the nucleus and the more active portion of the cytoplasm; the region that gives off the polar bodies. *Compare* vegetal pole.

animation. State of being active or alive.

anisodactyl. *ornith.* Having feet with three front toes and one hind toe, the hallux.

anisogamous. Having gametes differing in size and structure; heterogamous.

anisomerous. Having unequal numbers of structures in floral whorls, as four petals and six stamens.

anisophylly. Having leaves of two different types.

anisosporous. HETEROSPOROUS.

ankle. The joint between the leg and the foot; the region of this joint.

ankylosaur. An armored dinosaur.

ankylosis. **1.** A stiffness or immobility of a joint. **2.** The union or fusion of two separate bones.

anlage. A primordium or the first accumulation of embryonic cells which constitutes an identifiable beginning of a structure. *Compare* blastema.

annectant. Linking; connecting together.

annelid. Any member of the phylum Annelida, segmented, coelomate animals which comprise the polychaetes, mostly marine forms as the clam worm; oligochaetes, terrestrial or freshwater forms as the earthworm; hirudineans, parasitic forms as the leech.

annual. **1.** Yearly; occurring once a year or lasting for a year. **2.** *bot.* A plant which completes its life cycle within one year or one growing season.

annual ring. One of the concentric layers of xylem formed in the stem of a woody plant each year, each ring consisting of an inner layer of springwood and an outer layer of summerwood.

annular. Ring-shaped or forming a ring.

annularis. The fourth finger of the hand.

annular ligament. 1. A circular ligament, as that which encloses **a.** the head of the radius; **b.** the footplate of the stapes. **2.** A fibrous membrane separating the cartilage rings of the trachea.

Annulata. Old name for the phylum Annelida.

annulate. 1. Composed of rings or ringlike structures. **2.** An annelid.

annulus. 1. A ring or ring-shaped structure, marking, opening, or space. **2.** In mushrooms, a membranous ring that surrounds the stalk after expansion of the pileus. **3.** A ring of cells at the opening of the capsule in mosses or encircling the sporangium in ferns. **4.** The transverse groove or girdle of a dinoflagellate. **5.** A division of a somite in leeches. **6.** The seminal receptacle of a female crayfish. **7.** A concentric growth ring (circulus) on the scale of a fish.

anodontia. Absence of teeth.

anodynia. Absence of pain.

anole. A lizard of the genus *Anolis*, especially *A. carolinensis*, the common American chameleon.

anomaly. A deviation from normal; an abnormal or irregular structure; a malformation.

anopheline. Of or pertaining to mosquitoes of the genus *Anopheles* which serve as vectors of the causative organism of malaria in man.

anopluran. Any insect of the order Anoplura which includes the sucking lice. *See* louse.

anostracan. Any crustacean of the order Anostraca which includes the brine shrimp and fairy shrimp.

anoxia. HYPOXIA.

anserine. 1. Gooselike. **2.** Of or pertaining to the order Anseriformes which includes the screamers, swans, ducks, and geese.

ant. A hymenopterous insect of the family Formicidae most of which are social insects living in colonies, each colony having serveral castes.

antagonism. Opposition or counteraction as occurs between **a.** organisms, as a mold and inhibiting bacteria; **b.** chemical substances, as a vitamin and antivitamin; **c.** muscles, as a flexor and extensor.

ant bear. 1. The great anteater (*Myrmecophaga jubata*) of South America. **2.** The aardvark.

anteater. One of several mammals which feed principally upon ants, as the great anteater (*Myrmecophaga jubata*) of the order Edentata; the banded anteater (*Myrmecobius fasciatus*), a marsupial; the spiny anteater (*Tachyglossus aculeatus*), a monotreme; the scaly anteater or giant pangolin (*Manis gigantea*) of the order Pholidota.

antebrachium. The forearm from elbow to wrist.

antelope. Any of a number of graceful, rapid-running ruminants of the family Bovidae. Term is applied to all gazelles and related forms (hartebeestes, duikers, waterbucks, sables, bushbuck, kudu) restricted entirely to the Old World.

antenna. An elongated, movable, sensory structure, usually segmented, found on the head of various invertebrates, especially arthropods; a feeler.

antennal gland. A gland which opens on the antenna, as **a.** that of certain insects; **b.** the green gland of crustaceans.

antennule. A small antenna, especially the more anterior and smaller of the two antennae found in certain crustaceans as the crayfish.

antepenultimate. Second from the last segment, as in an insect antenna; third from the end.

anterior. 1. In front of or before. **2.** *zool.* **a.** Toward the forward or head end. **b.** In man and other erect animals, toward the front side of the body or ventral surface. **3.** *bot.* Facing away from the axis; inferior or lower, as the anterior side of a leaf. *Compare* posterior.

anterior cavity. In a vertebrate eye, the space containing aqueous humor located between the lens and cornea, subdivided into the anterior chamber, between cornea and iris, and posterior chamber, between iris and lens.

anterior root. The ventral root of a spinal nerve composed of efferent fibers.

anthelate. Having elongated, flower-bearing branches, as in certain rushes.

anthelmintic. A drug used to treat any worm infestation, especially one used to expel worms from the gastrointestinal tract. *See* vermicide.

anther. The part of a stamen in which pollen develops. *See* stamen.

antheridiophore. In hepatics, a specialized branch which bears antheridia.

antheridium pl. **antheridia.** In nonvascular plants, the structure (gametangium) which produces male gametes.

antheriferous. Bearing anthers.

antherozooid or **antherozoid.** A small, flagellated, male gamete found in certain algae, mosses, and ferns; a spermatozoid.

anthesis. The time of flowering or coming into full bloom.

anthocarpous. Pertaining to aggregated fruits with accessory structures called pseudocarps, as in the strawberry.

anthocaulus. In solitary corals, the stalk after separation of the anthocyathus.

anthocodium. The oral end of an alcyonarian coral.

anthocyanin. A water-soluble blue or purple pigment present in the cell sap of various plants. They are natural indicators, being red in acid solutions, blue in alkaline.

anthocyathus. In the young of solitary corals, the disklike terminal portion which separates from the stalk or anthocaulus.

anthodium. A composite flower; head; capitulum.

anthogenesis. The production of both males and females by parthenogenic females, as in aphids.

anthophilous. Attracted to flowers; feeding upon flowers.

anthophore. In certain flowers, the elongated portion of the receptacle between calyx and corolla, as in pinks.

anthophyte. Any member of the Anthophyta, a division of the plant kingdom which includes all flowering plants, vascular plants in which double fertilization occurs resulting in development of a fruit containing seeds. It comprises two classes, the Monocotyledonae and Dicotyledonae.

anthotaxy. The arrangement of flowers in a cluster. *See* inflorescence.

anthozoan. Any coelenterate of the class Anthozoa which includes the sea anemones and corals.

anthropogenic. Caused by or resulting from the acts of man.

[1]**anthropoid.** Manlike.

[2]**anthropoid.** Any primate of the suborder Anthropoidea which includes the monkeys, apes, and man.

27

anthropoid ape. One of the large, tailless apes of the family Pongidae, as the gibbon, chimpanzee, orangutan, or gorilla.

anthropology. The science which deals with man.

anthropomorphic. Attributing human attributes to nonhuman beings or objects.

antibiosis. An association between two organisms in which one produces a substance or a condition which is harmful to the other.

antibiotic. A substance which, in low concentrations, inhibits the growth of microorganisms. The term is generally restricted to substances produced by living organisms, as penicillin. *See* spectrum.

antibody. One of a group of immune substances present naturally or formed within an organism as the result of the presence of an antigen. Types include antitoxins, agglutinins, precipitins, lysins, and opsonins.

anticlinal. Radial; at right angles to the surface of a part.

anticoagulant. A substance which prevents or delays coagulation of blood, as sodium citrate.

antidiuretic. Opposing or preventing the formation and excretion of urine.

antidiuretic hormone. Vasopressin — also called ADH.

antidromic. Moving in a direction opposite to normal, as nerve impulses travelling peripherally in afferent fibers.

antidromous. *bot.* Spiraling in a reverse direction.

antigen. **1.** A substance which will stimulate the production of antibodies and in turn react with them. **2.** A substance which is not antigenic but will react visibly with antibodies *in vitro*.

antimere. One of a pair of corresponding parts in a bilaterally symmetrical animal.

antimetabolite. A substance which interferes with normal cellular metabolism, especially one which inhibits or blocks a specific step in a metabolic pathway. Antimetabolites include sulfonamides, antibiotics, antihistamines, and various herbicides and insecticides.

antipodal. Located on the opposite side.

antipodal cells. *bot.* Three cells in an embryo sac located opposite to the micropylar end.

antisepsis. Prevention of decay, putrefaction, or sepsis by exclusion of microorganisms or inhibition of their growth and multiplication.

¹**antiseptic.** Checking decay or putrefaction.

²**antiseptic.** An agent which inhibits growth of microorganisms. *See* bactericide, germicide, disinfectant.

antiserum. An immune serum prepared from the blood of man or an animal which has been immunized against bacteria or other antigenic agents and which contains antibodies specific for these agents.

antithrombin. A substance in blood plasma which neutralizes thrombin, thus preventing coagulation.

antitoxin. A substance which neutralizes a toxin, especially an antibody which neutralizes or inactivates the toxin that serves as its antigen.

antler. One of the paired, branched horns of members of the deer family, usually possessed by males only and shed annually.

ant lion. A neuropterous insect (*Dendroleon obsoletum*) whose larvae, called doodle bugs, conceal themselves at the bottom of a cone-shaped pit where they sieze their prey by long, sickle-shaped jaws.

antrorse. Directed upward and forward. *Compare* retrorse.

antrum. 1. A hollow space or cavity, especially one in bone. **2.** The cavity of an ovarian follicle.

antrum of Highmore. The maxillary sinus.

anuclear. Lacking a nucleus, as an erythrocyte.

anuran. A salientian; a frog or toad.

anuria. Inability to form or excrete urine.

anus. 1. The posterior opening of the alimentary canal. **2.** The cloacal aperture.

anvil. The incus; the middle ear bone.

aorta. 1. In vertebrates, the main trunk vessel which carries blood from the heart to the tissues. In fishes, the *ventral aorta* carries blood from the heart to the gills; the *dorsal aorta*, from the gills to the tissues. **2.** In various invertebrates, the main vessel carrying blood from the heart.

aortic arch. 1. In vertebrate embryos, one of a series of paired blood vessels which encircle the pharynx connecting the ventral and dorsal aortae. **2.** In aquatic vertebrates, one of a series of paired afferent and efferent branchial arteries.

aortic ring. A ring of dense fibrous connective tissue at the orifice of the aorta in the heart of certain mammals, sometimes containing a heterotopic bone, the ossa cordis, as in cattle.

apara. The three-banded armadillo (*Tolypeutes tricinctus*) of South America.

ape. 1. In a broad sense, any monkey. **2.** One of the large, tailless, semierect primates of the family Pongidae. *See* anthropoid ape.

aperture. An opening or orifice.

apetalous. Lacking petals.

apex. a. The summit or top; the vertex or summit; **b.** the tip of a conelike structure; the end opposite the base.

aphagia. Inability to swallow.

aphasia. Inability to use words as symbols of ideas.

aphid. A plant louse; a small, sucking, homopterous insect of the family Aphidae.

aphis lion. The larva of the common lacewing insect (*Chrysopa oculata*); feeds principally upon aphids. Also called aphid lion.

aphodus. A narrow canal connecting a flagellated chamber with an excurrent canal in certain leuconoid sponges.

aphonia. Loss of speech; voicelessness.

aphotic. Without light, as in the *aphotic zone* of the ocean, the region below 800 meters which receives little or no light.

aphyllous. Without leaves.

aphytal. Lacking plants.

apian. Of or pertaining to bees.

apiary. A place where bees are kept, especially a collection of hives maintained for honey production.

apical. At, near, or belonging to an apex, tip, or summit.

apical organ. A sensory structure at the apex of a trochophore which bears a tuft of cilia.

apiculate. With a short, abrupt point.

apiculture. Bee culture.

aplanospore. A nonmotile spore, as in the true fungi.

apneustic. Lacking spiracles, or with nonfunctional spiracles, as in aquatic insects.

apocarpous. With carpels separate, not united. *Compare* syncarpous.

apocrine gland. A gland in which a portion of the cytoplasm of the secretory cells is lost and becomes a part of the secretion, as the mammary gland.

apodal. 1. Lacking feet or legs; apodous. **2.** Lacking pelvic fins, as eels.

apodeme. An inward projecting portion of the exoskeleton of an arthropod serving primarily for the attachment of muscles.

apodous. 1. Lacking feet; apodal. **2.** Without thoracic or abdominal locomotor appendages, with reference to insect larvae.

apoenzyme. The protein portion of an enzyme which, when combined with coenzyme, forms a complete enzyme or holoenzyme.

apogamy. The development of a sporophyte directly from gametophyte without fusion of gametes; parthenogenesis; apomixis.

apolysis. The successive breaking off of gravid proglottids in tapeworms.

apomixis. APOGAMY.

aponeurosis. A flat, sheetlike tendon which attaches muscle to bone or other structures.

apopetalous. With separate and distinct petals; polypetalous.

apophysis. 1. *zool.* **a.** An outgrowth or projection of a structure, as of a bone. **b.** In insects, a tubular or spinelike apodeme. **2.** *bot.* **a.** The swollen portion at the base of a capsule, as in mosses. **b.** The rounded, exposed portion of a cone scale, as in pines.

apopyle. In sponges, the opening of a radial canal into the central cavity or spongocoel; in elasmobranchs, the anterior opening of the clasper tube. *Compare* hypopyle.

aporogamy. Fertilization in plants in which the pollen tube does not enter through micropyle. *Compare* porogamy.

aposematic. Warning or threatening, with reference to coloration.

aposepalous. With separate and distinct sepals; polysepalous.

apospory. The development of a gametophyte directly from sporophyte tissue rather than from a spore.

apothecium. The open disk or cup-shaped fruiting body or ascocarp of certain ascomycetes, as the cup fungi.

apparatus. A group of structures or parts which work together in performing a certain function, as the Golgi apparatus.

appendage. A structure attached to or appended to a larger structure; a projection of the body or of an organ; a limb.

appendicular. Consisting of, pertaining to, or possessing appendages.

appendiculate. Provided with or bearing an appendage or appendages.

appendix. 1. *anat.* An appendage or outgrowth, especially the vermiform appendix, a small, blind tube at the distal end of the cecum in man. **2.** *bot.* The sterile terminal portion of a spadix.

applanate. Flattened horizontally.

apple. The fleshy, pome fruit of a tree of the genus *Malus*, or the tree which produces it.

apposable. Capable of being arranged side by side or next to each other.

apposition. The act of fitting together; state of being in juxta-position.

appositional growth. Growth by deposition of material on previously formed layers, as in bone.

apposition image. A mosaic image, *q.v.*

appressed. Lying close to or flat against.

approximate. Close together but not united.

apterium pl. **apteria.** A bare or downy area between feather tracts of a bird.

apterous. Lacking wings or winglike processes.

apterygote. Any insect of the subclass Apterygota comprising primitive forms which undergo little or no metamorphosis, as bristletails and springtails. Sometimes proturans are included.

aquarium. 1. An artificial pond, pool, vessel, or container in which aquatic organisms are kept. **2.** A place for the exhibition and care of aquatic organisms, especially fishes.

aqueduct. A canal or duct for the passage of a fluid.

aqueduct of Sylvius. The cerebral aqueduct, a slender canal in the midbrain connecting the third and fourth ventricles.

aqueous. Consisting principally of or of the nature of water.

aqueous humor. A clear fluid occupying the anterior cavity of the eye.

ara. A large South American macaw.

aracari. A small, gregarious toucon (*Pteroglossus*).

arachnid. Any arthropod of the class Arachnida which includes the spiders, scorpions, ticks, and mites.

arachnoid. 1. Like a cobweb. 2. *anat.* The arachnoid membrane, the middle of the three meninges which enclose the brain and spinal cord. 3. *bot.* Covered with thin, loose hairs or fibers.

araneid. Any arachnid of the order Áraneae (Araneida) which includes the spiders.

araneiform. Like a spider.

arboreal. 1. Pertaining to or of the nature of trees. 2. Attached to or living in or among trees.

arboreous. Heavily wooded: ARBOREAL, ARBORESCENT.

arborescent. Treelike in structure, appearance, or manner of growth.

arboretum. A place where trees and shrubs are grown for educational or scientific purposes.

arborization. A treelike branching.

arbor virus. *See* arbovirus.

arbor vitae. 1. *bot.* An evergreen tree of the genus *Thuja*, a common ornamental shrub. 2. *anat.* A treelike structure composed of white matter seen in a median section of the cerebellum.

arbovirus. An arthropod-borne virus, as the causative agents of encephalitis and yellow fever.

arbuscula. A small shrub having the form of a tree.

archaeopteryx. A fossil bird of the genus *Archaeopteryx* with teeth embedded in sockets, separate metacarpal bones, and a lizardlike tail with feathers along the sides; of the late Jurassic.

archaeornis. A fossil bird of the genus *Archaeornis* similar to *Archaeopteryx*.

archegonial. Of, pertaining to, or bearing archegonia.

archegoniophore. A specialized branch bearing archegonia, as in hepatics.

archegonium pl. **archegonia.** The female sex organ of liverworts, mosses, and ferns, usually a flask-shaped structure with a swollen base or venter and a slender, elongated neck. Within the venter a single egg develops.

archenteron. The primitive gut or digestive cavity of the gastrula, lined with endoderm and opening to the outside by the blasto-pore; gastrocoel.

archeocyte. A large amebocyte in the mesenchyme of sponges.

archeology. In human evolution, the study of objects of cultural significance, as tools, weapons, ornaments, and utensils.

Archeozoic era. The earliest period of geological time, with an estimated duration of two billion years. Fossil remains are lacking, but there is evidence that primitive life forms existed.

archesporium. The spore-bearing region in the capsule of a bryophyte.

archetype. A basic model, plan, or prototype; a generalized type of structure assumed to have been possessed by ancestors of a group.

archinephros. A holonephros, a primitive type of kidney present in the larva of certain hagfishes and amphibians.

archipallium. The older, primitive portion of the cerebral cortex which, in mammals, comprises the hippocampal formation, a part of the rhinencephalon. *Compare* neopallium.

arctic. Pertaining to, or of the region of the North Pole; polar; frigid.

arcualia. Two pairs of cartilaginous structures (basidorsals and basiventrals) which develop in the mesodermal sheath surrounding the notochord in vertebrates. They give rise to the neural and hemal arches.

arcuate. Arched; curved like a bow.

arenaceous. Sandy; consisting of, or of the nature of sand; growing in sand.

arenicolous. Growing in sand or sandy areas; living in or on sand.

arenose. Sandy; gritty.

areola. 1. The colored ring surrounding a nipple. **2.** The part of the iris which encircles the pupil. **3.** A minute space in a tissue. **4.** A bare space between scales on the feet of birds. **5.** A median, longitudinal strip on the dorsum of a crayfish.

areolar tissue. Loose connective tissue.

areole. 1. An areola. **2.** A rounded or polygonal plate in the cuticle of nematomorphs.

argali. A wild sheep (*Ovis ammon*) of the mountains of Central Asia, noted for its massive horns.

argasid. Any acarid of the family Argasidae which includes the soft ticks, as *Ornithodoros*, transmitter of the infective agents of tickbite fever, Q fever, tularemia, and relapsing fever.

argentaffin. Having an affinity for silver or chromium salts, as *argentaffin* cells of the alimentary canal.

arginine. An essential amino acid present in both animal and plant cells.

argyrophilic. Readily impregnated or stained with silver.

arid. Dry; barren; lacking in life.

aril. A loose, fleshy covering which develops from the stalk or funiculus forming a fleshy bag enclosing a seed, sometimes brightly colored, as in the bittersweet.

arillate. Having an aril.

arillode. A false aril originating from the funicle or chalaza, as in the nutmeg.

arista. A bristlelike appendage, as **a.** that on the antenna of certain dipterous insects; **b.** the awn of cereal plants.

aristate. Slender and pointed; bearing an arista or awn.

Aristotle's lantern. A complicated, toothed structure consisting of several ossicles surrounding the mouth of a sea urchin.

arm. 1. In man, the portion of the upper extremity from shoulder to elbow. **2.** In common usage, the portion from shoulder to wrist. **3.** A structure of a corresponding nature, as **a.** the anterior limb of a vertebrate; **b.** the ray of a sea star; **c.** a tentacle of an octopus; **d.** an extension of the manubrium of a medusa.

armadillo. A nocturnal, burrowing mammal of the order Edentata, with a body encased in bony plates which enable the animal, when attacked, to roll up into a ball, as *Dasypus novencinctus*, the nine-banded armadillo of Texas.

armature. Collectively, defensive and protective structures, as teeth, horns, spines, or thorns.

armed. Bearing hooks, with reference to the scolex of tapeworms.

armored. Possessing a protective covering of heavy scales or bony plates.

armored scales. Scale insects of the family Diaspididae whose bodies are covered with a hard, waxy scale, as the San Jose scale.

35

armyworm. The larva of a noctuid moth (*Cirphis unipuncta*) which occurs in large numbers and is a serious pest to cereal crops. Also *Laphygma frugiperda*, the fall armyworm of Southern states, a pest to corn and cotton.

arna. The water buffalo (*Bubalus bubalis*).

arolium. In insects, a padlike structure at the tip of the tarsus.

arrector. An erector muscle, especially the arrector pili, consisting of smooth muscle fibers attached to the base of a hair follicle.

arrhenotoky. Female parthenogenesis in which only males are produced.

arrowworm. A small, marine worm (*Sagitta*) of the phylum Chaetognatha.

arteriole. A small artery less than 0.3 mm. (300 microns) in diameter.

artery. A vessel which conveys blood from the heart to the tissues.

arthrobranch. A gill which arises from the membrane joining an appendage to the body, as in the crayfish.

arthrodia. A joint which permits gliding or sliding movement, as that between articular processes of the vertebrae.

arthrodire. An extinct fish of the order Arthrodires with bony plates covering the anterior portion of the body.

arthromere. A somite or metamere; a segment of a segmented animal.

arthrophyte. A plant of the division Arthrophyta which includes the horsetails, characterized by a jointed stem; an articulate.

arthropod. An invertebrate of the phylum Arthropoda, which includes the crustaceans, insects, centipedes, millipedes, and arachnids which comprise the largest animal phylum (over 800,000 species); characterized by segmented body, jointed appendages, and exoskeleton of chitin.

arthrospore. A spore produced by certain fungi by fragmentation of the hyphae, as in some actinomycetes.

article. **1.** *bot.* A segment of a fruit or pod, as in legumes. **2.** *zool.* A serrated process at the tip of the chelicera of a tick.

articulamentum. The inner calcareous layer of the plate of a chiton.

articular bone. A cartilage bone in lower vertebrates which forms the posterior portion of the lower jaw articulating with the quadrate.

articular cartilage. A thin layer of hyaline cartilage which covers the end of a bone in a diarthrodial joint.

¹articulate. 1. Jointed; formed of segments; united by joints. **2.** Distinct, clear; formed of words or syllables, as *articulate* speech.

²articulate. 1. *zool.* A brachiopod with a hinged shell. **2.** *bot.* An arthrophyte.

³articulate. *anat.* To form a joint.

articulation. 1. *zool.* **a.** A joint between two bones or between a bone and cartilage. **b.** A point of union between any two parts or segments of an animal. **2.** *bot.* **a.** A point of union between two separable parts. **b.** A node or a joint.

articulator. A movable structure, as the lips, teeth, tongue, or soft palate, which functions in the production of distinct sounds in articulate speech.

artifact. 1. A structure seen in fixed preparations of cells and tissues which is not present in real life. **2.** An object, as a tool or utensil, indicating human workmanship.

artiodactyl. 1. An even-toed mammal. **2.** Any member of the order Artiodactyla, even-toed mammals, which include the pigs, peccaries, hippopotamuses, and all ruminants.

arytenoid. Ladle-shaped; pitcher-shaped.

arytenoid cartilage. One of a pair of cartilages in the larynx to which the vocal folds or cords are attached.

ascariasis. Infestation by an ascarid, especially *Ascaris*, a parasitic nematode inhabiting the intestine.

ascarid. Any nematode of the family Ascaridae to which *Ascaris lumbricoides*, a human intestinal parasite, belongs.

ascending. 1. Rising; going upward. **2.** *bot.* Rising obliquely upward from a prostrate position.

ascending series. A series of animals or plants arranged in order of complexity from the simplest to the most complex.

aschelminth. Any member of the phylum Aschelminthes, pseudocoelomate invertebrates comprising the rotifers, gastrotrichs, kinorhynchs, priapulids, nematodes, and nematomorphs.

ascidian. A sea squirt, a sessile tunicate of the class Ascidiacea.

ascidium. A flask- or pitcher-shaped structure, as the leaf of the pitcher plant.

ascocarp. A spherical or cup-shaped fruiting body containing asci found in Ascomycetes. Common types include: cleistothecia, perithecia, apothecia, and ascostromata.

ascogenous. Producing or giving rise to asci.

ascogonium. The female gametangium (oogonium) of Ascomycetes.

ascomycete. Any fungus of the class Ascomycetes which includes the sac fungi, characterized by the production of ascospores in a distinctive saclike structure, the ascus. Common representatives are the yeasts, blue and green molds, powdery mildews, ergot, cup fungi, sponge mushrooms, and lichens.

ascon. A type of sponge with a simple canal system in which openings in the body wall lead directly to the spongocoel, as in *Leucosolenia*.

asconoid. Pertaining to sponges with the ascon type of canal system.

ascorbic acid. A vitamin present in green vegetables, tomatoes, and citrus fruits which is necessary for the development of collagenous fibers essential in tissue repair. It plays a role in certain oxidation-reduction reactions and is essential for normal capillary permeadility. Extreme deficiency results in scurvy. Also called vitamin C, antiscorbutic factor.

ascospore. A spore produced by an ascus.

ascostroma. A stroma in which ascus-containing chambers or locules develop.

ascus pl. **asci.** A saclike sporangium characteristic of the Ascomycetes which is usually tubular in shape and produces typically eight spores.

asepalous. Lacking sepals.

aseptate. Without septa.

aseptic. Free of microorganisms, especially those causing decay, putrefaction, or disease.

asexual. Lacking sex; not involving sex.

asexual reproduction. Reproduction which does not involve the union of gametes or the sexual union of two individuals, as that by fission, fragmentation, spore formation, budding, and gemmule formation.

¹**ash.** A tree of the genus *Fraxinus*, a member of the olive family, Oleaceae.

²**ash.** The incombustible mineral matter which remains after a substance has been burned.

asity. A fruit-eating bird of Madagascar, as the velvet asity (*Philepitta castanea*).

asp. A small venomous snake of North Africa, especially *Cerastes cornutus*, the horned viper or a small cobra (*Naja haje*).

aspect. 1. The general appearance or view with respect to **a.** an object or individual; **b.** a species or population, especially its seasonal appearance. 2. An ecological season, as hiemal, vernal, aestival, autumnal.

asphyxia. Suffocation; coma or apparent death resulting from lack of oxygen.

aspidium. Male fern, the rhizome and stipes of *Dryopteris filix-mas*, the source of oleoresin of aspidium, a drug widely used for the expulsion of worms, especially tapeworms.

aspirator. An apparatus such as a suction pump in which a negative pressure is created, used for the withdrawal of liquids or gases from a cavity.

ass. A small quadruped of the genus *Equus*, especially *E. asinus*, the wild ass of Africa. The donkey and burro are domesticated forms.

assassin bug. A hemipteran insect of the family Reduviidae.

assimilation. Anabolism or constructive metabolism; the process of taking in and converting nutrient substances into components of the body; the conversion of nonliving matter into protoplasm

association. *ecol.* A division of a biome characterized by uniformity in species comprising plant dominants.

association neuron. An internuncial neuron, especially one which connects various regions or centers of the central nervous system.

associes. A seral plant community.

assumentum. One of the valves of a silique.

assurgent. Rising; ascending.

astasia. Inability to walk or stand due to motor incoordination.

astaxanthin. A xanthophyll pigment (a hematochrome) present in euglenoids and chlorophytes.

¹aster. *bot.* A fall-blooming flower of the genus *Aster.*

²aster. *cytol.* A star-shaped structure consisting of a centriole with radiating fibers appearing during mitosis at the two ends of the spindle.

asteroid. An echinoderm of the class Asteroidea which includes the sea stars (starfishes); asteroidean.

asthenia. Muscle weakness; loss of strength.

astigmatic. Lacking stigmas.

astomatous. 1. *zool.* **a.** Lacking a mouth or stoma. **b.** Lacking a cytostome, said of holotrich ciliates. 2. *bot.* Lacking stomata.

astragulus. The talus, a tarsal bone.

astrocyte. A type of neuroglial cell with many branching cytoplasmic processes.

asymmetrical. Lacking in symmetry; not capable of being divided into equal or corresponding parts.

atactostele. A type of stele in which the vascular bundles are scattered throughout the stem, as in maize.

atavism. The recurrence of traits possessed by a remote ancestor, usually the result of a recombination of genes.

ataxia. Muscular incoordination.

ateliosis or **ateleiosis.** Dwarfism with retention of infantile characteristics.

athecate. Lacking a theca.

atherosclerosis. A form of arteriosclerosis in which localized deposits of lipids form beneath the intima of an artery.

atlas. The first cervical vertebra with which the skull articulates.

atoke. In certain marine polychaetes, the anterior nonsexual form as distinguished from the posterior sexual form or epitoke, as in the palolo worm.

atokous. Lacking offspring.

atoll. A circular coral reef.

atom. The smallest particle of an element which **a.** can exist alone or in combination with other atoms; **b.** enters into the structure of a molecule; **c.** cannot be split into smaller particles by chemical means.

ATP. Adenosine triphosphate, *q.v.*

atresia. Imperforation or closure of a normal opening, canal, or cavity.

atrial. Of or pertaining to an atrium.

atrichous. 1. Without spines, with reference to nematocysts. **2.** Without flagella, with reference to bacteria.

atriopore. The opening of an atrial cavity to the outside.

atrioventricular. Pertaining to the atria and ventricles of the heart. Abbreviated AV, A-V.

atrium pl. **atria.** A cavity, entrance, or passageway, as **a.** the cavity surrounding the pharynx in lancelets and tunicates; **b.** the cavity in certain invertebrates which receives the genital ducts; **c.** in vertebrates, a chamber of the heart which receives blood from veins or the sinus venosus. *See* auricle.

atrocha. A ciliated larva of certain polychaete worms.

atrophy. The reduction in size of an organ or structure resulting from lack of nourishment or reduced functional activity.

atropine. An alkaloid obtained from *Atropa belladonna* and other plants of the family Solanaceae. It is a parasympathetic blocking agent and used as a respiratory stimulant, muscle relaxant, and mydriatic (dilator of pupils).

atropous. Designating an ovule which is not inverted with micropyle or nucellar apex directed away from the funiculus; orthotropous. *Compare* anatropous.

¹attenuate. 1. To make thin, dilute, or weaken. **2.** *bacteriol.* To reduce the virulence of, with reference to pathogenic organisms.

²attenuate. *bot.* With a long, gradual taper; accuminate.

attid. A jumping spider of the family Attidae.

attitude. The position or posture assumed by an organism.

atypical. Irregular; not according to type.

auditory. Of or pertaining to hearing or the sense of sound; acoustic.

auditory capsule. The portion of a chondrocranium which encloses the auditory vesicle.

auditory meatus. One of two canals in the temporal bone, the external conveying sound waves to the middle ear, the internal transmitting the vestibulocochlear and facial nerves.

auditory nerve. The vestibulocochlear nerve. Also called statoacoustic nerve.

41

auditory organ. An organ capable of being stimulated by sound vibrations, especially an organ of hearing as **a.** in vertebrates, the ear; **b.** in invertebrates, tympanal organs and auditory hairs, as in insects.

auditory ossicles. The ear bones, malleus, incus, and stapes, located in the middle ear of mammals. *See* columella.

auditory tube. A tube connecting the middle ear with the nasopharynx. Also called Eustachian tube.

auditory vesicle. *See* otocyst.

auger shell. A slender, many-whorled gastropod of the family Terebridae, common in the Indo-Pacific region. A common genus is *Terebra*.

auk. A sea bird of the family Alcidae, as the razor-billed auk (*Alca torda*) of the North Atlantic. The great auk (*Pinguinus* (*Plautus*) *impennis*) became extinct in 1844.

auklet. One of several species of small auks of the genera *Ptychoramphus* and *Cerorhinco* of the North Pacific.

auricle. 1. *zool.* An ear or earlike structure, as **a.** the pinna of the external ear; **b.** a receiving chamber of the heart in various invertebrates; **c.** the auricular appendage, a protruding portion of the atrium in the heart of mammals; **d.** a blunt sensory projection at the anterior end of turbellarians; **e.** a ciliated projection near the mouth in ctenophores and rotifers; **f.** a lip or plate forming a part of the pollen packer in honeybees. **2.** *bot.* An earlike lobe at the base of various leaves and petals.

¹**auricular. 1.** Of or pertaining to the ear or the sense of hearing. **2.** Shaped like an ear.

²**auricular.** *ornith.* One of the feathers which covers the external auditory opening.

auricularia. The free-swimming larva of holothuroidean echinoderms.

auriculate. Possessing an auricle or auricles.

auriform. Shaped like an ear.

aurochs. The European bison or wisent (*Bison bonasus*) now extinct in the wild state. It has been considered to be the ancestor of modern breeds of cattle.

aurophore. A constricted portion of the float in certain siphonophores.

auscultation. The art of listening to sounds produced within the body, as heart and respiratory sounds.

autacoid. Term formerly applied to a hormone which stimulated the target organ.

autecology. The study of an individual species. *Compare* synecology.

autocarp. A fruit developing from self-fertilization.

autocatalysis. A process in which the speed of a chemical reaction is accelerated by the presence of one or more products resulting from the reaction.

autochthonous. Local, indigenous, native.

autoclave. An apparatus for sterilizing with steam under pressure.

autoecious. Requiring but a single host to complete its life cycle, with reference to rust fungi.

autogamy. 1. *bot.* Self-fertilization; self-pollination. 2. In certain protozoans and algae, the fusion of two nuclei originating from a single cell.

autogenous. Arising from within; self-generated; endogenous.

autograph. Tissue removed from one part of the body and transplanted to another region of the same organism.

autologous. Obtained from or derived from an organism in contrast to homologous (from another of the same species) or heterologous (from one of a different species).

autolysis. 1. Self-digestion of tissues within a living body or within an organism that has just died. 2. The breakdown of tissues from the action of intracellular, proteolytic enzymes.

automatism. Spontaneous activity of tissues, organs, or organisms.

automixis. In certain protozoans, the fusion of two nuclei within a single cell; autogamy.

autonomic. Self-regulating; functioning independently; spontaneous.

autonomic nervous system. That portion of the peripheral nervous system which supplies fibers to structures not under voluntary control, as smooth and cardiac muscle, glands.

autonomous. Capable of existing or operating independently; self-regulating.

autophilous. Self-pollinated.

autophyte. A plant capable of manufacturing its own food. *Compare* heterophyte.

autoplastic. Of or pertaining to grafts transferred from one part of the body to another.

autoploid. An autopolyploid.

autopolyploid. A polyploid in which multiple genomes are identical, as an autotetraploid.

autoradiograph. A radioautograph.

autorhythmic. Capable of spontaneous, rhythmical self-excitation.

autosome. Any chromosome other than a sex (X or Y) chromosome.

autospore. A zoospore or aplanospore which has the same shape as the parent cell.

autostylic. A method of jaw suspension by the quadrate, common in tetrapods other than mammals.

autotomy. Self-mutilation; the automatic breaking off of a part, especially one that is injured, usually followed by regeneration. It occurs in starfishes, insects, and crustaceans.

autotroph. An autotrophic organism.

autotrophic. **1.** Self-nourishing, said of organisms which are capable of synthesizing food from inorganic compounds, as all chlorophyll-containing plants. **2.** *microbiol.* Obtaining carbon from carbon dioxide or carbonates, as nitrifying bacteria of the soil.

autozooid. A type of feeding polyp found in certain alcyonarians. *Compare* siphonozooid.

auxillary. That which complements, aids, or helps.

auxin. A phytohormone which regulates growth, especially one that induces longitudinal growth of shoots through cell elongation rather than through cell division, as indole-3-acetic acid (IAA) or heteroauxin.

auxospore. A spore formed in certain diatoms by the fusion of two protoplasts (gametes).

Aves. A class of vertebrates which includes the birds, warm-blooded animals with wings and feathers. About 8600 species.

avian. Of or pertaining to birds.

aviary. A house or enclosure where birds are kept.

avicularium. A modified protective zooid found on the surface of certain bryozoans, shaped like a bird's beak with a jaw which opens and closes.

avidin. A protein present in egg white which inactivates biotin.

avifauna. Collectively, the birds of an area or region.

avocado. A tropical tree of the genus *Persea*, or the edible fruit it produces; alligator pear.

avocet. A slender, graceful shorebird of the family Recurvirostridae, with long, slender legs and recurved bill, as *Recurvirostra americana*.

awn. A small, pointed process, especially one of the slender, barbed appendages constituting the beard of the spikelet of a cereal.

axenic. Germ-free; free from foreign organisms.

axial. Of or pertaining to an axis, especially the main axis of the body.

axial filament. *See* axial rod.

axial gland. A spongy mass forming part of the haemal system of echinoderms; axial organ.

axial rod. The central supporting structure of an axopodium; an axial filament.

axial skeleton. The skull, vertebral column, sternum, and ribs.

axil. *bot.* The angle between the stem axis and a leaf, branch, or other appendage attached to it.

axile. Of, or located in an axis, with reference to placentation in plants.

axilla. 1. *anat.* The armpit or axillary space. 2. *ornith.* The corresponding region under the wing of a bird.

axillar. *See* axillary.

¹axillary. 1. *anat.* Of or pertaining to the axilla. 2. *bot.* Of or located in an axil.

²axillary. *ornith.* An axillar, one of a group of feathers occupying the region between wing and body of a bird.

axis. 1. A line, real or imaginary, passing through the body about which the body revolves or theoretically is capable of revolving. 2. A line, real or imaginary, about which parts of an organism or of a structure are symmetrically arranged, as the spinal column of a vertebrate or the main stem of a plant. 3. *anat.* The second cervical vertebra; epistropheus. 4. *bot.* The stem of a plant.

axis cylinder. An axon, *q.v.*

axolotl. The larva of the tiger salamander (*Ambystoma tigrinum*), which is capable of breeding while still retaining its larval form. *See* neotony.

axon. A process of a neuron which usually conveys impulses away from the cell body; an axis cylinder. Term is also applied to the peripheral process of a unipolar or sensory neuron.

axoneme. The central, axial filament of a flagellum.

axoplasm. The cytoplasm or neuroplasm of an axon.

axopodium. A semipermanent type of pseudopodium which possesses a central, semirigid supporting rod, as in various heliozoans.

axostyle. A central supporting filament or rod in certain protozoans, as in trichomonads.

aye-aye. A nocturnal lemur (*Daubentonia madagascariensis*) of Madagascar.

azoic. Without life, especially with reference to geologic time antedating the origin of life on the earth.

azygous. Unpaired; existing singly.

B

baboon. An Old World ape of the genus *Papio*.

baccate. Berrylike; pulpy; fleshy; producing berries.

bacciferous. Producing berries.

bacciform. Shaped like a berry.

baccivorous. Feeding principally upon berries.

bacillary. 1. Rod-shaped; composed of rod-shaped structures. 2. Of, pertaining to, or caused by bacilli.

bacilliform. Rod-shaped.

bacillus pl. **bacilli.** A rod-shaped bacterium.

bacitracin. An antibiotic produced by *Bacillus subtilis*, effective especially against gram-positive bacteria.

backbone. The spinal or verterbal column.

backcross. The mating of a hybrid with one of the parental genotypes. *See* test cross.

backswimmer. An aquatic, hemipteran insect which swims ventral side up, as *Notonecta undulata* of the family Notonectidae.

bacteria. Minute, unicellular plant organisms of the division Schizomycophyta (Schizophyta), formerly considered a class Schizomycetes of the Fungi. They lack chlorophyll, most existing as parasites or saprophytes; a few are autotrophic. They are the chief agents of fermentation, putrefaction, and decay. Many are pathogenic.

bactericide. An agent which destroys or kills bacteria.

bacteriochlorophyll. A chlorophyll present in purple sulfur bacteria.

bacterioclasis. Disintegration of bacteria.

bacteriology. The science which deals with bacteria. *See* microbiology.

bacteriolysin. An antibody which, in the presence of the complement, causes the dissolution of the bacteria which are its antigen.

bacteriolysis. The dissolution of bacteria.

bacteriophage. A virus which infects bacteria, bringing about their destruction through lytic action.

bacteriostatic. Preventing or hindering the growth and multiplication of bacteria.

bacterium. *See* bacteria.

baculum. The penis bone (os penis, os priapi), a heterotopic bone present in the penis of various male mammals (insectivores, bats, rodents, carnivores, and primates, except man).

badger. A stout-bodied, burrowing carnivore of the family Mustelidae. A common American species is *Taxidea taxus*, European species is *Meles meles*.

bagworm. A caterpillar which constructs a silken bag about its body, characteristic of a number of moths of the family Psychidae.

bailer. The fused exopodite and epipodite of the second maxilla of a crayfish; scaphognathite.

balancer. 1. A halter. 2. One of a pair of lateral appendages extending from the angle of the mouth in certain immature salamanders.

balancing organ. A receptor which functions in the maintenance of posture and equilibrium. *See* statocyst, semicircular canal.

balata. A nonelastic rubber obtained from the latex of the bully tree (*Manilkara bidentata* or *Mimusops balata*).

balausta. The fruit of the pomegranate, a many-seeded, indehiscent fruit with a thick, tough rind.

bald. Lacking the usual covering, as absence of hair or feathers on the head.

baldhead. *See* baldpate.

baldpate. A surface-feeding duck (*Mareca americana*).

baleen. Whalebone, a horny substance growing from the upper jaw of whales forming a fringelike sieve used as a food-getting apparatus.

ballooning. A method of transport employed by young spiderlings in which secreted strands of silk are caught by air currents and the young are carried considerable distances.

balsa. The wood of *Ochroma pyramidale*, a tropical tree, the lightest wood known.

bamboo. Any of a number of woody, arborescent grasses of the family Gramineae, which grow in most tropical countries. Their hollow, woody stems are used extensively in furniture making and in various types of construction. An example is *Dendrocalamus*, the giant bamboo.

bamboo worm. *Axiothella*, a tube-dwelling polychaete worm, whose tube resembles the stem of a bamboo plant.

banana. A large herbaceous tropical plant (*Musa paradisiaca*) or the fruit it produces, borne in clusters on a single inflorescence.

bandar. *See* macaque.

bandicoot. 1. A small insectivorous and vegetarian marsupial of the genera *Perameles* and *Thylacomys* found in Australia. 2. A large rat of southern Asia of the genus *Nesokia*.

banner. The broad uppermost petal in an irregular flower, as that of members of the pea family. Also called standard, vexillum.

banteng. *Bibos sondaicus*, a large ox of Southeast Asia occurring both wild and domesticated.

barb. **1.** *bot.* A hair or bristle ending in a double hook. **2.** *zool.* One of a series of parallel projections from the shaft of a feather which collectively constitute the vane.

barbed. Having a barb or barbs; with short, pointed, and usually reflexed awns or bristles.

barbel. **1.** A slender, elongated sensory process on the head of various fishes, as the catfish. **2.** A European freshwater fish of the genus *Barbus*, family Cyprinidae.

barbellate. Finely barbed.

barberry. An ornamental shrub of the genus *Berberis*, family Berberidaceae. The common barberry (*B. vulgaris*) serves as intermediate host of the wheat stem rust.

barbet. A nonpasserine, tropical bird of the family Capitonidae, order Piciformes.

barbicel. A minute projection on the barbule of a feather.

barbulate. Finely bearded.

barbule. One of the numerous minute processes on each side of the barb of a feather.

bark. In woody plants, the tissues lying outside the cambium.

bark lice. Psocids, winged insects of the order Psocoptera. *See* book lice.

barley. A cereal grass of the genus *Hordeum*, its seed or grain widely used in malt beverages and as food for man and domestic animals.

barnacle. One of a number of sessile, marine crustaceans of the order Cirrepedia, as *Balanus*, the acorn or rock barnacle; *Lepas*, the goose barnacle.

baroceptor. A pressure receptor as that in the wall of certain blood vessels. Also called pressoreceptor or baroreceptor.

barracuda. A large, carnivorous marine fish of the family Sphyraenidae, order Percomorpha, as *Sphyraena barracuda*, the giant barracuda.

barrier. **1.** An obstacle or obstruction. **2.** *biol.* That which prevents the spread or distribution of animals or plants.

barrier reef. *See* coral reef.

barrow. A castrated male hog.

basal. **1.** Basic, fundamental. **2.** Located at or near the base of a structure.

basal disc. The flattened, adhesive disc of various coelenterate polyps or other sessile animals.

basal granule. A blepharoplast.

basalia. One of three basal cartilages in the fins of elasmobranch fishes, designated pro-, meso-, and metapterygium.

base. 1. *biol.* **a.** The lowest part of a body or structure; **b.** The portion upon which a structure rests; **c.** The broadest portion of a cone-shaped structure. **2.** *chem.* **a.** A substance which neutralizes an acid and turns litmus blue. **b.** A substance which furnishes hydroxyl (OH) ions and a positive ion, usually a metal. **c.** An ion or molecule which will combine with a proton. **d.** An alkali.

basement membrane. A layer of amorphous substance upon which epithelial cells rest.

basic stain. A biological dye whose ability to color resides in the basic radical or cation, as methylene blue, hematoxylin.

basidiocarp. The fruiting body of fungi belonging to the Basidiomycetes. *See* basidium.

basidiomycete. Any member of the Basidiomycetes, a class of true fungi which includes the mushrooms, toadstools, puffballs, rusts, and smuts, characterized by the production of basidia and basidiospores. About 25,000 known species.

basidiospore. A spore produced by a basidium.

basidium pl. **basidia.** A club-shaped, spore-producing structure characteristic of the Basidiomycetes.

basifixed. Fixed or attached by the base, said of ovules.

basilar membrane. A membrane in the floor of the cochlear duct upon which the organ of Corti rests.

basilisk. A tropical American lizard (*Basiliscus*) with an erectile crest.

basin. The concave, inner portion of the sternum in birds.

basinerved. Having veins extending from the base, said of leaves.

basipetal. Developing from apex toward the base, said of flowers and leaves. *Compare* acropetal.

basipodite. The second segment of a crustacean appendage. *See* protopodite.

basis. 1. The attached undersurface of a sessile barnacle. **2.** A basipodite, *q.v.*

basket. A type of phylactocarp. *See* corbula.

basket star. An echinoderm of the class Ophiuroidea, with arms which branch repeatedly and terminate in small flexible tendrils as *Gorgonocephalus*.

basophil or **basophile. 1.** A leukocyte with large cytoplasmic granules which stain readily with basic stains. **2.** A basophilic cell of the anterior lobe of the pituitary.

basophilic. Having an affinity for or staining readily with basic stains.

bass. 1. One of a number of spiny-finned food and game fishes of the family Centrarchidae, as *Micropterus dolomieui*, the smallmouth bass, and *M. salmoides*, the largemouth bass. **2.** One of a number of marine fishes, as the sea basses of the family Seranidae.

bassarisk. The ring-tailed cat (*Bassariscus astutus*) of the family Procyonidae; cacomistle.

basswood. Any of a number of species of the genus *Tilia* of the linden family (Tiliaceae). Also called whitewood.

bast. PHLOEM.

bastard wing. *See* alula.

bat. A flying mammal of the order Chiroptera.

batfish. One of a number of grotesque, marine fishes, as *Aetobatus*, the California batfish; *Dactylopterus*, the flying gurnard; *Ogcocephalus*, the longnose seabat of the batfish family, Ogcocephalidae.

bathyal. Of or pertaining to deeper portions of the oceans.

bathymetric. Vertical or altitudinal, with reference to distribution of organisms.

batrachian. A frog or toad.

battery. In coelenterates, a cluster of nematocysts.

beach flea. Any of a number of small, terrestrial amphipods of the family Talitridae, as *Talitrus*, *Orchestria*.

beak. 1. A projecting structure ending in a point. **2.** *zool.* **a.** The nib or bill of a bird or turtle. **b.** The projecting mouth parts of certain insects. **c.** The horny jaws of an octopus. **d.** The earliest part of the shell of a bivalve. **e.** The apex of the ventral valve of a brachiopod. **3.** *bot.* A prolongation of certain fruits or carpels.

beaker cell. A unicellular gland in the integument of a cyclostome.

51

bean. 1. The large, edible, kidney-shaped seed borne by various leguminous plants, especially those of the genus *Phaseolus*, as the kidney, lima, string, and soy *bean.* 2. The plants producing these seeds or the pods containing them. 3. Various other bean-like seeds or the plants producing them, as the castor *bean* or coffee *bean.*

bear. A large, carnivorous mammal of the family Ursidae, as the brown, polar, or grizzly *bear.*

bear animalcule. A tardigrade.

beard. 1. The hair that grows on the face of a man. 2. Any structure which resembles a man's beard, as **a.** the long hairs on the chin of a goat; **b.** hairlike feathers under the bill of a bird. 3. *bot.* A cluster of bristlelike hairs or awns.

beard worm. *See* pogonophoran.

beaver. An aquatic rodent of the genus *Castor*, with chisellike teeth, webbed hind feet, and a flat scaly tail; valued for its fur.

bedbug. *Cimex lectularius*, a flat, wingless, bloodsucking insect of the family Cimicidae, order Hemiptera.

beebread. Pollen stored by bees in the honeycomb and used for food by larvae and newly-emerged workers.

beech. A hardwood tree (*Fagus*) of the family Fagaceae.

bee louse. A minute, dipterous insect of the genus *Braula*, parasitic on bees.

bee moth. *See* wax moth.

beeswax. A wax produced by worker bees and used in the construction of the honeycomb. It is produced by glands on ventral surface of the abdomen.

beet. A biennial plant of the genus *Beta*, or its root, grown as a garden vegetable or cultivated commercially as a source of sugar. *See* mangel.

beetle. Any insect of the order Coleoptera.

beggar's lice. The fruit of the stickseed (*Hackelia*) or the plant that produces it.

beggar tick. A sticktight, the fruit or achene of *Bidens*, the burmarigold, or the plant which produces it, with recurved or hispid awns or bristles.

belemnite. A fossil cephalopod with a straight, slender, pointed shell.

bell. 1. The medusa form of a coelenterate; an umbrella. 2. The corolla of certain flowers.

bell toad. One of two genera of small frogs of the family Liopelmidae, *Liopelma* of New Zealand and *Ascaphus* of the Northwest United States.

belly. 1. The abdomen or abdominal cavity. **2.** The central, fleshy portion of a skeletal muscle. **3.** The underneath or ventral side of an animal.

bellying. Protruding on one side, said of the corolla of certain flowers.

benthic. Of or pertaining to the benthos.

benthoic. Bottom-dwelling.

benthos. 1. The bottom of the sea. **2.** The organisms living at the bottom of the sea or other body of water.

berry. 1. Any small juicy fruit. **2.** A simple fruit in which the inner portion of the ovary wall and the partitions between the locules are thick and juicy, as the grape, pepper, tomato, orange.

beta. The second letter of the Greek alphabet (β) used to designate the second of a series, as *beta* cell, *beta* carbon, *beta* oxidation.

betaceous. Beetlike.

bezoar. A concretion consisting principally of ingested hair found in the stomach or intestine of animals, especially ruminants.

bicarpellary. Composed of two carpels.

biceps. A muscle with two heads, as the *biceps* brachii.

biciliate. Having two cilia, flagella, or elaters.

bicipital. 1. Having two heads or points of origin, as a *bicipital* rib. **2.** Pertaining to the biceps muscle.

bicornuate. Having two horns, as a *bicornuate* uterus.

bicuspid. 1. Having two points or cusps, as a bicuspid tooth. **2.** Having two leaflets, as the *bicuspid* or mitral valve of the heart.

Bidder's organ. An ovarylike structure located anterior to the testis in male toads of the genus *Bufo*.

bidentate. Having two teeth.

¹biennial. Occurring once every two years.

²biennial. *bot.* A plant whose life cycle takes two years, seeds being produced the second year after germination.

biferous. Bearing two crops of fruit in one season.

bifid. Forked; divided into two parts by a median cleft.

bifurcated. Two-pronged; forked.

bighorn. The Rocky Mountain sheep (*Ovis canadensis*).

bilabiate. Having two lips, said of the corolla of certain flowers.

bilamellate. Consisting of two layers or lamella.

bilaminar. Consisting of two lamina or thin plates.

bilateral. Having two sides; pertaining to both sides of the body.

bilateral symmetry. Having a body consisting of two corresponding or complementary halves, with most organs occurring in pairs, one on each side of the median sagittal plane.

bile. A bitter, alkaline, digestive fluid secreted by the liver. It alkalinizes chyme, emulsifies fats, stimulates intestinal motility, and serves as a vehicle for waste products of metabolism.

bile acids. Glycocholic and taurocholic acids present in bile.

bile duct. In vertebrates, the common bile duct formed by a union of the hepatic duct from the liver and cystic duct from the gall bladder. Just before entering the duodenum, it receives the pancreatic duct.

bile pigments. Substances in bile primarily responsible for its color, principally bilirubin and biliverdin.

bilirubin. A bile pigment resulting from the reduction of biliverdin; the principal pigment in the bile of humans and carnivores.

biliverdin. A bile pigment resulting from the breakdown of hemoglobin and myoglobin. In the liver it is reduced to bilirubin.

bill. 1. The toothless jaws of a bird; the beak. **2.** A corresponding structure in other animals, as the turtle and duckbill platypus.

bilocular. Having two cells, cavities, or compartments.

binary. Composed of two elements or parts.

binary fission. A form of asexual reproduction in which an organism divides into approximately equal parts.

¹binocular. Having two eyes or employing two eyes at the same time.

²binocular. An optical instrument with two eyepieces or two sets of lenses which permits the use of both eyes simultaneously in viewing an object; a stereoscopic microscope.

binomial. Consisting of two words or names.

binomial nomenclature. A system of naming organisms established by Linnaeus in 1758 in which each species is given a scientific name consisting of two words in Latin, the first designating the genus, the second a subdivision of the genus. The genus name is capitalized, the second is not; both are italicized. The species comprising leopard frogs is designated *Rana pipiens;* the species comprising modern man is *Homo sapiens.*

binucleate. Having two nuclei.

bioassay. The determination of the potency of a drug by testing it upon a living organism.

biocenose. A community of organisms (plants and animals) which occupies a particular habitat.

biochemistry. The chemistry of living things; the study of the constituents of organisms, substances produced by them, and the changes which occur in all metabolic processes.

biochrome. A natural pigment or coloring matter in a plant or animal. *See* schemochrome.

biocies. The climax animal or biotic community of a sere.

bioclimatology. The study of the effects of climate on living things.

bioelectricity. Electric currents generated within living tissues.

biogenesis. The doctrine that living things originate only from preexisting living things. *Compare* abiogenesis.

biogenetic law. *See* recapitulation theory.

biogeography. The science which deals with the geographical distribution of organisms. *See* phytogeography, zoogeography.

biological clock. An innate physiological rhythm, such as metabolic rate, reproductive behavior, migration, feeding, cell division, or pigment dispersal, associated with and synchronized with solar, tidal, lunar, or other environmental cycles.

biological control. The use of parasitic, predaceous, or pathogenic organisms in the control of noxious or injurious plants or animals.

biology. 1. The science which deals with living things, comprising botany and zoology and all their subdivisions. 2. A treatise on living things. 3. All phenomena associated with a particular group of organisms, as the *biology* of protozoa.

bioluminescence. The production of light by living things, as occurs in fireflies, certain deepsea fishes, some bacteria and fungi. *See* phosphorescence, luciferin.

biomass. The total mass of organic material of a species per unit of area or volume, as 100 lbs. of perch per acre of pond surface. Term is used in expressing population density.

biome. A major biotic community or life zone characterized by distinctiveness of life-forms of the principal climax species. Terrestrial biomes are named after their plant dominants, as woodland, tundra, tropical forest; marine biomes, by their predominant animals, as coral reef.

biometry. The statistical study of biological phenomena; the application of mathematics to the study of living things. *See* biostatistics.

bionomics. Old term for ecology.

biophysics. The application of the laws of physics to the study of the properties of living systems. Principal fields include physiological physics, molecular biology, radiation biology, theoretical physics, instrumentation, and measurement.

biopsy. The study of tissues taken from a living organism, especially the microscopic examination of excised tissue for diagnostic purposes.

bioseston. The living components of seston.

biosphere. The portion of the earth inhabited by living organisms, including the land masses, oceans, and atmosphere.

biostatistics. The branch of biometry that deals with vital data.

biosynthesis. The formation of a chemical compound by an organism.

biota. The animal and plant life of a region.

biotic. Of or pertaining to life; biological.

biotin. Vitamin H, a member of the B-complex present in animals and plant tissues. In experimental animals, a deficiency results in skin disorders. Its role in human nutrition is uncertain.

biovulate. Having two ovules.

biparous. 1. *zool.* Producing two young at a single birth. **2.** *bot.* Having two branches or parts.

¹biped. 1. An animal with two feet, as man. **2.** Any two limbs of a quadruped, as the two forelimbs, the anterior biped.

²biped. Having two feet.

bipennate. Resembling a feather, as certain muscles; bipenniform.

bipinnaria. The bilaterally symmetrical, ciliated larva of asteroid echinoderms.

bipinnate. Twice pinnate, the condition of a leaf in which both primary and secondary divisions are pinnate.

bipolar. 1. Having two poles. **2.** Having two processes, as certain neurons.

bipotentiality. Having the potentiality of developing into two different structures, as an embryonic gonad developing into an ovary or testis.

biradial symmetry. A type of symmetry in which the body consists of radially arranged parts, half of which lie on each side of a median longitudinal plane, as a ctenophore.

biramous. Having two branches, an inner endopodite and an outer exopodite, with reference to appendages of various crustaceans.

birch. A tree (*Betula*) of the family Corylaceae (hazel family).

bird. Any warm-blooded, feathered vertebrate of the class Aves, with forelimbs modified into wings.

bird lice. Parasitic insects of the order Mallophaga.

bird of paradise. One of a number of brilliantly colored birds of the family Paradisaeidae inhabiting the forests of New Guinea.

bird's nest, edible. *See* swallow's nest.

bird's nest fungus. *Cyanthus stirecoreus*, a basidiomycete with a cup-shaped basidiocarp containing egglike spore-producing bodies.

birth. 1. The production of offspring; delivery, parturition, labor. **2.** That which is born or produced.

birth control. Control or limitation of the number of offspring; contraception.

birthmark. A nevus; an area of pigmentation or vascularization of the skin present at birth or shortly afterward.

birth rate. The number of live births reported in a calender year per 1000 of actual or estimated population, known as crude birth rate. In the United States, in 1960 the rate was 22.6.

birthwort. A species of low herbs or twining shrubs of the family Aristolochiaceae, especially those of the genus *Aristolochia*, whose aromatic roots were thought to aid in childbirth.

bisexual. 1. Of or pertaining to both sexes; having both male and female sex organs; hermaphroditic. **2.** *bot.* Having both stamens and pistils.

bison. 1. The wisent or European bison (*Bison bonasus*). **2.** The

57

American buffalo (*Bison bison*), a large, gregarious quadruped of western North America.

bitterling. *Rhodeus amarus*, a small European minnow recently introduced into the United States. The female deposits eggs in the mantle cavity of freshwater clams.

bittern. A wading bird of the heron family, Ardeidae, as *Botaurus lentiginosus*, the American bittern.

bittersweet. 1. A poisonous, climbing, solanaceous plant (*Solanum dulcamara*) with coral-red berries. **2.** An American shrub (*Celastrus scandens*).

bitunicate. Having a wall of two layers, with reference to an ascus.

biuret reaction. A reaction which occurs when proteins or other substances containing peptide bonds are treated with copper sulfate in an alkaline solution. A red, purple, or violet color results.

bivalent. 1. *biol.* Double or joined in pairs, said of chromosomes. **2.** *chem.* Having a valence of two.

¹bivalve. 1. *zool.* Having a shell consisting of two valves or valvelike parts. **2.** *bot.* Having two valves, as diatoms.

²bivalve. A mollusk of the class Pelecypoda.

bivium. The two rays of a starfish (sea star) between which lies the madreporite.

bivoltine. Having two broods a year, said of insects.

black bass. *See* bass.

black bear. The common American bear (*Ursus americanus*).

blackberry. The plant or fruit of any of the species of brambles of the genus *Rubus*.

blackbird. 1. Any of a number of species of birds in which individuals, especially males, are black or nearly so. **2.** In America, any of several birds of the family Icteridae, as the yellow-headed, red-winged, rusty, or Brewer's blackbird.

Black Death. *See* plague.

blackfish. One of a number of species of small, toothed whales of the genus *Globicephala*.

black fly. A small, dark-colored dipterous insect of the genus *Simulium*, family Simuliidae, a pest to man and domestic animals; also called buffalo gnat.

blackhead. 1. A scaup duck. **2.** An encysted glochidium in the tissues of a fish. **3.** An infectious disease of domestic fowl, especially turkeys, caused by a protozoan (*Histomonas melea-gridis*). **4.** A comedo, an enlarged, discolored sebaceous gland.

black mold. Any of the molds of the order Mucorales, as *Rhizopus nigricans*, the common bread mold.

blackout. Loss of vision and finally unconsciousness resulting from linear and centrifugal acceleration, due to reduced blood supply to the head.

black snake. One of a number of dark-colored snakes, especially *Coluber constrictor*, the blue racer, a nonpoisonous snake.

black spot grub. *See* yellow grub.

black widow spider. A poisonous spider (*Lactrodectus mactans*) identified by a reddish, hourglass-shaped spot on ventral surface of abdomen.

bladder. 1. *zool.* A membranous sac or vesicle which serves as a reservoir for a fluid or a gas, as a urinary *bladder*, swim *bladder*. **2.** *bot.* An expanded, gas-containing structure on the blade of various marine algae; a pneumatocyst.

bladder kelp. *See* bull kelp.

bladdernose. The hooded seal (*Cystophora cristata*).

bladdernut. A bladderlike seedpod of *Staphylea trifolia*, a shrub or small tree; also the plant which produces it.

bladder worm. A cysticercus.

bladderwort. An insect-eating aquatic or bog plant of the genus *Utricularia*.

bladder wrack. The common rockweed (*Fucus*), a brown alga.

bladdery. Inflated; having thin walls like a bladder.

blade. 1. The leaf of a plant, especially that of an herb. **2.** The thin expanded portion of a leaf, as distinguished from the petiole; lamina. **3.** The flat, leaflike thallus of certain brown algae, as in kelps.

blast. 1. To ruin or destroy, as by a noxious wind. **2.** To render incapable of producing fruit or seeds. **3.** To become withered or blighted.

blastema. 1. An undifferentiated part from which an organ or structure develops. **2.** The part of an organism which, in asexual reproduction, gives rise to a new organism. *Compare* anlage.

blastocoel. The cavity of a blastula or blastocyst; the segmentation or cleavage cavity; the subgerminal cavity.

blastocyst. The mammalian blastula, consisting of an outer layer of cells (trophoblast) enclosing a blastocoel. Attached to the inner surface of the trophoblast is an inner cell mass from which the embryo develops; also called blastodermic vesicle.

blastoderm. In eggs undergoing meroblastic cleavage, the disc of protoplasm in which cleavage takes place; in centrolecithal ova, the surface layer of cells enclosing the yolk.

blastodisc. In telolecithal ova, a small disc of yolk-free cytoplasm located at the animal pole from which the embryo and its membranes develop; germinal disc, blastodisk.

blastogenesis. The development of a new individual by asexual reproduction; the development of a new individual or a part from a blastema. *Compare* embryogenesis.

blastokinesis. Movement of an embryo within an egg, with reference to insects.

blastoma. A tumor, especially one that originates from embryonic cells.

blastomere. a. A cleavage or segmentation cell. **b.** Any of the cells resulting from cleavage prior to gastrulation. *See* macromere, micromere.

blastopore. In a gastrula, the opening of the archenteron or gastrocoel.

blastostyle. In colonial coelenterates, a modified reproductive polyp consisting of a central axis bearing numerous medusa buds.

blastozooid. In tunicates, a zooid which arises by budding; a blastozoite. *Compare* oozooid.

blastula. A stage in animal development following cleavage in which the cells are typically arranged in the form of a hollow sphere with a single layer, the blastoderm, surrounding a cavity, the blastocoel. In telolecithal ova, its form is greatly modified. Also called blastodermic vesicle, blastodisc.

bleed. 1. To lose blood, as from a wound. **2.** To exude water or sap.

bleeding. *See* exudation.

blenny. Any of a number of percomorph fishes of the family Blenniidae, as the butterfly fish.

60

blepharoplast. In flagellated protozoans, a deeply staining body from which the flagellum arises; basal granule.

blesbok. *Damaliscus albifrons*, a South African antelope; also blesbuck.

blight. Any plant disease which results in withering and cessation of growth, especially one caused by a fungus, as the chestnut *blight*.

blind. Sightless; lacking a sense of vision.

blind fish. One of several species of fishes of the family Amblyopsidae inhabiting caves in the United States. Common genera are *Amblyopsis* and *Typhlichthys*.

blind snake. A small, wormlike, burrowing reptile, as *Typhlops* of the family Typhlopidae, and *Leptotyphlops* of the family Leptotyphlopidae.

blind spot. The optic disk of the retina where optic nerve fibers make their exit from the eye. It is devoid of rods and cones, hence insensitive to light.

blindworm. *See* blind snake; slowworm.

blood. 1. In vertebrates, the fluid that circulates in the cardio-vascular system consisting of a fluid plasma containing corpuscles and platelets. 2. In invertebrates, a similar fluid, the plasma sometimes containing a respiratory pigment in solution.

blood clotting. Coagulation or the conversion of fluid blood into a solid mass or clot composed of fibers of fibrin in which blood cells are enmeshed.

blood corpuscle. A blood cell; an erythrocyte (red blood cell) or leukocyte (white blood cell).

blood count. The determination of the number of blood cells (erythrocytes or leukocytes) per cu. mm. of blood. In man, red cells average 5 million in males, $4\frac{1}{2}$ million in females; white cells average 5,000 to 10,000 per cu. mm.

blood fluke. A trematode of the genus *Schistosoma;* a schistosome.

blood groups. In man, four primary types of blood based on the presence or absence of antigens (agglutinogens) in red blood cells and antibodies (agglutinins) in the serum. The principal groups and their frequencies are: 0 (47%), A (41%), B (9%), and AB (3%). Other blood groups include M,N, and Rh groups based on the presence of corresponding antigens in red blood cells. *See* universal donor, universal recipient.

blood island. An isolated mass of mesodermal cells from which the first blood cells and blood vessels in vertebrates arise.

blood pigment. A respiratory pigment present in the blood, as hemoglobin.

blood plasma. The fluid portion of circulating blood consisting of water (91%) and solids (9%). Solids include proteins, nonprotein nitrogenous substances, nutrients, inorganic substances, and miscellaneous substances, as antibodies, enzymes, hormones, and vitamins. Gases are present in solution.

blood platelets. Minute bodies in blood plasma consisting of non-nucleated fragments of the cytoplasm of megakaryocytes. They are the source of thromboplastin. Also called thrombocytes.

blood pressure. The pressure exerted by the blood against the walls of the vessels containing it. In man, arterial blood pressure averages 120/80, the first figure representing systolic pressure, the second, diastolic.

blood proteins. Proteins present in the blood plasma, principally serum albumin, serum globulin, and serum fibrinogen.

blood serum. The clear fluid which exudes from a clot, consisting of plasma minus serum fibrinogen.

blood sinus. An irregular space or channel through which blood flows, as those in the tissues of arthropods or in the meninges of the brain of a vertebrate.

bloodsucker. An animal that sucks blood, especially a leech. Term is also applied to certain red-throated Asiatic and Australian lizards and true vampires.

blood sugar. A carbohydrate, chiefly glucose, present in blood plasma. Normal value in man — 70-120 mg. per 100 ml.

blood transfusion. The introduction of blood (whole or fraction) or plasma into a blood vessel.

blood typing. The procedure used to determine the blood group to which a person belongs, consisting of adding anti-A and anti-B sera to suspensions of unknown cells.

blood vessel. A vessel which conveys blood. *See* arteriole, artery, capillary, metarteriole, sinus, sinusoid, vein, venule.

bloodworm. The red, aquatic larva of dipterous insects of the genus *Chironomus*. *See* midge.

¹**bloom.** To yield or produce blossoms or flowers.

²**bloom.** **1.** The state of flowering, as roses are in *bloom*. **2.** The delicate, waxy or powdery coating of the fruit or leaves of certain plants. **3.** The sporadic seasonal occurrence of enormous numbers of algae in inland waters.

¹**blossom.** The flower of a seed plant, especially one which produces an edible fruit; a bloom.

²**blossom.** To bloom or put forth blossoms.

blow. *See* spout.

blowfish. *See* puffer.

blowfly. One of several species of flies, especially those of the family Calliphoridae, which deposit their eggs or larvae on meat or in wounds of animals. *See* myiasis.

blowhole. The nostril or spiracle on dorsal surface of the head of whales and other cetaceans through which spouting takes place.

blow snake. The hognose snake (*Heterodon*), noted for feigning death. Also called puff adder or spreading viper.

blubber. The fat of whales and other marine animals.

blue baby. A baby with bluish skin (cyanosis) resulting from imperfect oxygenation of the blood. Such may be temporary as from obstruction of respiratory passageways or permanent due to a cardiac anomaly such as a patent foramen ovale or ductus arteriosus or both.

bluebird. A North American songbird (*Sialia sialis*) of the thrush family, Turdidae.

blue crab. The edible crab (*Callinectes*) of the Atlantic coast.

blue fox. A color phase of the Arctic fox (*Alopex lagopus*).

blue gill. *Lepomis macrochirus*, a pan fish of the sunfish family, Centrarchidae, a highly esteemed sport and food fish.

blue jay. The common jay (*Cyanocitta cristata*) of the family Corvidae.

blue racer. *See* black snake.

BMR. Basal metabolic rate. *See* metabolic rate.

BNA. Basle Nomina Anatomica, a system of antomical nomenclature

boa. Any snake of the family Boidae.

boa constrictor. A large tropical American snake (*Constrictor constrictor*) which kills its prey by crushing it to death within its coils.

boar. 1. The uncastrated male of swine. **2.** The male of a guinea pig.

boat shell. A marine gastropod of the genera *Crepidula* and *Cymbium*.

bobcat. The wildcat or bay lynx (*Lynx rufus*).

bobolink. A songbird (*Dolichonyx oryzivorus*) of the family Icteridae. Also called reedbird or ricebird.

bobwhite. A species of American quail, especially *Colinus virginianus*, an upland game bird. Called partridge in southern United States.

body. 1. The total substance of an organism. **2.** The trunk or main part of an organism as distinguished from its limbs or appendages. **3.** The principal part of an organ or structure as distinguished from its ends or extremities, as of a bone.

body cavity. The principal cavity of an organism; a coelom, hemocoel, or pseudocoel.

body cells. Nonreproductive cells or soma; somatic cells.

body louse. A sucking, parasitic insect (*Pediculus humanus corporis*) of the order Anoplura, a transmitter of the causative organisms of epidemic typhus, trench fever, and relapsing fever. *See* louse, nit.

body stalk. In mammalian development, a narrow stalk of mesoderm which connects the caudal end of the embryo to the chorion.

body whorl. The largest coil of a gastropod shell.

bog. Wet, spongy earth consisting principally of decayed vegetable matter; a morass.

bog moss. Any moss which grows in bogs, especially those of the genus *Sphagnum*. *See* peat.

bole. The main trunk or stem of a tree; a caudex.

boll. The pod or capsule of a plant, as a cotton *boll*.

boll weevil. *Anthonomus grandis*, a serious pest of cotton.

bollworm. The larva of a moth (*Heliothis armigera*) which feeds on cotton bolls. Also called corn earworm and tomato fruitworm.

bolus. A round mass, especially a moist mass of food formed in the mouth and prepared for swallowing.

bone. **1.** Osseous tissue, a form of hard, rigid connective tissue consisting of bone cells (osteocytes) embedded in a matrix of organic matter in which mineral matter, principally calcium and phosphorus compounds, has been deposited. *See* haversian system. **2.** An individual unit or element of the skeletal system, as the femur.

bonebreaker. A bird which breaks bones to feed on the marrow, as the giant fulmer, lammergeier, and osprey.

bone cell. A cell in bone tissue, as an osteoblast, osteocyte, osteoclast.

bongo. *Boocercus euryceros*, a large antelope of equatorial Africa.

bonito. One of several species of percomorph fishes of the tuna family, Thunnidae.

bonnet. A protuberance on the snout of certain whales.

bontebok. *Damaliscus pygargus*, a South African antelope.

bony fishes. The Osteichthyes, a class of vertebrates whose representatives have an ossified or partly ossified skeleton.

booby. A large tropical sea bird related to the gannets, as the Atlantic blue-faced booby (*Sala dactylatra*) of the family Salidae.

book gill. A gill composed of thin plates or lamella, as that of the king crab (*Xiphosura* or *Limulus*).

book lice. Wingless insects of the order Psocoptera, family Lipsocelidae.

book lung. A respiratory structure composed of a number of thin plates or lamella enclosed within an air chamber, as in spiders.

boomer. A large male kangaroo hunted for sport.

boreal. Northern.

boring clam. Any of a number of marine clams of the family Pholadidae which bore into wood, coral, or soft rock, as *Pholas*.

boring sponge. A sponge which bores into the shells of oysters, clams, and other mollusks, as *Cliona*.

¹boss. A knoblike protuberance or umbo.

²boss. In the United States, a cow or calf.

bossed. Having a rounded surface with a projection in the center.

bostryx. A helicoid cyme.

bosvark. The bushpig of South Africa.

bot. The larva of a botfly.

botanize. To study plants in the field or to collect plants for botanical purposes.

botany. 1. The science which deals with plants and plant life. 2. A book which deals with plants.

botfly. One of a number of flies belonging to the families Gasterophilidae (horse bots), Guterebridae (skin bots), and Oestridae (head bots), whose larvae called bots or warbles infest man and various animals causing myiasis. *See* warble.

bothridium. One of four lappetlike outgrowths on the scolex of certain tapeworms.

bothrium. A slitlike groove on the scolex of certain tapeworms, as the fish tapeworm.

botryoid. Having many rounded prominences, thus resembling a cluster of grapes.

botryoidal tissue. A form of connective tissue present in the coelomic spaces of a leech.

bottle. A unistratose involucre which encloses the sex organs of certain bryophytes.

botulism. A form of food poisoning resulting from ingestion of a toxin (botulin) produced by *Clostridium botulinum*.

bouton. The lappetlike terminal portion of the glossa in certain hymenopterans as the honeybee.

bouton terminal. The expanded end of the terminal branch of an axon which makes a synaptic connection with the cell body or dendrites of another neuron; an end bulb, end foot, or neuropodium.

bovid. Any member of the Bovidae, a family of hollow-horned ruminants which includes cattle, sheep, goats, bison, water buffalo, musk ox, and certain antelopes.

bovine. Cattlelike; of, pertaining to, or derived from cattle.

bowel. The gut or intestine.

bowfin. A carnivorous, freshwater fish (*Amia calva*) of central and eastern North America. Commonly called dogfish or mudfish.

Bowman's capsule. The expanded end of a renal tubule which encloses a glomerulus. *See* renal corpuscle.

brachial. Of or pertaining to the brachium or arm.

brachial plexus. An interconnecting network of cervical and thoracic nerves which innervate the forelimb of vertebrates or the arm of man.

brachiate. 1. *bot.* Possessing widely spreading branches resembling arms, as the maple. **2.** *zool.* Possessing arms.

brachiation. Moving through trees by swinging from limb to limb suspended by the forelimbs, as the anthropoid apes.

brachiolaria. The free-swimming, ciliated larva of asteroid echinoderms which develops from a bipinnaria.

brachiopod. Any member of the Brachiopoda, a phylum of marine, lophophorate invertebrates which includes the lamp shells, sessile forms with an external shell of two valves (dorsal and ventral), as *Lingula, Terebratulina*.

brachium. 1. The arm; the portion between the shoulder and elbow. **2.** An armlike structure or process.

brachycephalous. Short-headed or broad-headed.

brachycerous. Possessing short antennae.

brachydactyly. Having abnormally short fingers and toes.

brachydont. Possessing or denoting teeth with short, low crowns, long roots, and reduced pulp cavity.

brachypterous. Having short wings, especially wings that do not cover the abdomen.

brachyurous. Having a short, reduced abdomen, as in crabs.

brachyury. Having short tails, as in certain mutant strains of mice.

bracken fern. The brake fern (*Pteris aquilina*).

bracket fungi. The pore fungi, basidiomycete fungi whose sporophores or basidiocarps grow horizontally on tree trunks forming shelflike structures (brackets or conks).

brackish. Salty, with saline content less than that of sea water.

braconid. A hymenopterous insect of the family Braconidae, whose larvae parasitize other insects, as *Microgaster tibialis*, a parasite of the European corn borer. *See* ichneumon fly.

bract. 1. *bot.* **a.** A small, specialized leaf in the axil from which a flower or floral axis develops; a modified leaf on the axis itself. **b.** A modified, scalelike leaf of a cone or strobilus. **2.** *zool.* A modified medusoid (hydrophyllium, phyllozooid) serving a protective function, as in siphonophores.

bracteal. Of or pertaining to bracts.

bracteate. Possessing bracts.

bracteolate. Possessing small bracts or bractlets.

bracteole. A small secondary bract or bractlet.

bracteose. Having many bracts.

bractlet. A small bract situated in the axis of a larger bract; bracteole.

bradycardia. Abnormal slowness of heart beat.

bradytelic. Evolving slowly.

brain. 1. In vertebrates, the encephalon; the portion of the central nervous system lying within the cranial cavity, consisting of the telencephalon (olfactory lobes and cerebral hemispheres), diencephalon (pineal body, thalamus, hypothalamus, hypophysis), mesencephalon (midbrain), metencephalon (pons, cerebellum), and myelencephalon (medulla oblongata). **2.** In invertebrates, a large ganglion located in the anterior region from which a nerve cord or cords pass posteriorly.

brain case. The cranium or portion of the skull that encloses the brain.

brain coral. A coral (*Meandrina* or *Meandra*) which resembles the human brain in appearance.

brain membranes. The meninges, *q.v.*

brain sand. Corpora aranacea, sandlike particles present in the pineal body.

brain stem. The portions of the vertebrate brain other than the cerebrum and cerebellum. It includes the medulla oblongata, pons, midbrain, and diencephalon.

branchia pl. **branchiae.** A gill; a ctenidium.

branchial. Of or pertaining to gills or branchiae.

branchial aperture. The opening of a gill chamber to the exterior.

branchial arch. 1. A gill-supporting structure between the gill clefts of fishes; a gill arch or visceral arch. **2.** In a vertebrate embryo, the structure that develops in the wall of the pharynx between successive branchial grooves.

branchial basket. The cartilaginous framework supporting the gill pouches of cyclostomes.

branchial chamber. A cavity which contains gills, as that in crustaceans, amphibians, and fishes.

branchial cleft. 1. A gill slit. **2.** In a vertebrate embryo, the opening between two visceral arches.

branchial cyst. A cervical cyst, a closed epithelial sac located in the neck region, resulting from incomplete obliteration of a branchial groove or pouch.

branchial fistula. A cervical fistula, a canal or diverticulum in the neck region resulting from incomplete closure of a branchial groove or pouch.

branchial groove. One of a series of ectodermal invaginations in the neck region of vertebrate embryos corresponding in position to the pharyngeal pouches.

branchial heart. A supplementary, contractile structure located at the base of the gill in a squid.

branchial musculature. The muscles which move the visceral arches and jaws of lower vertebrates or homologues of these structures in higher vertebrates; muscles which are derived from splanchnic mesoderm of the hypomere instead of from myotomes.

branchial ray. A gill ray.

branchial skeleton. *See* visceral skeleton.

branchiate. Possessing branchiae or gills.

branchiocardiac. Pertaining to gills and to the heart.

branchiocardiac sinus. In crustaceans, one of the several channels through which blood passes from the gills to the pericardial sinus and thence to the heart.

branchiopod. Any crustacean of the subclass Branchiopoda which includes various shrimps (brine, fairy, tadpole, clam) and water fleas, mostly freshwater forms with leaflike thoracic appendages which bear branchia (gills).

branchiostegal membrane. A membrane supported by bony rods (branchiostegal rays) which, in most fishes, extends from the inner surface of the operculum to the body wall.

branchiostegite. In crustaceans, the portion of the carapace that covers the gills.

branchiuran. Any crustacean of the subclass Branchiura, which includes the fish lice.

brandling. The stinking earthworm (*Eisenia foetida*).

brant. A wild goose of the genus *Branta*.

Brazil nut. A large, three-angled nut (actually a seed) of a brazilian tree (*Bertholletia excelsa*), developing in a large globular fruit containing 18 to 24 closely packed nuts.

breakage plane. In certain decapod crustaceans which exhibit autotomy, a region traversing one of the basal segments of an appendage at which self-amputation occurs.

breast. 1. The fore part of the body between neck and abdomen; the anterior portion of the chest or thorax. **2.** One of the human mammary glands located in the thoracic region.

breastbone. The sternum.

breathing. The inhalation and exhalation of air; inspiration and expiration; external respiration.

breech. The lower, posterior portion of the body; the buttocks or nates.

¹breed. To produce or bring forth young; to propagate.

²breed. A race or variety of animals or plants, especially one developed through the influence of man.

brier or **briar.** A plant bearing thorns or prickles, especially those of the genera *Rosa*, *Rubus*, and others; a bramble.

brine. Water saturated with salt; a strong, alkaline solution.

brine shrimp. *Artemia salina*, a branchiopod crustacean which lives in waters of high salt concentration.

bristle. A stiff, coarse hair or hairlike structure.

bristletail. A small, wingless insect of the order Thysanura, as the silverfish and firebrat.

bristly. Bearing stiff hairs or bristles.

brit. Dense swarms of minute, copepod crustaceans, especially *Calanus finmarchicus*, an important food for marine fishes and whales.

brittle star. An ophiuroid with long, slender, flexible arms set off sharply from the central disc, as *Ophiothrix*.

brocket. A small South American deer of the genus *Mazama*.

bronchiole. One of the numerous small divisions of a bronchus.

bronchus. One of the two divisions of the trachea or windpipe leading to a lung.

brontosaur. A large, herbivorous dinosaur of the genus *Brontosaurus*.

¹brood. The young of animals, especially birds, which are hatched and taken care of together, as a brood of chickens.

²brood. To sit on and incubate eggs.

brood capsule. A hollow cyst which develops from the inner wall of a hydatid cyst or from the wall of a secondary cyst of *Echinococcus granulosus*, a tapeworm of carnivores. From its inner wall, scolices develop.

brood chamber. A brood pouch, a marsupium.

brood parasitism. Behavior exhibited by certain birds which deposit their eggs in the nests of other birds which incubate, hatch, and care for the young, as European species of cuckoos, cowbird.

brood pouch. A space or cavity other than the uterus in which young develop, as the modified outer gill of a freshwater clam; a brood chamber or marsupium.

broomrape. One of a number of leafless herbs of the genera *Orobanche* and *Phelipaea* which live as parasites on the roots of other plants, as tomato, tobacco.

brotochore. A plant spread or dispersed by man.

brown body. 1. In bryozoans, a dark brown mass formed by a degenerating polypide. 2. In holothurians, a mass of amebocytes present in the coelom.

Brownian movement. A jiggling or vibratory movement exhibited by microscopic particles, especially those of a colloidal system when suspended in a fluid.

browse line. In wooded regions, a line usually at a height of about 8 feet below which all available vegetation has been consumed, as that seen in overpopulated deer areas. Also called deer line.

brucellosis. A febrile disease resulting from infection by bacteria of the genus *Brucella*. Also called undulant fever, Malta fever, Mediterranean fever, contagious abortion.

bryophyte. Any plant of the division Bryophyta which includes the mosses, liverworts, and hornworts.

bryozoan. An invertebrate of the group Bryozoa (Polyzoa) which included organisms commonly called moss animals or zoophytes. The phylum has now been divided into two phyla, Entoprocta and Ectoprocta, *q.v.*

71

bubble shell. A tropical marine gastropod with a thin, globose, or oval shell, of the family Bullidae, as *Bulla, Hydatina.*

bubo pl. **buboes.** Inflammation and swelling of one or more lymph nodes, especially those of the groin and axilla, as occurs in infectious diseases as syphilis and bubonic plague.

bubonic plague. A form of plague characterized by buboes, primarily a disease of rodents but transmissible to man through the agency of fleas. *See* plague.

buccal. 1. Of or pertaining to the cheeks 2. Of or pertaining to the sides of the mouth cavity.

buccal cavity. 1. The mouth cavity of gastrotrichs, nematodes, and other invertebrates. 2. The mouth cavity or oral cavity of most vertebrates, especially the region lying within the dental arches.

buccal cirri. The oral tentacles surrounding the vestibule of a lancelet.

buccal field. The area surrounding the mouth of a rotifer.

buccal funnel. The oral funnel, a suctorial structure by which the sea lamprey attaches itself to and secures food from its host.

buccal mass. A mass of tissue in the head of gastropods and other mollusks consisting principally of the radula and associated structures.

buck. The male of various mammals (deer, antelopes, goats, hares, rabbits, rats).

buckeye. A tree of the genus *Aesculus*, especially *A. octandra*, the yellow or sweet buckeye. Also called horse chestnut.

buckthorn. One of a number of trees or shrubs of the genus *Rhamnus*, of the buckthorn family, Rhamnaceae. *R. purshiana* is the source of cascara, a laxative.

buckwheat. *Fagopyrum sagittatum*, an herb of the buckwheat family, Polygonaceae, cultivated as a food plant.

bud. 1. *bot.* **a.** An undeveloped shoot usually containing rudimentary floral or foliage leaves and enclosed within protective bud scales. **b.** A protrusion of a bacterium or yeast cell capable of developing into a new organism. 2. *zool.* **a.** A protuberance or outgrowth which is capable of developing asexually into a new organism. **b.** *embryol.* A protuberance which forms the primordium of a structure, as a limb *bud.*

budding. 1. *biol.* A method of asexual reproduction in which a new individual arises from an outgrowth or bud. **2.** *horticulture*. A method of grafting in which a scion composed of a single bud with a small amount of adjacent tissue is inserted into a slit in the bark of the receiving stock.

bud scales. Scalelike structures which form a protective covering for winter buds.

buffalo. One of several species of heavily-built bovine animals, as the water buffalo (*Bubalus bubalis*), cape buffalo (*Syncerus caffer*), anoa (*Anoa depressicornis*), and North American buffalo or bison (*Bison bison*).

buffalo fish. One of several species of freshwater food fishes of the sucker family, Catostomidae, as the large-mouth buffalo fish (*Ictiobus cyprinella*).

buffalo gnat. A black fly.

buffer. A substance which prevents a change in the hydrogen ion concentration (pH) of a solution upon the addition of an acid or base.

buffered. Treated in such a way as to resist changes in pH, said of solutions, bacteriological media, etc.

buffer solution. A solution consisting of a weak acid and a highly ionized salt of the same acid or base.

buffer species. An alternative food for a predator, which reduces pressure on the primary prey.

bufflehead. *Bucephala* (*Glaucionetta*, *Charitonetta*) *albeola*, a small diving duck of the family Anatidae.

bug. 1. *entomol.* **a.** Any of a number of insects of the order Hemiptera which includes the true bugs, as the giant water bug, bedbug, chinch bug, squash bug. **b.** Any of a number of insects of various orders, as the mealy bug (Homoptera), lightningbug (Coleoptera) crotonbug (Orthoptera). **2.** Any small creeping or flying organism.

bulb. 1. *bot.* A single, large and usually globose bud, generally subterranean, consisting of a short stem bearing fleshy, overlapping, scalelike leaves, as that formed by many monocots, as the onion, lily, tulip. **2.** *anat.* **a.** The expanded oval portion of a tube or structure, as the bulb of the aorta or of the penis. **b.** Formerly, the medulla oblongata.

bulbil. A small, secondary bulb or bulblike reproductive structure, as that which develops on the tips of new stems in *Lycopodium* or in the axils of leaves in the tiger lily.

bulbillus. A contractile enlargement at the base of each of the lateral arteries leading to the gills in a lancelet.

bulblet. A small bulb that develops in a leaf axil inflorescence or some other unusual location.

bulbose, bulbous, bulbar. 1. Having the structure of or resembling a bulb. 2. Producing or containing bulbs. 3. Growing from bulbs.

bulbourethral gland. Cowper's gland, one of two compound, tubuloalveolar glands located near the bulb of the urethra which secrete a viscid, alkalinizing fluid into the urethra.

bulbus arteriosus. The expanded proximal portion of the ventral aorta in lower vertebrates.

bull. The male of any bovine mammal; also the male of most large mammals, as the moose, elephant, seal, sea lion.

bulla. 1. A large bleb or blister. 2. In certain mammals, **a.** the expanded portion (auditory bulla) of the tympanic bone; **b.** an enlarged ethmoid cell which forms a rounded projection into the middle meatus.

bullate. Having a blistered or puckered appearance, as the surface of certain leaves, as the Savoy cabbage.

bullfrog. *Rana catesbeiana*, a large frog common in the southern United States.

bullhead. Any of a number of species of fishes having a large, blunt head, especially a catfish of the genus *Ameiurus*.

bulliform cells. Large, thin-walled cells present in the leaves of most monocots which function in the unrolling of developing leaves and in opening and closing movements of mature leaves.

bull kelp. *Nereocystis*, the sea otter's cabbage or bladder kelp, a giant kelp sometimes reaching a length of 150 feet.

bull snake. One of several species of snakes of the genus *Pituophis*. They are large, harmless but pugnacious, killing their prey by constriction, as the pine snake and gopher snake.

bully tree. *See* balata.

bulrush. *Juncus effusus*, a large rush growing in marshy land or in water. Term is also applied to various species of the genus *Scirpus*.

bumblebee. One of several species of heavy-bodied bees of the genera *Bombus* or *Psithyrus*, the latter, social parasites living in the nests of the former.

bundle. 1. *anat.* A group of elongated structures as muscle or nerve fibers; a fasciculus. **2.** *bot.* An elongated group of conducting vessels in the stem of a vascular plant. *See* vascular bundle.

bundle scar. A scar marking the end of a vascular bundle, one or more being present in a leaf scar.

bundle sheath. A layer of lignified cells surrounding a vascular bundle in monocots.

bundle tip. The end of a vascular bundle in a leaf, usually a single vessel or tracheid.

bunodont. Having low, rounded cusps, as a molar tooth in man.

bunt. Stinking smut of wheat caused by various species of *Tilletia*.

bunting. One of several species of seed-eating birds of the family Fringillidae, as the indigo bunting (*Passerina cyanea*).

buprestid. A beetle of the family Buprestidae which includes the metallic wood-boring beetles, as *Chrysobothris femorata*, the flat-headed apple tree borer.

bur. A rough and prickly covering of a fruit, as that of the chestnut or burdock.

burbot. A freshwater food fish (*Lota lota*) of the family Gadidae.

burdock. A composite of the genus *Arctium*, especially *A. lappa*, a coarse biennial weed.

burl. An abnormal outgrowth or excrescence on a tree trunk, usually due to stimulation of the cambium by insects or other agents.

burro. A small donkey, especially one used as a pack animal.

¹burrow. A hole or excavation used as a breeding place, shelter, runway, or habitat.

²burrow. To make a passageway beneath a surface.

burrowing. FOSSORIAL.

bursa. 1. *anat.* A sac or saclike cavity, as **a.** one in connective tissue filled with synovia for minimizing friction; **b.** the cavity of the greater omentum. **2.** *zool.* **a.** The flared, posterior end of male nematodes. **b.** An eversible, cuplike copulatory structure

75

in male acanthocephalans. **c.** A sac in female turbellarians which stores sperm. **d.** An invagination on oral surface of ophiuroids.

bursa Fabricii. The bursa of Fabricius, a saclike lymphatic organ in young birds communicating dorsally with the cloaca.

bush. 1. A shrub, especially one with many branches and dense foliage. **2.** Uncultivated land covered with scrubby vegetation, as in Australia.

bush baby. A lemur (*Galago senegalensis*), an arboreal, nocturnal primate of Africa.

bushbuck. The harnessed antelope (*Tragelaphus scriptus*) of South Africa; also called guib.

bushman. One of a race of primitive peoples of Central and Southern Africa, related to the pygmies.

bushmaster. *Lachesis mutus*, a large, venomous snake of Central and South America.

bushpig. *Potamochoerus porcus*, a wild pig of Africa, called the red river hog in West Africa, the bosvark in South Africa.

bustard. A game bird of the genus *Otis*, family Otididae, related to the cranes and plovers. The great bustard (*Otis tarda*) is the largest European game bird.

butcher-bird. *See* shrike.

buteo. A buzzard hawk, one of a number of broadwinged hawks of the genus *Buteo*, as *B. jamaicensis*, the red-tailed hawk.

butterball. A bufflehead, *q.v.*

buttercup. The crowfoot, a plant of the genus *Ranunculus*, possessing bright yellow flowers.

butterfly. One of a large number of slender-bodied, diurnal insects of the order Lepidoptera, suborder Rhopalocera, usually possessing knobbed antennae and brightly colored wings held vertically over the back when at rest. *Compare* moth.

button. 1. *zool.* The cone-shaped tip of the rattle of a rattlesnake. **2.** *bot.* Any of a number of structures which resemble a button, as a bud, a rose hip, an immature fruiting body of a mushroom.

buzzard. 1. Term applied to one of several hawks of the genus *Buteo*. **2.** A vulture, *q.v.*

byssal gland. A glandular structure in bivalve mollusks which secretes thrèads of a viscid substance which harden on contact with sea water, forming byssal threads. Also called byssal pit. *See* byssus.

byssus. A filamentous structure composed of byssal threads by which certain mollusks, especially those of the genera *Mytilus* and *Pinna*, attach themselves to rocks and other objects.

C

cabbage. A leafy, garden plant (*Brassica oleracea*) of the mustard family, Cruciferae.

cacao. A South American tree (*Theobroma cacao*) extensively cultivated for its seeds, cocoa beans, the source of commercial cocoa and chocolate.

cactus. Any plant of the cactus family, Cactaceae, characterized by thick, fleshy, green stems and branches bearing scales, spines or prickles; common inhabitants of warm, dry areas.

caddis fly. A small to medium-sized insect of the order Trichoptera. Their aquatic larvae, called caddisworms, construct portable cases in which they live and pupate.

caducibranch. An amphibian which loses its gills at metamorphosis.

caducous. Not persistent; falling off early or prematurely.

cae-. For words beginning with CAE- not found here, see words beginning with CE-.

caecilian. A wormlike, burrowing amphibian of the order Gymnophiona (Apoda).

caeoma. A spore-producing structure in rust fungi.

caerulescent. Bluish.

caffeine. An alkaloid, $C_8H_{10}N_4O_2$, present in tea and coffee, mate, kola, and other plants. It is a central nervous system, cardiac, and respiratory stimulant and a diuretic.

caiman or **cayman.** A tropical American alligator of the genus *Caiman*.

calabash. 1. A gourd, especially one shaped like a flask. **2.** The hard-shelled fruit of the calabash tree (*Crescentia cujete*).

calamus. 1. The quill of a feather. **2.** The sweet, aromatic rhizome of the sweet flag (*Acorus calamus*).

calathiform. Cup-shaped.

calcaneus. The heel bone, os calcis.

calcar. A spur or spurlike process.

calcarate. Bearing a spur or calcar.

calcareous. Of the nature of, consisting of, or containing calcium carbonate.

calcicolous. Preferring soils heavily impregnated with lime.

calciferol. Vitamin D_2 produced by irradiation of ergosterol.

calciferous. Producing or containing calcite or carbonate of lime.

calcification. The deposition of calcium or lime salts within an organ or tissue.

calcifuge. A plant which does not thrive on calcareous soils.

calcite. CALCIUM CARBONATE.

calcium. A soft, silver-white metallic element abundant in nature but occurring only in combination with other elements. Symbol Ca; at. wt. 40.08.

calcium carbonate. $CaCO_3$, the most abundant and widely distributed compound of calcium. It occurs in nature as calcite (limestone, marble, chalk) and appears as a white precipitate when CO_2 is passed through limewater. It is an important constituent of bone, animal shells, and skeletal structures of invertebrates.

calculus. A hard concretion composed principally of mineral matter which forms in hollow organs or ducts, as a gallstone.

calf. 1. The young of a cow or other bovine quadruped. **2.** The young of most large mammals. **3.** The fleshy portion of the leg below the knee.

call. A simple, brief, vocal sound produced by birds usually for a distinct purpose, as an alarm *call*, begging *call*, or a *call* to bring the brood together.

callose. A carbohydrate present especially in the sieve plates of vascular plants.

callosity. A callus.

callous. Having a callus or callosities; hardened.

callus. **1.** *bot.* **a.** A mass of tissue which forms as a result of injury to a plant. **b.** In certain grasses, an enlargement at the base of a spikelet or flowering glume. **2.** *anat.* **a.** A callosity, an area of thickened and hardened skin resulting from excessive pressure. **b.** A mass of tissue which forms between and around the ends of fragments of a fractured bone. **3.** *zool.* **a.** A growth of the inner surface of the umbilicus of a gastropod shell. **b.** A knoblike structure on the exoskeleton of dipterous insects.

caloric. Of or pertaining to heat.

Calorie. A kilocalorie or large calorie (Cal.), a unit of heat used in the study of metabolism. It is the amount of heat required to raise the temperature of one kilogram of water 1° C. (from 15° to 16°). A small calorie (cal.) is the amount of heat required to raise the temperature of one gram of water 1° C.

calorigenic. Heat-producing, said of certain foods.

calorimetry. The determination of heat exchange in an organism or in a system.

calycanthemy. Abnormal development of a calyx resulting in formation of petallike structures.

calyces. Plural of CALYX.

calyciflorous. With stamens and petals adnate to the calyx.

calyciform. Calyxlike.

calyculate. Resembling a calyx in shape.

calyculus. A calyxlike structure composed of bracts or bractlets.

calypter. In dipterous insects, a lobelike structure at the base of the wing; also called alula or squama.

calyptopis. The protozoea larva of *Euphausia* and other shrimp-like crustaceans.

calyptra. A hoodlike covering, as **a.** in liverworts, a structure enclosing a developing sporangium; **b.** in mosses, a caplike structure which partially encloses the developing capsule; **c.** in angiosperms, the covering of certain fruits, as in the eucalyptus.

calyx pl. **calyces.** **1.** *bot.* Collectively, the sepals, the outermost floral whorl **2.** *zool.* **a.** The central cuplike portion of the crown of a crinoid. **b.** The body of certain pedunculate entoprocts. **c.** The thickened basal portion of certain corals. **3.** *anat.* **a.** A division of the renal pelvis. **b.** The expanded cuplike termination of an axon.

cambium. A layer of persistent meristematic tissue in the stems and roots of dicotyledonous plants which gives rise to secondary tissues (xylem, phloem, parenchyma).

Cambrian period. The first or earliest division of the Paleozoic era; its rocks contain fossils of most invertebrate phyla, trilobites being abundant.

camel. A large, domesticated ruminant common in dry regions of Asia and Africa, as *Camelus dromedarius*, Arabian camel or the dromedary, and *C. bactrianus*, the Bactrian camel.

camelopard. The giraffe.

campanula. A small bell.

campanulate. Bell-shaped.

campestrian. Pertaining to the plains or open spaces.

camptodromous. Designating a type of leaf venation in which secondary veins curve toward the margins.

campylotropous. Curved so that the funiculus appears to be attached at its side, said of an ovule.

Canada balsam. A yellowish, viscid turpentine produced by the balsam fir (*Abies balsamea*).

Canada goose. *Branta canadensis*, the common wild goose of North America.

canal. A tubular passageway or channel.

canaliculate. Bearing longitudinal grooves or channels.

canaliculus pl. **canaliculi.** A small canal, especially one of the minute canals extending from lacunae into the matrix of bone.

canal system. In sponges, the passageways through which water passes in its course from the surface pores to the osculum or excurrent opening. Types include asconoid, syconoid, and leuconoid. *Also see* rhagon.

canary. *Serinus canarius*, a common domesticated songbird of the finch family.

cancellate. CANCELLOUS.

cancellous. Having a porous or spongy structure; cancellate.

cancer. A malignant tumor; a carcinoma or sarcoma.

cancroid. 1. Resembling a crab. **2.** Cancerlike, with reference to growth characteristics.

cane. 1. Any slender, jointed, hollow or pithy, more or less flexible stem, as that of various grasses and bamboo. **2.** Rattan, especially the split stem used in wickerwork. **3.** The stem of certain berries, as raspberry.

canescent. Hoary; covered with a gray-white pubescence.

cane sugar. SUCROSE.

¹canine. Any member of the dog family, a group of digitigrade carnivores which includes the dog, wolf, coyote, jackal, and fox.

²canine. Of, pertaining to, or resembling a dog or doglike animal.

canine tooth. A sharp, pointed tooth lying lateral to the incisors, especially prominent in canines.

canities. Hoariness; grayness or whiteness of hair.

cankerworm. The larva of a moth of the family Geometridae.

cannabis. *See* marijuana.

cannibalism. Eating the flesh of one's own species.

canopy. In a forest, the uppermost layer of branches with their twigs and leaves.

cantharophilous. Pollinated by beetles.

canvasback. A North American wild duck (*Nyroca* (*Aythya*) *valisineria*).

cap. **1.** *bot.* **a.** The pileus of a mushroom. **b.** The calpytra of a moss capsule. **2.** *zool.* The acrosome of the head of a spermatozoan.

¹capillary. Resembling a hair; very slender; having a small bore, as a *capillary* tube or vessel.

²capillary. A small, thin-walled tube, as a bile, lymph, or blood *capillary*.

capillary water. Water in the soil which is held by surface or capillary forces about soil particles or in the spaces between them. It is the principal source of moisture for plants.

capillitium. A delicate, branched network within the sporangium of slime molds and certain fungi.

capillus. A hair, especially one of the head.

capitate. **1.** Having a head or headlike structure. **2.** *bot.* Gathered into a dense cluster or head, with reference to flowers.

capitellum. **1.** A capitulum. **2.** *zool.* The expanded terminal portion of a halter.

capitulum. **1.** *anat.* A small head or bony prominence, as that

81

on a bone. **2.** *zool.* **a.** The expanded portion of a polyp which bears the tentacles. **b.** The body of a gooseneck barnacle. **c.** The expanded end of a hair, tentacle, antenna, or proboscis. **d.** The gnathostome of a tick. **3.** *bot.* A head, a type of inflorescence.

capon. A castrated cock, especially one fattened for use as food.

capreolate. Having one or more tendrils.

capreole. A tendril.

caprification. A method of pollination employed in the cultivation of figs in which fruits of the wild fig (caprifig) containing the fig wasp (*Blastophaga*) are placed in trees of cultivated figs. The insects on emerging enter the flowers of edible figs, thus bringing about cross-pollination.

capsicum. Cayenne or red pepper, the dried fruit of *Capsicum frutescens*, known in commerce as Tabasco or long pepper.

capsid. The protein coat or envelope of a virion. *See* virus.

capsomere. A morphological subunit of a specific viral protein.

capsulate. Enclosed within a capsule.

capsule. **1.** *anat.* **a.** A sheathlike structure which encloses an organ or structure. **b.** A thickened portion of the chondrocranium which encloses a sense organ. **2.** *zool.* **a.** A membranous sac which encloses eggs, as in annelids, mollusks, and insects. **b.** The central, nucleated portion of a radiolarian. **3.** *bot.* **a.** The spore case of various liverworts, mosses, and ferns. **b.** A simple, dry, dehiscent fruit which develops from a compound ovary. *See* pyxis, silique, silicle. **c.** A thin, gelatinous envelope which surrounds certain microorganisms.

capuchin. A long-tailed, South American monkey of the genus *Cebus*, family Cebidae.

capybara. The water pig or water cavy (*Hydrochoerus hydrochaeris*) of South America, the largest of all rodents. Also called carpincho.

carapace. **1.** The portion of the exoskeleton which covers the cephalothorax of various arthropods. **2.** The fleshy mantle which encloses the body of a gooseneck barnacle. **3.** The dorsal, convex portion of the exoskeleton of a turtle. **4.** The armor of an armadillo.

carbohydrate. Any of a number of compounds containing carbon, hydrogen, and oxygen, the latter in the ratio of 2:1; a sugar or starch. *See* monosaccharide, disaccharide, polysaccharide.

carbon. Symbol C, atomic weight 12. A nonmetallic element present in all organic compounds, occurring in three elementary forms: amorphous carbon, diamond, and graphite. It has a positive and negative valence of four, hence readily combines with other carbon atoms forming long chains as in alcohols or a ring as in benzene.

carbon clock. *See* radioactive carbon.

carbon dioxide. Carbonic acid gas, CO_2, an odorless, colorless, incombustible gas produced by the action of acids on carbonates. It is used by plants in photosynthesis and is an end product in the processes of fermentation, respiration, combustion, and decomposition of organic substances.

carbonic acid. H_2CO_3, a weak acid formed when carbon dioxide is dissolved in water. Its salts are carbonates.

carbonic anhydrase. An enzyme which catalyzes the reaction, $CO_2 + H_2O \rightarrow H_2CO_3$.

Carboniferous period. The Mississippian (Lower Carboniferous) and Pennsylvanian (Upper Carboniferous) periods of the Paleozoic era, during which time the great coal deposits were formed.

carbon monoxide. A colorless, odorless gas, CO, formed when organic substances are burned in an inadequate supply of oxygen. It is poisonous in that it forms a stable compound with hemoglobin.

carboxyl group. The univalent, acidic radical, —COOH, characteristic of organic acids, as formic, acetic, lactic.

carcinogenic. Inducing cancer formation.

[1]**carcinoid.** Of, pertaining to, or resembling crabs.

[2]**carcinoid.** A type of benign tumor derived from argentaffin cells.

carcinoma. A malignant, epithelial tumor; a cancer.

cardiac. Of, pertaining to, or near the heart.

cardiac muscle. A type of involuntary, striated muscle tissue found in the myocardium of the heart.

cardiac stomach. 1. The large, anterior portion of the stomach of a decapod crustacean; the gastric mill. 2. The folded, adoral portion of the stomach of a sea star, the portion everted when feeding.

cardiac valves. The valves of a mammalian heart, comprising two atrioventricular valves (bicuspid and tricuspid) and two sets of semilunar valves (pulmonary and aortic).

cardinal. *Richmondena cardinalis*, a brightly-colored songbird of the family Fringillidae; the redbird or cardinal grosbeak.

cardinal vein. One of three pairs of veins (anterior, posterior, and common cardinal veins) present in fishes and the embryos of tetrapods. The common cardinal or duct of Cuvier empties into the sinus venosus.

cardiovascular. Of or pertaining to the heart and blood vessels.

cardo. A hinge or hingelike structure, as **a.** the basal segment of the maxilla of certain insects; **b.** the hinge in bivalve mollusks.

caribou. One of several species of reindeer inhabiting North America and Greenland, best known of which is *Rangifer caribou*, the woodland caribou.

carina. 1. Any keellike structure. **2.** *zool.* **a.** The prominent ridge on the breastbone of a bird. **b.** The median, dorsal plate of a barnacle. **3.** *bot.* **a.** In irregular flowers, the structure which encloses the stamens and pistil, usually consisting of two fused petals lying opposite the banner. **b.** A longitudinal projection on the glumes of certain grasses.

carinal. Pertaining to or possessing a keel or keellike ridge.

carinate. *ornith.* Having a keeled sternum.

carmine. A crimson dye substance obtained from cochineal.

carneous. Resembling flesh in color.

carnivore. A flesh-eating animal, especially a mammal of the order Carnivora, which comprises predatory, flesh-eating animals as dogs, cats, weasels, bears, seals, walruses.

carnivorous. Flesh-eating; preying or feeding upon animals. *Compare* herbivorous.

carotene or **carotin.** A hydrocarbon, $C_{40}H_{56}$, synthesized by plants, the precursor of vitamin A.

[1]**carotenoid.** Resembling carotene.

[2]**carotenoid.** One of a group of yellow and orange plant pigments associated with chlorophylls. The group includes carotenes and xanthophylls.

carotid artery. One of two large arteries which supply the neck and head regions of vertebrates.

carp. A freshwater food fish (*Cyprinus carpio*) inhabiting ponds and sluggish streams.

carpal. Of or pertaining to the wrist or carpus.

carpel. *bot.* **1.** A single pistil or a division of a compound pistil; one of the innermost whorl of modified leaves which bear the megaspores. **2.** A modified, ovule-bearing megasporophyll. **3.** In conifers, a scale leaf of a female or carpellate cone.

carpellate. Pertaining to, bearing, or possessing carpels.

carpet beetle. One of a number of small, dermestid beetles whose larvae feed on woolens, furs, rugs, and other materials, as the black carpet beetle (*Attegenus piceus*) and buffalo carpet beetle (*Anthrenus scrophulariae*).

carpincho. *See* capybara.

carpocephalon. An erect, sporangium-bearing structure in certain liverworts.

carpogonium. The female gametangium in certain red algae, as *Polysiphonia*.

carpometacarpus. The fused carpal and metacarpal bones in the wing of a bird.

carpophore. **1.** A structure which supports the sporocarp in certain fungi. **2.** In the Umbelliferae, a slender stalk which supports a ripe carpel.

carpospore. A nonmotile, diploid spore produced by red algae. *See* cystocarp.

carpus. The wrist or the group of bones supporting the wrist.

carrageen. *See* Irish moss.

carrier. 1. *med.* **a.** A healthy individual who transmits pathogenic organisms but does not show symptoms of the disease. **b.** An organism which passively transmits pathogenic organisms, as the housefly. *Compare* vector. **2.** *physiol.* A substance with which another substance combines in its transport through a cell membrane. *See* active transport. **3.** *genetics.* A heterozygous individual which transmits a recessive gene for an abnormal or pathologic condition.

carrion. A dead body or dead and decaying flesh.

carrion fungi. Gasteromycete fungi of the order Phalles, whose odor attracts carrion-loving insects.

carrot. A biennial garden plant of the genus *Daucus*, or its yellow root, widely used as food.

cartilage. A form of connective tissue consisting of a flexible and resilient matrix containing cells (chondrocytes) lodged in lacunae.

cartilage bone. *See* endochondral ossification.

caruncle. 1. *zool.* **a.** A small, fleshy outgrowth or excrescence, as the lacrimal caruncle of the eye. **b.** The wattles or comb of certain birds. **c.** An area of contact between a placental cotyledon and uterine mucosa. **d.** A disclike structure on the tarsus of ticks and mites. **e.** A horny process on the beak of a newly hatched chick or turtle. **2.** *bot.* A fleshy, water-absorbing structure attached to certain seeds, as the castor bean.

caryopsis. A dry, one-seeded indehiscent fruit characteristic of grasses, consisting of a single seed in which the seed coat and pericarp are fused, as in corn, wheat, oats, rice.

casein. *See* caseinogen.

caseinogen. A soluble phosphoprotein in milk which, through the action of enzymes (rennin, pepsin), is converted into insoluble casein, the basis of the curd of milk.

cassava. A tropical plant (*Manihot esculenta*) of the spurge family, Euphorbiaceae, an important root crop widely used for food. It is the source of tapioca.

cassia. A tropical plant of the genus *Cassia* which yields senna, a drug with laxative properties.

cassowary. A large, flightless bird of the genus *Casuarius* of Australia and neighboring islands.

cast. A type of fossil in which the original organic substance has disappeared but counterparts of the organism are formed by the filling of the molds by uncrystallized substance, as *casts* of leaves, footprints.

castaneous. Dark brown or chestnut-colored.

caste. In social insects as bees or ants, a group of individuals which performs a special function in the economy of the colony, as a worker *caste* or soldier *caste*. Members of castes usually differ in structure, physiology, and habits.

casting. A mass of material discarded or thrown off by an animal, as the excrement of an earthworm or regurgitated food of a hawk.

castor bean. The seed of *Ricinus communis*, the source of castor oil, a cathartic.

castration. 1. *zool.* **a.** Removal of the testes (orchiectomy) or ovaries (ovariectomy, oophorectomy, spaying). **b.** Inactivation of the testes or ovaries by irradiation or through the action of hormones, parasites, or infections. **2.** *bot.* Removal of the androecium of a flower.

cat. 1. *Felis domestica*, the domestic cat. **2.** Any mammal of the cat family, Felidae, as the lion, tiger, leopard, puma, and related forms.

catabolism. The destructive phase of metabolism; the opposite of anabolism. It includes all processes involved in the break-down of complex substances into simpler substances, especially those involved in the release of energy. Also katabolism.

catadromous. Migrating downstream and into the sea to spawn, as the freshwater eel. *Compare* anadromous.

catalase. 1. An oxidizing enzyme. **2.** An enzyme present in tissues which decomposes hydrogen peroxide.

catalyst. A substance or agent which increases the velocity of a chemical reaction but is not itself changed in the process. *See* enzyme.

catamount. A cougar or lynx.

cataphyll. A small, scalelike leaf, as a bud scale.

catarrhine. A primate of the suborder Catarrhini (now super-family Cercopithecoidea) which includes the Old World monkeys.

catbird. *Dumetella carolinensis*, a passerine bird of the family Mimidae.

catfish. Any of a number of scaleless, mostly freshwater fishes, the upper jaw bearing whiskerlike barbels. North American forms belong to the family Ameiuridae and include the bull-head or horned pout, channel cat, and madtom.

catenoid. Linear or end to end, with reference to the arrange-ment of individuals in a colony.

caterpillar. An eruciform larva, as the wormlike larva of a butterfly, moth, sawfly, or scorpionfly.

cathepsin. An intracellular enzyme which catalyzes the hydrolysis of proteins; responsible for autolysis or liquification of tissues which occurs after death or in pathological conditions.

cathode. The negative electrode or pole of a battery or source of electricity. *See* anode.

cation. The positive ion which is attracted to and moves toward the cathode. *See* ion.

catkin. An ament; a special type of spikelike inflorescence which bears either pistillate or staminate flowers (but not both) and falls as a whole from the plant.

catodont. **1.** Having teeth in the lower jaw only. **2.** Pertaining to the sperm whale (*Physeter catodon*).

cattail. A tall marsh plant of the genus *Typha* whose stem terminates in a dense, cylindrical spike.

cattle. Large quadrupeds of the bovine family held as property and raised for some purpose, as the production of food.

cattle tick. *Boophilus annulatus*, a transmitter of *Babesia*, causative agent of Texas fever.

caudad. Toward the tail or posterior end. *Compare* cephalad.

caudal. Of, pertaining to, or resembling a tail; near or in the region of the tail; posterior.

caudal fin. The terminal or tail fin of an aquatic vertebrate; the primary organ of locomotion.

caudate. Bearing a tail or taillike appendage.

caudex. A stem, especially the stem of a palm tree or tree fern. *See* bole.

caudicle. The structure which connects a pollen mass or pollonium to the stigma, as in orchids.

caul. **1.** The embryonic membranes which cover the head of a fetus at birth when the amnion fails to rupture. **2.** The greater omentum.

caulescent. Having a stem above the ground.

caulicle. A rudimentary stem, as the region between the radicle and cotyledons in a plant embryo; the hypocotyl.

caulid. The main shoot of a leafy liverwort.

cauliflower. A variety of cabbage (*Brassica oleracea botrytis*) in which the head is modified, consisting of a thickened flower cluster.

cauline. Growing on or belonging to a stem.

caulis. The stalk or stem of a plant.

¹**caustic.** Corrosive, burning; capable of destroying by corroding or eating away.

²**caustic.** A substance which destroys tissue by corrosive action.

caustic potash. Potassium hydroxide, KOH.

caustic soda. Sodium hydroxide, NaOH.

cauterize. To burn or sear.

cave biology. The study of cave-dwelling organisms (troglobionts).

cave dweller. A troglobiont; cavernicole.

cavernicolous. Inhabiting caves.

cavernous. Full of cavities or hollow spaces.

caviar. The prepared and salted roe of the sturgeon or other fishes.

cavy pl. **cavies.** One of several species of tailless or short-tailed rodents of the family Caviidae, as the guinea pig, capybara, and mara.

cayman. *See* caiman.

cecidium. An abnormal plant growth, as a gall.

cecropia moth. The giant silkworm moth (*Samia cecropia*), the largest native moth of the United States.

cecum. A cavity open at one end; a blind pouch or cul-de-sac, as the first portion of the large intestine in mammals.

cedar. One of a number of coniferous trees characterized by fragrant and durable wood, as **a.** cedars of Lebanon of the genus *Cedrus;* **b.** white cedar or cypress of the genus *Chamaecyparis;* **c.** arbor vitae of the genus *Thuja.*

cedar-apple rust. A fungus disease of apples caused by *Gymnosporangium juniperi*, which also infects cedars causing swellings on branches called cedar apples.

celiac. COELIAC.

cell. 1. *biol.* The fundamental unit of which all organisms are composed, consisting of a mass of protoplasm composed of cytoplasm and a nucleus or nuclear material all enclosed within a cell membrane. *See* inclusion, organelle. **2.** *zool.* A compartment, as **a.** an air cell within certain bones; **b.** an area bounded by veins in the wing of an insect; **c.** a compartment of a honeycomb; **d.** a cavity in the ground which contains an insect pupa. **3.** *bot.* The cavity within an anther or ovary.

cell body. *See* neuron.

cell division. The process by which two new cells are formed from a single parent cell. *See* amitosis, meiosis, mitosis.

cell doctrine. *See* cell theory.

cell lineage. Tracing the developmental history of individual blastomeres to their ultimate fate in the formation of definite parts of the organism.

cell membrane. The surface layer of a cell or protoplasmic mass consisting of a layer of lipoid molecules separating two layers of portein molecules; the plasma membrane or plasmalemma.

cell plate. The structure which develops in the equatorial plane of a plant cell during telophase of mitosis. Cellulose cell walls are formed on each side of the plate, the plate remaining as the middle lamella cementing the cells together.

cell sap. The fluid contained within the vacuole of a plant cell, consisting of water and various substances in solution.

cell theory. The theory attributed to M. J. Schleiden, botanist, and T. Schwann, zoologist, announced in 1838–39 that all plants and animals are made up of microscopic units called cells. From it developed the cell doctrine which holds that all organisms are composed of cells or the products of cells and that cells constitute the structural and functional units of an organism.

cellular. Of or pertaining to cells; composed of or derived from cells.

cellulose. A complex carbohydrate which forms the chief constituent of the cell walls of plant cells. *See* tunicine.

cell wall. **1.** *bot.* A rigid wall lying outside of and secreted by the cytoplasm of a plant cell. It is nonliving and composed principally of cellulose. **2.** *zool.* A thin, nonliving sheath or pellicle which lies outside the plasma membrane of certain animal cells, as in various protozoans.

cenanthy. A condition in which stamens and pistil fail to develop.

cenogenetic or **coenogenetic.** Pertaining to characters appearing during development which are not of phylogenetic significance. *Compare* palingenetic.

Cenozoic era. The most recent of geologic eras, characterized by the rise and spread of modern mammals and the rise of land floras. It is called the "Age of Mammals" and covers about 60 million years.

censer action. The release of seeds from censerlike capsules from which seeds are shaken out through small openings, as in the poppy.

center of dispersal. An area on the earth's surface where a species arose and from which it spread to occupy its present geographic range. Also called center of origin.

centigrade thermometer. A thermometer having 100 divisions or degrees between freezing (0°) and boiling (100°). Abbreviated C.

centimeter. One hundredth part of a meter, approximately 2/5 of an inch. Abbreviated cm.

centipede. An elongated, many-segmented, swift-moving arthropod of the class Chilopoda, its first body segment bearing poison claws (maxillipeds).

central body. A colorless central region in certain blue-green algae.

central capsule. The inner, multinucleated portion of a radiolarian.

centrifugal. 1. Proceeding from or being thrown away from the center. *Compare* centripetal. **2.** Moving toward the periphery, said of nerve impulses. **3.** *bot.* Blooming outwardly from the inside or downward from the top.

centriole. A small, spherical body in a cell center which, in mitosis, forms the center of the astral rays. *See* diplosome.

centripetal. 1. Proceeding inwardly from the outside or periphery. *Compare* centrifugal. **2.** *bot.* Blooming inwardly from the outside or from the base upward.

centrolecithal. Having yolk centrally located and surrounded by a narrow layer of clear cytoplasm, as the eggs of insects.

centromere. A clear, constricted region of a chromosome which marks the junction of its two arms and serves as a point of attachment of the spindle fibers during mitosis.

centroplasm. The inner, colorless region (central body) of a blue-green algal cell. *Compare* chromoplasm.

centroplast. In certain heliozoans, a centrally located granule from which axopodia radiate.

centrosome. The dense zone surrounding the centriole in a cell center; also called microcentrum.

centrosphere. The clear region which surrounds a centriole.

centrum. 1. *anat.* The cylindrical mass forming the body of a vertebra. **2.** *bot.* A large central air space in certain stems, as in *Equisetum.*

century plant. A Mexican plant (*Agave americana*) of the Amaryllis family, a longlived, monocarpic plant which flowers and produces seeds only once and then dies.

cephalad. Toward the head or anterior end.

cephalic. Of or pertaining to the head or anterior end.

cephalin. One of a group of phospholipids found in animal and plant tissues, present in the brain and spinal cord of mammals.

cephalization. In the evolution of animals, the development of a head, that is the concentration of nervous tissue (brain and sense organs) and the development of a mouth and food-getting apparatus at the anterior end.

cephalochordate. Any chordate of the subphylum Cephalochordata which includes the lancelets.

cephalodium. An epiphytic lichen which grows on the surface of another lichen forming a wartlike protuberance.

cephalon. The anterior body region of a trilobite or pycnogonid.

cephalopod. Any mollusk of the class Cephalopoda which includes the nautili, squids, and octopi.

cephalothorax. The anterior portion of the body of various arthropods, consisting of the fused head and thorax, as in crustaceans and arachnids.

ceraceous. Waxy.

ceral. 1. Of or pertaining to wax. **2.** Of or pertaining to the cere of a bird's beak.

cerata. Filamentous or club-shaped respiratory structures on the backs of nudibranch mollusks.

cercaria pl. **cercariae.** The free-swimming, larval form of a fluke which develops from a sporocyst or redia within the tissues of a mollusk (snail or bivalve).

cercariaeum. A cercaria in which a tail is lacking; sometimes found within the molluscan host.

cerci. Plural of CERCUS.

cercocystis. A cysticercoid with a tail.

cercopod. A segmented terminal process on the telson of branchiopods.

cercus pl. **cerci.** One of a pair of sensory appendages, usually segmented, found at the posterior end of insects and other arthropods.

cere. A cushionlike protuberance at the base of the beak in certain birds, as in the eagle, hawk, parrot.

cereal. 1. Any grass plant which produces seeds which are used for food. **2.** The seed or grain of such a plant, as wheat, corn, rice, oats, rye, barley.

cerebellum. A part of the vertebrate brain which lies dorsal to the pons and medulla oblongata. It coordinates muscular movements, especially reflex activities involving maintenance of muscle tonus, posture, and equilibrium.

cerebral. 1. Of or pertaining to the brain. **2.** Of or pertaining to the cerebral hemispheres.

cerebral hemispheres. One of the two halves of the cerebrum.

cerebrospinal fluid. The fluid contained within the ventricles of the brain and the subarachnoid spaces of the meninges.

cerebrum. The portion of the vertebrate brain which lies anterior to and above the brain stem. It contains important sensory and motor centers and is the seat of higher mental activities involving association and learning.

ceriferous. Waxy; bearing or producing wax.

cernuous. Pendulous, drooping, nodding.

ceroma. A cere.

certation. Agonisis; competition, as between pollen grains in rate of growth.

cerumen. 1. Earwax, a waxlike substance formed from the secretions of ceruminous and sebaceous glands of the external auditory meatus. **2.** A substance produced by stingless bees of the genus *Melipoma* and used in nest construction.

cervical. 1. Of or pertaining to the neck. **2.** Of or pertaining to the cervix of an organ.

cervine. Of or pertaining to deer or the deer family, Cervidae.

cervix pl. **cervices.** A neck or necklike part.

cespitose. Growing in dense clumps or tuffs; matted.

cestodarian. A tapeworm of the subclass Cestodaria characterized by lack of scolex and a body not divided into proglottids, as *Amphilina*, which lives in a sturgeon.

cestode. A tapeworm, especially any member of the subclass Cestoda (Eucestoda) which includes the true tapeworms, parasitic worms with body consisting of a scolex and a linear series of proglottids, as *Taenia*.

cetacean. Any marine mammal of the order Cetacea which includes the whales, dolphins, and porpoises. Forelimbs are modified into flippers, hindlimbs are lacking.

chaeta pl. **chaetae.** A spine or bristle; in annelids, a seta.

chaetognath. Any invertebrate of the phylum Chaetognatha, a group of small, free-swimming, marine worms which includes the arrowworm (*Sagitta*).

chaff. 1. The husks or outer envelopes of the seeds of various cereal grains or grasses. 2. Small, degenerate bracts on the receptacle of flowers of various composites.

chalaza. 1. *bot.* The region in a seed where the funiculus spreads out and unites with the ovule. 2. *embryol.* One of two thickened strands of albuminous material in a hen's egg by which yolk is suspended from the shell.

chalazogamy. Fertilization in plants in which the pollen tube enters ovule through the chalaza.

chalcid. A small, hymenopterous insect of the superfamily Chalcidoidea, parasitic on other insects.

chalcidfly. A chalcid.

chalimus. A stage in the development of *Lernaeocera*, a copepod parasitic on the gills of fishes.

chalk. A soft limestone of marine origin formed from the deposition of shells of foraminiferans.

chambered nautilus. A cephalopod mollusk with a shell coiled like a watchspring and divided into a number of chambers or compartments.

chameleon. A small arboreal lizard common in warm regions of the Old World, most belonging to the genus *Chameleon*.

chamois. A small, goatlike antelope (*Rupicapra rupicapra*) of the mountain ranges of Europe and Asia.

channelled. Bearing longitudinal grooves.

chaparral. Vegetation consisting of thickets of dwarfed, drought-resistant and often thorny shrubs and bushes characteristic of much of the western United States.

char or **charr.** A trout of the genus *Salvelinus*.

character. A feature, trait, or characteristic, either structural or physiological, possessed by individuals of a group by means of which the members of the group can be recognized or distinguished from those of other groups.

chartaceous. Like paper; with a paperlike texture.

chasmogamous. Designating pollination which takes place in open flowers. *Compare* cleistogamous.

chasmogamy. The opening of a flower at maturity.

cheek. The region of the face below the eye and lateral to the mouth. *See* gena.

cheetah. *Acinonyx jubatus*, of the Old World, a catlike mammal noted for its speed. When trained to hunt, it is called the hunting leopard.

chela. **1.** A pincherlike organ, as that borne on the tips of the limbs of various crustaceans and arachnids. **2.** A type of sponge spicule with recurved hooks or plates at each end.

chelate. Possessing or resembling chelae.

chelicera pl. **chelicerae.** One of the first pair of appendages used as feeding structures in chelicerates. In spiders, each bears a terminal fang on which a poison duct opens.

chelicerate. Any arthropod of the subphylum Chelicerata which includes the classes Merostomata, Arachnida, and Pycnogonida.

cheliform. Pincherlike.

cheliped. A walking leg which bears at its tip a chela, especially the first walking leg of a lobster or crayfish.

¹chelonian. Of or pertaining to turtles or tortoises.

²chelonian. Any reptile of the order Chelonia (Testudinata), which includes the turtles, tortoises, and terrapins.

chemistry. The science which deals with the composition and structure of matter. Organic chemistry deals with compounds of carbon; inorganic chemistry with all other compounds.

chemolithotroph. An autotrophic organism which can obtain its energy from inorganic compounds, as the iron, sulfur, hydrogen, and nitrifying bacteria.

chemoorganotroph. A heterotrophic organism which uses prefabricated foods as a source of nutrition, as saprotrophic decay organisms, all metazoans.

95

chemoreceptor. A sensory receptor which is stimulated by chemical substances, as taste cells, olfactory cells, and cells in the carotid and aortic bodies.

chemosynthetic. Designating a type of autotrophic nutrition in which organisms obtain their energy from the oxidation of various inorganic compounds instead of from light.

chemotaxis. The turning of a motile organism or cell in response to a chemical stimulus. *Compare* chemotropism.

chemotherapy. The treatment of disease or the prevention of infection through the use of drugs which suppress growth or which kill pathogenic organisms in the body. Chemotherapeutic agents include synthetic substances (sulfonamides, arsenic compounds) and antibiotics (streptomycin, penicillin).

chemotroph. An organism which is capable of obtaining its energy from chemical compounds in the absence of light, as certain bacteria. *Compare* phototroph.

chemotropism. The turning or growth of a plant or a part of it in response to a chemical stimulus. *Compare* chemotaxis.

cheradad. A plant which inhabits wet sand bars.

cherry. Any of several species of trees of the genus *Prunus* or its fruit, a globose drupe with a smooth stone.

chersad. A chersophyte.

chersophyte. A plant which inhabits dry wastelands.

chestnut. A tree of the genus *Castanea* or the sweet, edible nut it produces.

chestnut blight. Canker blight, a destructive disease of the chestnut tree, caused by an ascomycete fungus, (*Endotheia parasitica*).

chevrotain. The mouse deer, a small, hornless ruminant (*Tragulus javanicus*) of Indonesia.

chiasma or **chiasm** pl. **chiasmata.** A crossing, as **a.** the crossing of the optic nerves, the optic *chiasma;* **b.** the interchange of parts of two chromosomes during meiosis. *See* crossing-over.

chickadee. A titmouse, a northern, passerine bird of the genus *Parus*, family Paridae.

chickaree. The North American red squirrel (*Tamiasciurus hudsonicus*).

chigger. A red bug, the six-legged, parasitic larva of mites of the family Trombiculidae. Also called harvest mite.

chigoe. A chigger or sand flea. *See* sand flea.

chilopod. Any arthropod of the class Chilopoda which includes the centipedes.

chimera. An individual with a mixture of genetically different cells.

chimpanzee. An anthropoid ape (*Pan troglodytes*) of tropical Africa.

chinch bug. *Blissus leucopterus*, a small hemipteran insect destructive to grasses and cereal crops.

chinchilla. A small, squirrellike rodent (*Chinchilla laniger*) of South America, now extensively raised for its fine, silky fur.

chinquapin. The dwarf chestnut (*Castanea pumila*), which produces a small edible nut contained in a bur.

chionad. A plant inhabiting snow-covered areas.

chipmunk. A small rodent common throughout the United States, as *Tamias striatus*, the Eastern chipmunk or hacker.

chironomid. *See* midge, a.

chiropter. Any mammal of the order Chiroptera which includes the bats.

chitin. A nitrogenous polysaccharide comprising the principal structural material in the exoskeleton of arthropods.

chiton. A mollusk of the class Amphineura.

chloranthus. Possessing green and usually inconspicuous flowers.

chlorenchyma. Parenchyma cells which contain chloroplasts, seen in the cortex of various stems.

chlorophyll. One of a number of green pigments present in plant cells which are essential in the utilization of light energy in photosynthesis. There are ten or more chlorophylls, the principal ones being chlorophylls *a* and *b*. Except in blue-green algae and certain bacteria, chlorophyll always occurs in chloroplasts.

chloroplast. A plastid which contains chlorophyll, abundant in the cells of green plants. They function in photosynthesis.

chlorosis. 1. Yellowness in normally green plants resulting from failure of chlorophyll to develop. **2.** An iron-deficiency anemia in man.

choana pl. **choanae.** One of the two posterior or internal nares, the openings of the nasal cavity into the mouth cavity or pharynx.

choanosome. The inner portion of a leuconoid sponge consisting of groups of flagellated chambers. *See* endosome.

choanocyte. A collar cell, a flagellated cell with a transparent collar surrounding the base of the flagellum, found in the gastral epithelium of sponges.

choanoflagellate. A protozoan of the order Choanoflagellida which resembles a choanocyte, as *Proterospongia*.

cholesterol. A sterol widely distributed in animal tissues being present in the brain and spinal cord, in animal fats and oils, egg yolk, and bile. It is a normal constituent of the blood and the main constituent of gallstones. It is the precursor of corticosteroids, androgens, and estrogens.

choline. A nitrogenous substance, a member of the vitamin B complex, essential for the synthesis of amino acids.

cholinergic fibers. Nerve fibers which release acetylcholine at their terminations.

chondrification. Conversion to cartilage.

chondriocont. A rod-shaped or filamentous chondriosome.

chondriome. Collective term for mitochondria and chondrioconts.

chondriosome. A mitochondrion or chondriocont.

chondroblast. A cartilage-forming cell.

chondroclast. A cartilage-destroying cell.

chondrocranium. In vertebrates, the cartilaginous portion of the embryonic skull forming a part of the base of the skull and the capsules enclosing the sense organs. In cyclostomes and elasmobranchs, it persists as the permanent skull; in higher vertebrates, it is partially or completely replaced by bone.

chondrocyte. A mature cartilage cell.

chordal. Of or pertaining to the notochord.

[1]**chordate.** Of, pertaining to, or possessing a notochord.

[2]**chordate.** Any member of the phylum Chordata which includes the tunicates, lancelets, and vertebrates, animals characterized by possession, at some stage during life, of a notochord.

chordotonal organ. A sense organ in insects which is stimulated by body movements. *See* sensillum.

chorion. 1. *entomol.* The outermost covering or shell of an insect egg. 2. *embryol.* An extraembryonic membrane which, in birds and reptiles lines the shell and functions in respiration and excretion; in mammals it is involved in the formation of the placenta.

choroid. The middle, vascular, pigmented layer of the vertebrate eye, anteriorly continuous with the ciliary body and iris, the three comprising the vascular tunic or uvea.

choroid plexus. A vascular structure which projects into a ventricle of the brain. It is the source of cerebrospinal fluid.

chorology. The science dealing with the distribution of organisms over the surface of the earth.

chromaffin reaction. Turning brown in the presence of chrome salts, a characteristic of cells containing epinephrine (adrenalin) or its precursor.

chromaffin system. Collectively, structures in higher vertebrates which contain chromaffin cells. Includes the carotid bodies, paraganglia proper, aortic bodies, and the medulla of the adrenal gland.

chromatic. Pertaining to, produced by, or consisting of a color or colors.

chromatid. 1. One of two spiral filaments comprising a chromosome during the prophase and metaphase of mitosis. **2.** In meiosis, one of the four elements comprising a tetrad.

chromatin. A substance present in the nuclei of cells which usually stains intensely with basic dyes. It is Feulgen-positive material (DNA) present in the interphase nucleus in the form of flakes or granules; during mitosis and meiosis, it forms the substance of the chromosomes. *See* euchromatin, heterochromatin.

chromatography. A method of chemical analysis based on the selective absorption of components of a mixture in a column of absorbent (column chromatography) or on a strip of paper (paper chromatography).

chromatophore. 1. *zool.* **a.** A pigment cell, especially one with branched radiating processes, as seen in crustaceans, amphibians, and reptiles. **b.** A chlorophyll-containing body present in protozoans, especially the phytoflagellates. **2.** *bot.* **a.** A structure in certain bacteria which contains photosynthetic pigments. **b.** The chloroplast of green algae.

chromidial substance. Basophilic material containing RNA present in the cytoplasm of neurons and gland cells. Also called chromophil substance. *See* Nissl bodies.

chromocenter. A karyosome; also called false, nucleinic, or chromatin nucleus.

chromogen. A colorless substance which, under suitable conditions, can become colored; a precursor of a pigment.

chromogenesis. The production of pigment, as by a microorganism.

chromomere. One of the beadlike granules of chromatin distributed linearly along a filiform chromosome.

chromonema pl. **chromonemata.** A fine, spiral filament which lies within the matrix of a chromosome.

chromopexy. The ingestion and storage of vital dyes, as by macrophages.

chromophil or **chromaphil.** A cell whose cytoplasm stains readily, as certain cells in the anterior lobe of the hypophysis.

chromophil substance. Chromidial substance. Also called ergastoplasm.

chromophobe. A cell whose cytoplasm does not stain readily, as certain cells in the anterior lobe of the hypophysis. *Compare* chromophil.

chromoplasm. The outer, pigmented portion of a blue-green algal cell. *Compare* centroplasm.

chromoplast or **chromoplastid.** A plastid which contains a colored pigment other than green, as the red, orange, or yellow plastids responsible for the colors of various plant structures, as leaves, flowers, and fruits.

chromosome. A self-duplicating body present in the cells of higher plants and animals, especially observable during stages of mitosis when each appears as a rod-shaped structure which stains intensely. In somatic cells, they occur as homologous pairs, the number varying with species but remaining constant within a species. Chromosomes are composed of DNA, the repository for genetic information transmitted from cell to cell. *See* meiosis.

chromosome map. A map or diagram showing the relative position of genes, prepared from crossing-over data and from the study of chromosomal aberrations correlated with phenotypic changes.

chrysalis or **chrysalid.** The pupa of a butterfly encased within a tough integument, often brightly colored.

chrysanthine. Having yellow flowers.

chrysophyllous. Having leaves of a golden color.

chub. One of a number of small, freshwater fishes of the genera *Hybopsis, Gila,* and *Leuciscus.*

chuckwalla. A large lizard (*Sauromalus obesus*) of the southern United States.

chukar partridge. *Alectoris graeca,* an Old World partridge introduced into the western United States as a game bird.

chyle. The milklike fluid containing fat globules present in lymph vessels draining the intestine during and following digestion.

chyme. The semifluid contents of the stomach and intestines.

chymotrypsin. A proteolytic enzyme present in the intestine formed from chymotrypsinogen secreted by the pancreas.

chytrid. A phycomycete fungus of the order Chytridiales. Most are minute intracellular parasites, as *Clytridium lagenaria* in algae.

cicada. A heavy-bodied insect of the order Homoptera, the males producing a characteristic buzzing sound. Common species are the periodical cicada (17-year locust) and the dog-day cicada (harvest fly).

cicada killer. A large wasp (*Sphecius speciosus*) which provisions its nest with cicadas stung to insensibility.

cicatrix pl. **cicatrices.** 1. A scar or scarlike mark. 2. *zool.* **a.** The fibrous connective tissue which replaces injured or destroyed tissue. **b.** A nonvascular area on the surface of an ovary which marks the point of rupture of a follicle, as in birds and reptiles. 3. *bot.* A mark or scar left after abscission of a leaf or bract.

cilia sing. **cilium.** 1. *anat.* The eyelashes. 2. *biol.* Minute, hairlike cytoplasmic processes of a cell which beat rhythmically. In lower forms, they serve as locomotor organelles; on ciliated surfaces they propel substances, as mucus. 3. *bot.* Minute hairs along a margin forming a fringe.

ciliary body. In a vertebrate eye, a thickened portion of the vascular tunic consisting of ciliary processes to which fibers of the suspensory ligament of the lens are attached and ciliary muscle which functions in accommodation.

ciliary feeding. Feeding accomplished by a current of water produced by ciliary action, as in certain protozoans, rotifers, and bivalves.

[1]ciliate. Provided with or bearing cilia.

[2]ciliate. A protozoan of the taxon Ciliata, characterized by possession of cilia, as *Paramecium*.

ciliolate. Minutely ciliated.

cilium. Singular of CILIA.

cinchona. The dried bark of the stem or root of several species of *Cinchona*, a South American tree, the source of quinine and other alkaloids. Now extensively cultivated in the East Indies.

cincinnus. A helicoid cyme.

cineraceous or **cinereous.** Of the nature of or resembling ashes; ash-colored.

cingulate. 1. Resembling a girdle. 2. Possessing a girdle or girdlelike band.

cingulum. 1. The waist. 2. A band or girdlelike structure, as **a.** the outer ciliated band of a trochal disc of a rotifer; **b.** a ridge of enamel about the margin of certain teeth; **c.** a bundle of association fibers which arches over the corpus callosum of the brain.

circadian. Occurring at or about at twenty-four hour intervals, as *circadian* rhythms.

circinate. Ring-shaped or rolled up in a coil, as *circinate* vernation exhibited by the leaves of ferns.

circulation. The movement of fluid throughout a circuit, as the flow of blood or lymph within the body or the movement of cytoplasm within a cell.

circulatory system. The cardiovascular system; the heart and blood vessels. A closed system is one in which the blood is confined to tubes throughout its entire course; an open system is one in which the blood leaves the arteries and circulates through body spaces (hemocoel) before reentering the heart.

circumflex. Bent or winding, said of blood vessels or nerves.

circumnutation. An irregular spiral movement of a plant resulting from unequal growth rates. *See* nutation.

circumscissile. Designating the opening or dehiscing of a fruit or anther in which the valve comes off like a lid.

circumvallate. Surrounded by a deep groove, as the *circumvallate* papillae of the tongue.

cirrate. Bearing cirri.

cirri. Plural of CIRRUS.

cirriferous. Bearing a curl or tendril.

cirriped. Any crustacean of the subclass Cirripedia which includes the barnacles.

cirrose. Bearing tendrils or cirri; resembling a cirrus.

cirrus pl. **cirri.** **1.** *bot.* A tendril. **2.** *zool.* A slender filamentous structure, as **a.** the fused cilia of certain protozoans, or hairs on the antennae or legs of certain insects; **b.** a blunt process or processes on the parapodia of polychaete worms, on the oral hood of lancelets, or about the mouth of fishes; **c.** rootlike processes of a feather star or whorled processes on the stalk of a crinoid; **d.** the protrusible copulatory organ of certain flatworms and mollusks.

cisco. A whitefish of the genus *Coregonus*, especially *C. artedi*, a food fish of northern United States and Canada; also called lake herring.

cisterna. A reservoir or cavity, as the *cisterna* chyli.

cistron. *genetics.* A hereditary unit which carries information for the synthesis of a single enzyme or protein molecule.

citreous. Lemon-yellow in color.

citric acid cycle. Kreb's cycle; tricarboxylic acid cycle. A complex series of reactions in which pyruvic acid, a product of carbohydrate metabolism, is broken down, under aerobic conditions, to CO_2 and H_2O with the release of energy which is utilized principally in the synthesis of adenosine triphosphate (ATP). In this cycle, an organism can also utilize fatty acids and amino acids, products of fat and protein metabolism; consequently it is considered to be the common pathway for the metabolism of the three primary classes of foods.

citron. A citrus fruit of a thorny tree (*Citrus medica*) or the tree producing it. Commercial citron is its candied rind.

citrus fruit. A fruit produced by one of several species of trees or shrubs of the genus *Citrus*, family Rutaceae. The principal fruits are the orange, lemon, lime, grapefruit, citron, and kumquat; minor fruits are bergamot, calamondin, and shaddock (pomelo).

civet. A substance with a musky odor found in a pouch near the sex organs of civet cats. It is used in the manufacture of perfumes.

103

civet cat. One of a number of catlike carnivores of the genera *Civettictis* and *Viverra* of Africa and Southeast Asia.

cladoceran. Any member of the Cladocera, an order of branchiopod crustaceans which includes the water fleas, as *Daphnia*.

cladode. A cladophyll.

cladophyll. A stem which resembles a leaf in form and color; a leaflike branch.

clam. A mussel; a bivalve mollusk, especially one of the edible species of the class Pelecypoda, as the hard-shell clam or quahog (*Venus mercenaria*) and the soft-shell clam (*Mya arenaria*).

clamber. 1. *zool.* To crawl or climb over a surface using hands and feet for holding. 2. *bot.* To grow over a surface using tendrils for attachment.

clam shrimp. A crustacean of the suborder Conchostraca with a bivalve carapace with concentric markings, as *Cyzicus*.

clam worm. A marine polychaete (*Neanthes* (*Nereis*) *virens*).

clasper. A structure which enables the male to hold to the female during copulation, as in insects and elasmobranch fishes.

class. A taxon ranking below a division or phylum but above an order.

classification. Taxonomy; the systematic arrangement of animals and plants into groups or categories based on characteristics common to a group, especially those due to relationship and common descent. The categories or taxa in common use, from the largest and most inclusive to the smallest and most limited, are: kingdom, phylum or division, class, order, family, genus, and species. *See* binomial nomenclature.

clavate. Club-shaped.

clavicle. A bone of the pectoral girdle extending from sternum to scapula; the collar bone in man. It is rudimentary in ungulates and carnivores.

claviculate. Possessing tendrils or hooks.

cleaner. An organism which removes ectoparasites, fungi, and diseased and injured tissue from the surface of another, as birds, fishes, and crustaceans which feed off the surface of other animals.

cleavage. Cell division, especially the series of mitotic divisions of a fertilized ovum resulting in multiplication of cells and formation of a blastula; segmentation.

¹**cleft.** A fissure; a narrow, cracklike slit or opening.

²**cleft. 1.** Divided or split, as a cleft palate. **2.** *bot.* Divided into lobes or divisions by narrow sinuses which extend more than halfway to the midrib, with reference to leaves.

cleistogamous. Occurring in unopened flowers, with reference to fertilization. *Compare* chasmogamous.

cleistogamy. The production of small, inconspicuous, self-pollinating flowers which do not open, as in the violet or pansy.

cleistothecium. A hollow, completely closed ascocarp.

click beetle. A beetle of the family Elateridae which is capable of righting itself from an overturned position, during which it makes a clicking sound. Larvae are called wire worms. Also called snapping beetle or skipjack.

climax community. The final and most stable of a series of communities in a succession, remaining relatively unchanged as long as climatic and physiographic factors remain constant.

climber. A plant which grows upward using other plants or objects for support.

cline. A gradual and more or less continuous change in a character in a series of continuous populations or throughout the range of a species in which individuals at the two extremes differ markedly.

clingfish. One of several fishes of the order Gobiesocida (Xenopterygii) which possess a ventral sucker by which they cling to stones, as *Gobiesox*.

clitellum. A swollen glandular region in the anterior portion of certain annelids.

clitoris. A small, erectile structure in female vertebrates, a homologue of the male penis.

cloaca. The terminal portion of the digestive tract which serves as a common passageway for the products of the digestive, urinary, and reproductive systems. It is present in many invertebrates and most adult vertebrates except mammals.

clone. All the descendants of a single individual arising by asexual reproduction or parthenogenesis.

clonus. Alternate reflex contraction and relaxation of a voluntary muscle.

clot. A semisolid mass of blood or lymph; a coagulum.

clotting. Coagulation.

cloven. Split into two or more parts, as a cloven hoof.

105

clubbed. Shaped like a club, as an insect antenna.

club fungi. The Basidiomycetes, which possess club-shaped basidia.

club moss. A vascular plant of the order Lycopodiales, so called because of their club-shaped strobili, as *Lycopodium*, *Selaginella*.

clypeate. Scutate; shaped like a shield.

clypeus. A median sclerite on the head of an insect to which the labrum is attached.

cnidarian. A coelenterate.

cnidoblast. A cell in coelenterates which produces a nematocyst.

cnidophore. A stalked structure bearing nematocysts present on the tentacles of certain medusae.

cnidosac cells. Cells in cerata of nudibranch mollusks in which ingested nematocysts from coelenterates are stored and later used for defense.

coachwhip. A long, slender snake (*Masticophis flagellum*) of the family Colubridae, common in southern United States and Mexico.

coagulation. The formation of a clot or coagulum as occurs in blood or milk.

coagulum. A semisolid mass; a clot or curd.

coarctate. Crowded together; closely packed.

coati. A tropical, raccoonlike mammal (*Nasua narica*); also called coatimundi.

cob. A male swan. *See* pen, cygnet.

cobalamine. Generic name for vitamin B_{12}, a cobalt-containing vitamin.

cobra. One of several venomous snakes, as *Naja naja*, the Indian cobra, and *N. hannah*, the King cobra of Southeast Asia. Also *Hemachates haemachatus*, the ringhals of South Africa.

cocci. Plural of COCCUS.

coccidiosis. Any of a number of sporozoan infections caused by protozoans of the class Telosporidia, subclass Coccidia. It includes a number of serious diseases of poultry and domestic animals.

coccus pl. **cocci. 1.** A spherical bacterium. **2.** A one-seeded locule of a lobed fruit.

coccyx. A small bone terminating the vertebral column in man and apes, formed by a fusion of four rudimentary vertebrae.

cochineal. A crimson dye obtained from homopterous insects of the genera *Epicoccus* and *Dactylopius*, family Coccidae, which feed upon cacti of Mexico.

cochlea. 1. *anat.* The spiral-shaped portion of the osseous labyrinth of the internal ear in which are located the receptors for hearing. **2.** *bot.* A coiled legume.

cochlear. 1. Of or pertaining to a cochlea. **2.** Designating a type of estivation in which the external leaf entirely covers an internal one.

cochleate. Spiral or screw-shaped; coiled like a snail shell.

cock. The male of barnyard fowls, also the male of other birds, especially gallinaceous birds; a rooster.

cockatoo. One of a number of parrotlike birds of the Australian region, as *Kakatoe galerita*, the sulfur-crested cockatoo.

cockerel. A young domestic cock, especially one less than a year old.

cockle. A bivalve mollusk of the family Cardiidae, most possessing a ribbed shell with prominent umbo, as *Cardium edule*, an edible species of Europe.

cockroach. One of several species of nocturnal, dark-colored, foul-smelling insects of the order Blatteria (formerly Orthoptera), as *Blattella germanica*, the German cockroach or croton bug; *Blatta orientalis*, the Oriental cockroach; *Periplaneta americana*, the American cockroach.

cocoa. The pulverized seeds of cacao.

coconut. The large, hard-shelled, oval fruit of the coconut palm (*Cocos nucifera*), its shell lined with an edible meat and containing a milklike fluid. *See* coir.

cocoon. In general, any tough, protective covering which encloses the eggs or young, and sometimes adults of animals, as **a.** the silken envelope enclosing the larva or pupa of an insect; **b.** the membranous sac enclosing the developing larvae of earthworms and leeches; **c.** the silken sac enclosing the eggs of spiders; **d.** the mud case which, during the dry season, encloses a lungfish.

codfish. The cod (*Gaddus callarias* and *G. morrhua*), an important North Atlantic food fish of the family Gadidae. Also *G. macrocephalus* of the North Pacific.

codling moth. *Carpocapsa pomonella*, a small moth whose larva is a serious pest of apples and other fruits.

codon. A group of three successive nucleotides in a DNA molecule which act as a code in the placing of an amino acid in a protein molecule.

coe-. For words beginning with COE- not found here, see CE-.

coelacanth. A primitive, lobe-finned, crossopterygian fish, abundant as fossils and considered extinct until a living species (*Latimeria chalumnae*) was discovered in 1938 in the Indian Ocean.

coelenterate. An invertebrate of the phylum Coelenterata (Cnidaria) which includes bydroids, sea anemones, jellyfishes, and corals. *See* polyp, medusa.

coelenteron. The gastrovascular cavity of coelenterates.

coeliac. Abdominal; of or pertaining to the belly.

coelom. The true body cavity or perivisceral space present in most of the higher metazoans. It develops as a cavity of the mesoderm, and is lined with epithelium. *See* pseudocoel.

coelomate. Any member of the Coelomata which includes all animals possessing a true coelom.

coelozoic. Inhabiting a coelom or body cavity, said of parasites.

coenenchyme. In alcyonarian corals, the mass of tissue from which polyps project, consisting primarily of a gelatinous mesoglea containing skeletal elements.

coenobium. A nonfilamentous, spherical or discoid colony as seen in certain algae and protozoans, as *Pediastrum*, *Volvox*.

coenocarpium. A fruit formed of an entire inflorescence, as a fig.

coenocyte. A multinucleate plant body as that of certain slime molds, nonseptate algae and certain fungi in which cell walls are absent. *Compare* syncytium.

coenogamete. A multinucleate gamete, as in *Rhizopus*, the bread mold.

coenosarc. The hollow, soft tissues in the stem of a colonial hydrozoan or coral which unites the individuals of a colony. *See* perisarc.

coenosteum. The surface layer of the calcareous mass forming the skeleton of various corals.

coenozygote. A multinucleate zygote, as in *Mucor*.

coenurus. A type of tapeworm larva consisting of a bladderlike sac containing many scolices. *Compare* cysticercus, hydatid. *See* gid.

coenzyme. A substance associated with an enzyme and necessary for its activation; the prosthetic or nonprotein part of an enzyme system.

coffee. The green or roasted seeds of various species of a shrub or small tree of the genus *Coffea* or the plant which produces them.

coherent. 1. Sticking together by adhesion. **2.** Touching and adhering but not fused.

cohesion. 1. The force which holds molecules of a substance together. **2.** *bot.* The union of similar parts as those of a floral whorl.

cohort. A taxon between class and order.

coir. The coarse, short fibers from the husk of a coconut.

coition. Sexual intercourse; coitus.

colchicine. A poisonous alkaloid obtained from *Colchicum autumnale*, the meadow crocus. When applied to the tips of growing plants, it interferes with mitotic divisions resulting in polyploidy.

cold-blooded. POIKILOTHERMOUS.

cold light. Light emitted by bioluminescent organisms.

coleopteran. Any insect of the order Coleoptera which contains the beetles and weevils; the largest insect order including over 250,000 known species.

coleoptile. A sheath enclosing the shoot apex or plumule in the embryos of grass plants, as wheat, corn.

coleorhiza. A sheath enclosing the radicle or rudimentary root in the embryos of grass plants, as in wheat, corn.

colicin. A protein or polypeptide released by *Escherichia coli* into the surrounding medium.

coliphage. A phage of *Escherichia coli*.

collagen. A protein forming the principal constituent of white fibers of connective tissue which upon boiling or hydrolysis, is converted into gelatin.

collagenous fibers. Wavy, white, nonelastic fibers found in connective tissue, especially tendons and ligaments.

collar. 1. *bot.* **a.** A line of junction between a root and its stem. **b.** The annulus or ring of a mushroom. **2.** *zool.* A structure or marking which resembles a collar, as **a.** a colored band of feathers about the neck of a bird; **b.** a circular fold or parapet at the junction of column and oral disc in a sea anemone; **c.** the funnel-shaped structure surrounding the base of a flagellum in choanocytes or choanoflagellates.

collar bone. The clavicle.

collar cell. A choanocyte.

[1]**collateral.** Subsidiary, subordinate, indirect; acting secondarily.

[2]**collateral.** A branch of an axon other than one at the terminal arborization.

collateral bundle. A vascular bundle in which the xylem and phloem are in contact on one side only.

collembolan. A primitive insect of the order Collembola comprising the springtails, as *Achoretes*, the snow flea.

collenchyma. **1.** *bot.* A strengthening tissue in the stems of plants composed of closely placed cells with cell walls thickened at their angles. **2.** *zool.* Mesenchyme when cells are few in number and there is much gelatinous intercellular material.

collencyte. An amebocyte with slender, branching pseudopodia, as in sponges.

colletereal glands. Glands in a female insect which secrete an adhesive substance or provide material for the egg case.

colloblast. An adhesive or lasso cell found on the tentacles of ctenophores.

colloid. **1.** A colloidal system. **2.** A gluelike substance such as a protein whose particles (molecules or aggregates of molecules) when dispersed in a solvent remain uniformly suspended and do not form a true solution. **3.** A homogeneous, gelatinous substance present in the follicles of the thyroid gland.

colloidal state. A state of matter in which one substance in the form of many small particles (discontinuous phase or dispersion) is suspended in or dispersed in another (continuous phase).

colloidal system. A mixture of two substances which behave as two separate phases, as oil in water to form an emulsion or water in oil to form a butter Various types of colloidal systems and the resulting product are: gas in a liquid — foam; gas in a solid — floating soap; liquid in a gas — cloud, mist, or aerosol; liquid in an immiscible liquid — emulsoid (skim milk) or an emulsion (cream); liquid in a solid — cheese, gelatin; solid in a gas — smoke; solid in a liquid — glue; solid in a solid — stained glass.

collophore. A tubelike structure on the ventral surface of the abdomen of springtails which functions as an adhesive structure.

collum. **1.** A neck, collar, or collarlike structure. **2.** The first, legless segment behind the head in millipedes and pauropods.

colon. **1.** In vertebrates, the portion of the large intestine extending from the cecum to the rectum. **2.** In insects, the region of the digestive tract between the ileum and rectum.

colony. A group of organisms of the same kind living in close association with each other.

colugo. A flying lemur.

columella. **1.** *bot.* The central axis of a sporangium in mosses, liverworts, and certain fungi. **2.** *zool.* **a.** A bone of the middle ear in amphibians, reptiles, and some birds. **b.** The central portion of a snail shell. **c.** The central axis of the cochlea of the ear. **d.** The central skeletal mass of a coralite.

column. **1.** *zool.* Any column-shaped structure, as **a.** the body of an anthozoan; **b.** the central bony axis (spinal column) of a vertebrate; **c.** a longitudinal division of the spinal cord; **d.** an extension of the renal cortex into the medulla; **e.** a thickened strand of the ventricle of the heart or the mucosa of the vagina or anal canal. **2.** *bot.* **a.** The united stamens and style forming a solid structure as in orchids; **b.** The twisted portion of an awn in grasses.

¹**coma.** *physiol.* A state of unconsciousness from which a person cannot be aroused by ordinary stimuli.

²**coma.** *bot.* **1.** A leafy crown or head, as in palms. **2.** A tuft of hairs at the end of a seed, as in the milkweed. **3.** A tuft of leaflike structures at the end of a fruit, as in the pineapple.

comb. A serrated structure bearing teeth or teethlike parts, as **a.** the fleshy crest on the head of domestic fowl and other birds, especially the male; **b.** the pecten in the eye of a bird; **c.** the pecten on the third leg of a honeybee; **d.** a row of fused cilia on the comb plate of a ctenophore; **e.** the pecten of a scorpion.

comb jelly. An animal belonging to the phylum Ctenophora; a sea walnut.

comb plate. A locomotor organ in a ctenophore. Also called plate row, comb row, comb column, rib or costa, paddle plate.

111

comet. A single arm of an asteroid which is in the process of regenerating a new disc and arms, as in *Linckia*, a Pacific sea star.

commensalism. A form of symbiosis in which two species live in close association in such a manner that one, the commensal, usually the smaller of the two, benefits but the other, the host, does not.

commissure. 1. *zool.* **a.** A band of nerve fibers which connects paired ganglia or the two halves of the brain or spinal cord. **b.** A seam or line of junction marking the union of two parts, as the labia. 2. *bot.* A line of junction between two united structures, as carpels.

communal. Organized into societies, as man, social insects.

community. An association of plants and animals in a given area or region in which the various species are more or less interdependent upon each other, as a desert *community*, pond *community*.

comose. Having a tuft or tufts of hair.

companion cells. Cells in phloem associated with sieve tube cells.

compensation point. The light intensity at which oxygen production during photosynthesis is equal to oxygen consumption by respiration.

competence. In development, the physiological state or condition of a tissue which enables it to react morphologically in a specific way to determinative stimuli, as inductors.

complement. A heat-labile substance in blood plasma or serum which has the ability to react with and destroy pathogenic organisms or foreign substances but only after these agents have been acted upon and become sensitized by their antibodies.

complementary gene. One of a pair of genes which interact to produce a character or trait different from that produced by either acting alone.

complement fixation. An antigen-antibody-complement reaction in which the complement is destroyed or rendered unavailable, thus preventing certain reactions, such as hemolysis or bacteriolysis.

complicate. Folded longitudinally.

composite. Any member of the Compositae, the largest family of dicotyledonous plants consisting of over 1000 genera containing some 20,000 species characterized by possession of a composite flower, as the daisy, dandelion, sunflower.

composite flower. One consisting of a group of many small flowers forming a compact head, characteristic of plants of the Compositae; anthodium, capitulum.

compound. 1. *biol.* Composed of several elements or parts united into a single structure, as a *compound* gland or leaf. **2.** *chem.* A substance composed of two or more elements combined in definite proportions by weight.

compound eye. An eye composed of many individual units or ommatidia, characteristic of arthropods. *See* mosaic image.

compressed. Flattened laterally or from side to side.

compression. A fossil consisting of the carbonized remains of an organism brought about by compressive forces; an impression.

conaria. The early larva of certain siphonophores, as *Velella*.

concave. Having a curved, depressed surface. *Compare* convex.

concentric. A vascular bundle in which xylem surrounds phloem or phloem surrounds xylem.

conceptacle. A concavity in the thallus of certain algae in which gametangia develop.

conception. Fertilization of an ovum; the act of becoming pregnant; impregnation.

conch. A large, marine gastropod of the family Strombidae, especially those of the genera *Strombus* and *Cassis* or their shells, used in the production of jewelry.

concha. A structure shaped like a shell, as **a.** the deeper portion of the external ear; **b.** one of three scroll-like structures on the lateral wall of the nasal fossa; a turbinate bone.

conchiolin. A nitrogenous substance present in the shells of mollusks.

conchology. The science which deals with shells or the animals which produce them; the study of mollusks.

concrescence. 1. The growing together of parts originally separate, as the roots of teeth. **2.** The movement of masses of cells toward each other and their eventual fusion, as in gastrulation.

concretion. A solidified mass; a calculus.

condor. A large, South American vulture (*Vultur gryphus*), one of the largest of living birds.

conduction. The passage of electrons, atoms, or molecules through a suitable medium, as **a.** *bot.* the passage of water, minerals, or foods through conducting tissues; **b.** *zool.* the movement of heat, sound waves, or nerve impulses through cells or tissues.

conduplicate. Folded together lengthwise.

condyle. **1.** A rounded process on a bone which serves for articulation. **2.** A similar process in the joint of an arthropod appendage.

cone. **1.** *zool.* A cone-shaped dendritic process of a cone cell which functions as a photoreceptor in the retina of a vertebrate eye. **2.** *bot.* **a.** A strobilus, a fruiting structure present in club mosses, conifers, and cycads consisting of a group of sporophylls bearing sporangia, as a pine *cone.* **b.** An inflorescence or fruit with a central axis bearing overlapping scales.

cone cell. **1.** A photoreceptor cell in the retina of the eye. **2.** In an ommatidium, one of four cells which produce the crystalline cone.

conelet. A small cone, especially one of the first year's growth.

conenose. A bloodsucking assassin bug of the genus *Triatoma*, a vector of trypanosomes.

cone shell. A poisonous, carnivorous marine gastropod of the genus *Conus*.

coney. A small, rodentlike mammal of the order Hyracoidea found in Africa, Arabia, and Syria. Also called hyrax, dassia, rock rabbit, and tree bear.

conferted. Densely or closely packed; crowded.

confluent. Merging; coming together; blended together as one.

congenital. Present at birth.

congested. Overcrowded; full.

conglobate. Spherical; grouped together into a ball or rounded structure.

conglomerate. Grouped together into a dense cluster or mass, as a conglomerate fruit.

congression. In mitosis, the movement of chromosomes so that their centromeres occupy a position midway between the two poles of the spindle.

conical. Cone-shaped.

conidia. Plural of CONIDIUM.

conidiophore. A specialized hypha which produces conidia.

conidiosporangium. A conidiophore, *q.v.*

conidiospore. A conidium, *q.v.*

conidium pl. **conidia. 1.** An asexual spore produced by certain fungi, usually formed by abstriction at the tip of a sporophore. **2.** *bacteriol.* A specialized cell or spore produced at the tips of certain filamentous bacteria.

conifer. A cone-bearing tree or shrub, especially one of the order Coniferales which includes the Pinaceae (pines) and Taxaceae (yews). Most bear true cones and are evergreen.

coniferous. 1. Producing or bearing cones. **2.** Composed of cone-bearing trees, as a *coniferous* forest.

conjoined. United or joined together, as *conjoined* twins.

conjugant. One of a pair of individuals engaged in conjugation.

conjugate. 1. To undergo conjugation. **2.** *bot.* To join in pairs.

conjugation. 1. The act of uniting or joining together, especially the temporary union of two individuals in which there is a transfer of nuclear material from one cell to the other, as occurs in various algae, fungi, and protozoans. **2.** In certain algae, the union of isogametes.

conjunctiva. The mucous membrane which lines the eyelids and covers the anterior surface of the eyeball.

conk. The basidiocarp of a bracket fungus.

connate. 1. United at birth. **2.** *bot.* Firmly united or joined.

connate-perfoliate. United at the base and surrounding a stem, said of certain leaves.

connective. A structure which binds together or connects, as **a.** *bot.* the band of tissue connecting two lobes of an anther; **b.** *zool.* a strand of nervous tissue which connects two ipsilateral ganglia, as the circumpharyngeal connectives in annelids.

connective tissue. A primary tissue of mesodermal origin which supports and binds structures together but also functions in food storage, hemopoietic, and reticuloendothelial activity. Intercellular substance consists principally of collagenous and elastic fibers; cells few and widely separated. It is the principal tissue of fascia, aponeuroses, tendons, ligaments, capsules, fibrous membranes, cartilage, and bone.

conoid. Conoidal; cone-shaped.

consortium. The symbiotic relationship between an alga and a fungus to form a lichen.

constitutive. Constantly present, with reference to enzymes in bacteria. *Compare* inductive.

constrictor. **1.** A muscle which narrows an opening, reduces the size of a cavity, or compresses an organ. **2.** A snake which kills its prey by coiling its body tightly around it, inducing suffocation. *See* boa constrictor.

consumer. In an ecosystem, a heterotrophic organism which feeds on other organisms. A primary consumer, as a herbivore, obtains its nutrition directly from plants; a secondary consumer, as a carnivore, obtains its energy indirectly by feeding upon herbivores.

contagious. Readily transmissible from one individual to another, said of diseases.

contiguous. Adjacent to; in contact with; adjoining.

contorted. Twisted; bent; convolute (in estivation).

contoured. Having an irregular but smooth, undulating surface, as a bacterial colony.

contraception. Prevention of conception; birth control; planned parenthood.

contract. To shorten; to narrow or reduce in size; to draw parts together; to shrivel or shrink.

contracted. **1.** Shortened. **2.** *bot.* With short, appressed branches, said of an inflorescence.

contractile fibrils. Myofibrils; myonemes.

contractile vacuole. A small excretory vesicle present in certain protozoans which fills with a fluid and then discharges its contents through the plasma membrane.

contralateral. Pertaining to or located on the opposite side. *Compare* ipsilateral.

control. In experimental work, a standard against which observations and results can be checked in order to determine their validity.

conure. A long-tailed New World parrot, especially one of the genus *Aratinga*.

convergence. **1.** In phylogeny, the appearance of similar structures in unrelated organisms. **2.** *physiol.* **a.** The condition in which several axons synapse with one or a few motor neurons.

b. The inclination of the visual axes toward a common point. **3.** *embryol.* The movement of cells from lateral areas toward the midline resulting in formation of the primitive streak, as in the chick.

convergent. Coming together, as veins in a leaf which pass from base to apex in a curving fashion.

convergent evolution. The tendency of unrelated animals in a particular habitat to acquire similar adaptive structures, as the streamline form in an aquatic environment.

convex. Having a curved, rounded exterior surface, as that of a sphere. *Compare* concave.

convolute. *bot.* **1.** Rolled up or bound together so that the edge of one part overlaps the next, as in floral buds. **2.** Rolled up longitudinally from the sides.

convoluted. Coiled or rolled up; having convolutions.

convolution. **1.** A turn, twist, bend, or fold. **2.** A permanent fold on the surface of the brain; a gyrus or folium.

cony. A coney, rabbit, or pika.

coon. A raccoon.

coot. Any of a number of species of aquatic birds of the genus *Fulica*, family Rallidae, as *F. americana*, the American coot, commonly called mud hen.

cooter. A turtle (*Pseudomys floridana*).

cootie. Common name for human body louse. *See* louse.

copal. A hard resin of plant origin used in varnishes.

copepod. Any crustacean of the subclass Copepoda, comprising mostly microscopic free-living forms, as *Cyclops, Calanus;* important as fish food. Some are parasitic. *See* fish louse.

copper. A metallic element, symbol Cu; atomic weight 63.54. It is a constituent of certain respiratory enzymes and is essential for the production of chlorophyll and hemoglobin. It is toxic to fungi.

copperhead. A poisonous snake (*Ancistrodon* (*Agkistrodon*) *contortrix*), a pit viper of eastern and southern United States.

copra. The dried meat of the coconut.

coprodeum. The anterior, dorsal portion of the cloaca into which the intestine empties.

coprolite. A petrified mass of fecal matter.

coprophagous. Feeding upon feces, as the dung beetle.

copulation. Sexual union of a male and female; coition.

coracidium. The ciliated embryo of the fish tapeworm and related species.

coracoid bone. A bone of the pectoral girdle of amphibians, reptiles, and birds; absent in most mammals.

coral. 1. The calcareous skeleton formed by certain anthozoan coelenterates. **2.** In a collective sense, masses of coral skeletons forming reefs, islands, etc. **3.** A coelenterate of the class Anthozoa, a small polyp which secretes a calcareous cup about its base. They are solitary or colonial animals inhabiting warm, shallow waters. **4.** The egg mass of a lobster, which when boiled resembles in color red coral.

¹coralline. *zool.* Of or resembling coral.

²coralline. 1. An animal which resembles a coral, as certain hydrozoans and bryozoans. **2.** *bot.* A red alga with thallus heavily impregnated with lime, encrusting forms forming an important component of coral reefs.

corallite. The skeleton of an individual coral polyp.

corallum. The entire skeletal mass of a colony of corals.

coral reef. A calcareous mass formed by the deposition of coral skeletons over a long period of time. Types include: fringing reef, one that is close to land; barrier reef, one separated from the shore by a navigable channel; circular reef or atoll, one which encloses a lagoon.

coral snake. One of several species of venomous snakes of the family Elapidae marked by distinctive black, red, and yellow bands, as the harlequin or Eastern coral snake (*Micrurus fulvius*) or the Western coral snake (*Micruroides euryxanthus*).

corbicula. The pollen basket of a honeybee.

corbina. *Menticirrhus undulatus,* of the croaker family, a game and food fish of the Pacific coast.

corbula. A phylactocarp, a basketlike structure enclosing the gonangium of certain hydroids.

cord. An elongated, cordlike structure, as the spinal *cord*, umbilical *cord*.

cordate. Heart-shaped, as a *cordate* shell or leaf.

cordiform. CORDATE.

core. The central portion of a fleshy fruit, especially pome fruits, consisting of ripened carpels containing seeds, as in an apple.

coremium. A cluster of hyphae forming a column which bears asexual spores (conidia).

coriaceous. Tough, leatherlike.

Cori cycle. The process in which muscle glycogen is broken down to lactic acid which is carried to the liver and reconverted to glycogen, which in turn is broken down to glucose and returned to muscle where it is utilized.

corium. 1. The dermis or derma, the layer of skin immediately beneath the epidermis; the true skin. **2.** The thickened, basal portion of a hemelytron.

cork. 1. *bot.* The outer protective covering of stems and roots of woody plants consisting of dead cells, the walls of which have been impregnated with suberin. **2.** The thickened outer bark of the cork oak (*Quercus suber*), used for bottle stoppers, floats, etc.

corm. 1. *bot.* A short, upright, underground stem, usually bulblike, as in the crocus. **2.** *zool.* The axial portion of a phyllopodium.

cormatose. Producing corms.

cormorant. A large, marine, fish-eating bird of the genus *Phalacrocorax*, family Pelicaniformes.

corn. 1. a. In the United States, Indian corn or maize (*Zea mays*), a tall cereal plant which produces seeds on large ears. **b.** The grain, seeds, or kernels of *Zea mays* used principally as food for livestock. **2.** In various countries, the seeds of any of the cereal grasses which are used for food. In England the term is applied to wheat, in Scotland, oats.

corn borer. A moth (*Pyrausta nubialis*), the European corn borer of the family Pyralidae, whose larvae live in cornstalks, introduced into the United States about 1917.

corncob. The central axis of an ear of Indian corn or maize upon which the grains or kernels develop.

cornea. 1. The anterior, transparent portion of the sclera of the eye of a vertebrate. **2.** The outermost, transparent layer of a compound eye of an arthropod. It is divided into facets, each facet covering the end of an ommatidium and functioning as a lens.

corn earworm. A moth (*Heliothis armigera*) of the family Noctuidae, whose larvae feed on developing corn grains. It is

119

also called tomato fruitworm or cotton bollworm as the larvae also feed on these plants.

cornicle. One of a pair of tubes at the posterior end of an aphid through which a waxy fluid is discharged.

corniculate. 1. Bearing horns or hornlike structures. **2.** Shaped like a horn.

cornification. Conversion into horn or hornlike material.

cornu pl. **cornua.** A horn or horn-shaped structure.

corolla. The petals of a flower taken collectively.

corollate. Possessing a corolla.

corona. 1. A crown or crownlike structure. **2.** *zool.* **a.** The ciliated disc at the anterior end of a rotifer. **b.** The pentamerous body of a crinoid. **3.** *bot.* **a.** A ring of scalelike appendages between the corolla and stamens in certain flowers, as the daffodil. **b.** A group of cells enclosing an oogonium in stoneworts.

coronal. Designating a plane or section through the coronal suture, which separates the frontal and parietal bones.

corona radiata. A layer of follicle cells which surrounds a mature mammalian ovum.

coronary. 1. Resembling a crown. **2.** Encircling, as the coronary arteries which supply the heart.

coronate. Possessing a crown, corona, or similar structure.

coronoid. Pointed, beaklike.

corpora. Plural of CORPUS.

corpora allata. A pair of neuroendocrine structures attached to the corpora cardiaca in certain insects; the source of the juvenile hormone (JH).

corpora cardiaca. A pair of neuroendocrine structures attached to the brain of certain insects.

corpora cavernosa. A pair of cylindrical bodies of erectile tissue present in the penis and clitoris.

corpus pl. **corpora.** A body or structure.

corpus callosum. A broad band of commissural fibers connecting the two cerebral hemispheres.

corpuscle. 1. A small body, as a renal or thymic *corpuscle*. **2.** A cell, especially one floating in a fluid, as blood, coelomic fluid, or saliva. **3.** An encapsulated sensory nerve ending, as Meissner's corpuscle.

corpus luteum. A small, yellow body which develops in a ruptured graafian follicle. It is the principal source of progesterone, also a possible source of estrogens.

cortex. **1.** *bot.* Parenchymal tissue of a stem or root which lies between the vascular tissue and epidermis. **2.** *zool.* The outer portion of an organism, as in certain protozoans. **3.** *anat.* The outermost layer of certain organs as the adrenal gland, ovary, kidney, brain.

corticate. Possessing a cortex.

corticosteroid. A steroid hormone secreted by the adrenal cortex. *See* glucocorticoid, mineralocorticoid.

corticotrophin. A hormone which stimulates the adrenal cortex, as ACTH.

cortisone. Compound E, an anti-inflammatory glucocorticoid secreted by the adrenal cortex.

corymb. A raceme type of inflorescence in which the lower flowers are borne on long stalks so that the entire flower is flat-topped, as in the cherry.

corymbiform. Shaped like a corymb.

corymbose. Corymbiform; corymbous.

cosmine. A hard, dentinelike substance present in certain types of fish scales.

cosmoid scale. A type of scale found in primitive fishes, as in crossopterygians and dipnoans.

cosmopolitan. Worldwide in distribution.

cosmozoic. Pertaining to life on other planets.

costa. **1.** A rib. **2.** *zool.* A riblike structure, as **a.** a ridge on the shell of certain mollusks; **b.** a vein in the wing of certain insects usually forming the anterior margin; **c.** a rodlike structure in the base of the undulating membrane of trichomonads; **d.** a comb row in ctenophores; **e.** an extension of the sclerosepta of corals connecting the theca to the epitheca. **3.** *bot.* **a.** The midvein of a simple leaf. **b.** The rachis of a pinnately compound leaf.

costal. Of or pertaining to a rib or costa.

costate. Bearing a longitudinal rib or ribs, said of leaves.

cotton. A soft, white, fibrous material obtained from fibers enclosing seeds of various plants of the family Malvaceae, especially that from the cotton plant (*Gossypium hirsutum* or *G. barbadense*). *See* boll.

cottonmouth. The water moccasin. *See* moccasin.

cottontail. Any of several species of North American rabbits of the genus *Sylvilagus*, especially *S. floridanus*, the Eastern cottontail.

cotyledon. 1. *bot.* A seed leaf, the first leaf that develops in the embryo sporophyte of a seed plant. **2.** *zool.* A cluster of villi on the placenta of certain ungulates, as the cow, deer.

cotylosaur. A primitive reptile of the order Cotylosauria, a group considered to have given rise to modern reptiles, as *Seymouria*.

cotype. *See* syntype.

cougar. A large feline quadruped (*Felis cougar, F. concolor, Puma concolor*) common in the western United States. Also called puma, mountain lion, panther, painter, catamount.

coupling. LINKAGE.

covert. One of the small feathers which covers the bases of the wing and tail feathers in birds; a tectrix.

covey. A brood, hatch, or small flock of birds, especially of quail and partridges.

cow. The mature female of any bovine animal or of any animal, the male of which is designated bull.

cowbird. *Molothrus ater* of the family Icteridae, a small blackbird which lays its eggs in the nests of other birds.

coxa. 1. The first or proximal segment of the leg of various arthropods. **2.** The hip or hip joint.

coxite. The basal segment of certain abdominal appendages in insects.

coxopodite. In crustacean appendages, the proximal or basal segment of the protopodite. *Compare* basipodite.

coyote. The prairie wolf (*Canis latrans*), a wolflike carnivore of the western United States.

coypu. The swamp beaver or beaver rat (*Myocastor coypus*), a large, aquatic rodent of South America, the source of a fur, nutria.

crab. 1. One of a large number of crustaceans of the order Decapoda, true crabs possessing a broad, flat cephalothorax and a short abdomen flexed under the body proper. *See* blue crab. **2.** One of a number of related crustaceans, as the hermit crab, sand crab, mole crab, or other arthropods, as the king crab which resemble true crabs. **3.** A crab louse.

crab louse. *Phthirus pubis*, a sucking louse which frequents the pubic and perianal regions of man.

crampon. A hooklike, aerial, adventitious root which serves as a supporting structure, as in the ivy.

crane. A tall wading bird of the family Gruidae, as the sandhill crane (*Grus canadensis*) and the whooping crane (*G. americana*), the latter rare and threatened with extinction.

crane fly. A dipterous insect of the family Tipulidae with extremely long, slender legs, as *Tipula simplex*, the range crane fly. Most of their larvae are aquatic or semiaquatic, but some, known as leather jackets, are terrestrial.

craniad. Toward the head or anterior end.

cranial. Of or pertaining to the cranium or the skull.

¹craniate. 1. Possessing a cranium **2.** Of, pertaining to, or belonging to the subphylum Vertebrata (Craniata).

²craniate. An animal with a skull and vertebral column.

cranium. The portion of the vertebrate skull which encloses the brain. *See* chondrocranium, dermatocranium, skull.

crappie. A freshwater pan fish of the family Centrarchidae, as *Pomoxis annularis*, the white crappie, and *P. nigromaculatus*, the black crappie.

craspedote. Possessing a velum, as in medusae.

crateriform. Shaped like a saucer or hollow, shallow bowl.

craw. 1. The crop of an insect or bird. **2.** The stomach of an animal.

crawdad. 1. In the United States, a crayfish. **2.** In England, the spiny lobster.

crawler. 1. An animal which crawls, as the night *crawler;* an annelid. **2.** The first instar of a scale insect.

crayfish. A freshwater, decapod crustacean, resembling a lobster but smaller in size. American forms belong principally to the genus *Cambarus*, European forms to the genus *Astacus*.

creatine. $C_4H_9N_3O_2$, a nitrogenous compound in muscle and other tissues. It combines readily with phosphate to form phosphocreatine (PC) which serves as a reservoir for high-energy phosphate used in the formation of adenosine triphosphate (ATP).

creatospore. A plant which produces nuts.

creeper. 1. *bot.* **a.** A plant which spreads along the surface of the ground. **b.** A plant which climbs by means of aerial roots, as the Virginia creeper. 2. *zool.* **a.** A small bird of the family Certhiidae, as the brown creeper (*Certhia familiaris*). **b.** Any small creeping animal.

creeper fowl. A heterozygous fowl with short, crooked legs. Homozygous recessives die due to a lethal factor.

crenad. A plant which grows near a spring.

crenate. Having a notched or scalloped border.

crenicolus. Inhabiting brooks fed by springs.

crenulate. Minutely or finely crenate.

creodont. A primitive, fossil mammal.

creophagous. Carnivorous, said of insect-eating plants.

crepitation. Production of a grating or crackling sound.

crepuscular. 1. Pertaining to dusk or twilight. 2. Active at twilight or in dim light, as bats.

crest. 1. A tuft of hair or a process on the head of a bird or mammal, as the comb of a cock. 2. A prominent ridge, as **a.** that along the neck of newts; **b.** that on certain bones or seeds.

crested. Bearing a crest or crestlike structure.

Cretaceous period. The third and final period of the Mesozoic era characterized by great swamps, formation of the Rocky Mountains, the rise of flowering plants, modern insects, archaic mammals, and birds, and the extinction of dinosaurs, pterodactyls, and toothed birds.

cribriform. Sievelike, as the *cribriform* plate of the ethmoid bone with openings through which pass the olfactory nerves.

cricket. One of a number of leaping, orthopterous insects of the family Gryllidae, noted for their chirping sounds produced by stridulating organs on the forewings of the males.

crinoid. Any echinoderm of the class Crinoidea which includes the sea lilies and feather stars, abundant during the Paleozoic era. Most were sessile forms, attached by a jointed stalk. Living species include *Antedon*, a feather star of the Atlantic coast.

crisp. Curled; in curls or ringlets; brittle.

crissum. Undertail coverts constituting the circumanal plumage of birds.

crista. 1. A crest or ridge. 2. A fold on the inner wall of a mitochondrion.

crista ampullaris. A sensory structure on the inner surface of the ampulla of a semicircular duct containing hair cells which are stimulated by movement of the endolymph.

cristate. Bearing a crest or crista.

cristulate. Bearing small crests.

crithridia. A type of trypanosome.

croaker. 1. An animal which croaks, as a frog. 2. One of a number of fishes which produce a croaking or grunting sound, as the Atlantic croaker (*Micropogon undulatus*) of the family Sciaenidae. *See* drumfish.

crochets. 1. Minute curved hooklets on the tips of the prolegs of a caterpillar. 2. Small double-pointed chaetae in certain oligochaete worms.

crocodile. A large, long-tailed, aquatic reptile of the order Loricata (Crocodilia), as *Crocodylus americana*, the American crocodile.

crocus. A bulbous plant of the genus *Crocus*, an early spring flower.

Cro-Magnon man. A race whose remains are found in caves of southern France; considered to be the earliest type of modern man. They lived in the upper Pleistocene 20,000 to 50,000 years ago.

crop. 1. In birds, a dilated portion of the esophagus in which food is stored and moistened; the ingluvies, craw. 2. In various invertebrates, a dilated region of the foregut in which food is stored, as in annelids and insects.

crop-milk. PIGEON MILK.

crosier. A curled structure, as a young frond of a fern.

cross. To interbreed; to mate two individuals of different breeds, races, or species; the product of such a mating.

crossbill. A small bird of the family Fringillidae with curved mandibles which cross each other, as *Loxia leucoptera*, the white-winged crossbill.

crossbreed. To mate two unrelated individuals, especially those of different breeds.

125

cross-fertilization. The union of gametes of two different individuals. *See* cross-pollination.

crossing-over. In genetics, a process occurring during synapsis in which pairs of homologous chromosomes bearing linked genes mutually exchange corresponding parts. *See* linkage, crossover.

crossopterygian. A primitive lobe-finned fish of the superorder Crossopterygii, a group considered ancestral to amphibians and higher vertebrates. *See* coelacanth.

crossover. An individual resulting from a new combination of linked genes brought about by crossing-over.

cross-pollination. The transfer of pollen from a flower to the stigma of another or, more commonly, the transfer of pollen from a flower of one plant to that of another.

crotalid. A pit viper.

croton bug. The German cockroach (*Blattella germanica*).

crow. A glossy, corvine blackbird of the genus *Corvus*, as the common crow (*C. brachyrhynchos*).

crown. 1. A corona; the top portion of a structure. **2.** *bot.* **a.** The junction between stem and root in a seed plant. **b.** A portion of a rhizome with a large bud, used in propagation. **c.** A circle of processes on the inner surface of a corolla. **d.** The branches of a tree with their leaves. **3.** *zool.* **a.** The body of a crinoid with its projecting arms. **b.** The crest of a bird. **4.** *anat.* **a.** The exposed portion of a tooth covered by enamel. **b.** The top portion of the head.

crown gall. A malignant growth at the crown of many broad-leaved plants caused by a bacterium (*Agrobacterium tumefaciens*).

crown-of-thorns. A sixteen-armed, spiny starfish, *Acanthaster planci*, destructive to living coral, especially in the islands of Oceania.

cruciate. Resembling or shaped like a cross; bearing the mark of a cross.

cruciferous. Of or pertaining to the mustard family, Cruciferae.

cruciform. CRUCIATE.

crura. Plural of CRUS.

crus pl. **crura. 1.** The leg or shank. **2.** A structure which re-

126

sembles a pair of legs or roots, as the crura cerebri (cerebral peduncles).

crustacean. Any arthropod of the class Crustacea which includes some 28,000 species of mostly aquatic forms, as lobsters, crayfishes, shrimps, crabs, barnacles, water fleas, copepods, barnacles, and fossil trilobites.

crustaceous. Hard; brittle, with a crustlike surface or shell.

crustose. Lying flat and close to the substrate, said of lichens.

cryophilous. Inhabiting cold regions; thriving in low temperatures.

cryophyte. A plant which grows on ice or snow.

cryoplankton. The plankton of glacial or polar waters.

crypt. A slender sac or cavity; an intestinal gland (crypt of Lieberkuhn).

cryptanthous. Cleistogamous; having hidden flowers or stamens.

cryptic. **1.** Hidden or concealed. **2.** Adapted for concealing, as *cryptic* coloration.

cryptogam. Any member of the Cryptogamia in older classifications, a group which included plants which did not produce true flowers or seeds, as thallophytes, mosses, and ferns. *Compare* phanerogam.

cryptomitosis. Mitosis in which distinct chromosomes are not formed.

cryptophyte. A plant which reproduces by underground or underwater structures, as by bulbs, corms, or rhizomes.

cryptorchism. Failure of the testes to descend; cryptorchidism.

cryptozoic. Living in concealed places, as under stones, in crevices.

cryptozoite. The preerythrocyte stage of the malarial organism found in liver cells.

crystalline. **1.** Of, resembling, or pertaining to crystals. **2.** Clear or transparent like a crystal, as the *crystalline* lens. *See* lens.

crystalline cone. A cylindrical, refractive structure in an ommatidium.

crystalline style. *See* style.

¹crystalloid. Like a crystal.

127

²**crystalloid.** **1.** A substance which forms a true solution, as salt, sugar. **2.** A protein crystal, a solid nitrogenous particle present in various seeds.

ctenidium. **1.** A gill, especially in a mollusk. **2.** A row of stiff bristles forming a comblike structure, as that on the head of a flea.

ctenoid. Resembling a comb; pectinate.

ctenoid scale. A type of dermal scale with an edge bearing comblike projections, as that in most teleost fishes.

ctenophore. Any member of the phylum Ctenophora which includes the comb jellies or sea walnuts, transparent, gelatinous, free-swimming marine animals with a locomotor organ consisting of eight rows of comb plates.

cub. The young of various animals, as the lion, tiger, bear.

cubical. Cube-shaped.

¹**cubital.** Of or pertaining to the forearm or the elbow.

²**cubital.** A secondary remex of a bird.

cubitus. **1.** The forearm. **2.** A vein in an insect wing.

cuboid. Resembling a cube in shape.

cuckoo. A common European bird (*Cuculus canorus*), a social parasite which lays its eggs in the nests of other birds.

cucullate. **1.** Hooded. **2.** Shaped like a hood.

cucumber. A vine (*Cucumis sativus*) or its long, succulent fruit.

cucurbit. Any member of the melon or gourd family, Cucurbitaceae, as the gourd, pumpkin, squash, cucumber, muskmelon, and watermelon.

cud. In ruminants, a mass of food which is regurgitated from the rumen, chewed, and swallowed a second time.

culm. The jointed stem of a grass, usually hollow except at the nodes.

culmen. **1.** A median, longitudinal ridge on the beak of a bird. **2.** The median, superior portion of the cerebellum.

culmicolous. Growing on grass stems, said of fungi.

cultch or **culch.** Broken stones and shells placed in oyster beds which serve for attachment of young oysters.

cultellus. A sharp, lancetlike mouth part of certain flies.

cultigen. A plant or group of plants which has arisen from or is known only in cultivation. *Compare* indigen.

culture. **1.** The cultivation of microorganisms (bacteria, yeasts, molds) or tissue cells on specially prepared media. **2.** A mass of organisms or cells cultivated on such a medium. **3.** The rearing of animals or the cultivation of plants, as a crop.

cultured. Grown or produced artificially, as cultured pearls.

cuneate. 1. Wedge-shaped. **2.** Triangular and tapering to the point of attachment, said of leaves.

cuneiform. Cuneate; wedge-shaped.

cuneiform bones. Three tarsal bones lying proximal to the first three metatarsal bones.

cuneus. 1. A convolution on the medial surface of the occipital lobe of the cerebrum. **2.** A triangular-shaped portion of the corium of the forewing of a hemipterous insect.

cunner. A small, marine percomorph fish (*Tautogolabrus adspersus*).

cup. A shallow, open organ or structure, as the involucre of an acorn.

cup fungi. Ascomycetes of the order Pezizales which possess cup-shaped fruiting bodies or apothecia, as *Peziza.*

cupola. CUPULA.

cupula. 1. A gelatinous mass surmounting **a.** the neuromasts of a lateral line system; **b.** a crista ampullaris of the inner ear. **2.** The rounded tip of the osseous cochlea. **3.** The dome of the diaphragm.

cupulate. Cup-shaped; bearing cupules.

cupule. A cup-shaped structure, as **a.** the involucre of an acorn consisting of bracts fused at their bases; **b.** the structure in liverworts which produces gemmae; **c.** the apothecium of a cup fungus.

cupuliform. Cup-shaped.

curare. An extract of tropical plants of the genera *Chondrodendron* and *Strychnos* of the Amazon valley used by South American Indians for poisoning arrows. It contains alkaloids which block impulses at the myoneural junction. It is used medicinally as a muscle relaxant.

curassow. One of several gallinaceous birds of the genera *Mitu* and *Crax*, family Cracidae, inhabiting forests of Central and South America.

curculio. Any snout beetle of the family Curculionidae, as the plum curculio (*Conotrachelus nenuphar*).

129

curlew. One of several large shorebirds of the family Scolopacidae with long, downcurved bills, especially those of the genus *Numenius*.

currant. 1. A small, dried grape grown extensively in Greece. **2.** A small, acid berry of the shrub *Ribes* of the gooseberry family, used extensively in jams, sauces, jellies, and wines.

cursorial. Adapted for running.

cuscus. A phalanger, *q.v.*

cusp. 1. A pointed structure, peak, or projection, as that on the crown of a tooth. **2.** One of the leaflets of a heart valve.

cuspid. A tooth with one cusp, as a canine tooth.

cuspidate. Terminating in a sharp, pointed tip, said of leaves.

cutaneous. Of or pertaining to the skin.

cuticle. 1. *zool.* **a.** A noncellular pellicle, membrane, or covering of the body of various invertebrates. **b.** The outermost layer of a hair shaft. **c.** The epidermis of the skin. **2.** *bot.* The noncellular surface layer covering the aerial parts of a vascular plant.

cuticula. CUTICLE.

cuticularization. The formation of a cuticle.

cutin. A waxy, waterproofing substance present in plant cuticles and in the walls of certain cells, as epidermal cells of leaves, flowers, and fruits.

cutinization. The impregnation of the walls of epidermal cells with cutin.

cutis. The skin.

cutlass fish. A long, slender marine fish of the family Trichiuridae; hairtail.

cutting. A part of a plant removed for vegetative propagation.

cuttlebone. The calcareous internal shell of the cuttlefish.

cuttlefish. A dibranchiate, decapod cephalopod of the genus *Sepia*.

cutworm. The larva of various moths of the family Phalaenidae (Noctuidae) which feed upon roots and shoots of various plants, cutting them off at the surface of the ground.

Cuvierian organs. Slender tubules attached at the base of a respiratory tree in sea cucumbers which, when discharged through the anus, form a sticky, entangling mass.

cyanophyte. Any member of the Cyanophyta, a division of the plant kingdom which includes the blue-green algae, simple organisms lacking a distinct nucleus and plastids, as *Gleocapsa*, *Nostoc*.

cyanosis. Blueness of the skin due to an excess quantity of reduced hemoglobin in the blood. *See* blue baby.

cyathiform. Cup-shaped.

cyathium. A type of inflorescence in which modified unisexual flowers are borne on a cup-shaped involucre, as in *Euphorbia* (spurge).

cybernetics. The application of statistical mechanics to communication systems, especially in relation to human control systems.

cycad. A primitive, gymnospermous plant of the subclass Cycadophytae, with large divided leaves borne on the tip of a short, unbranched stem, abundant during the Mesozoic era. Living forms comprise about 100 species.

¹cycle. A complete course of operations or a series of events which occurs repeatedly in the same sequence, as a life *cycle*.

²cycle. 1. *zool.* An alternating circle of tentacles on the oral disc of a sea anemone. **2.** *bot.* **a.** One turn of a helix or spire, with reference to leaf arrangement. **b.** A whorl of floral parts.

cyclic. 1. Pertaining to or occurring in cycles. **2.** *bot.* Whorled, said of floral parts.

cycloid scale. A type of fish scale which is thin, lacks enamel, is roughly circular in outline, and bears concentric lines of growth.

cyclomorphosis. Seasonal changes in form which occur in various plankton organisms.

cyclopia. Possessing a single, median eye.

cyclosis. The streaming of protoplasm, as seen in the cells of *Elodea* and various protozoans.

cyclostome. A lamprey or hagfish.

cydippid larva. The free-swimming larva of certain ctenophores.

cyesis. Pregnancy.

cygnet. A young swan. *See* cob; pen.

cymba. A woody, boatlike spathe which encloses the inflorescence in certain palms.

cyme. A determinate type of inflorescence in which flowers form a flat-topped or convex cluster in which the terminal flower of each axis blooms first.

cymose. 1. Of or pertaining to a cyme. **2.** Bearing cymes.

cymule. A small cyme.

cynarrhodium. A type of fruit characteristic of roses, as a rose hip.

cynipid. A gall-producing insect of the family Cynipidae, order Hymenoptera. *See* gall wasp.

cyphonautes. A ciliated larva which develops from a non-brooded egg in ectoprocts.

cypris larva. The free-swimming larva of certain crustaceans, as barnacles.

cyrtopia stage. The postlarval stage of euphausiaceans, shrimp-like crustaceans.

cyst. 1. A small, closed sac or vesicle; a bladder. **2.** *zool.* A thick-walled structure containing an organism in an inactive stage, as that of various protozoans and flukes. **3.** *bot.* A resistant, sporelike body formed by various algae, bacteria, fungi, and myxomycetes. **4.** *med.* An abnormal, saclike structure or closed cavity containing fluid or other material, as a hydatid *cyst*.

cystacanth. The infective form of an acanthocephalan.

cystic duct. The duct leading from the gall bladder to the common bile duct.

cysticercoid. A larval form of a tapeworm resembling a cysticercus but with a solid body and usually a taillike appendage, as in *Hymenolepis*.

cysticercus. A bladder worm, the larva of a tapeworm consisting of a fluid-filled sac into which a scolex is invaginated; found in the intermediate host.

cystid. In bryozoans, the exoskeleton and body wall.

cystidium. A sterile structure on the basidium of various basidiomycetes.

cystocarp. In red algae, a spore case containing diploid spores (carpospores).

[1]cystoid. Bladderlike; resembling a cyst.

[2]cystoid. A fossil echinoderm of the class Cystoidea.

cystolith. *bot.* A crystalline body which develops within the epidermal cells of certain plants, as the stinging nettles.

cytase. An enzyme in seeds which acts on hemicelluloses.

cytochemistry. The science which deals with the constituents of cells.

cytochrome. One. of a group of enzymes, cytochromes a, b, and c, widely distributed in animal and plant tissues which play an important role in cellular respiration. Each is a complex pigmented compound containing iron.

cytochrome oxidase. One of several enzymes which catalyze the final steps in Krebs cycle in which water is an end product.

cytogamy. In protozoa, a process resembling conjugation but differing in that the synkaryon is formed from fusion of the male and female nuclei of the same cell.

cytogenetics. The science which deals with cytological changes occurring in connection with or related to hereditary phenomena, especially changes occurring in the chromosomes.

cytokinesis. The phase in mitosis or meiosis which involves the division of the cytoplasm.

cytology. The division of biology which deals with cells, their origin, structure, and physiology.

cytolymph. The fluid matrix of cytoplasm.

cytolysin. A substance which brings about the dissolution of cells, as an antibody (bacteriolysin or hemolysin) or a nonspecific substance, as snake venom or a hypotonic saline solution.

cytolysis. The dissolution or disintegration of cells.

cytomegalovirus. A virus which causes abnormal enlargement of host cells.

cytomorphosis. *bacteriol.* A succession of morphological types from large, young cells through smaller, uniformly staining, mature forms, to senescent, degenerating forms, as may appear in a bacterial culture.

cyton. The cell body of a neuron; perikaryon.

cytopathic. Causing injury to cells.

cytopharynx. A cavity in certain protozoans which leads from the cytostome (mouth) into the endoplasm; the gullet.

cytoplasm. The protoplasm of a cell surrounding the nucleus. It includes the cytoplymph, organelles, inclusions, and the plasma membrane.

cytoproct. The anal opening of a protozoan; cytopyge.

cytopyge. CYTOPROCT.

cytorrhysis. Wrinkling of a schizophyte when placed in a hypertonic solution.

cytosine. A water-soluble pyrimidine base, $C_4H_5N_3O$, present in nucleotides and nucleic acids.

cytosome. The cytoplasm of a cell.

cytostome. In protozoans, the mouth or opening leading to the cytopharynx.

cytotaxonomy. The branch of taxonomy which uses cytology as an aid in determining relationships.

cytozoan. An animal which lives as an intracellular parasite, as *Plasmodium*.

cytozoic. Living within cells, said of animal parasites.

D

dab. A flatfish, especially a European flounder of the genus *Limanda*.

dabbler. A surface-feeding or dabbling duck.

dace. 1. A small European fish (*Leuciscus*). 2. In the United States, one of a number of small, freshwater fishes, especially those of the genus *Rhinichthys*.

dactyl. A digit (finger or toe).

dactylogram. A fingerprint.

dactylozooid. In colonial coelenterates, a defensive or protective zooid; a zooid which produces nematocysts.

daddy longlegs. *See* harvestman.

dahlia. A plant of the genus *Dahlia*, propagated by root tubers.

dalmanite. A fossil trilobite of the genus *Dalmanites*.

dam. A female parent, especially of a quadruped.

daman. *See* hyrax.

damping-off. A disease of seedlings in which plants die just before or after emergence from the soil; caused by a phycomycete fungus, *Pythium*.

damselfly. Any of a number of slender-bodied, rapid-flying insects of the order Odonata, suborder Zygoptera. They differ from dragonflies in that wings are held vertically over the body when at rest.

dart. *See* dart sac.

darter. 1. One of a large number of small, freshwater fishes of the family Percidae characterized by swift, sudden movements, as *Etheostoma* and *Percina*. **2.** WATER TURKEY.

dart sac. In snails, an outpocketing of the vagina which produces calcareous spicules or darts which are discharged reciprocally during copulation.

dassie. *See* hyrax.

dasyphyllous. Having thick, hairy leaves.

date. The fruit of the date palm (*Phoenix dactylifera*).

¹daughter. A female child.

²daughter. Designating first generation offspring without regard to sex, as a *daughter* chromosome.

Dawn Stone Age. The Eolithic Age, the first of the cultural ages of man based on the existence of crude stone instruments called eoliths. It lasted from the Miocene through most of the Pliocene epoch.

day neutral plant. An indeterminate plant, or one which produces flowers irrespective of whether the days are long or short, as corn, tomato.

dead. Lifeless; not alive.

dead-leaf butterfly. An Indo-Malayan butterfly of the genus *Kallima* which, when resting, resembles a dead leaf.

dead-man's-fingers. A soft coral (*Alcyonium digitatum*) found on beaches after storms.

¹dealate. Having lost or been deprived of wings.

²dealate. An insect which has lost its wings, as a termite.

deamination. Removal of the amino (NH_2) group from an organic compound, especially an amino acid, accomplished by oxidation, reduction, or hydrolysis.

death. Cessation of life or vital activities beyond the possibility of resuscitation; termination of the existence of an organism. *See* necrosis.

death cup. *See* volva.

decacanth. The ciliated larva of *Amphilina foliacea*, a cestodarian tapeworm. Also called lycophore.

decalcification. Removal of lime or calcareous matter.

decapitate. To cut off or remove the head.

decapod. 1. Any crustacean of the order Decapoda which includes the lobsters, crayfishes, shrimps, and crabs. **2.** Any cephalopod mollusk of the order Decapoda which includes the squids and cuttlefishes.

decarboxylation. The loss or removal of carbon dioxide from an organic acid, especially an amino acid.

decay. 1. Decomposition, putrefaction, fermentation, rotting; the transformation from a more or less solid or perfect state to a less perfect state. **2.** *bacteriol.* The decomposition of organic matter, especially carbohydrates through the action of aerobic organisms such as bacteria, yeasts, and molds.

decerebration. Removal of the brain, especially the cerebrum.

decibel. A unit of sound intensity equivalent to one tenth of a bel. Abbreviated db.

decidua. The mucosa or endometrium of a pregnant uterus.

deciduous. Falling off or shed at maturity, with reference to leaves or fruit in plants, or antlers, teeth, or uterine endometrium in mammals.

decomposer. An organism which converts the bodies or excreta of other organisms into simpler substances, as bacteria, yeasts, molds, and other fungi; a reducer.

decompound. Compound more than once, said of leaves.

decortication. 1. *bot.* Removal of the bark or outer covering of a plant or plant product. **2.** *zool.* Removal of the cortex of an organ.

decumbent. Lying on the ground but with end ascending.

decurrent. Prolonged downward, said of a leaf whose base extends downward beyond its insertion forming a wing or ridge along the stem.

decussate. 1. *anat.* Crossed or intersected so as to form an X-shaped structure, as the optic chiasma. **2.** *bot.* Arranged in pairs alternately crossing at right angles, as *decussate* leaves or branches.

deep. Situated or located below the surface; not superficial.

deer. In a broad sense, any of a number of ruminants of the family Cervidae whose males possess solid antlers which are shed and grown anew each year, as the white-tailed deer (*Odocoileus virginianus*).

deerfly. A dipterous insect of the genus *Chrysops*, a pest to domestic and wild animals and a transmitter of disease organisms.

deer line. *See* browse line.

deer mouse. A rodent (*Peromyscus leucopus*) and related species, of the family Muridae. Also called white-footed mouse.

defassa. *Kobus defassa*, a bovine animal of Africa, resembling the waterbuck. Also called sing-sing.

defaunate. To make free of animal inhabitants.

defecation. Evacuation of the bowel; passage of feces or a stool.

deficiency. *See* deletion.

definite. 1. Precise, exact, fixed. 2. *bot.* **a.** Not exceeding twenty, with reference to stamens. **b.** Cymose, with reference to inflorescence.

definitive. Complete, fully developed, final or last.

deflexed. Bent abruptly downward; reflexed.

deflorate. To deprive of flowers.

defoliation. The loss of leaves, as that occurring in the natural shedding or that resulting from the depredations of insects.

degeneration. 1. *biol.* Reversion or retrogression from a highly specialized type to a simpler type. 2. *pathol.* Retrogressive changes in cells, tissues, or organs involving chemical or physical changes usually resulting in impaired function and reduced vitality. May be physiological, as occurs in aging, or may be the result of disease processes.

deglutition. The act of swallowing.

dehiscence. The bursting or splitting open of a part of a plant at maturity, as a capsule, seed pod, fruit, or anther.

dehydrate. To render free of water.

dehydration. Removal of or loss of water.

dehydrogenase. An enzyme which catalyzes the oxidation of a substance by removal of hydrogen atoms.

delamination. The process of splitting into layers.

deletion. *genetics.* The loss of a portion of a chromosome with its contained genes. Also called deficiency.

¹**deliquescent.** *chem.* Capable of absorbing atmospheric moisture and going into solution in the water absorbed.

²**deliquescent.** *bot.* **1.** Becoming semiliquid, applied to the perianth of certain flowers, as the water hyacinth, or to certain fungi upon maturing. **2.** Designating a type of branching in which the main stem divides and then divides repeatedly forming many small branches. *See* excurrent.

¹**deltoid.** Triangular in shape like the Greek letter delta.

²**deltoid.** In man, a large muscle which covers the shoulder joint, an abductor of the humerus.

deme. **1.** A genetic population; one of two interbreeding races of a population which resemble each other morphologically but differ physiologically. **2.** A local breeding population.

demology. The study of population genetics.

denatured. Deprived of natural qualities or attributes; rendered unfit for eating or drinking.

dendrite. A short and usually branched process of a neuron which conducts impulses to the cell body.

dendritic. Branching like a tree: ARBORESCENT.

dendrogram. A treelike representation of taxonomic relationships.

dendroid. Treelike; resembling a tree in shape.

dendrology. **1.** The study of trees and shrubs **2.** A treatise upon trees.

dendron. A dendrite.

dendrophilous. Tree-loving; living in, on, or among trees: ARBOREAL.

denitrification. The process of freeing nitrogen from its compounds or of reducing nitrates to simpler compounds (nitrites, oxides of nitrogen, ammonia) and eventually free nitrogen. It occurs especially in waterlogged soils under anaerobic conditions through the action of denitrifying bacteria.

denizen. **1.** A resident or inhabitant. **2.** A naturalized animal or plant in a region to which it is not indigenous.

dens. A tooth or toothlike process; the odontoid process of the axis.

dental. Of or pertaining to teeth or dentistry.

dental formula. In mammals, a method of expressing for a given species, the number of various kinds of teeth, that is, incisors, canines, premolars, and molars. The numbers are given in the form of a fraction, the upper teeth being designated by the denominator above the line, the lower teeth by the numerator below. The dental formula for man is $i\frac{2}{2}$, $c\frac{1}{1}$, $p\frac{2}{2}$, $m\frac{3}{3}$, or, simplified, $\frac{2.1.2.3}{2.1.2.3}$.

dentalium. A tooth shell of the genus *Dentalium*.

dentary. One of several bones of the lower jaw in vertebrates. In man, it alone forms the mandible.

dentate. Having a notched or serrated margin, as a *dentate* leaf.

denticle. Any small tooth or toothlike projection.

denticulate. Having minute, toothlike projections.

dentine. A hard, dense calcareous tissue present in dermal scales and forming the major portion of a tooth. It surrounds the pulp cavity and underlies the enamel of the crown and cementum of the root.

dentition. 1. Teething or the eruption of teeth. 2. The teeth of an individual or species taken collectively or as expressed by a dental formula.

dentoid. Toothlike.

denudate. Bare or naked.

denude. To make bare or naked; to strip; to remove the surface layer.

deoxy compound. A substance which has one less oxygen atom than the reference substance.

deoxyribonuclease. An enzyme which catalyzes the hydrolysis of deoxyribonucleic acid (DNA).

deoxyribonucleic acid. DNA, a substance present in nuclei of cells consisting of phosphoric acid, a sugar, and nitrogenous bases. The sugar is deoxyribose, a pentose sugar; the bases are purine bases, adenine and guanine, and pyrimidine bases, thymine and cytosine. It is a nucleic acid localized principally in the chromosomes. It regulates protein synthesis and is the key molecule of the genes which give each cell its hereditary qualities and characteristics.

depauperate. Dwarfed, stunted, impoverished; poorly developed.

dependent. 1. Hanging downward. 2. Relying on something else for support or sustenance.

deplasmolysis. The return to a normal state following plasmolysis.

depolarization. The process of depriving of polarity, said of cell membranes when movement of ions across the plasma membrane results in inactivation of electrical charges and a drop in potential. Such occurs in a nerve cell upon application of an adequate stimulus.

depressant. An agent which lowers vital activity.

depressed. Pressed down; flattened from above.

depression. A state in which functional acitivity is below normal.

depressor. 1. A muscle which depresses or draws a part downward. 2. A nerve which, when stimulated, lowers blood pressure by reflexly inducing vasodilation or a slowing of heart beat.

dermal. Of or pertaining to the skin or, more specifically, the dermis.

dermal bone. A membrane bone.

dermal branchia. A papula.

dermal scales. Typical scales of fishes present in the epidermis but of mesodermal origin.

dermal skeleton. Skeletal structures which originate from the dermis, as the dermal scales of fishes, dermal plates of reptiles and, in birds and mammals, certain membrane bones of the skull.

dermapteran. Any insect of the order Dermaptera which includes the earwigs, nocturnal insects with large, forcepslike cerci, as *Forficula auricularia*, the European earwig, now common in the United States.

dermatocranium. The portion of the skull or cranium consisting of membrane bones which forms a roof over the chondrocranium.

dermatogen. Meristem tissue in plants which gives rise to the epidermis.

dermatome. 1. The portion of the lateral wall of a somite which gives rise to the dermis. 2. An area of the skin supplied by a single spinal nerve.

dermatophyte. A parasitic fungus which lives in or on the skin, the cause of various skin diseases (dermatomycoses), as ringworm.

dermestid. Any beetle of the family Dermestidae, small dark beetles, some of which are serious household pests, as the black carpet beetle (*Attagenus piceus*).

dermis. *See* corium.

dermopteran. Any mammal of the order Dermoptera which includes the flying lemurs, nocturnal, arboreal animals with webbed feet and a patagium. Comprises one genus (*Galeopithecus*) of Southeast Asia.

derris. The plant or root (tuba root) of various species of *Derris*, a leguminous plant of Southeast Asia. It is the source of rotenone, an insecticide.

desaturation. The conversion of a saturated compound to an unsaturated compound by the removal of hydrogen, as stearin to olein.

descending. Directed downward or tending to grow downward.

desiccation. The act of drying thoroughly; exhausting or depriving of moisture.

desman. An aquatic, molelike mammal of the family Talpidae, order Insectivora.

desmid. A green alga of the family Desmidiaceae, common in fresh waters. Each consists of two semicells connected by an isthmus, as *Cosmarium*.

desmoneme. A nematocyst with a threadlike tube coiled like a corkscrew and closed at one end. Also called volvent.

desmosome. A condensation of protoplasm at points of contact between cells, as those of the epidermis of the skin.

Desor's larva. The ciliated larva of *Lineus*, a nemertine.

despeciate. To remove all species of organisms from a specific habitat.

determinate. Definite, fixed, established; with specific limits.

determinate cleavage. Spiral cleavage, as in protostomes, in which the early blastomeres have specific fates.

determination. *embryol.* The assumption by an embryonic region of a specific course of development and irrevocable structural determination.

detorsion. A reversal in rotation as occurs in some gastropods whose ancestors had undergone complete torsion.

deuteromycete. Any fungus of the Deuteromycetes which comprises the Fungi Imperfecti, fungi whose sexual stages are lacking or unknown.

deuterostome. Any member of the Deuterostomia, animals in which the blastopore becomes the anal opening, the mouth being formed later as an independent opening, as in the echinoderms and chordates.

deuterotoky. Parthenogenesis in which individuals of both sexes are produced from female gametes.

deutocerebrum. The second or middle portion of the brain of an arthropod.

deutomerite. The third or posterior body segment of a gregarine.

deutonymph. The second stage in the development of certain trombiculid mites.

deutoplasm. Yolk or food material within the cytoplasm of an ovum.

development. 1. The series of orderly changes by which a mature, functional cell, tissue, organ, organ system, or an organism comes into existence. 2. The evolution or phylogenesis of a group of organisms, as the *development* of mammals.

deviation. A departure from a standard or norm; an alteration in the regular course or path of development.

devilfish. 1. Any of the large rays of the genus *Manta*, especially the greater devilfish (*Manta birostris*). 2. An octopus.

Devonian period. A period of the Paleozoic era following the Silurian and preceding the Carboniferous during which primitive lycopods, bryophytes, and amphibians came into existence.

dewclaw. A vestigial first digit of the hind foot of certain mammals, as the dog.

dewlap. 1. A pendulous fold of skin hanging from the neck in most bovine animals. 2. The wattles of a bird.

dextral. 1. Pertaining to or on the right side; right-handed. 2. Coiling from left to right, as in gastropods. *Compare* sinistral.

dextrin. A polysaccharide resulting from incomplete hydrolysis of starch.

dextrorse. 1. *bot.* In a plant arising vertically, twining from left to right. 2. To the right or clockwise.

dextrose. $C_6H_{12}O_6$, d-glucose or grape sugar. *See* glucose.

dextrosinistral. From right to left.

dhole. The red dog or Indian wild dog (*Cuon dukhunensis*).

diabetes insipidis. A disorder of the pituitary gland or hypothalamus characterized by excessive production of urine. It is caused by insufficient production of vasopressin, the antidiuretic hormone (ADH).

diabetes mellitus. A disorder of carbohydrate metabolism resulting from inadequate production of insulin by the pancreas. It is characterized by presence of sugar in the urine and excessive production of urine.

diadelphous. United into two bundles or fascicles, said of stamens.

diageotropic. Transversely geotropic, said of leaves or roots which grow horizontally.

diagnosis. **1.** *biol.* A concise, technical description of a taxonomic group. **2.** *med.* The recognition or determination of the existence of a disease or pathological condition. **3.** The scientific examination and analysis of a situation or condition.

dialycarpic. Composed of distinct carpels, said of fruits.

dialysis. The separation of crystalloids from colloids by the use of selectively permeable membranes.

diandrous. Having two stamens.

diapause. A period of rest or delayed development as occurs in an insect in which growth started in the fall ceases and is not resumed until spring.

diapedesis. The passage of ameboid white blood cells through an unruptured capillary wall.

diaphragm. **1.** A thin dividing membrane or partition. **2.** *anat.* The musculotendinous sheet separating the thoracic and abdominal cavities. **3.** *microscopy.* A device for regulating the amount of light passing to and through an object.

diaphysis. **1.** *anat.* The shaft of a long bone. **2.** *bot.* The elongated axis of an inflorescence.

diapsid. A type of vertebrate skull in which two temporal fossae are present, as in *Sphenodon* and crocodiles.

diarch. Having two protoxylem groups or poles, with reference to roots.

diarthrosis. A type of articulation in which there is considerable freedom of movement. The joint possesses a joint cavity lined with synovial membrane.

diaspore. A disseminule.

diastase. An enzyme which hydrolyzes starch to maltose, as ptyalin.

diastema. A natural gap or space between two types of teeth.

diastole. 1. The relaxation phase following contraction during which the heart chambers dilate and fill with blood. *Compare* systole. **2.** The filling of a contractile vacuole of a protozoan.

diatomaceous earth. Earth composed principally of the siliceous skeletons of diatoms, used as an abrasive, for insulation, and for filtration.

diatoms. Golden-brown algae of the division Chrysophyta, class Bacillariophyceae. They possess a siliceous cell wall consisting of two valves which fit together, one overlapping the other.

dibranchiate. A cephalopod of the subclass Coleoidea (formerly Dibranchia) with a single pair of gills, as the squid and octopus.

dicentric. Possessing two centromeres, with reference to a chromatid or chromosome.

dicarpellary. Composed of two carpels.

dichasium. A cyme with two lateral branches.

dichlamydeous. With both a calyx and corolla.

dichogamous. Not maturing at the same time, with reference to stamens and pistil.

dichotomous. Branching repeatedly in pairs; forking.

dichotomous key. In taxonomy, a key in which each division is divided into two subdivisions.

dichromatic. Having two color phases independent of age or sex, applied to certain birds and insects.

diclinous. Having separate staminate and pistillate flowers borne either on the same plant or on separate plants.

dicoelous. Having two cavities.

dicot. A dicotyledon.

dicotyledon. Any member of the Dicotyledoneae, a taxon (class or subclass) of the flowering plants characterized by two cotyledons in the embryo, leaves with netted venation, flower parts in groups of four or five, vascular tissue in the form of a cylinder, a cambium present. Comprises over 150,000 species including the most advanced and highly developed plants, as the bean, pea, oak, maple.

dictyopteran. An insect of the order Dictyoptera which includes the cockroaches and mantids.

dictyosome. The Golgi complex in invertebrates.

dictyostele. A type of siphonostele in which the vascular bundles are separate, as in *Polypodium*.

dicyclic. 1. Arranged in two whorls. **2.** BIENNIAL. **3.** Having two sexual cycles a year.

didactyl. Having two digits.

didelphic. Having two uteri, as marsupials.

didymous. In pairs, twofold, bilobed.

didynamous. Having four stamens occurring in pairs of unequal length, as in the flowers of certain Labiatae.

diencephalon. The portion of the vertebrate brain between the cerebral hemispheres and the midbrain. It includes the epithalamus (pineal body), thalamus, hypothalamus, and neural lobe of the hypophysis.

diestrus. Sexual inactivity between two estrous periods.

differentiation. The process by which unspecialized structures become modified and specialized for the performance of specific functions.

diffuse. 1. Spread out; not localized. **2.** To spread about or be disseminated.

diffusion. 1. The process by which substances in a solution tend to become uniformly distributed. **2.** The movement of atoms, ions, or molecules from a region of higher concentration to one of lower concentration.

digenetic. 1. Pertaining to or exhibiting alternation of generations. **2.** Pertaining to flukes of the subclass Digenea in which one or more asexual generations are interposed between sexual generations. *Compare* monogenetic.

digestion. The process by which large, complex food molecules are broken down into simpler molecules capable of being absorbed and assimilated by an organism.

digestive system. The alimentary canal and its accessory organs (salivary glands, liver, pancreas).

digit. 1. A finger or toe. **2.** In ticks, an article.

digitate. 1. Having or resembling fingers or digits. **2.** *bot.* Having parts resembling the fingers of a hand, as a palmately compound leaf.

145

digitiform. Shaped like a finger.

digitigrade. A type of foot posture in which the animal walks upon the ventral surfaces of the digits with posterior portion of the foot raised, as a dog or cat. *Compare* plantigrade, unguligrade.

dihybrid. One of the offspring of parents differing in two unit characters.

dik-dik. A small African antelope of the genera *Madoqua* and *Rhynchotragus*.

dilatation. The state of being expanded or dilated.

dilate. To enlarge, expand, or widen.

dilation. DILATATION.

dilute. To diminish in strength or flavor; to weaken or decrease the concentration.

dimastigate. Having two flagella.

dimerous. 1. *bot.* Having parts in twos, with reference to flowers. 2. *zool.* In two parts, said of the tarsi of certain insects.

dimidiate. 1. Reduced to one half. 2. Condition in which half of an organ is markedly reduced or wanting.

dimorphic. Of or pertaining to dimorphism.

dimorphism. 1. *zool.* The existence of two forms among individuals of a species or colony. 2. *bot.* The occurrence of two types of leaves, flowers, or other structures on a single plant or on different plants of the same species.

dingo. *Canis dingo*, the wild dog of Australia, presumably introduced by man.

dinoflagellate. A protozoan of the order Dinoflagellata possessing two flagella, one longitudinal and one transverse, as *Noctiluca*, a luminescent, marine form.

dinosaur. A fossil reptile of the subclass Archosauria which comprised a dominant group during the Mesozoic era. Some attained tremendous size, as *Brontosaurus*, the largest terrestrial animal known, reaching a length of 90 feet.

dioecious. 1. *zool.* Unisexual; having separate sexes. 2. *bot.* **a.** With sex organs on separate gametophytes. **b.** With staminate and pistillate flowers borne on separate plants.

diphotic. With two surfaces receiving equal illumination.

diphtheria. An acute, infectious disease of the respiratory passageways caused by a bacterium (*Corynebacterium diphtheriae*).

diphycercal. Designating a type of caudal fin in which the vertebral column extends to the tip of the tail dividing it into dorsal and ventral portions which are approximately equal, as in dipnoan fishes.

diphygenic. Producing embryos of two types.

dipleurula. The hypothetical, bilaterally symmetrical, ciliated, ancestral larva of echinoderms or an existing larva resembling this form.

diploblastic. Having two germ layers, ectoderm and endoderm.

diplohaplontic. Having a life cycle in which a haploid generation alternates with a diploid generation, sporogenic meiosis occurring, as in bryophytes.

diplospondyly. The condition in which a vertebra possesses two centra, as in *Amia*. *Compare* monospondyly.

diploid. Having double the haploid number of chromosomes or double the number of chromosomes present in gametes.

diplontic. Designating a type of life cycle in which adults are diploid and gametes are haploid.

diplosegment. A double segment resulting from the fusion of two segments.

diplosome. The two centrioles of a cell center or centrosome.

dipnoan. Of or pertaining to the Dipnoi, the lung fishes.

dipper. Common name applied to a number of water birds, as the grebe, ruddy duck, water ouzel.

dipteran. Any insect of the order Diptera which includes the true flies, gnats, midges, and mosquitoes, characterized by a single pair of wings, the second pair represented by a pair of halteres.

dipterous. 1. *zool.* Of or pertaining to the order Diptera. 2. *bot.* Possessing two wings or winglike processes.

disaccharide. A double sugar, one of a group of carbohydrates which upon hydrolysis yields two monosaccharide molecules, as lactose, maltose, and sucrose. Basic formula is $C_{12}H_{22}O_{11}$.

disc or disk. *bot.* A flat, circular or oblong structure, as **a.** the central portion of the head in various composites; **b.** a flat, elevated receptacle about the base of a pistil; **c.** the flattened tip of a tendril.

disc flower. One of the tubular flowers comprising the central portion of the head in composites. *Compare* ray flower.

disciform. Circular in outline and shallow, like a disc.

147

discoblastula. A disc-shaped blastula; a germinal disc or blastoderm.

discoid. 1. Shaped like a disc. 2. Having only disc flowers, with reference to the head of composites.

discontinuous. Not continuous; having gaps or interruptions.

discontinuous distribution. The occurrence of related organisms in widely separated localities, as tapirs in Malaya and Central America.

discrete. Separate, distinct.

disease. 1. A sickness, illness, malady, or ailment. 2. A condition of the body characterized by a group of symptoms peculiar to it which differentiates it from other bodily states, either normal or pathological. 3. A disordered state of metabolism or bodily functioning. 4. *bot.* A condition in which a plant or a part of it is affected continuously by some factor which interferes with normal growth or development or the normal functioning of its organs or tissues.

disinfectant. An agent which has marked antimicrobial activity, killing all or nearly all pathogenic organisms but whose use is largely restricted to inanimate matter because of toxic properties; a bactericide or germicide.

disintegration. Breaking up or falling apart.

disjunction. 1. The act of disjoining or separating. 2. *genetics.* The separation of homologous chromosomes during meiosis.

disk. A disc; a flat platelike structure.

dissect. To divide into separate parts, especially to cut into and expose and separate the various parts of a plant or animal for study purposes.

dissected. Deeply cut into or divided into many lobes or divisions, as a *dissected* leaf.

disseminate. To spread about widely; to disperse.

disseminule. 1. A part of a plant which gives rise to a new plant. 2. Any special structure that aids in the dissemination of species, as a thick-walled spore or cyst.

dissepiment. A septum or partition, as **a.** that separating the segments of an annelid; **b.** that separating the locules of the ovary of a flower.

dissilient. Bursting open suddenly.

dissimilation. CATABOLISM.

148

dissociation. 1. The act of disunion or separation, as the breakdown of a substance into simpler substances or of a molecule into its ions. **2.** *bacteriol.* A form of bacterial variation in which a change in the form of a colony occurs usually associated with a change in virulence or other physiological properties.

distal. Remote; away from point of attachment or origin.

distant. Not closely aggregated; remote; not approximate.

distichous. In two rows or ranks.

distinct. Separate; apart; clearly evident.

distome. A fluke with two suckers.

distribution. The natural range or area inhabited by a species or a group of organisms.

distromatic. Two cells in thickness, with reference to a thallus.

ditrematous. Having two openings.

diuresis. Increased production and excretion of urine.

diurnal. 1. Pertaining to the day; occurring daily. **2.** *zool.* Active during the day, as certain birds. **3.** *bot.* Opening during the day and closing at night, as certain flowers. *Compare* nocturnal, crepuscular.

divaricate. Diverging widely; forking.

divaricator. A muscle which opens the beak of an avicularium or the valves of a brachiopod.

divergence. Moving or spreading apart from a central point.

diverticulum. A blind pouch or sac arising from a tubular structure or cavity.

divided. 1. Separated into parts. **2.** *bot.* Separated into distinct parts by cuts or incisions extending to or nearly to the base or midrib, said of leaves.

division. 1. The act or process of dividing or separating into parts. **2.** *taxonomy.* **a.** In plants, a major taxon made up of classes, comparable to a phylum in animals. **b.** In animals, a neutral term used to designate a subdivision of a taxonomic unit or a group of related units.

division of labor. Specialization of function resulting from differentiation during development.

dizygotic. Originating from two zygotes, as fraternal twins.

DNA. Deoxyribonucleic acid, *q.v.*

dobsonfly. A large insect (*Corydalis cornutus*) of the order Megaloptera. Larvae are called hellgramites.

dodder. The love vine or strangleweed (*Cuscuta*), a leafless annual herb which lives as a parasite on various shrubs and herbs.

dodecamerous. In twelve parts.

dodo. *Raphus cucullatus*, a large, heavy, flightless bird of the family Raphidae, formerly inhabiting Mauritius and neighboring islands of the Indian Ocean. It became extinct about 1681.

doe. The female of the deer, hare, and other mammals whose males are referred to as bucks.

dog. A domesticated carnivorous mammal (*Canis familiaris*) of the family Canidae.

dogfish. **1.** One of several species of small sharks of the genera *Squalus* and *Mustelus*. **2.** The freshwater bowfin (*Amia calva*).

dogwood. A tree or shrub of the family Cornaceae, especially *Cornus florida*, the flowering dogwood or cornel.

dolabriform. Hatchet-shaped.

dolioform. Barrel-shaped.

doliolaria. The barrel-shaped, ciliated larva of crinoids and holothuroids.

dolphin. **1.** One of several species of beaked cetaceans of the family Delphinidae, as the common dolphin (*Delphinus delphis*). **2.** A slender marine fish (*Coryphaena hippurus*) of the family Coryphaenidae.

dominant. **1.** Prevailing, governing, controlling. **2.** *genetics.* Designating a character or its gene which, in a heterozygous condition manifests itself to the exclusion of its allelomorph, the recessive character. **3.** *ecol.* Designating an organism or a group of organisms which, by their size and numbers or both, determine the character of a community.

donor. **1.** One who gives or donates. **2.** *physiol.* The individual which is the source of blood for transfusion, organs or tissues for grafting, etc. **3.** A substance which is capable of transferring hydrogen atoms to another substance. *See* acceptor.

doodlebug. The larva of the ant lion.

DOPA. Dihydroxyphenylalanine, an intermediate product in the metabolism of tyrosine. It is the precursor of melanin.

dormancy. A dormant or inactive state. *See* estivation, desiccation, encyst, inanition, hibernate, torpor.

dormant. **1.** *zool.* Quiescent, not active; in a state of suspended animation. **2.** *bot.* In a resting or nonvegetative state, as in buds, seeds, spores.

dormouse. An Old World rodent (*Muscardinus avellanarius*) of the family Muscardinidae, resembling small squirrels. Other genera are *Glis, Graphiurus*, and *Typhlomys*.

dorsal. 1. *anat.* **a.** Pertaining to the back (dorsum) or upper surface of the body or one of its parts. **b.** In man, pertaining to or towards the back or posterior portion of the body. **2.** *bot.* **a.** Pertaining to the upper surface of a plant having dorsiventral surfaces, as a liverwort. **b.** Pertaining to or designating the surface which is away from the axis, as the outer or under surface of a leaf: ABAXIAL.

dorsifixed. Attached by the back or dorsal surface.

dorsiventral. Having distinct dorsal and ventral surfaces, as the thallus of a liverwort, or distinct upper and lower surfaces, as a leaf.

dorsolateral. Pertaining to the back and side.

dorsum. 1. The back or upper surface of an animal. **2.** The upper surface of a part, as of the foot or tongue.

dotterel. An Old World plover (*Eudromias morinellus*) or one of several related species.

double. Twofold; twice as many or as large.

double fertilization. *See* triple fusion.

dove. A pigeon or any of the numerous birds of the family Columbidae, especially the eastern mourning dove (*Zenaidura macroura*).

dovekie. A small seabird (*Plautus alle*) of the North Atlantic coast. It belongs to the family Alcidae.

dowitcher. One of several snipelike shorebirds of the family Scolopacidae, especially *Limnodromus griseus*, the eastern dowitcher.

down. 1. *zool.* **a.** The covering of soft plumage possessed by young birds, especially precocial birds. **b.** In adult birds, soft down feathers which underlie the contour feathers. **c.** Fine soft hair on the skin. **2.** *bot.* A soft, fine pubescence, as on certain leaves and fruits: PAPPUS.

down feather. A plumule; a small, soft feather lacking a vane but with a short, vestigial rachis or quill from which numerous barbs bearing barbules arise.

downy mildew. One of a number of phycomycete fungi which form a downy growth on the surface of their host or substrate. They are the causative agents of a number of plant diseases, as downy mildew of grapes, caused by *Plasmopara viticola*.

dragonfly. A stout-bodied, carnivorous insect of the order Odonata, suborder Anisoptera. Commonly called devil's darning needle, snake doctor, or snake feeder.

drake. An adult male duck. *See* cob, gander.

drepaniform. Sickle-shaped.

drill. **1.** A small, carnivorous, marine gastropod which feeds on other mollusks by drilling a hole in their shell, especially *Urosalpinx cinerea*, the oyster drill. **2.** An African baboon (*Mandrillus leucophaeus*).

drillworm. The larva of the spotted cucumber beetle (*Diabrotica undecimpuntata*); the corn rootworm.

dromedary. The Arabian or one-humped camel (*Camelus dromedarius*) of the family Camelidae.

drone. A male honeybee.

drosophila. A pomace or fruit fly of the genus *Drosophila*, especially *D. melanogaster*, widely used in genetic research.

drum. **1.** The tympanic or middle ear cavity; the eardrum. **2.** A drumfish.

drumfish. A fish of the family Sciaenidae capable of producing grunts or booming sounds, as *Pogonias chromis*, the common drumfish; a grunt.

drumming. The sound produced by the movement of the wings of the males of ruffed grouse.

drumstick. A minute mass of chromatin attached by a fine strand to a lobe of the nucleus of a neutrophil leukocyte which is present only in the female.

drupaceous. Producing drupes; consisting of drupes.

drupe. A stone fruit, as that of the plum, peach, or cherry, consisting of the outermost skin or exocarp, a fleshy mesocarp, and an inner hard and stony endocarp which encloses the seed.

drupelet. A small drupe, as that in an aggregate fruit as the blackberry.

duck. One of a number of species of waterfowl of the order Anseriformes, family Anatidae, characterized by short neck and legs, webbed feet, bill with serrated edges, and sexes differing in plumage. Types include diving ducks (redhead, canvasback, scaup), surface-feeding ducks or dabblers (mallard, pintail, teal), and fish-eating ducks (mergansers).

duckbill. 1. The duckbilled platypus (*Ornithorhynchus anatinus*), a primitive egg-laying mammal (monotreme) of Australia and Tasmania; duckmole or water mole. **2.** A bipedal, beaked dinosaur of the order Ornithischia.

duckweed. Any of a number of small, flowering plants of the family Lemnaceae, especially those of the genus *Lemna*. They are the smallest angiosperms.

duct. 1. *anat.* **a.** A small vessel which conveys a fluid, especially one from a gland. **b.** One of the main lymph vessels, as the thoracic duct. **2.** *bot.* A vessel for conducting specialized substances as latex or various resins.

ductless gland. *See* endocrine gland.

ductule. A small duct.

ductus deferens. In a male, the main duct which transports spermatozoa: VAS DEFERENS. In man, it extends from the epididymis to the ejaculatory duct.

dugong. A large, herbivorous mammal (*Dugong dugon*) found in waters off Southern Asia. It belongs to the order Sirenia.

duiker. One of a number of small, South African antelopes of the genera *Sylvicapra, Cephalophus,* and *Guevei*.

dulse. Red algae of the genera *Dilsea* and *Rhodymenia* used as food.

dumetose. Bushy.

dumose. Consisting principally of bushes; shrublike.

dung. Animal excrement.

dung beetle. *See* scarab.

dunlin. The red-backed sandpiper (*Erolia alpina*).

duodenum. In most vertebrates, the first portion of the small intestine.

duplicate genes. The existence of two genes each capable of producing the same phenotypic effect.

duplication. 1. A doubling of parts. **2.** *genetics.* A type of chromosomal aberration in which a chromosome bears two groups of identical genes.

dura mater. A tough, fibrous membrane forming the outermost of the three meninges enveloping the brain and spinal cord.

duramen. The hard central wood of a tree stem; the heartwood.

durango. A poisonous scorpion (*Centruroides*) of the southwest United States.

Dutch elm disease. A disease of American elm trees caused by a fungus (*Ceratostomella ulmi*), transmitted by elm bark beetles of the genera *Scolytus* and *Hylurgopinus*.

dwarf. An abnormally small individual.

dyad. A pair or a couple, especially one of the two pairs of homologous chromatids present in a tetrad in meiosis.

dye. A stain or coloring substance. *See* stain.

dyscrasia. An abnormal state or condition of the body.

dysentery. A disease of the intestine, especially the colon, characterized by inflammation, griping pains, diarrhea, mucus and blood in stools, and sometimes systemic symptoms. Common types are amebic, balantidial, and bacillary.

dysfunction. Any impairment in normal functioning.

dysgenic. Detrimental to the race from the standpoint of hereditary makeup or constitution.

dysphotic. Dim; with little light.

dyspnea. Difficult or labored breathing.

dystrophic lake. One of low productivity in which the content of organic matter is high and the amount of dissolved oxygen low.

dystrophy. Impaired or defective nutrition; defective or abnormal development; degeneration or atrophy.

E

eagle. A large bird of prey of the family Falconidae, as the golden eagle (*Aquila chrysaetos*) and bald eagle (*Haliaeetus leucocephalus*).

ear. 1. *anat.* The organ of hearing and equilibrium; the acoustic or auditory organ. In mammals, it consists of three divisions, the external, middle, and internal ear, the latter containing the maculae and cristae, receptors for equilibrium and position, and the organ of Corti, the receptor for hearing. **2.** *zool.* The auditory organ (tympanal organ or eardrum) of various insects. **3.** *bot.* A fruiting structure consisting of a cluster of pistillate flowers borne on an elongated peduncle or cob, as in maize.

ear bones. The auditory or ear ossicles, the malleus, incus, and stapes (hammer, anvil, stirrup) which transmit vibrations from the tympanic membrane to the internal ear. *See* columella, 2a.

eardrum. 1. The tympanum or tympanic cavity; the middle ear. **2.** The tympanic membrane, especially in amphibians and reptiles. **3.** The tympanum in insects.

ear dust. Otoconia or otoliths.

eared. Having external ears or earlike processes.

earlet. *See* tragus, 2.

ear ossicle. One of the ear bones.

ear pit. In reptiles, a shallow pit leading to the tympanic membrane.

ear shell. An abalone.

ear stone. A large otolith present in the ear of certain fishes.

earthworm. A soil-inhabiting oligochaete, especially one of the genus *Lumbricus;* a night crawler.

earwax. CERUMEN.

earwig. An insect of the order Dermaptera.

ebony. A hard durable wood obtained from trees of the genus *Diospryos* grown principally in Asia and Africa.

ebracteate. Lacking bracts.

ecblastesis. The development of buds within a flower.

eccrine gland. A merocrine sweat gland.

ecdysis. Molting; the shedding of the outer exoskeleton as occurs in various arthropods; exuviation.

ecdysone. The molting hormone (MH) secreted by the prothoracic gland of insects.

ecesis. Successful establishment in a new habitat; oikesis.

echidna. The spiny anteater, an insectivorous monotreme.

echinate. Possessing prickles or spines.

echinoderm. Any marine invertebrate of the phylum Echinodermata which includes the sea stars, sea urchins, serpent stars, sea cucumbers, crinoids, and related forms characterized by radial symmetry, an endoskeleton with external spines, and possession of a water-vascular system.

echinoid. An echinoderm of the class Echinoidea which includes the sea urchins, sand dollars, heart and cake urchins.

echiuroid. A burrowing, marine worm of the phylum Echiuroidea, related to the sipunculids and annelids.

echolocation. The process by which an animal as the bat or dolphin orientates itself through the emission of high-frequency sounds.

eclosion. *entomol.* **1.** Hatching or escape of the young from the egg. **2.** Emergence of the imago or adult from the pupa case.

ecology. The branch of biology which deals with the interrelationships between living things and the relationships between organisms and their environment; environmental biology.

economic botany. The study of plants of importance to man, especially plants as sources of food, clothing, drugs, and products used in industry.

economic zoology. The study of animals of importance to man, especially animals as sources of food, clothing, and various commercial products.

ecosystem. An ecological system, a natural unit of living and nonliving components which interact to form a stable system in which a cyclic interchange of materials takes place between living and nonliving units, as in a balanced aquarium or in a large lake or forest. *See* consumer, decomposer, producer.

ecotone. A transition zone, as that between two biomes.

ecotype. In a species having a wide geographical distribution, a subgroup which has developed specific adaptations to local conditions such as temperature, light, humidity.

ectoblast. Term formerly applied to ectoderm.

ectocommensal. An organism which lives as a commensal on the surface of another organism, as certain ciliates on *Hydra*.

ectocrine. An ectohormone.

ectoderm. The outermost of the three primary germ layers. It gives rise to epidermis and its derivatives, all nervous tissue and sense organs, and the linings of the nasal cavity, mouth, and anal canal.

ectohormone. A chemical substance produced by an organism which has an effect on nearby organisms; an ectocrine. *See* pheromone.

ectolecithal. A type of egg in which the yolk lies outside the ovum, as in triclad turbellarians.

ectoparasite. A parasite which lives on the surface of an organism, as a mite, tick, louse, or flea.

ectophyte. An ectoparasitic plant.

ectopic. Occupying an abnormal position.

ectoplasm. The peripheral layer of protoplasm of a cell, usually more rigid and lacking granules. *See* plasmagel.

ectoplast. The plasma membrane of a cell.

ectoproct. Any invertebrate belonging to the Ectoprocta, a phylum of sessile, bilaterally symmetrical, lophophorate metazoans which form arborescent and encrusting colonies. Formerly comprised a class of the phylum Bryozoa. Examples are *Plumatella, Bugula*.

ectosome. The outer layer of a sponge enclosing the inner choanosome.

ectotrophic. Securing nourishment from without.

ectozoan. An ectoparasitic animal.

edaphic. Pertaining to or influenced by soil conditions.

edaphon. The fauna and flora of the soil.

edeagus. The male copulatory organ of an insect: AEDEAGUS.

edema. Excessive accumulation of fluid in tissue spaces.

edentate. 1. Any member of the Edentata, an order of placental mammals which includes the sloths, armadillos, and anteaters. 2. Lacking teeth.

eel. 1. An elongated, snakelike fish of the order Apodes (Anguilliformes), as *Anguilla rostrata* of eastern North America. Adults inhabit freshwater streams but spawn in the Atlantic. Larvae are called leptocephali. 2. Any of a number of animals shaped like an eel, as the electric eel, lamprey eel, moray eel, vinegar eel.

eelpout. A marine European fish of the family Zoarcidae, as *Zoarces viviparus*.

eelworm. A small nematode worm, as **a.** the vinegar eel (*Turbatrix* (*Anguillula*) *aceti*); **b.** the wheat eelworm (*Anguina tritici*).

effector. 1. a. A specialized motor nerve ending in a muscle; a motor end plate; **b.** a secretory nerve ending in a gland. **2.** An effector organ.

effector organ. A structure which, when stimulated, responds, as a muscle or gland.

efferent. Conducting or conveying away from. *Compare* afferent.

effloresce. 1. To blossom or bloom. **2.** To become covered with a powdery crust.

efflorescence. The period of flowering: ANTHESIS.

effuse. 1. *bot.* Diffuse, expanded, spread out. **2.** *zool.* With edges of valves separated by a gap.

eft. A newt, as the American newt (*Diemictylus viridescens*),· especially its terrestrial stage in development.

egest. To discharge or cast from the body. *Compare* ingest.

egesta. Matter discharged from the body: EXCREMENT.

egg. 1. *zool.* **a.** An ovum; a female gamete or germ cell. **b.** A reproductive body produced by females of various animals (flat and round worms, reptiles, birds) from which, after a period of development or incubation, the young emerge or hatch. **2.** *bot.* A female gamete, variously called oosphere, ovum, macrogamete, megagamete.

egg apparatus. *bot.* The egg cell and two synergid cells located in the micropylar end of the embryo sac (female gametophyte) of angiosperms.

egg burster. A sharp process on the dorsal surface of the head of a developing insect which aids in rupture of the eggshell.

egg case. An ootheca, a protective covering or capsule enclosing one or more eggs, as in insects (cockroach), some mollusks (cephalopods), and vertebrates (sharks, rays).

egg membrane. A membrane surrounding an egg or ovum as the cell or vitelline membrane, the zona pellucida, or the shell membrane.

eggplant. A large, branching herb (*Solanum melongena*) cultivated for its fruit, a large, ovoid white or purple berry about the size of a coconut.

egg pod. A structure formed by several eggs glued together, as that formed by a grasshopper.

egg tooth. A calcareous structure on the tip of the upper mandible of a developing bird or reptile which serves to cut the egg membrane and shell, thus aiding in hatching.

egret. A white heron (*Egretta thula*), the snowy egret, and *Casmerodius alba*, the American egret of the family Ardeidae, order Ciconiiformes. They produce long white nuptial plumes or aigrettes during the breeding season.

eider. One of two species of diving ducks, the common eider (*Somateria mollissima*) and the king eider (*S. spectabilis*).

ejaculate. To discharge forcibly; to eject seminal fluid.

elaioplast. A plastid in plant cells which stores oil.

eland. One of two species of large African antelopes of the genus *Taurotragus*.

elasmobranch. Of or pertaining to fishes of the class Chondrichthyes, or any member of this class which includes the sharks, skates, and rays, characterized by a cartilaginous skeleton.

elastic. Capable of resuming its original length, size, or shape after being stretched or deformed.

elater. 1. *bot.* A hygroscopic structure present in the sporangia of liverworts or attached to the spores of horsetails which changes shape in response to changes in humidity. 2. *zool.* A click beetle of the family Elateridae.

electric catfish. *Malapterurus*, a large catfish of the river Nile.

electric eel. A large cyprinoid fish (*Electrophorus electricus*) of the rivers of South America, capable of producing a powerful electric shock from its electric organ.

electric organ. A structure present in various vertebrates capable of producing electric shocks of considerable intensity. It is composed of muscle tissue in the form of plates or discs (electroplaxes).

electric ray. An elasmobranch fish (*Torpedo*) which possesses large electric organs located dorsally on each side of the head.

electrolysis. Decomposition of a substance by passing an electric current through it. Positively charged ions (cations) migrate to the negative electrode, negatively charged ions (anions) to the positive electrode.

electrolyte. A substance, as an acid, base, or salt, whose molecules, when in solution, separate into ions. Such a solution is capable of conducting an electric current.

159

electron. A fundamental particle of matter, the smallest known particle bearing a negative electric charge. Each atom consists of a nucleus and its encircling electrons.

electron microscope. A microscope that utilizes electrons instead of light rays in the magnification of objects. Electrons emitted from a cathode pass through a magnetic field, then through the object, and through another magnetic field. An image is formed which can be observed on a fluorescent screen or photographed. Magnifications of 300,000 and higher are possible.

electrophoresis. Cataphoresis; the migration of charged colloidal particles through their dispersion medium, a method used in the analysis of proteins.

electroplaxes. Electric plates or discs arranged in piles in the electric organ of fishes.

element. 1. The ultimate unit of which anything is composed. 2. *chem.* One of some 100 elementary units of which all matter is composed; a substance which cannot be broken down into simpler substances by chemical means. 3. *bot. See* vessel member.

elementary bodies. Minute bodies within inclusion bodies present in cells infected with certain viruses.

elephant. A large, five-toed mammal of the order Proboscidea characterized by possession of an elongated, muscular proboscis or trunk. Common species are the Indian elephant (*Elephas maximus*) and African elephant (*Loxodonta africana*).

elephantiasis. Pathological enlargement of parts of the body resulting from blockage of lymph vessels, as that caused by a nematode worm (*Wuchereria bancrofti*). Also called filariasis.

elk. 1. *Cervus canadensis*, the American elk or wapiti, a member of the deer family of large size and possessing massive antlers shed annually. 2. The European elk (*Alces alces*), the largest living deer.

elliptic. Shaped like an ellipse, with an oval body rounded at both ends.

elm. One of several species of lumber and shade trees of the genus *Ulmus*, especially the American elm (*U. americanus*).

elver. A young eel, especially one migrating from the ocean upstream.

elytron pl. **elytra. 1.** The thickened, hardened forewing of a beetle. **2.** A platelike respiratory structure on the dorsal surface of a scaleworm.

emarginate. 1. Having a notched border or margin. **2.** Bearing a notch at its tip, said of leaves.

emasculate. 1. *zool.* To castrate. **2.** *bot.* To remove anthers from a bud or flower.

embiid. An insect of the order Embioptera which includes the web spinners.

embolism. The blocking of a blood vessel by an embolus.

embolus. 1. *pathol.* A mass of material, as a thrombus, oil droplet, air bubble, or tissue debris which is transported by the bloodstream to a point where it lodges, forming an obstruction. **2.** *zool.* A penislike projection on the pedipalp of certain male spiders.

emboly. The process of invagination, as the formation of a two-layered gastrula from a blastula.

embracing. Encircling, enclosing, clasping, amplectant.

embryo. 1. *zool.* **a.** The young of an organism in its early stages of development. **b.** The young before hatching from an egg. **c.** In mammals, the stage between blastocyst and fetus, when organs and organ systems are coming into existence. **2.** *bot.* **a.** In vascular plants, the young sporophyte before period of rapid growth. **b.** In seed plants, the rudimentary plant within a seed consisting of plumule, cotyledons, hypocotyl, and radicle.

embryogenesis. The development of an embryo; embryogeny. *Compare* blastogenesis.

embryology. The science which deals with the development of an organism; developmental anatomy.

embryophore. The protective covering enclosing the embryo (onchosphere, coracidium) of certain tapeworms.

embryo sac. The female gametophyte (gynogametophyte) of an angiosperm, usually an eight-celled structure which develops within the ovule of a flower. It consists typically of two synergids and an ovum at the micropylar end, three antipodal cells at the opposite end, and two polar or endosperm nuclei in the center.

embryotroph. The nutrient material which nourishes a mam-

161

malian embryo during early pregnancy before establishment of the placenta; histotroph.

emergence. 1. *entomol.* The escape of an insect from a cell, cocoon, or puparium. **2.** The coming into existence from a simpler group, as the *emergence* of land vertebrates. **3.** *bot.* An outgrowth which contains some epidermal tissue, as the prickle of a rose.

emersed. Rising out of or raised above a surface, as of water.

eminence. A protuberance or projection.

¹emission. The act of emitting or of sending forth; to emanate.

²emission. A discharge, especially a discharge of seminal fluid.

emulsion. A colloidal system consisting of minute droplets of a liquid suspended in another liquid, as oil in water.

enamel. The hard, calcareous substance forming the outer layer of the crown of a tooth.

enation. An outgrowth on the surface of a plant.

encapsulated. Enclosed within a capsule.

encephalon. The brain.

encrusted. Incrusted.

encyst. 1. To form a cyst. **2.** To become enclosed within a cyst.

endarch. Designating the protoxylem or first-formed wood when it is centrally located next to the pith.

endartery. One without branches or anastomoses.

end bulb of Krause. An encapsulated receptor for cold present in the skin.

endemic. 1. Peculiar to a particular region or locality. **2.** Native, in contrast to that which is foreign or introduced: INDIGENOUS.

endergonic. *chem.* Designating a reaction in which energy is absorbed.

endite. 1. The expanded coxa of the pedipalp of a spider. **2.** An appendage on the endopodite of a crustacean appendage.

endobiotic. Living within another organism.

endocardium. The innermost layer of the heart wall.

endocarp. The inner layer of the pericarp.

endochondral. Within cartilage.

endochondral ossification. Bone formation in which a model of hyaline cartilage is formed and then replaced by bone except at joint surfaces as in development of a long bone, as the femur or humerus.

162

endocrine. Secreting internally.

endocrine gland. A ductless gland; any of a number of glands whose secretions called hormones are secreted into the blood or lymph.

endocrine system. Collectively, the endocrine glands, which, in vertebrates, include the pituitary, thyroid, parathyroid, pancreas, adrenal, ovary, and testis. Other structures which secrete hormones are the placenta, intestinal mucosa, and hypothalamu.

endocrinology. The science dealing with endocrine glands and their secretions; the study of hormones.

endocyst. The inner layer of the wall of a cyst.

endoderm. The innermost of the three primary germ layers; entoderm.

endodermis. A layer of cells on the inner surface of the cortex present in most vascular plants.

endoenzyme. An intracellular or endocellular enzyme; one that acts within a cell.

endogamy. 1. Conjugation of gametes with the same genetic composition. **2.** Marriage within the same clan or tribe.

endogenous. Originating within; due to or resulting from internal causes or factors.

endolymph. The fluid filling the spaces of the membranous labyrinth of the ear.

endolymphatic duct. 1. In elasmobranchs, the invagination canal. **2.** A blind duct which grows out of the sacculus terminating in the endolymphatic sac.

endometrium. The mucous membrane which lines the uterus.

endomitosis. Reproduction of nuclear elements which is not followed by chromosome movements and division of the cytoplasm.

endomixis. A process occurring in certain protozoans which involves nuclear disintegration, reorganization, and reconstruction.

endomysium. A delicate network of reticular fibers which envelopes individual muscle fibers within a fasciculus.

endoneurium. A delicate covering of connective tissue which encloses individual nerve fibers.

endoparasite. A parasite which lives within its host.

endophoresis. Phoresis in which one of the two organisms lives within a cavity of the other, as the fierasfier, a fish within the cloaca of a sea cucumber.

endophyte. A plant which grows within the tissues of another plant.

endoplasm. The inner, more granular portion of the protoplasm of a cell.

endoplasmic reticulum. A cell organelle consisting of a complicated network of fine, branching, and anastomosing tubules or spaces (cisternae) or isolated vesicles present in the cytoplasm of most cells. It forms a structural framework for the cell and a circulation pathway between the plasma membrane and nuclear membrane. Its surface may bear ribosomes, the site of protein synthesis.

endopodite. The inner ramus or division of a biramous appendage; endopod.

endoral. Within the mouth or buccal cavity.

end organ. The expanded terminal portion of a nerve fiber (motor or sensory).

endoscopic. With the apical pole of the embryo directed towards the base of the archegonium, as in *Selaginella*.

endoskeleton. An internal skeleton, as the bony skeleton of vertebrates.

endosome. 1. A solid body present in the nucleus of certain protozoans. **2.** CHOANOSOME.

endosperm. Nutritive tissue formed within the ovule or embryo sac of a seed plant, constituting a reserve food in seeds of many plants, especially cereals. It is triploid, arising from the union of a sperm cell with two endosperm (polar) nuclei.

endospore. A spore formed within a parent cell as the thick-walled spore of certain bacteria.

endosteum. A fibrous membrane lining the marrow cavity of compact bone and covering trabeculae of cancellous bone.

endostyle. A ciliated, glandular groove on the inner ventral surface of the pharynx of tunicates, lancelets, and the ammocoetes larva of cyclostomes.

endothecium. 1. A layer of cells underlying the epidermis of an anther. **2.** The inner tissue in a developing moss capsule from which the columella and archesporium develop.

endothelium. A form of simple squamous epithelium which lines the heart, blood and lymph vessels, and comprises the wall of a capillary.

endothermic. Designating a chemical reaction in which heat is absorbed: ENDERGONIC.

endothermous. Warm-blooded: HOMOIOTHERMOUS.

endotoxin. A toxin produced by a bacterial cell which is released only after death and disintegration of the cell.

endotrophic. Nourished from within, said of mycorrhiza which obtain their nourishment from their host.

endozoic. Living within an animal; entozoic.

end plate. *See* motor end plate.

endysis. Development of a new coat of hair or feathers.

energy. The capacity to perform work manifested in various ways, as chemical, mechanical, electrical, thermal, or radiant energy, forms which can under suitable conditions be converted one to another.

ensheathed. Enclosed within a sheath, as certain larval nematodes.

ensiform. Shaped like a sword, with sharp edges and tapering to a point.

enteric. Of or pertaining to the intestine.

enterobiasis. Infestation of the intestine by *Enterobius vermicularis*, a nematode worm commonly called the pinworm or seatworm; oxyuriasis.

enterocoel. The cavity of a mesodermal somite which has arisen as an outpocketing of the archenteron.

enteron. The alimentary canal or digestive tract.

enteronephric. Emptying into the intestine, said of certain types of nephridia.

enteropneustan. An invertebrate of the class Enteropneusta, phylum Hemichordata which includes the acorn worms.

entire. **1.** Complete; whole; without parts or divisions. **2.** With a continuous edge or margin, said of leaves or bacterial colonies.

entoderm. *See* endoderm.

entomology. The study of insects.

entomophagous. Feeding upon insects; insectivorous.

entomophilous. Pollinated by insects.

entomostracan. Any of a large number of small, aquatic crustaceans formerly comprising the Entomostraca, a subclass which included the copepods, ostracods, and others.

entophyte. A plant which lives within another plant; an endophyte.

entoproct. A pseudocoelomate, marine animal of the phylum Entoprocta characterized by possession of a lophophore, as *Urnatella*. Formerly considered as a class of the phylum Bryozoa.

entozoan. An animal which lives as a parasite within another animal.

[1]**enucleate.** Without a nucleus.

[2]**enuleate.** To remove a nucleus.

environment. 1. That which surrounds a cell or organism. **2.** All the factors, forces, or conditions which affect or influence the growth and development or the life of an organism.

enzyme. An organic catalyst; a substance produced by living cells which is capable of catalyzing a specific reaction; a ferment.

Eocene epoch. The second epoch of the Tertiary period characterized by the rise of modern mammals.

eolith. A crude stone instrument of the Dawn Stone Age.

eosin. 1. A reddish dye obtained from fluorescin. **2.** A yellowish-red acid stain widely used in histology and bacteriology.

eosinophil. A type of granular leukocyte with coarse cytoplasmic granules which stain readily with eosin.

eosinophilic. Having an affinity for or staining readily with eosin.

epaxial. Above or dorsal to an axis.

ependyma. A layer of neuroglia cells which lines the ventricles of the brain and central canal of the spinal cord.

ephedrine. A white crystalline alkaloid whose effects are similar to epinephrine; obtained from *Ephedra equisetina* or produced synthetically.

ephemeral. 1. Existing for a short time; short-lived. **2.** Lasting only for a day, said of flowers.

ephemerid. An insect of the order Ephemeroptera (Ephemerida) which includes the mayflies.

ephippium. In cladocerans, a modified portion of the carapace which forms a protective capsule for fertilized eggs.

ephyra pl. **ephyrula, ephyrae.** The minute, free-swimming medusa of a scyphozoan jellyfish formed by transverse fission (strobilization) of the scyphistoma.

epibenthos. The organisms of the sea bottom between low water mark and 100 fathoms.

epibiotic. 1. Surviving, with reference to endemic species which are relics or survivors of a preceding age. **2.** Living on the surface of another organism.

epiblast. 1. *bot.* A small appendage along the plumule in the embryo of grasses. **2.** *zool.* The surface layer of a blastodisc; the ectoderm.

epiboly. Expansion and growth of one part over another, as occurs in the development of certain gastrulae.

epibranchial. Lying above the gills.

epicalyx. A whorl of bracts resembling a calyx.

epicanthus. A fold above the inner canthus of the eye, characteristic of mongolians.

epicardium. 1. In vertebrates, the visceral pericardium forming the outer layer of the wall of the heart. **2.** In tunicates, a tubular structure surrounding the stomach and intestine.

epicone. In a dinoflagellate, the portion of the body anterior to the annulus.

epicotyl. The portion of a plant embryo or seedling which lies above the cotyledons. It gives rise to the shoot (stem and leaves).

epidemic. Affecting large numbers of a population at the same time, said of certain diseases.

epidermis. 1. *bot.* The outermost layer of cells of the primary plant body; the superficial layer of cells of various parts, as the root, stem, leaf. **2.** *zool.* The outermost layer of cells of various invertebrates. **3.** *anat.* The surface layer of the skin of a vertebrate.

epididymis. In vertebrates, a coiled mass of tubules located between the efferent ductules of the testis and the vas deferens.

epifauna. Benthic animals living on the surface of bottom material.

epigamic. Pertaining to factors other than those of a sexual nature which serve to attract the opposite sex, as colors and songs of birds.

epigeal. Living near the surface of the soil, said of insects.

epigenesis. A theory of development proposed by Wolff in 1759

167

which maintained that an embryo did not exist preformed within the egg but developed out of unformed material. *Compare* preformation.

epigenous. Growing on a surface, as that of a leaf.

epigeous 1. Lying close to or growing on the surface of the ground. **2.** Appearing above the surface of the ground, with reference to the cotyledons of a seedling.

epiglottis. In mammals, a platelike cartilaginous structure at the base of the tongue, the free end of which is directed posteriorly over the opening to the larynx.

epignathous. Condition in which the upper jaw or mandible is longer than the lower jaw.

epigynous. Designating a type of flower in which the ovary is embedded in the receptacle with the other parts (sepals, petals, stamens) attached at or near its tip, as in the evening primrose.

epigynum. A sclerotized plate on the ventral surface of a female spider bordering the genital opening.

epilimnion. In a lake, the upper layer of warm, circulating water.

epimerite. The hooked, anchoring structure on the protomerite of a gregarine.

epimorphic. Designating a mode of development in which larval stages are suppressed and the individual hatches as a postlarva, as in certain crustaceans.

epinasty. Condition in which the upper surface of a structure, as a leaf, grows more rapidly than the lower surface, causing a downward curvature.

epinephrine. A hormone, one of two catecholamines secreted by the adrenal medulla. It raises blood pressure and brings about other changes necessary for the quick mobilization and release of energy as required in emergency situations: ADRENALIN. *See* norepinephrine.

epineurium. A sheath of connective tissue which **a.** in vertebrates, surrounds a nerve; **b.** in arthropods, surrounds a ganglion.

epipelagic. Pertaining to suspended organisms living in water between surface and a depth of 200 meters.

epipetalous. Having stamens attached to or inserted on petals.

epipharynx. A lobelike or pointed structure on the inner surface of the labrum or clypeus of an insect.

epiphragm. A membrane which closes an opening, as **a.** *bot.* the platelike structure which closes the capsule in certain mosses; **b.** *zool.* a calcareous plate which closes the aperture of the shell in certain hibernating gastropods.

epiphyll. A plant which grows on the upper surface of a leaf.

epiphysis. 1. A portion of a bone which develops from a separate center of ossification and later becomes the terminal portion of a long bone. **2.** A movable process on the tibia of the front leg of certain insects. **3.** A barlike structure in Aristotle's lantern. **4.** The pineal body.

epiphyte. A plant which grows upon another plant but is not parasitic upon it, securing its moisture through aerial roots, as an orchid.

epiplasm. The protoplasm remaining in an ascus after the spores have been liberated.

epipleuron. The turned-down, lateral margin of the elytron of a beetle.

epipodite. A flat, leaflike structure attached to the coxa of certain appendages in various crustaceans which functions as a gill separator.

epipodium. A lateral fold or ridge along the edge of the foot in certain gastropods.

episematic. Pertaining to recognition marks or colors.

episepalous. Attached to or adnate to the sepals, with reference to floral parts.

episome. A replicon in a bacterial cell, either attached to the bacterial chromosome or existing in the cytoplasm and replicating independently.

epistasis. 1. Cessation of a flow or discharge. **2.** A scum or film on the surface of urine. **3.** *genetics.* The suppression or masking of the effects of a gene through the action of a nonallelic gene.

epistome. A structure lying anterior to or overhanging the mouth in various invertebrates.

epithalamus. The dorsal portion of the diencephalon comprising the epiphysis (pineal body) and habenular nuclei.

epithecium. A compact membrane formed by the tips of the paraphyses in the apothecia of ascomycetous fungi.

epithelial tissue. A primary tissue consisting of cells arranged in one or a few layers. It is found on surfaces, lines tubes and

169

cavities, and forms the secretory portions and ducts of glands; epithelium.

epitheliomuscular cell. A type of cell present in the epidermis of coelenterates whose basal end bears projections containing contractile fibrils or myonemes.

epitoky. A reproductive phenomenon occurring in certain polychaete worms in which the anterior asexual portion or atoke differs from the posterior sexual portion or epitoke, as in nereids.

epizoon. An animal which lives on the surface of another animal; *Compare* ectozoan.

¹epizootic. Attacking many animals at the same time, said of a disease.

²epizootic. A disease which is prevalent among animals.

epoch. In geologic time, a subdivision of a period.

eponym. A person for whom or from whom the name of an animal or plant or a part of it takes its name, as Malpighi (Malpighian corpuscle).

equatorial plate. The structure formed in the metaphase of mitosis or meiosis consisting of chromosomes arranged in the equatorial plane.

equine. Of, pertaining to, or resembling a horse.

equisetum. *See* horsetail.

equitant. Overlapping, said of leaves whose bases overlap the bases of the leaves above them, as in the iris.

era. A major division of geologic time.

erect. Upright; not reclining or leaning.

erectile. Capable of being erected, as *erectile* tissue of the penis or clitoris.

erection. The act of becoming erect; becoming swollen and turgid, with special reference to the penis or clitoris.

eremad. A desert plant.

eremophilous. Desert-loving.

eremophyte. A desert plant: EREMAD.

ergastoplasm. Chromidial substance in the cytoplasm of a cell; the chromatophil or basophil material, as Nissl bodies.

ergosterol. A sterol present in yeast and other plants which, upon irradiation by ultraviolet rays, is converted into vitamin D_2 or calciferol.

ergot. 1. A fungus disease of grasses and cereal plants, especially rye, caused by species of *Claviceps*, an ascomycete. **2.** The dried sclerotia of *Claviceps purpura* which contains a number of alkaloids, some of medical importance. It induces powerful uterine contractions.

erichthus. The ciliated larva (pseudozoea) of certain stomatopod crustaceans.

ermine. 1. The stoat (*Mustela erminea*), a species of weasels whose fur, in northern latitudes, turns white in winter. **2.** The fur of these animals in winter phase.

erose. Having an irregularly notched margin.

eruciform larva. A caterpillarlike larva with well-defined segmentation, prolegs, and stigmata along sides of body, as in Lepidoptera.

erumpent. Bursting or breaking out suddenly, as a spore mass.

eruption. A sudden emergence or breaking out.

erythroblast. A nucleated cell of the red bone marrow which gives rise to an erythrocyte.

erythrocyte. A red blood cell or corpuscle.

erythrophore. A melanophore with red pigment.

erythropoiesis. The formation of red blood cells.

escape. 1. *bot.* **a.** The breaking away from, as spores or seeds from a capsule. **b.** The ability of a susceptible plant to avoid becoming infected through some condition or factor other than natural, inherent resistance factors. **2.** *physiol.* A release from control, as the *escape* of the heart from vagus control.

escargot. A snail, especially one prepared for eating.

esculent. Edible; eatable.

esophagus. The tube that leads from the buccal cavity or pharynx to the crop or stomach.

espalier. A framework or trellis upon which plants are trained to grow, or the plant that is so trained.

ester. A compound formed from the reaction of an alcohol with an acid.

esterase. An enzyme which catalyzes the hydrolysis of an ester to an alcohol and an acid.

esthetasc. A long, delicate hair or a minute plate which functions as a chemoreceptor in crustaceans.

estipitate. Lacking a stipe.

estipulate. Lacking a stipule.

estival. Of or pertaining to summer.

estivation. 1. *bot.* The arrangement of floral parts within a floral bud; also aestivation. **2.** *zool.* A state of dormancy or inactivity induced by the heat of summer.

estrogen. A substance, natural or synthetic, capable of inducing estrus and the development of female secondary sexual characteristics; a female sex hormone. In man the principal estrogens are estradiol, estrone, and estriol, steroid hormones secreted by the ovarian follicle, corpus luteum, placenta, adrenal cortex, and testis.

estrogenic. Inducing estrus.

estrous cycle. The sequence of events from the beginning of one estrus to the beginning of the next. Stages include estrus, metestrus, diestrus, and proestrus.

estrus. Among mammals other than primates, the period of sexual excitement and receptivity in the adult female during which mating usually occurs; the period of heat or rut; oestrus.

estuary. The lower end of a river where fresh water meets and mixes with sea water.

ethmoid. A bone, located anteriorly at the base of the cranium which forms the supporting structure for the walls and roof of the nasal cavity. It contains openings for the olfactory nerves.

ethnology. The science which deals with the races of man.

ethology. The study of animal behavior.

etiolation. Blanching; becoming white or pale, as when plants are grown in the absence of sunlight.

etiology. The study of the causes of disease.

eucaryotic. Possessing a true nucleus.

eucephalous. Possessing a distinct head, with reference to insect larvae.

euchromatin. A type of chromatin present in chromosomes which stains densely during metaphase but weakly during interphase. It is active genetically, containing the genes. *Compare* heterochromatin.

eugenics. The science which deals with improvement of the human race through application of the principles of genetics.

euglenoid. Resembling *Euglena*, a flagellated one-celled organism with plant-animal characteristics.

eunuch. A castrated male person.

euphotic zone. The surface layer of a body of water in which light is sufficient for photosynthesis. Average thickness in the sea is 100 fathoms.

euplankton. True plankton minus debris.

eupnea. Normal, quiet breathing.

eurybathic. Having a wide depth range in ocean waters.

eurychoric. Having a wide geographic range.

euryhaline. Capable of withstanding wide variations in osmotic pressure or salinity.

euryhygric. Having a wide range of tolerance with reference to humidity of the atmosphere.

euryphagous. Capable of utilizing or existing on a wide variety of foods.

eurypterid. A scorpionlike, aquatic arachnid of the extinct order Eurypterida, which included the largest known arthropods.

eurythermic. Capable of withstanding wide extremes in temperature.

eurytopic. Tolerating wide extremes in environmental conditions, hence having a wide range of distribution.

eusporangiate. Having a sporangium formed from a group of cells which divide parallel to the surface, the inner cells forming sporogenous tissue, the outer forming the wall, as in most vascular plants.

eustachian tube. The auditory tube.

eusymbiosis. MUTUALISM.

eutely. Having a constant characteristic number of nuclei in various somatic organs as in rotifers.

euthenics. The science which deals with improvement of the human race through improvement of the environment. *Compare* eugenics.

eutherian. Any mammal of the group Eutheria which comprises the placental mammals.

eutrophic lake. One whose waters contain a minimal amount of dissolved oxygen, usually a shallow lake with abundant organic matter.

eutropic. Twining in a clockwise direction.

evagination. An outgrowth or outpouching.

evanescent. Tending to vanish or disappear quickly.

evaporation. Conversion to a gaseous state or vapor.

¹**evergreen.** Remaining green throughout the winter.

²**evergreen.** A tree which retains its leaves through the year, as various conifers.

eversion. The act of turning outward or inside out.

evisceration. 1. Removal of thoracic or abdominal organs. **2.** Protrusion of the viscera through an opening in body wall. **3.** Removal of the contents of an organ, as the eye. **4.** Reflex discharge of internal organs through the anal opening, as occurs in holothuroids.

evocation. The process by which a specific morphogenic effect is brought about by a chemical substance (evocator) transmitted from a region (inductor or organizer) to another part of the embryo: INDUCTION.

evolution. 1. The process by which something complex develops in an orderly fashion from something simpler. **2.** *biol.* The process by which different kinds of organisms have developed from simpler forms as a result of change and adaptation; the development of complex, highly specialized forms from simpler, more generalized types; descent with modification.

evolve. To undergo evolution; to become more specialized.

evulsion. The act of forcibly tearing away or forcing out a part.

ewe. The female of the sheep.

exarate. With legs and wings free of secondary attachment to the body, with reference to insect pupae.

exarch. Designating protoxylem when it is located outside of the older xylem or wood.

exasperate. Having stiff, hard, projecting points, as a leaf.

excentric. Off center; one-sided; eccentric.

excise. To cut out or remove a part.

excitation. Increasing the activity of; stimulation.

exconjugant. Either of a pair of organisms after conjugation.

excrement. Excreted waste material: FECES.

excrescence. An outgrowth of the body. May be normal, as a hair or horn, or abnormal, as a wart.

excrete. To discharge or eliminate; to expel from a cell, tissue, or organ.

¹**excretion.** The process of excreting, especially the elimination or discharge of waste products or substances present in excess from the body.

²**excretion.** That which is excreted.

excretory organ. An organ or structure which functions in excretion, as a contractile vacuole, flame bulb, nephridium, Malpighian tubule, green gland, or kidney.

excurrent. **1.** *zool.* Flowing outward; pertaining to an outflowing current or flow. **2.** *bot.* **a.** Extending beyond the apex or margin of a leaf, with reference to the midrib or lateral veins. **b.** Having a continuous axis or main stem, with reference to branching of stems.

excyst. To break out of a cyst or capsule.

excystation. The escape of an organism from a cyst; excystment.

exergonic. Liberating energy, with reference to chemical reactions.

exfoliation. Shedding in flakes or thin layers, as **a.** *zool.* the shedding of the epidermis of the skin; **b.** *bot.* shedding of the bark of certain trees.

exhale. To breathe out; to emit as a vapor.

exine. The outer wall of a pollen grain.

exobiology. The study of extraterrestrial life.

exocarp. The outer layer of the pericarp; epicarp.

exocrine. Secreting externally.

exocrine gland. A gland which delivers its secretion onto an epithelial surface usually through a duct.

exogamy. **1.** The union of gametes of different genetic constitution: CROSS-FERTILIZATION; CROSS-POLLINATION; CROSSBREED. **2.** Marriage outside the tribe or clan.

exogenous. **1.** Resulting from or due to external causes or factors; originating outside of an organism. **2.** Designating a type of protein metabolism which is dependent upon dietary composition. **3.** *bot.* Growing or developing on the outside, said of stems which add a layer of wood each year.

exophthalmos. Abnormal protrusion of the eyeballs.

exopodite. The outer or lateral branch of a biramous appendage in crustaceans; exopod.

exopterygote. Any insect of the Exopterygota which includes all winged insects with simple metamorphosis, as orthopterans and hemipterans.

exoscopic. With apex of the embryo directed toward the neck of the archegonium, as in bryophytes.

exoskeleton. **1.** A protective, external skeleton as in most

175

arthropods. **2.** The protective covering of certain vertebrates, as the dermal plates of an armadillo or the dermal scales of fishes.

exosmosis. In osmosis, the passage of water outward.

exothermic. Designating chemical reactions in which heat is liberated.

exotic. Of foreign origin; not native.

exotoxin. A toxin produced by a microorganism which diffuses through the cell membrane into the surrounding medium.

expiration. The expulsion of air from the lungs; breathing out; exhalation.

expire. **1.** To breathe out or exhale. **2.** To die.

explanate. Spread out flat.

explant. The tissue removed in explantation.

explantation. The removal of living tissue from an organism and its cultivation in an artificial culture medium: TISSUE CULTURE.

explosive fruit. One which bursts suddenly and violently, scattering seeds over a considerable area, as in the squirting cucumber.

express. To force out by the application of pressure.

expressivity. The extent or degree to which a character or trait manifests itself in individuals possessing its gene.

exsert. To thrust out.

exserted. Protruding; sticking out or projecting beyond the usual enclosing structure.

exsiccate. To deprive of moisture; to dry up.

exstipulate. Lacking stipules.

extant. Now living or existing.

extension. *anat.* Straightening of a limb or a part. In man, the backward or posterior movement of the head, trunk, or limbs.

extensor. A muscle which extends or straightens a part.

external. Outside of, from the outside, pertaining to the exterior; toward the outside or away from the center.

exteroceptor. A receptor which responds to stimuli from the outside, as the eye, ear, olfactory cells, and various cutaneous receptors.

extinct. No longer in existence.

extirpation. **1.** Complete removal of a part. **2.** Eradication of a species from a given area.

extracellular. Outside of a cell or cells; occurring in intercellular spaces or in a cavity surrounded by cells.

extracorporeal. Outside the body.

extraembryonic membranes. The yolk sac, amnion, allantois, and chorion, auxillary structures concerned with protection, nutrition, respiration, and excretion in the embryo. Also called embryonic membranes.

extremity. 1. The terminal portion of a structure. **2.** *anat.* A limb of the body and the bones to which it is attached. The upper *extremity* includes the pectoral girdle and upper limb; lower *extremity* includes pelvic girdle and lower limb.

extrinsic. External to or outside of the body or a structure.

extrorse. Facing outward.

extrude. To expel or force out.

exudate. Exuded material.

exudation. 1. The passage of fluids, such as sweat, sap, or gums through minute openings from an organism; oozing out; bleeding. **2.** The passage of certain constituents of the blood into tissue spaces, as occurs in inflammation.

exude. To discharge slowly through small pores.

exumbrella. The convex aboral surface of a medusa.

exuviae. Cast-off materials from the surface of an organism as shed exoskeleton, skin, scales, hair, etc.

exuviation. The act of shedding a surface covering; ecdysis; molting.

eye. 1. *zool.* **a.** The organ of sight or vision. **b.** An organ which contains photoreceptors. **c.** A structure which resembles an eye in appearance. *See* ocellus, ommatidium. **2.** *bot.* **a.** The center of a flower. **b.** A bud of a tuber, as the *eye* of potato. **c.** The bud of a cutting.

eyeball. The eye proper; the globe of the eye as distinguished from the accessory structures.

eye brush. A group of stiff hairs on the prothoracic legs of the honeybee and other insects used for cleaning the surface of the compound eye.

eyed. 1. Having eyes or eyelike structures. **2.** Resembling an eye in appearance.

eye gnat. A frit fly or eye fly, a small dipterous insect of the families Chloropidae and Oscinidae, of importance in the transmission of eye diseases.

eyespot. 1. A pigmented, light-sensitive structure present in algae and various invertebrates; a stigma. **2.** One of a number of fungus diseases of plants.

eyetooth. A canine tooth of the upper jaw.

eye worm. The loa worm (*Loa loa*), a filarial worm living normally in subcutaneous tissues but often migrating to the eye region. Common in west and central Africa. It is transmitted by the mango fly (*Chrysops*).

F

F₁. The offspring of a cross in which homozygous parents differ in one or more traits; the first filial generation.

face. The anterior part of the head; the region bearing the mouth, nasal openings, and the orbits.

facet. 1. *anat.* A small, smooth flat surface, as that on a bone forming an articulating surface. **2.** *zool.* The surface of an ommatidium.

factor. 1. A circumstance, fact, condition, or influence which tends to produce, alter, or modify a result; a constituent. **2.** *genetics.* A gene or hereditary determiner.

facultative. Not limited to a specific condition; having the ability to live under varying conditions; optional.

fairy ring. A circular figure formed by the fruiting bodies of mushrooms, especially *Marasmius oreades*.

fairy shrimp. One of a number of branchiopod crustaceans occurring in temporary ponds in the early spring, as *Branchinecta*, *Eubranchipus*.

falcate. Curved or shaped like a sickle.

falciform. FALCATE.

falcon. One of a number of birds of prey of the family Falconidae, especially the peregrine falcon or duck hawk (*Falco peregrinus*) used in the sport of hawking or falconry.

Fallopian tube. In mammals, the uterine tube or oviduct.

fallow deer. A European deer (*Dama dama*), a common park deer of Great Britain and western Europe.

family. 1. A group of closely related persons, consisting of parents and their children. 2. A group of persons related by common ancestry, as the Smith *family*. 3. *biol.* A taxonomic group which comprises the principal division of an order and in turn is divided into genera. In zoology, the family name ends in *-idae;* in botany, the family name usually ends in *-aceae.*

famulus. A microsensory seta on the legs of mites.

fan. The tail feathers of a bird; a flabellum.

fang. A long, sharp, pointed tooth or toothlike structure, as **a.** the canine tooth of a carnivore; **b.** the long, hollow, curved tooth of a venomous reptile; **c.** a chelicera of a spider.

fanworm. One of a number of sedentary polychaete worms which possess a crown of pinnate processes (radioles), as *Sabella*, a feather duster.

farsightedness. HYPERMETROPIA.

fascia. A layer of loose connective tissue containing much adipose tissue which lies under the skin (superficial fascia) and continues internally as the deep fascia. It binds organs together and forms investing sheaths for muscles, nerves, and blood vessels.

fasciate. Bound or compressed into a bundle.

fascicle. A group, cluster, or compact bundle, as a *fascicle* of nerve fibers, muscle fibers, leaves, or roots; a fasciculus.

fascicled. Bearing bundles or fascicles; arranged in fascicles.

fascicular. Of or pertaining to fascicles.

fasciculate. FASCICLED.

fasciculus. A fascicle.

fat. 1. A fatty, greasy substance present in adipose tissue and in certain seeds as those of cotton, soy beans, and corn. 2. A lipid, one of a class of substances containing carbon, hydrogen, and oxygen, which is soluble in alcohol or ether but not in water and which, upon saponification or alkaline hydrolysis yields glycerol and soap. True fats are esters of glycerol and fatty acids.

fat body. A structure containing stored fat which is utilized during periods of metamorphosis, hibernation, or dormancy. Present in amphibians, reptiles, and insects.

father. The male parent of a child.

fatigue. 1. Weariness, exhaustion. **2.** Inability or reduced capacity of muscle cells to perform work or nerve cells to respond to stimulation. **3.** The sensation or feeling of weariness or exhaustion.

fat tissue. ADIPOSE TISSUE.

fatty acid. One of a number of organic acids, as palmitic, oleic, and stearic acids, components of neutral fats. Saturated fatty acids have the general formula $C_nH_{2n}O_2$.

fauces. The passageway from mouth or buccal cavity to the pharynx.

fauna. The animal life of a locality or region or that existing during a specific geological period or time.

faveolate. Having cavities or cells like those of a honeycomb; favose: ALVEOLATE.

fawn. A young deer.

feather. One of the specialized epidermal outgrowths which form the surface covering or plumage of birds. Each consists of a central shaft composed of a proximal hollow portion (quill or calamus) and a distal rachis. The latter bears laterally extending barbs which collectively comprise the vane.

feather duster. A fanworm.

feather star. A living crinoid of the genus *Antedon*.

feather tract. *See* pteryla.

fecal. Of or pertaining to feces.

feces. Excrement, especially material discharged from the anus.

fecundation. Fertilization or impregnation.

fecundity. The capacity or power of producing abundant fruit or numerous offspring.

feedback mechanism. A controlling mechanism in a cell or organism by which functional activity is regulated through factors which are returned or fed back to the controlling center with resultant depression or stimulation.

feeler. A tactile organ, as an antenna, tentacle, or vibrissa.

feeling. 1. The sense or sensation of touch. **2.** Bodily consciousness or awareness, as a *feeling* of pain or warmth. **3.** An emotion or emotional state, as a *feeling* of joy.

feline. 1. Of or pertaining to the cat family, Felidae, a group of carnivores which includes the lion, tiger, leopard, puma, jaguar, wildcat, lynx, cheetah, and the domestic cat. Most belong to the genus *Felis*. **2.** Any member of the family Felidae.

felted. Closely matted with intertwining fibers or hairs.

female. 1. *zool.* **a.** Pertaining to or designating the individual which produces ova or which conceives and brings forth young. **b.** Pertaining to or characteristic of a woman. **2.** *bot.* **a.** Pertaining to the plant or structure which produces the ovum or female gamete. **b.** PISTILLATE.

femoral. Of or pertaining to the femur or thigh.

femur. 1. In vertebrates, the thigh bone. **2.** In arthropods, the third segment of the leg located immediately distal to the trochanter.

fen. Marshy, swamp land, especially in Britain.

fenestra. 1. A small opening. **2.** *entomol.* A transparent spot or clear area in the wing of an insect.

fenestrated. Possessing small openings.

feral. Wild, savage, untamed, undomesticated.

fer-de-lance. *Bothrops atrox*, a large, venomous snake of tropical America.

ferment. An enzyme.

fermentation. The decomposition under anaerobic conditions of complex organic substances, especially carbohydrates, through the action of enzymes produced by microorganisms, as yeasts, molds, and bacteria.

fern. A simple, vascular plant of the class Filicineae, division Pteridophyta, with an underground stem or rhizome from which leaves and roots develop.

ferret. 1. A small mammal of the weasel family, considered to be an albino form of the polecat, used in hunting rabbits and rats. Sometimes erroneously given the scientific name *Putorius furo*. **2.** A North American weasel (*Mustela nigripes*), the black-footed ferret.

fertile. 1. Rich, productive, fruitful. **2.** Capable of growing, with reference to seeds or eggs. **3.** *bot.* Capable of producing fruit. **4.** *zool.* Fertilized and capable of developing, said of eggs.

fertilization. 1. Impregnation, fecundation, syngamy. **2.** The union of a male gamete with a female gamete with resultant formation of a zygote. **3.** Enrichment of the soil to increase crop productivity.

fertilization membrane. A membrane at the surface of a fertilized ovum which prevents other sperm from entering.

fertilizin. A substance present on the surface of certain ova which causes sperm to adhere. Separate from the egg, it agglutinates sperm.

fetus. The unborn young of a viviparous mammal especially in later stages of development. In man, the young from the end of the second month to birth.

Feulgen reaction. A test for determining the presence of DNA based on the formation of a purple colored substance following treatment with Schiff's reagent (decolorized fuchsin) after acid hydrolysis.

fiber. 1. *zool.* **a.** A fine, threadlike structure, as a nerve or muscle *fiber.* **b.** A slender filament, as the *fiber* of a spider's web. **2.** *bot.* **a.** An elongated, thick-walled supporting or strengthening cell (sclerenchyma) in various plant structures. **b.** Any of a number of elongated structures obtained from plants and utilized in the manufacture of textiles, netting, cordage, paper, etc., as cotton, flax, or hemp *fibers.*

fibril. 1. A minute, threadlike structure within the cytoplasm of a cell, as a myofibril or neurofibril. **2.** A minute component of a fiber.

fibrillation. A local, quivering movement of muscle fibers.

fibrin. A fibrous protein which develops from fibrinogen in the formation of a blood clot.

fibrinogen. A protein present in blood plasma which, in coagulation, is converted into fibrin.

fibroblast. A cell present in connective tissue which functions in the formation of fibers and the amorphous ground substance.

fibrous. Containing or consisting principally of fibers or fibrous structures.

fibrovascular. *bot.* Composed of conducting cells and their fibrous sheaths.

fibula. The smaller of the two bones of the hind or lower limb extending from knee to ankle. It lies lateral to the tibia.

ficin. A proteolytic enzyme present in fresh fig latex, used as an anthelminthic against nematodes.

fiddler crab. A burrowing crab of the genus *Uca*, with a greatly enlarged chela.

field mouse. *See* vole.

fig. A tree or shrub of the genus *Ficus* or the fruit it produces. Types include common figs, caprifigs, Smyrna figs, and San Pedro figs.

fig insects. Small, hymenopterous insects of the family Agaontidae essential for the pollination and commercial production of the Smyrna fig. The two principal species are *Blastophaga psenes* and *Secundeisenia mexicana*. *See* caprification.

filament. 1. A minute, threadlike structure, process, or growth form. 2. The stalk of an anther.

filaria. A slender, threadlike nematode worm of the superfamily Filaroidea. Most are parasites of vertebrates; larvae or microfilariae are usually transmitted by bloodsucking arthropods. *See* elephantiasis.

file. The notched portion of a stridulating organ in insects.

filial. Of or pertaining to offspring. *See* F_1.

filical. Fernlike.

filiform. Filamentous; threadlike; long and slender.

filipendulous. Hanging by a thread.

filly. A young mare or female horse.

filoplume. A minute, hairlike feather.

filopodium. A filamentous pseudopodium consisting entirely of ectoplasm.

filose. Terminating in a threadlike structure.

filter. *bacteriol.* A special type of filter, as one of diatomaceous earth, porcelain, or asbestos through which bacteria cannot pass.

filterable. Capable of passing through the pores of a bacterial filter, with reference to viruses.

filter feeding. Obtaining food by passing water through a filtering mechanism, as the fine processes on the appendages of certain crustaceans.

filum. A threadlike structure.

fimbria. A fringe or fringelike border.

fimbriate. Having a fringe or fringelike border.

fimbrillate. Bearing a fringe of minute processes.

fin. A thin winglike or paddlelike process as that present on

183

fishes and other aquatic animals which functions in propulsion, guiding, balancing, and braking.

finback. The finner whale (*Balaenoptera physalus*), also called the rorqual or razorback.

finch. Any of a number of songbirds of the family Fringillidae which includes the sparrows, grosbeaks, towhees, buntings, crossbills, goldfinches, and linnets. Most are seedeaters with bills specially adapted for crushing seeds; many are fine singers, as the canary.

finger. A digit of the forelimb other than the thumb.

fingerling. A young fish from time of disappearance of the yolk sac to the end of the first year of growth.

fin rays. Slender supporting structures of dermal origin in the fins of fishes. Types include ceratotrichia, actinotrichia, lepidotrichia, and camptotrichia.

fir. A coniferous tree of the genus *Abies*, especially the balsam fir (*Abies balsamea*). Term is also applied to certain other coniferous trees, as the Douglas fir (*Pseudotsuga taxifolia*).

fire algae. Algae of the phylum Pyrrophyta which includes the cryptomonads and dinoflagellates.

fire ant. *Solenopsis geminata* and *S. saevissima*, the imported fire ant of the southern states, noted for the severity of their stings.

fire blight. A bacterial disease of pears, apples, and quinces caused by *Erwina amylovora*.

firebrat. A bristletail.

firefly. A lightning bug, as *Photinus pyralis*, one of several species of soft-bodied beetles of the family Lampyridae capable of emitting cold light from their abdominal segments. *See* glowworm, luciferin.

fish. An aquatic vertebrate of the superclass Pisces. Living fishes comprise the classes Chondrichthyes (cartilaginous fishes) and Osteichthyes (bony fishes).

fish duck. A merganser.

fisher. A carnivorous, fur-bearing mammal (*Martes pennanti*) of the family Mustelidae. Also called pekan.

fish hawk. An osprey.

fish louse. A parasitic crustacean of the genus *Argulus*, subclass

Branchiura, which lives on the skin and gills of both fresh and salt water fishes.

fish tapeworm. *Dibothriocephalus latus* (*Diphyllobothrium latum*) which lives in fish-eating mammals including man. Intermediate hosts are copepods and fishes. *See* procercoid, plerocercoid.

fissile. Easily split or separated into two parts.

fission. **1.** A splitting, cleavage, or division. **2.** A method of asexual reproduction in which an organism divides into two or more parts. *See* binary fission.

fissure. A deep groove or cleft.

fistula. An abnormal tube or opening forming a communication between **a.** two hollow organs; **b.** a hollow organ or cavity and the outside.

fistular. Cylindrical; elongated and hollow.

fixation. **1.** In microscopy, the process of treating cells and tissues so that their structure is maintained and altered as little as possible by subsequent procedures. **2.** *chem.* Converting from a gaseous to a usable solid form, as nitrogen *fixation.*

fixator. A muscle which stabilizes a part, thus preventing undesired movement.

fixed. **1.** Not free; securely attached or combined with; stationary. **2.** *chem.* Combined in a chemical compound, as *fixed* nitrogen.

flabellate. Shaped like a fan; bearing a fanlike process or projection.

flabellum. A fanlike structure or process.

flaccid. Limp, flabby; lacking rigidity.

¹**flagellate.** Possessing or bearing flagella.

²**flagellate.** **1.** An organism with one or more flagella. **2.** A protozoan of the class Mastigophora (Flagellata).

flagellum. **1.** A slender, whiplike process of a cell, as that possessed by certain bacteria, algae, protozoans, choanocytes, and reproductive cells. It functions as a locomotor organ, for propulsion of fluids, and as a sensory organelle. **2.** A slender process, as **a.** the penis of gastropods; **b.** the terminal portion of the abdomen in various arthropods; **c.** the exopodite of certain crustacean appendages; **d.** the portion of an insect's antenna beyond the second segment.

flame bulb. A ciliated structure consisting of a part of a cell, a flame cell, or a group of cells found at the terminations of excretory tubules in a number of invertebrate groups.

flame cell. A cell containing a tuft of cilia found at the ends of excretory tubules in certain invertebrates, especially flat worms.

flamingo. One of several species of large, gregarious, aquatic birds with long, slender legs, long, curved neck and beak, and brilliantly colored plumage, as *Phoenicopterus ruber*, the American species.

flank. The fleshy region of the side between the ribs and hip.

flatfish. One of several species of fishes of the order Heterosomata which includes the halibut, flounder, and sole. Body is markedly compressed laterally with both eyes on upper side.

flatworm. Any member of the phylum Platyhelminthes; a turbellarian, fluke, or tapeworm.

flavin. One of a group of yellow, water-soluble pigments isolated from various plant and animal sources, an important constituent of certain yellow enzymes, as flavoprotein.

flavoprotein. A yellow enzyme, one of a group of dehydrogenases each consisting of a protein carrier and a prosthetic group, a flavin.

flax. A plant of the genus *Linum*, especially *L. usitatissimum*, cultivated for its fiber used in the manufacture of linen and its seed (flaxseed), the source of linseed oil.

flea. A wingless, bloodsucking insect of the order Siphonaptera, as *Pulex irritans*, the human flea. *See* bubonic plague, typhus.

fledgling. A young bird which has just acquired feathers necessary for flight.

flesh. The soft parts of an animal, especially muscle tissue.

flesh fly. Any of a number of dipterous insects of the family Sarcophagidae which deposit their eggs or larvae in carrion or in wounds or body openings. Important genera are *Sarcophaga* and *Wohlfahrtia*.

fleshy. 1. *bot.* Succulent, pulpy, with reference to fruits; plump, tender, with reference to leaves. 2. *zool.* Of, pertaining to, or of the nature of flesh; plump; fat.

flexion. 1. The bending of a part in which the angle of the joint is reduced. 2. In man, the forward movement of the head, vertebral column, arm, or thigh.

flexor. A muscle which bends or flexes a part.

flexuous. Sinuous, undulating, wavy.

flexure. A bend or fold.

flicker. 1. A wavering, fluttering, or unsteady movement. 2. One of several species of woodpeckers of the genus *Colaptes*, as *C. auratus*, of eastern North America.

flier. A small sunfish (*Centrarchus macropterus*) of southeastern United States.

flight. The act of flying or passing through the air with the aid of wings or winglike structures.

flight feathers. Special contour feathers on the wings and tail of a bird used especially in flying; remiges and rectrices.

flimmer. A mastigoneme.

flipper. A broad, flat limb adapted for swimming, as that of a seal or whale.

flittermouse. An insectivorous bat.

float. A gas-filled structure which supports an organism in water, as the pneumatophore of coelenterates or the air bladder of seaweeds.

floccose. Woolly; bearing tufts of soft, woolly hair: FLOCCULOSE.

flocculation. The coalescence of finely divided or colloidal particles into larger masses which form a precipitate.

flocculose. FLOCCOSE.

flocculus. 1. A small, detached mass of wool or hair. 2. A small body forming part of the cerebellum.

flora. 1. The plant life of an area or locality or of a geological era or period. 2. Plants taken collectively as distinguished from animals. *Compare* fauna.

floral. Of, pertaining to, or consisting of flowers.

floral diagram. A diagram which shows the relative position and number of the floral parts.

floral formula. A simple method of expressing in shorthand the information given in a floral diagram.

floral parts. The sepals, petals, stamens, and pistil(s) of a flower.

floral tube. The fused bases of the sepals, petals, and stamens which surround the ovary in certain flowers.

florescence. The time of flowering: ANTHESIS.

floret. A small flower, especially one of the small flowers which make up the head of a composite flower; the spikelet of a spike.

floricane. A flowering and fruiting stem, as in the bramble, *Rubus*.

floriculture. The growth and cultivation of flowers for ornamental purposes.

florid. Bright red in color; ruddy.

floridian starch. A form of polysaccharide found in red algae, which upon treatment with iodine, turns yellow, then reddish purple.

floriferous. Bearing flowers, especially in abundance.

florigen. A flower-forming substance or hormone produced where flowers are forming which, when transported to other parts of the plant, is capable of inducing flower formation.

flotoblast. A floating statoblast in an ectoproct.

flounder. One of a number of flatfishes of the families Bothidae and Pleuronectidae, as the European flounder, *Platichthys*. *See* plaice.

flower. 1. The bloom or blossom of a plant. 2. A plant which is grown for its blossoms. 3. *bot*. A specialized reproductive structure in higher plants consisting of a stem or branch bearing modified leaves. A typical flower bears the following parts: sepals, petals, stamens, and pistil, all borne on a receptacle.

flower bud. A bud which develops into a flower; an unopened flower. Also called a fruit bud, if the flower develops into a fruit.

flowering. The act or state of producing flowers; blossoming.

fluctuation. A variation.

fluke. 1. A parasitic flatworm of the class Trematoda. 2. A flatfish or flounder. 3. One of the horizontal lobes of the tail of a whale.

fluorescence. The property of absorbing light of a particular wave length and then emitting light of a different color and wavelength.

fluorine. Symbol F. A gaseous element of the halogen group. Atomic weight 19.00. It is a highly reactive, poisonous gas.

fluted. Bearing alternating ridges and grooves.

fly pl. **flies.** 1. A two-winged insect of the order Diptera, especially those of the family Muscidae. 2. Any of a number of insects of various orders, as a mayfly, butterfly, dragonfly, sawfly.

fly agaric. A poisonous mushroom (*Amanita muscaria*).

flycatcher. One of a number of insect-eating passerine birds of the family Tyrannidae.

flying fish. One of several fishes of the family Exocoetidae with greatly enlarged pectoral fins which enable them to glide through the air after leaping from the water, as *Cypselurus cyanopterus*.

flying fox. A fruit-eating bat, especially one of the genera *Pteropus* and *Epomophorus*.

flying lemur. The colugo or cobego, a herbivorous mammal of southeast Asia comprising a single species, *Cynocephalus variegatus* of the order Dermoptera.

flying phalanger. One of several species of small marsupials of the genera *Petaurus*, *Petsuroides*, and *Acrobates* inhabiting Australia and neighboring islands.

flying reptiles. Pterosaurs, fossil reptiles of the order Pterosauria, abundant in the late Mesozoic era.

flying squirrel. *Glaucomys volans*, an arboreal rodent of the family Sciuridae.

flytrap. An insectivorous plant, as Venus's-flytrap.

flyway. A definite migratory route followed by birds in flying to and from their breeding grounds.

[1]foal. A colt or the young of any of the horse family, Equidae.

[2]foal. To give birth to a colt.

foliaceous. **1.** *bot.* Resembling a foliage leaf in shape, form, texture, or color. **2.** *zool.* Leaflike, with reference to crustacean appendages.

foliage. Collectively, the leaves of a plant.

foliar. Of or pertaining to leaves.

foliate. **1.** Furnished with or possessing leaves. **2.** Resembling a leaf or possessing leaflike parts.

foliation. The formation of a leaf or leaves; the state of being in leaf.

foliferous. Bearing leaves or leaflike structures.

foliicolous. Growing upon leaves.

foliolate. Pertaining to or consisting of leaflets.

foliole. A small leaflet or leaflike structure.

foliose. Having many leaves; leafy.

follicle. 1. *anat.* A small, closed, or nearly closed cavity, as that found in the ovary or thyroid gland. **2.** *bot.* A simple, dry, dehiscent fruit consisting of one carpel and splitting along one suture, as that of the milkweed or columbine.

follicle-stimulating hormone. FSH, a gonadotrophic hormone secreted by the anterior lobe of the pituitary which promotes growth and maturation of ovarian follicles. In the male, it stimulates the seminiferous tubules.

fontanel or **fontanelle. 1.** *anat.* An unossified, membranous area in the skull of a fetus or infant. **2.** A depression on the dorsum of the head of a termite.

food. A nutritive substance which serves as a source of energy or is utilized in growth and repair of tissues or in the regulation of bodily processes. The term is generally restricted to energy sources, namely carbohydrates, fats, proteins.

food accessory. A nutrient substance which is utilized by the body but does not provide energy, as water, mineral salts, vitamins.

food chain. A group of organisms involved in the transfer of energy from its primary source, plants, as algae, insects, small fishes, larger fishes, fish-eating birds or mammals.

food web. A complex pattern of several interlocking food chains in a complex community or between several communities.

foot. 1. *anat.* The terminal portion of a limb upon which the body rests when standing. **2.** *zool.* **a.** One of a number of locomotor structures in various invertebrates. **b.** The tail of a rotifer. **3.** *bot.* In liverworts, mosses, club mosses, and ferns, the basal portion of a developing sporophyte by which it is attached to the gametophyte.

foram. A foraminiferan.

foramen pl. **foramina.** An opening or perforation, as that in a bone.

foramen magnum. An opening in the occipital bone through which the spinal cord passes to the cranial cavity.

foramen ovale. An opening **a.** in the interatrial septum in the developing heart of birds and mammals; **b.** in the sphenoid bone for transmission of the mandibular branch of the trigeminal nerve.

190

foraminiferan. Any member of the order Foraminiferida, rhizopod protozoans with a calcareous shell usually perforated by small openings through which reticulopodia extend, as *Globigerina*.

forceps. A pincerlike structure formed of cerci at the posterior end of an earwig.

forcipate. Deeply forked, resembling forceps.

forcipulate. Resembling small pincers.

forearm. The portion of the upper extremity between elbow and wrist.

forebrain. The prosencephalon, the most anterior of the three primary vesicles of the vertebrate brain. It gives rise to the telencephalon and diencephalon.

forefoot. 1. The anterior foot of a quadruped. **2.** In man, the portion of the foot lying distal to the cuboid and cuneiform bones.

foregut. 1. *invert.* The portion of the digestive tract lying anterior to the midgut; the stomodeum. **2.** *embryol.* The anterior portion of the enteron from which the alimentary canal from mouth to midportion of the intestine develops.

forelimb. The anterior limb of a tetrapod.

foreskin. The prepuce.

forest. A large area of land covered by trees and underbrush.

forester. 1. One trained in forestry; one engaged in forest management. **2.** A moth of the family Agaristidae. **3.** The great gray kangaroo.

forestry. The science and art of managing forest resources; the growth and cultivation of trees.

forewing. One of the anterior wings of an insect; the mesothoracic wing.

formalin. An aqueous solution of formaldehyde containing not less than 37% by weight. It is an antiseptic, disinfectant, deodorant, and preservative.

form group. A taxonomic group, as a form family or form genus, so designated to indicate that its members do not necessarily form a natural group, as the Fungi Imperfecti.

formic. Of, pertaining to, or derived from ants.

formicary. An ant nest.

fornicate. 1. Arched. **2.** Bearing arched, scalelike appendages.

191

fossa. A depression; a pit or concavity, especially one on the surface of a bone.

fossil. Any remains, impression, cast, or trace of an animal or plant of a past geological period. The term is generally restricted to parts which have been petrified or converted to stone.

fossiliferous. Containing fossils.

fossorial. Adapted for burrowing or digging.

fosterer. The host in nest parasitism.

fovea. A small pit or depression.

fowl. 1. Any bird, especially a large edible bird. 2. Collective term for birds of one type or an area, as waterfowl, wildfowl. 3. The domestic cock or hen of the species *Gallus domestica*, especially when used for food.

fox. A carnivorous, fur-bearing mammal of the family Canidae noted for its speed, cunning, and resourcefulness, as *Vulpes fulva*, the common red fox and its color mutations, the black and silver foxes.

fragmentation. A method of asexual reproduction in which the organism breaks up into several parts each capable of developing into a new organism, as occurs in certain algae, bacteria, fungi, and liverworts.

free. Distinct, separate; not adnate, adherent, or united to other organs or structures.

free-living. Capable of living independently; not parasitic; not sessile.

freemartin. The sterile, masculinized female member of a pair of unlike twins in cattle.

frenulum. 1. *anat.* A thin, membranous fold of skin or mucous membrane which attaches a structure and checks its movement, as that of the tongue, prepuce, or clitoris; a frenum. 2. *zool.* A structure on the hind wing of a moth or butterfly which locks it to the front wing.

frenum. *See* frenulum.

frigate bird. The man-o'-war bird (*Fregata*), a large, rapacious seabird with extraordinary powers of flight.

fringillid. Of, pertaining to, or belonging to the family Fringillidae, which includes the grosbeaks, finches, sparrows, and buntings.

frog. 1. Any of a number of tailless, leaping amphibians of the subclass Anura (Salientia). True frogs belong to the genus

Rana, family Ranidae. **2.** An elastic, horny pad in the middle of the sole of a horse's foot.

froghopper. *See* spittlebug.

frond. 1. *bot.* **a.** The broad, flattened thallus of certain algae. **b.** A thalluslike stem of duckweeds. **c.** The leaf of a fern. **d.** The fernlike leaf of certain cycads. **2.** *zool.* A hollow, foliose enlargement on the parapet of certain sea anemones.

frondose. Bearing fronds; resembling a frond in structure or appearance.

frontal. Of or pertaining to the anterior portion of the body or a part.

frontal bone. A bone forming the anterior dorsal portion of the cranium.

frontal section. A coronal section; one which divides the body or a part into anterior and posterior portions.

fructiferous. Bearing fruit.

fructification. *bot.* **1.** The act of producing fruit. **2.** The production of a reproductive structure. **3.** In fungi, the production of spores.

fructose. Levulose or fruit sugar, $C_6H_{12}O_6$.

frugivorous. Fruit-eating.

fruit. *bot.* A structure consisting of one or more ripened ovaries, with or without seeds, and sometimes with accessory structures derived from other parts of the flower, as an apple, bean pod, nut, berry, melon, grain.

fruit bat. A bat of the suborder Megachiroptera, as *Pteropus*, the flying fox.

fruit dot. The sorus of a fern.

fruit fly. 1. A pomace fly, any of a number of species of flies which frequent ripe and decaying fruits and vegetables. Most belong to the genus *Drosophila*, family Drosophilidae. **2.** A trypetid, a small to medium sized fly of the family Trypetidae, some of which are serious pests, as *Ceratitis capitata*, the Mediterranean fruit fly.

fruiting body. A reproductive structure in plants, especially the complex structure in fungi which produces spores.

fruitlet. One of the small fruits comprising an aggregate fruit.

frustule. One of the two halves of the cell wall of a diatom.

fruticose. 1. Pertaining to or resembling a shrub. **2.** Erect and branched, with reference to lichens.

fucoxanthin. Fucoxanthol, $C_{40}H_{60}O_6$, a brown pigment present in brown algae.

fugacious. Falling off soon or fading away; temporary; not permanent, as the petals of a flower.

fumigant. Any substance used for fumigation.

fumigation. Disinfection by use of a vapor, smoke, or gas.

fumitory. 1. A plant or plant product which is smoked for pleasure, physiological effects, or for use in religious ceremonies, as tobacco, opium, cannabis. **2.** Any plant of the fumitory family, Fumariaceae, as *Fumaria*, the earth-smoke.

function. The characteristic action or the normal activities of an organ or structure.

fundus. The bottom or base of an organ; the part farthest from the aperture.

fungating. Growing rapidly like a fungus, said of bacterial colonies or a tumor.

fungi. Plural of FUNGUS.

fungicide. An agent which kills or inhibits the growth of fungi.

fungiform. Shaped like a mushroom.

Fungi Imperfecti. An artificial group which includes a large number of fungi whose sexual stages are lacking or unknown.

fungistatic. Inhibiting the growth of fungi.

fungoid. Resembling or having the characteristics of a fungus.

fungous. 1. Spongy or like a fungus in texture. **2.** Springing up suddenly and growing rapidly, like a fungus.

fungus pl. **fungi.** Any of a large group of simple plants characterized by lack of chlorophyll, as the molds, mildews, mushrooms, rusts, and smuts. Most have a filamentous body or mycelium and subsist on organic matter as saprophytes or parasites. Many are pathogenic.

fungus garden. A mass of fungi cultivated within the nests of termites or fungus ants for use as food.

funicle. A slender cord or stalk; a funiculus.

funiculus. 1. *anat.* A division of the white matter of the spinal cord. **2.** *zool.* A strand of tissue which connects stomach to the body wall in bryozoans. **3.** *bot.* The stalk by which an ovule is attached to the placenta.

funnel. 1. *bot.* A space beneath the outer coat of a macrospore in aquatic plants of the waterclover family. **2.** *zool.* The excurrent siphon of a squid or octopus: HYPONOME.

funnelform. 1. Shaped like a funnel. **2.** Designating flowers in which the tube gradually widens into an expanded corolla.

funori. A phycocolloid obtained from red algae, used as a sizing, thickening, and adhesive agent.

fur. The soft, fine, hairy coat of a mammal.

furbearer. A fur-bearing mammal of the family Mustelidae.

furca. A forked process.

furcate. Forked or Y-shaped.

furcilia. The zoea larva of crustaceans of the order Euphausiacea.

furcocercous. Fork-tailed, as the cercariae of blood flukes.

furcula. A fork-shaped process or structure, as **a.** the wishbone of birds; **b.** the springing structure in springtails.

furrowed. Bearing longitudinal grooves or channels: SULCATE.

fur seal. One of several species of eared seals, especially *Callorhinus ursinus*, the Alaska fur seal noted for its summer migrations to the Pribilof Islands; also called sea bear.

fusarium disease. Any of a number of rots, blights, and wilts caused by parasitic phycomycete fungi of the genus *Fusarium*.

fusiform. Spindle-shaped; tapered at each end.

G

gadwall. A surface-feeding duck (*Anas strepera*).

gaggle. A flock of geese.

galactose. A hexose sugar, $C_6H_{12}O_6$, a constituent of many complex sugars.

galago. *See* bush baby.

galea. 1. *bot.* A helmet-shaped structure forming a part of the calyx or corolla in certain flowers. **2.** *zool.* A lobelike structure on the maxilla of certain insects, as the grasshopper.

galeate. Helmet-shaped.

gall. 1. Bile, especially that of an ox. **2.** *bot.* A localized growth on a plant induced by various organisms, as fungi, mites, roundworms, and gall insects. **3.** *zool.* **a.** A growth on certain corals induced by gall-forming crabs and shrimps. **b.** A sore on a horse's back caused by friction.

gallbladder. A saclike structure lying between the lobes of the liver which concentrates and stores bile. It is drained by the cystic duct.

gallfly. A gall-producing dipterous insect (*Eurosta*) of the family Trypetidae, the cause of stem galls on goldenrod.

gall gnat. A gall midge, a small dipterous insect of the family Cecidomyiidae, the larvae of which live in plants, forming galls.

gallinaceous. 1. Of or resembling domestic fowl. **2.** Of or pertaining to the order Galliformes, which includes most upland game birds, as the grouse, prairie chicken, quail, Hungarian partridge, pheasant, and wild turkey.

gall insect. Any of a number of insects which cause galls to develop on plants.

gallinule. A henlike bird of the family Rallidae, as *Gallinula chloropus*, the Florida gallinule.

gall midge. A gall gnat.

gall mite. An elongated, four-legged mite of the family Eriophyidae which causes galls to form on various plants.

gallstone. A concretion or calculus found in the gallbladder or bile ducts.

gall wasp. One of a number of small hymenopterous insects of the family Cynipidae which cause galls to form on plants.

gam. A group of whales or porpoises swimming together.

game. Various animals, principally birds and mammals which are hunted for sport and for their value as food.

game birds. Birds considered worthy of pursuit and capture by sportsmen, principal ones of which are upland game birds and waterfowl.

gamecock. A male game fowl.

game fishes. Fishes sought by anglers because of their cunning or wariness or difficulty in landing when caught.

game fowl. One of a breed of poultry bred for its fighting qualities.

game mammals. Mammals which are hunted for sport. Small game includes the cottontail, jackrabbit, and squirrel; big game includes deer, elk, moose, antelope, caribou, bighorn sheep, mountain goat, bear, and peccary. The term is also applied to animals as the lion, tiger, rhinoceros, and elephant.

gametangium. A gamete-producing structure in plants. In algae and fungi, it is usually unicellular; in higher plants, multi-cellular. *See* antheridium, oogonium, and archegonium.

gamete. A mature haploid reproductive or sex cell; an ovum or spermatozoon; any cell which upon union with another cell results in the development of a new individual.

gametocyte. A cell which develops into a gamete. *See* oocyte, spermatocyte.

gametogenesis. The origin and development of mature gametes. *See* oogenesis, spermatogenesis.

gametophore. A structure upon which gametangia are borne.

gametophyte. A gamete-producing plant; in a plant exhibiting alternation of generations, the generation which produces gametes or sex cells; the sexual generation, usually haploid.

gamic. SEXUAL.

gamogenesis. SEXUAL REPRODUCTION.

gamone. An ectohormone which acts to bring about the union of gametes.

gamopetalous. A condition in which the petals of a flower are united: SYMPETALOUS.

gamophyllous. With united leaves or leaflike parts.

gamosepalous. Synsepalous; a condition in which sepals of a flower are united.

gander. An adult male goose.

ganglion. 1. A mass of nervous tissue consisting principally of nerve cell bodies. **2. a.** In invertebrates, the brain or one of the masses of nervous tissue comprising the ventral nerve cord. **b.** In vertebrates, a mass of nervous tissue lying outside the brain and spinal cord. **3.** A cystic tumor involving a tendon sheath.

gannet. A large, fish-eating seabird of the family Sulidae, as *Sula leucogaster*, the booby gannet.

ganoid fishes. The garpikes, sturgeons, and their allies, primitive fishes with ganoid scales.

ganoid scale. A type of fish scale consisting of layers of ganoin deposited on a layer of lamellate bone.

gap. A break in the continuity of the vascular cylinder of a stem, as leaf gap or bud gap.

¹**gape.** To open the mouth widely.

²**gape.** A wide opening as that between **a.** the opened jaws of a

vertebrate; **b.** the two valves of a bivalve mollusk; **c.** the scutes of a barnacle.

gapeworm. *Syngamus trachea,* a parasitic nematode worm which infects the respiratory passages of fowl, causing gapes.

gar. The garpike, one of several species of primitive, ray-finned fishes of the genus *Lepisosteus* inhabiting central United States.

garfish. *See* gar.

garpike. *See* gar.

garter snake. A harmless, viviparous snake (*Thamnophis sertalis*).

gas. A fluid which exists in the form of a vapor.

gaseous. In the form of or of the nature of a gas or gases; pertaining to gases.

gas gland. A gland which secretes a gas, as that of a siphonophore or the red gland of fishes.

gaster. The portion of the abdomen lying posterior to the petiole, as in various hymenopterans, for example, ants. *See* propodium.

gastralia. Ventral, riblike, dermal bones present in the alligator and other long-bodied reptiles.

gastric. Of, pertaining to, or originating in the stomach.

gastric ceca. Blind evaginations from the stomach or midgut of insects.

gastric filaments. Nematocyst-bearing threads borne on the septa of the gastrovascular cavity of scyphozoans.

gastric glands. Tubular glands of the stomach which secrete gastric juice, a digestive fluid containing hydrochloric acid (0.4%), pepsin, and a lipase.

gastric mill. 1. A grinding structure in the stomach of various crustaceans. **2.** The gizzard of an insect.

gastric pouch. One of four pouches extending from the central stomach of a scyphozoan jellyfish.

gastric shield. In various mollusks, a platelike structure in the dorsal wall of the stomach with which the tip of the crystalline style comes in contact.

gastrocoel. The cavity of the gastrula: ARCHENTERON.

gastrodermis. The layer of epithelium which lines the gastrovascular cavity of a coelenterate.

gastrolith. A calcareous body in the lateral wall of the stomach of a crayfish.

gastropod. Any mollusk of the class Gastropoda which includes the snails, slugs, limpets, periwinkles, and whelks.

gastrotrich. Any member of the Gastrotricha, a phylum or class of pseudocoelomate animals, mostly minute, bottom-dwelling, marine and freshwater animals, as *Chaetognathus.*

gastrovascular. Pertaining to both digestion and circulation.

gastrozooid. A nutritive polyp of a colonial coelenterate: TROPHOZOOID.

gastrula. The stage in embryonic development which follows the blastula. In its simplest form, it is cup-shaped with a wall of two layers, ectoderm and endoderm, enclosing a cavity, the gastrocoel or archenteron with a single opening, the blastopore.

gastrulation. The processes involved in the formation of a gastrula.

gavial. A large Indian crocodile (*Gavialis gangeticus*) with long slender jaws.

gazelle. One of a number of small, graceful, swift-running antelopes inhabiting Asia and Africa, especially those of the genus *Gazella.*

gecko. Any of a number of small, swift-moving lizards of the family Gekkonidae. Most are tropical; some are sound-producing.

gel. A phase or state of a colloidal solution in which the colloid is of a semisolid or jellylike consistency, as gelatin when cool.

gelatin. A protein resulting from the partial hydrolysis of collagen, obtained by boiling white connective tissue fibers.

gelation. The conversion of a colloid from a fluid or sol state to a solid or gel state.

gelding. A castrated animal, especially a horse.

gemma. 1. A bud or budlike body. 2. A small, asexual reproductive body which develops in gemma cups on the thalli of liverworts, hornworts, and some mosses.

gemmation. The process of producing gemmae.

gemmule. 1. An asexual, internal reproductive body produced by certain sponges. 2. One of the numerous, spinelike processes on the dendrite of a neuron.

gemsbuck. A large, African antelope of the genus *Oryx;* gemsbok.

gena. The cheek or side of the face.

gene. 1. A hereditary factor or unit of heredity; a determiner. **2.** A specific region of a chromosome which is capable of determining the development of a specific trait, composed wholly or in part of DNA. It is a self-duplicating particle involved in the transmission of genetic information from one generation to the next. *See* muton.

genealogy. An account or diagram of the descent of an individual, family, or race from its progenitors or ancestors; pedigree: LINEAGE.

genera. Plural of GENUS.

generalized. Of a general nature; not specific; not highly differentiated.

generate. To originate, produce, propagate, beget, or bring forth.

¹generation. The act or process of producing offspring.

²generation. All the individuals comprising a stage in natural descent, as the F_1 *generation* or gametophyte *generation*.

generative. Having the power to bring forth or produce.

generative cell. The cell in a pollen grain which gives rise to two male gametes.

generative organ. REPRODUCTIVE ORGAN.

generic. Pertaining to or having the rank of a genus.

genesis. The beginning or coming into existence; the origin of anything.

genet. A small Old World carnivore of the genus *Genneta*, related to the civet and mongoose.

genetic. Of or pertaining to genetics or heredity.

genetic code. The arrangement of bases within a DNA or RNA molecule which determines the arrangement of amino acids within a protein molecule.

genetic drift. The tendency in small populations for gene pairs which are heterozygous to become homozygous for one or the other allele by chance, regardless of whether the gene is advantageous or not, resulting in small interbreeding groups which tend to become genetically stable, homozygous populations. Sometimes referred to as "bottle-neck effect."

genetic load. The number of recessive lethal or deleterious genes carried within a population.

genetics. The science which deals with heredity and variation.

Special fields include molecular, population, evolutionary, and medical genetics.

genic. Of or pertaining to genes.

geniculate. Bent sharply, as at the knee.

genital. Of or pertaining to the sex organs or the organs of reproduction.

genitalia. The reproductive organs, male or female, especially the external reproductive structures.

genome. A complete haploid set of chromosomes, as that in a gamete.

genotype. 1. *genetics.* **a.** A group of individuals with identical genes. **b.** The genetic constitution of an individual usually represented by a formula such as AABB or AaBb, in contrast to the appearance of an individual. *Compare* phenotype. **2.** *taxonomy.* The type species of a genus.

genu. 1. The knee. **2.** A segment in the leg of a tick between femur and tibia.

genus. *taxonomy.* A subdivision of a family which includes one or more closely related species, as *Canis* which includes dogs and *Trifolium* which includes the clovers. Genus name is always capitalized and italicized.

geocarpy. The development of a fruit beneath the ground which originated from a flower above the ground, as the peanut.

geoduck. A large, edible clam (*Panope generosa*) of the Pacific coast; gweduc.

geographical distribution. The distribution of organisms over the surface of the earth.

geographical isolation. Isolation of species resulting from geographic factors acting as barriers, as oceans or high mountains acting as a barrier to distribution of terrestrial forms.

geographic range. The area of the earth's surface occupied by a species, outside of which it does not normally occur.

geography. The science which deals with the earth, its natural features (terrain, climate, water, soils, minerals), the plants and animals living on it, including man, and those features which man has added through his utilization of earth resources, such as communities, towns, cities, farms, factories, mines, transportation facilities, etc.

geological distribution. The distribution of animals and plants in time as determined by presence of their fossil remains in sedimentary rocks.

geologic timetable. A table listing the geological eras and their subdivisions with the principal forms of life occurring in each.

geology. The science which deals with the structure and history of the earth. Structural geology deals with the form, arrangement, and structure of the rocks; dynamic geology with changes in structure of the rocks, causes, processes involved, etc.; historical geology with changes as revealed by study of fossils. *See* paleontology.

geometer. A moth of the family Geometridae whose larvae (inchworms or measuring worms) move by looping movements, as the spring cankerworm.

geonasty. Curving towards the earth.

geophilous. 1. *bot.* Fruiting underground. 2. *zool.* Living in or on the ground; terrestrial.

geophyte. A plant propagated by underground buds.

geotaxis. Orientation of a motile organism with respect to gravity.

geotropism. The response of a plant to the force of gravity.

gerbil. The sand rat, a nocturnal, burrowing rodent of Africa and Asia as *Gerbillus, Tatera.*

gerenuk. *Litocranius walleri,* a long-necked gazelle of East Africa.

germ. 1. A small bit of protoplasm capable of developing into a new organism, as a spore, egg, or seed. 2. The embryo of a seed, as wheat *germ.* 3. Any microorganism, especially a pathogenic bacterium. 4. Any minute disease-causing agent, as a virus, rickettsia, spirochaete.

germ ball. A mass of germinal cells in a developmental stage of a fluke.

germ band. A thickened area on the ventral surface of an insect egg from which the embryo develops.

germ cell. A mature sex cell or gamete (ovum or spermatozoan) or one of its progenitors.

germ-free. GNOTOBIOTIC.

germicide. An agent which kills microorganisms.

germinal. Of or pertaining to reproductive cells.

germinal disc. A blastodisc.

germinal epithelium. An epithelium which gives rise to sex cells, as **a.** that covering the surface of an indifferent gonad in an embryo; **b.** that lining a seminiferous tubule or covering the surface of the ovary in mammals.

germination. The beginning of growth of a spore, seed, or reproductive body after a period of dormancy.

germ layer. One of the three primary layers of an embryo from which the definitive tissues and organs develop. *See* ectoderm, mesoderm, endoderm.

germ plasm. The reproductive tissues of the body as distinguished from the soma (nonreproductive tissues).

germ tube. A threadlike filament that grows out of a germinating spore, as in various fungi.

gerontology. The study of old age and various phenomena associated with aging.

gestation. Pregnancy; the act of carrying young within the uterus.

gestation period. The time from conception to birth.

giant. 1. Of unusual size, power, or strength. **2.** A person of extraordinary stature or size.

giant chromosomes. Extremely large chromosomes present in the salivary glands of *Drosophila*, characterized by distinct banding.

giant fibers. Extremely large axons of neurons found in various invertebrates (annelids, crustaceans, mollusks) and some vertebrates (fishes, amphibians).

giantism. GIGANTISM.

gibberellic acid. A gibberellin which, when sprayed on plants or added to the soil, brings about excessive increase in height and length of leaves.

gibberellin. One of a number of growth-promoting substances produced by a fungus (*Gibberella fujikuroi*).

gibbon. One of several species of small long-armed, anthropoid apes of the genera *Hylobates* and *Symphalangus* of southeast Asia, noted for their powerful voices.

gibbosity. An asymmetrical swelling or protuberance usually near the base of an organ.

gibbous. Enlarged or swollen on one side.

gid. A disease principally of sheep caused by development within the brain of the larva (coenurus) of a tapeworm (*Multiceps multiceps*).

gigantism. Abnormal size or tallness resulting from over-production of the growth hormone of the anterior pituitary.

Gila monster. A large venomous lizard (*Heloderma suspectum*) of southwestern United States, with orange and black markings.

gill. 1. *zool.* A respiratory structure in aquatic organisms through which gaseous exchange takes place: BRANCHIA; CTENIDIUM. **2.** *bot.* A thin, platelike, spore-producing structure on the under surface of the pileus of a mushroom.

gill aperture. An external gill cleft or opening, as in cyclostomes.

gill arch. A branchial arch.

gill bar. 1. A supporting structure separating gill clefts in lancelets. **2.** A vertical supporting structure in the gill filament of a bivalve.

gill bypass. A blood vessel in the larvae of caudate amphibians through which blood passes directly from the ventral aorta to dorsal aorta without passing through the gills.

gill chamber. A cavity or space which contains gills.

gill cleft. A branchial cleft or gill slit.

gill fungi. Mushrooms or toadstools of the family Agaricaceae whose fruiting body bears leaflike plates or gills upon which basidiospores develop.

gill lamella. A thin plate forming a part of a gill, as in bivalves.

gill maggot. A wormlike parasitic copepod, as **a.** *Lernaeocera* on the gills of a codfish; **b.** *Salmincola* on the gills of salmon.

gill pouch. A cavity which contains gills, as in the lamprey.

gill rakers. Projections on the inner borders of gill arches which in fishes prevent food particles from passing outward through the gill slits.

gill rays. Branchial rays, slender, cartilaginous, supporting structures in the interbranchial septum of a gill.

gill slit. 1. In chordates, one of a series of openings in the wall of the pharynx. In aquatic vertebrates, they are separated by arches which bear gills; in terrestrial forms, they are transitory structures in the embryo. **2.** A branchial cleft or gill cleft; in tunicates, a stigma.

gilt. A young female swine.

ginger. A spice and condiment obtained from the rhizomes of the ginger plant (*Zingiber officinale*) an erect perennial herb which is also used medicinally.

gingival. Of or pertaining to the gums of the oral cavity.

ginkgo. The maidenhair tree (*Ginkgo biloba*), an ornamental gymnospermous tree, the only living representative of the order Ginkgogales.

giraffe. A large ruminant (*Giraffa camelopardalis*) with an extremely long neck, mobile, hairy lips and an extensible tongue, a native of Africa. Also called camelopard.

girdle. **1.** An encircling or confining structure. **2.** *bot.* The region where the two valves of a diatom overlap. **3.** *zool.* **a.** The mantle of a chiton, especially the peripheral portion. **b.** The annulus of a dinoflagellate. **c.** A supporting structure in vertebrates, as the pectoral or pelvic *girdle*.

girdler. An insect which cuts a groove about a twig causing it to fall to the ground, as the beetle, *Oncideres cingulata*.

¹**girdling.** Making a cut encircling a part of a plant.

²**girdling.** The arrangement of leaf traces in cycads.

gizzard. The proventriculus, a thick-walled grinding structure, as **a.** that between the crop and intestine in various annelids and arthropods; **b.** the second portion of the stomach in birds.

glabella. A smooth region of the frontal bone between the superciliary arches.

glabrate. Approaching a glabrous condition, especially at maturity.

glabrous. Lacking hair or pubescence.

gladiate. Shaped like a sword: ENSIFORM.

gladiolus. **1.** *anat.* The middle portion of the sternum. **2.** *bot.* A plant (*Gladiolus*) of the iris family, or its corm or flower.

gland. **1.** *zool.* A secreting structure; a cell or organ which manufactures or elaborates a substance which is discharged directly or through ducts onto a surface or into the bloodstream for utilization in other parts of the body or elimination. **2.** *bot.* **a.** A secreting structure, as a nectary, hydathode, lysigenous cavity, resin, oil, gum, or lactiferous duct of a plant or the digestive gland of an insectivorous plant. **b.** A structure or an appendage which has the appearance of a gland.

glandular. Of or pertaining to glands; bearing glands or gland cells.

glandule. A retinactulum of orchids.

glans. 1. *anat.* A conical structure at the tip of the penis or clitoris. **2.** *bot.* A nut enclosed within a cup-shaped involucre, as the acorn.

glass snake. A limbless lizard (*Ophisaurus ventralis*) with an extremely long, fragile tail.

glass sponge. A sponge of the class Hexactinellida (Hyalospongiae) with a siliceous skeleton which resembles glass when dried, as *Euplectella aspergillum,* Venus's flower basket.

glaucescent. Slightly glaucous.

glaucous. 1. Yellowish-green. **2. a.** Having a surface with a waxy bloom. **b.** Possessing a whitish coating which rubs off, as that of a plum or cabbage leaf.

gleba. A soft, fleshy, spore-producing mass of hyphae and basidia within the basidiocarp of puffballs and stinkhorns.

glenoid. Having the shape of a shallow depression.

glenoid cavity. A depression on the pectoral girdle which receives the head of the humerus.

glia. NEUROGLIA.

gliadin. A protein present in wheat.

glioma. A tumor composed principally of neuroglia cells.

globefish. *See* puffer.

globigerina ooze. Vast marine deposits consisting principally of the calcareous skeletons of foraminiferans, especially those of the genus *Globigerina.*

globin. One of a class of proteins obtained from the breakdown of hemoglobin.

¹globoid. Globe-shaped or globular.

²globoid. A minute globular structure as that in the aleurone grains of certain seeds, as in the castor bean.

globose. GLOBOID.

globular. Shaped like a globe; spherical.

globule. A small spherical body.

globulin. Any of a group of animal and plant proteins which are insoluble in water, soluble in salt solutions and coagulable by heat, as serum globulin of the blood and edestin from hemp seed.

glochid. A barbed hair or bristle.

glochidiate. 1. Having a barbed tip. **2.** Covered with barbed hairs or bristles.

glochidium. The larva of freshwater clams of the family Unionidae which on leaving the gill of the mother, attaches to the gills or skin of fishes where it lives as a temporary parasite for 3 to 12 weeks.

glomerate. 1. Grouped into a compact cluster. **2.** Rolled into a ball-like structure.

glomerule. A small, compact cluster, especially an inflorescence of this type.

glomerulus. A small convoluted mass, as **a.** the capillary mass within a renal capsule; **b.** the coiled proximal portion of the process of a spinal ganglion cell.

glomus. A small globular mass or body.

glossa. 1. *anat.* The tongue. **2.** One of a pair of median lobes at the tip of the labium of certain insects.

glossal. Pertaining to the tongue.

glottis. 1. In most vertebrates, the opening of the larynx between the vocal cords. **2.** In man, the sound-producing apparatus of the larynx, comprising the vocal folds (cords) and the opening between them, the rima glottidis.

glowworm. Any of a number of larvae or wingless beetles of the family Lampyridae, which are capable of producing cold light. *See* firefly.

glucagon. A hyperglycemic-glycogenolytic factor (HGF) present in pancreatic extracts which causes a rise in blood sugar.

glucocorticoid. One of a group of 11-oxygenated corticoids secreted by the adrenal cortex which markedly influence carbohydrate metabolism. Also called GC.

glucose. Dextrose, $C_6H_{12}O_6$, a crystalline monosaccharide formed in the hydrolysis of starch. It is the first stable product formed in photosynthesis and the final product of digestion of carbohydrates in vertebrates. Also called dextrose, d-glucose, grape sugar, corn sugar.

glucoside. A compound such as digitonin or indican which upon hydrolysis yields a sugar, usually d-glucose, and another product. Also called glycoside.

glumaceous. Bearing glumes; composed of or resembling glumes.

glume. One of a pair of the lowermost bracts of a grass spikelet.

gluten. A protein present in dough, especially that made from wheat flour, which gives it cohesiveness. It is highly nutritious.

glutinant. A form of nematocyst with an open sticky tube.

glutinous. 1. Like glue; viscid, sticky. **2.** *bot.* Covered with a sticky exudate.

glycerine. GLYCEROL.

glycerol. A viscous, hygroscopic liquid, miscible with water, obtained as a by-product in the manufacture of soaps as a result of the hydrolysis of fats.

glycogen. Animal starch, a polysaccharide, $(C_6H_{10}O_5)_n$, found in the liver and muscles of vertebrates.

glycogenesis. The formation of glycogen.

glycogenolysis. The conversion of glycogen to glucose.

glycolysis. In animal metabolism, the breakdown of carbohydrates to pyruvic and lactic acids.

glyconeogenesis. The formation of glycogen from noncarbohydrate sources, as fats and proteins.

glycoprotein. A mucoprotein, one of a group of conjugated proteins composed of a protein and a carbohydrate, as mucin.

glyptodon. A large, fossil, edentate mammal, related to the armadillo.

gnat. Any of a number of small, dipterous insects, as the buffalo gnat, eye gnat, fungus gnat, gall gnat, or wood gnat.

gnathic. Of or pertaining to the jaw.

gnathobase. The basal segment of an appendage which is used as a grinding structure, as that present on the legs of a king crab.

gnathopod. A thoracic appendage which is used for prehension or for excavating, as in certain isopods and amphipods.

gnathostome. Any jawed vertebrate.

gnathostomulid. Any member of phylum Gnathostomulida, the newest of animal phyla, discovered in 1928, and comprising microscopic forms related to the Turbellaria.

gnaur. A burl or excrescence on a tree trunk or root.

gnotobiotic. Sterile; germ-free, with reference to the culture of animals.

gnu. The wildebeest, an ungainly appearing antelope of Africa, as the white-tailed gnu (*Connochaetes gnu*) and brindled gnu (*Gorgon taurinus*).

goat. Any of a number of hollow-horned ruminants of the genus *Capra*, family Bovidae. True goats include the wild goat (*Capra hircus*), ibex (*C. ibex*), angora goat (*C. angorensis*), and milch goat (*C. hircus*). *See* Rocky Mountain goat.

goatsucker. An insect-eating bird of the family Caprimulgidae, as the nighthawk, whippoorwill, and oil bird.

goblet cell. An epithelial cell which secretes mucinogen, the precursor of mucin.

godwit. A long-billed wading bird of the genus *Limosa*, as the Hudsonian godwit (*L. haemastica*).

goiter. Enlargement of the thyroid gland, commonly the result of iodine deficiency.

goldeneye. A diving duck of the genus *Glaucionetta*, as the American goldeneye or whistler (*G. clangula*) and Barrow's goldeneye (*G. islandica*).

goldfinch. The wild canary or yellowbird, a small, brightly colored finch (*Spinus tristis*), the Eastern goldfinch; also *Carduelis carduelis*, the British goldfinch.

goldfish. A small, golden-colored fish (*Carassius auratus*) of the family Cyprinidae, a native of Asia but long domesticated and kept in aquaria and ponds.

Golgi apparatus. A cytoplasmic organelle especially well-developed in neurons and secretory cells, thought to play a role in the process of secretion; Golgi complex, internal reticular apparatus.

gonad. 1. A reproductive organ; an organ which produces gametes; an ovary or testis. **2.** An undifferentiated embryonic sex organ.

gonadotrophic. Gonad-stimulating. Also gonadotropic.

gonadotrophic hormone. A gonadotrophin, one of several hormones secreted by the anterior lobe of the pituitary or the placenta which act on or stimulate the gonads.

gonadotrophin. A gonad-stimulating hormone. Also gonadotropin.

gonangium. In colonial coelenterates, a reproductive polyp consisting of a blastostyle bearing gonophores (medusa buds) enclosed within a gonotheca.

gonidium. 1. A minute reproductive body in certain bacteria. **2.** An asexual reproductive cell, as in *Volvox*.

gonococcus. *Neisseria gonorrhoeae*, the causative organism of gonorrhea.

gonocoel. A cavity of the coelom within which a gonad lies.

gonoduct. A duct which conveys gametes from a gonad.

gonophore. *zool.* An asexual bud in hydrozoan coelenterates which gives rise to a medusa, a sessile medusoid, or in some cases sex cells directly.

gonopod. A modified appendage which functions as a copulatory structure, as in certain millipedes and insects.

gonopodium. The modified portion of an anal fin which functions as a copulatory organ, as in certain teleost fishes.

gonopore. 1. The external opening of a reproductive duct. **2.** In hydrozoans, the external opening of a gonotheca.

gonorrhea. A contagious venereal disease characterized by inflammation of the mucous membrane of the urinogenital tract, caused by a bacterium (*Neisseria gonorrhoeae*).

gonosome. Collectively, the reproductive structures of a hydrozoan coelenterate. *See* trophosome.

gonotheca. The vaselike covering of a blastostyle in coelenterates.

gonotyl. A retractile, genital sucker in certain heterophyid flukes.

gonozooid. In colonial coelenterates and tunicates, a reproductive zooid.

gooney bird. The black-footed albatross (*Diomedra nigripes*).

goosander. A merganser.

goose pl. **geese.** Any of a number of species of waterfowl of the family Anatidae, intermediate between ducks and swans. Common species are those of the genera *Branta*, *Anser*, and *Chen*, as the Canada goose (*Branta canadensis*).

gooseberry. An acid berry produced by the shrub, *Ribes*.

gopher. 1. One of a number of small, burrowing rodents of the genera *Geomys*, *Cratogeomys* and *Thomomys*, called *pocket gophers* because of possession of fur-lined cheek pouches used in carrying food. **2.** Common name for ground squirrels of the genus *Citellus*.

gopher snake. 1. The indigo snake (*Drymarchon corais*). **2.** The bull snake (*Pituophis catenifer*).

gopher tortoise. *Gopherus polyphemus* of southern United States.

goral. A shaggy, goatlike ruminant (*Naemorhedus goral*) of Asia.

gorge. The throat of a flower.

gorgonian. A horny coral, of the order Gorgonacea which

includes the sea whips, sea feathers, and sea fans, character-ized by the possession of a skeleton containing a hornlike material, gorgonin.

gorilla. *Gorilla gorilla,* the largest and most powerful of the anthropoid apes, an inhabitant of West Africa.

goshawk. A large, long-tailed hawk (*Accipiter gentilis*) of the family Accipitridae.

gosling. A young goose.

gourd. 1. Any plant of the genus *Cucurbita,* a tendril-bearing vine of the gourd family, Cucurbitaceae, including the cucum-ber, melon, squash, and pumpkin. **2.** The hard-shelled fruit of certain species of the gourd family especially those of the genera *Coccinea* (ivy gourd), *Lagenaria* (calabash gourd), and *Trichosanthes* (snake gourd).

grackle. 1. One of a number of blackbirds of the family Icteridae, especially *Quiscalus quiscula,* the purple grackle. **2.** One of a number of Old World starlings.

gradient. 1. A gradual ascent or descent or a curve which represents such. **2.** *biol.* A system of relationships within an organism which involves changes of increasing and decreasing magnitude with reference to metabolism, growth, pressure, velocity, or other physiological activities.

graduated. 1. Marked or divided into equal units by lines, as a thermometer. **2.** Tapered, with reference to a bird's tail, in which the length of the feathers increases gradually from the outside inward.

¹graft. 1. *bot.* A scion or shoot that is grafted on to a stock. **2.** *zool.* A piece of tissue, a part of the body, or an entire organ which is transferred to another region of the body or to another individual.

²graft. 1. To insert a bud or scion from one plant onto another so that a permanent union is effected. **2.** To transfer a piece of tissue, a part of the body, or an entire organ to another region of the body or to another individual.

grain. 1. A simple, single, dehiscent fruit produced by cereal grasses: CARYOPSIS. **2.** Any single, small, hard seed. **3.** Col-lectively, the seeds or fruits of cereal grasses, as wheat, oats, maize, rice, millet, or the plants which produce them. **4.** A

211

single hard particle, as a *grain* of salt. **5.** A unit of weight of the troy, avoirdupois, and apothecaries' system. Abbreviated gr.

gram. A unit of weight of the metric system; the weight of a milliliter of water at the temperature of maximum density (4°C); one thousandth of a kilogram. Abbreviated g., gm.

gramineous. Of or pertaining to grasses or plants of the grass family.

grampus. 1. The killer whale. **2.** Risso's dolphin (*Grampus griseus*), related to the killer whale.

Gram's method. A procedure by which bacteria are divided into two large groups, gram-negative and gram-positive bacteria.

granivorous. Feeding on small seeds or grain.

granular. Consisting of, containing, or of the nature of granules.

granule. A minute grain or particle.

granulocyte. A polymorphonuclear leukocyte containing granules in its cytoplasm, as a neutrophil, basophil, or eosinophil.

granum pl. **grána.** A disc-shaped unit of a chloroplast.

grape. A thin-skinned, juicy berry forming the fruit of a vine of the genus *Vitis* or the vine producing the fruit.

grape fern. One of a number of ferns of the genus *Botrychium*, whose sporophylls resemble a cluster of grapes.

grapefruit. The pomelo, a large thin-skinned citrus fruit or the tree (*Citrus maxima* or *C. parodisa*) which produces it.

grape sugar. D-glucose or dextrose.

graptolite. A fossil of an extinct group, the Graptolita (Graptolitoidea) of uncertain taxonomic position, now considered to be a primitive chordate.

grass. 1. A monocotyledonous plant of the grass family, Gramineae. **2.** Green plants, especially those with narrow leaves, used for food by grazing animals.

grass flower. A spikelet.

grass frog. *Rana pipiens*, also called leopard frog.

grasshopper. A herbivorous, leaping, orthopteran insect of the families Acrididae (short-horned grasshoppers or locusts), Tettigoniidae and Gryllacrididae (long-horned grasshoppers).

grassland. An area in which grasses constitute the dominant vegetation, as in the prairies or plains region of the United

States, the pampas of South America, and the steppes of Russia.

gravid. Pregnant; bearing eggs or young.

grayling. A freshwater game fish (*Thymallus signifer*) of northern United States and Canada.

gray matter. The gray substance of the spinal cord and brain consisting principally of the cell bodies of neurons and neuroglia cells.

grebe. One of several species of aquatic, ducklike birds of the family Podicipedidae.

green. **1.** Of a color characteristic of growing herbage, as grass. **2.** Immature, not ripe, said of fruits.

green algae. Algae of the division Chlorophyta, as *Spirogyra*, *Oedogonium*.

green gland. An excretory organ located in the base of the antenna of the crayfish and other crustaceans.

green pigments. The chlorophylls.

green turtle. A large sea turtle (*Chelonia mydas*), highly valued for its flesh.

gregarine. A parasitic sporozoan of the class Telosporea, inhabiting the digestive tract and body cavity of insects and other animals.

gregarious. **1.** *zool.* Living together in groups, as flocks or herds. **2.** *bot.* Growing in clusters.

gribble. An isopod crustacean (*Limnoria lignorum*) which burrows into marine timbers.

grilse. The young adults of salmon upon their return from the sea and entry into freshwater streams.

gristle. CARTILAGE.

grit cell. A stone cell in a fleshy fruit.

grizzly bear. A large, ferocious bear (*Ursus horribilis*) of western North America, becoming rare in the wild state.

grosbeak. One of a number of seed-eating birds of the finch family, as the rose-breasted grosbeak (*Pheuticus ludovicianus*).

groundhog. The woodchuck (*Marmota monax*), a heavy-bodied, burrowing rodent.

ground meristem. The primary meristem from which the pith, pith rays, and cortex develop.

ground pine. Common name for *Lycopodium*, a club moss.

213

ground squirrel. *See* gopher, 2.

grouper. One of a number of large marine fishes of the family Serranidae, especially the genera *Epinephelus* and *Mycteroperca* inhabiting warm waters.

grouse. One of several species of gallinaceous, upland game birds of the family Tetraonidae, especially the ruffed grouse (*Bonasa umbellus*) and sharp-tailed grouse (*Pedioecetes phasianellus*).

growth. 1. Increase in mass; increase in size of an organism or one of its parts. *See* hypertrophy, hyperplasia. **2.** The processes involved in coming to maturity. **3.** Increase in numbers, as *growth* in population.

growth hormone. 1. *zool.* Somatotrophin (STH) produced by the anterior lobe of the pituitary. **2.** *bot.* An auxin.

growth ring. One of the numerous concentric rings of wood seen in a cross section of a stem of a dicot or gymnosperm. *See* annual ring.

grub. 1. The thick-bodied larva of various insects, especially beetles. **2.** The encysted metacercaria of certain flukes. *See* yellow grub.

grubby. 1. Infested with warbles or bots, with reference to domestic animals. **2.** Infested with metacercaria of flukes, with reference to fishes.

guan. A large, gallinaceous bird of tropical America, of the family Cracidae.

guanaco. A South American ruminant (*Lama guanacoe*) related to the llama and vicuna.

guanine. A purine, nitrogenous base, $C_5H_5N_5O$, present in DNA and RNA. It is a crystalline substance present in the scales of fishes, various animal and plant tissues, and in excreta.

guano. A phosphorus-rich substance found along the coasts of South America and on Pacific Islands consisting principally of the excreta of seabirds, used as a fertilizer.

guanophore. A pigment cell containing crystals of guanine, present in the integument of fishes and amphibians; iridocyte.

guard cell. One of two chlorophyll-containing cells bordering a stoma in the epidermis of a plant.

guava. A tropical tree or shrub (*Psidium guajava*) or the yellow, berrylike fruit it produces, used extensively in preserves and jellies.

guayule. A semishrubby plant (*Parthenium argentatum*) of Mexico and southwestern United States, the source of rubber.

guest. An organism which lives within the nest or den of another, as those inhabiting the nests of termites, ants, or bees. *See* inquiline.

guillemot. An auklike, open-sea bird of the family Alcidae, as *Cepphus grylle*, the black guillemot.

guinea fowl. The guinea, a gallinaceous bird (*Numida meleagris*), a native of west Africa but now domesticated. Females are called guinea hens.

guinea pig. A small, short-eared, short-tailed rodent (*Cavia porcellus*) of the family Cavidae, widely used for experimental purposes. *See* cavy.

guinea worm. A parasitic, nematode worm (*Dracunculus medinensis*) of northern Africa and Asia; the fiery serpent or serpent worm inhabiting the subcutaneous tissue of man.

gular. Of or pertaining to the upper part of the throat.

gular pouch. **1.** An inflatable extension of the larynx in chameleons. **2.** An expansible pouch attached to the mandible in fish-eating birds, as the pelican.

gulfweed. An olive-brown seaweed (*Sargassum*), a brown alga widely distributed in tropical waters and the Gulf Stream.

gull. A heavy-bodied sea bird of the family Laridae, especially those of the genus *Larus*, as *L. argentatus*, the herring gull.

gullet. A food passageway, as **a.** the esophagus in vertebrates; **b.** the cytopharynx in protozoans; **c.** the tube from mouth to gastrovascular cavity in anthozoans.

gulper. An eellike, deep-sea fish of the order Lyomeri.

¹gum. *anat.* The soft tissue covering the alveolar processes of the mandible and maxilla and surrounding the necks of the teeth; gingiva.

²gum. *bot.* An amorphous substance exuded from or extracted from plants which hardens in air or dissolves in water (a true gum) or swells up forming a viscid mass (mucilage).

gundi. A small, short-tailed African rodent of the family Ctenodactylidae.

guppy. A small, ovoviparous fish (*Lebistes reticulatis*) raised extensively in aquaria.

gustatory. Of or pertaining to the sense of taste.

gut. 1. The alimentary canal, especially the intestine. **2.** The embryonic digestive tube or cavity.

gutta. A drop or droplike marking.

guttate. Shaped like a drop; with droplike markings.

guttation. The exudation of water in liquid form by uninjured plants usually on the edges or tips of leaves from hydathodes.

gutter. A groove, as that on the ventral surface of the proboscis of an echiurid.

gymnanthous. Lacking a floral covering.

gymnoblastic. Lacking a hydrotheca or gonotheca, with reference to buds in colonial coelenterates.

gymnocarpous. Having a naked or uncovered fruiting body, as lichens with exposed apothecia or fungi with exposed hymenia.

gymnorhinal. With nostrils not covered by feathers.

gymnosperm. Any plant of the class Gymnospermae, vascular plants which produce naked seeds, as the conifers.

gymnospore. 1. *bot.* A naked spore, that is, one not produced within a sporangium. **2.** *zool.* A spore formed by gregarines in the hindgut of crustaceans.

gynander. A gynandromorph.

gynandromorph. A sex mosaic, an individual in which sex characteristics are intermixed, as in certain insects.

gynandrous. Having stamens and pistils united, as in the orchid.

gynecandrous. Having staminate and pistillate flowers on the same spike, the pistillate being located at the apex, as in sedges.

gynecogenic. Inducing the development of female characteristics; estrogenic.

gynecology. The branch of medicine which deals with diseases of women, especially those involving the reproductive organs.

gynecomastia. Excessive development of the breast in males.

gynecophoric canal. A groove on the ventral side of a male blood fluke which encloses the female during copulation.

gynobase. The expanded or enlarged basal portion of the receptacle bearing the gynoecium.

gynodioecious. With some flowers perfect, others pistillate, but on different plants.

216

gynoecium. The pistil or pistils taken collectively; the female parts of a flower.

gynoecy. A condition in which only females are known to exist in a species.

gynogametophyte. The mega- or macrogametophyte.

gynogenesis. PSEUDOGAMY.

gynomonoecious. With perfect and pistillate flowers on the same plant.

gynophore. The elongated stipe or stalk of a pistil.

gynosperm. A female-producing sperm; one containing the X-chromosome.

gynospore. The megaspore of seed plants or certain plants of the Isoetales.

gynostegium. A structure covering the gynoecium.

gynostemium. The united stamens and pistil in orchids.

gypsy moth. A European moth (*Porthetria dispar*) introduced into the United States about 1869, its larvae causing extensive damage to trees throughout the northeastern states.

¹gyrate. Moving in a circular or spiral fashion.

²gyrate. 1. *zool.* Of or pertaining to a gyrus. **2.** *bot.* Winding or coiled; curved so as to form a circle. CIRCINATE.

gyration. 1. A turning, whirling, or rotating movement. **2.** One of the whorls of a spiral shell of a gastropod.

gyroscopic organs. Balancing organs, as the halteres of dipterans.

gyrose. *bot.* Possessing or marked by wavy, undulating lines.

gyrus. A convolution or fold, as those on the cerebral hemispheres of the brain.

H

habit. 1. *bot.* A characteristic form or mode of growth. **2.** *zool.* **a.** A body build or constitution. **b.** A fixed, repetitive behavior pattern.

habitat. The natural home or dwelling place of an organism.

hacker. The Eastern chipmunk (*Tamias striatus*).

hackle. A long, slender feather on the neck of gallinaceous birds.

haddock. An important food fish (*Melanogrammus aeglefinus*) of the cod family.

hae-. For words beginning with HAE- not found here, consult HE-.

haemosporidian. A sporozoan of the order Haemosporidia which includes a number of parasitic protozoans inhabiting the blood of reptiles, birds, and mammals. Includes the causative agents of malaria and Texas fever.

hagfish. An eellike cyclostome, as *Myxine glutinosa*, the Atlantic hagfish.

hair. 1. *bot.* A slender outgrowth of the epidermis, serving glandular, protective, or absorptive functions. 2. *zool.* **a.** A horny, threadlike outgrowth of the epidermis in man and other mammals. **b.** Collectively the entire hairy coat of an animal or a part of it, as the *hair* of the head.

hair ball. A mass of hair and other undigested material which forms in the stomach of cattle, cats, and other mammals who lick themselves.

hair cell. A neuroepithelial cell whose free end bears fine hairlike processes, as those in a macula, crista, or organ of Corti.

hair follicle. A cylindrical invagination of the epidermis from which a hair develops.

hair snake. *See* horsehair worm.

hairworm. Any of a number of minute, slender nematodes, **a.** of the family Trichostrongylidae, parasitic in the intestines of various mammals; **b.** of the genus *Capillaria*, parasitic in rats and other mammals.

hake. One of several food fishes, especially those of the genera *Merluccius* and *Urophycis* of the cod family, Gadidae.

halibut. A large, marine flatfish of the order Heterosoma, an important food fish, as the Atlantic halibut *Hippoglossus hippoglossus*) and Pacific halibut (*H. stenolepis*).

haliplankton. Saltwater plankton.

hallux. The innermost digit in a pentadactyl hind limb; in birds, the toe directed backward; in man, the great toe.

halobiont. An organism which lives in an environment in which salt concentration is high.

halophile. An organism which cannot live in a salt-free environment.

halophyte. A plant which grows where salt concentration is high, as in salt marshes or on alkaline soils.

halter pl. **halteres.** One of a pair of club-shaped balancing organs on the metathorax of dipterous insects.

¹hamate. 1. Shaped like a hook. **2.** Bearing a hooklike process.

²hamate. One of the carpal bones in man.

hammer. The malleus, an ear bone.

hammerhead. 1. A shark of the genus *Sphyrna* with lateral processes extending from the sides of the head. **2.** A heronlike African bird of the genus *Scopus*, family Scopidae.

hamster. An Old World rodent of the genus *Cricetus* and related genera. The golden hamster (*Mesocricetus auratus*) is widely used in biological research.

hamstring. One of the tendons on the posterior side of the knee.

hamulate. Pertaining to or involving hooks or hooklike structures.

hamulus pl. **hamuli.** A hook or hooklike process.

hand. 1. In man, the terminal portion of the forearm consisting of the wrist, metacarpus, and digits; the manus. **2.** In tetrapods, the corresponding portion of the forelimb. **3.** The terminal portion of any appendage when modified and functioning as a prehensile organ.

hangul. An Asian deer (*Cervus cashmiriensis*).

hanuman. The entellus monkey (*Presbytis entellus*), sacred to the Hindus.

haplodont. Pertaining to or possessing molar teeth with simple crowns.

haploid. Possessing half the diploid or somatic number of chromosomes; having one of each of the pairs of chromosomes characteristic for a species.

haplontic. Designating a type of life cycle in which adults are haploid, meiosis occurring in the zygote following fertilization, as in blue-green algae.

hapteron pl. **haptera.** A rootlike attaching structure forming a part of the holdfast of certain brown algae.

haptonema. A coiled attaching structure in certain golden algae.

haptor. An organ of attachment in a monogenetic fluke.

hardening. Increasing the ability of plants to withstand freezing by gradually subjecting them to lower temperatures.

harderian gland. A sebaceous gland in amphibians, reptiles, and birds which secretes an oily fluid which lubricates the nictitating membrane.

hard seeds. Seeds impervious to water, as alfalfa, sweet clover.

hardwood. 1. Any dense, heavy, close-grained wood. **2.** *forestry.* Wood produced by broad-leafed, deciduous trees.

Hardy-Weinberg law. In a large population in which random mating occurs, with respect to a particular pair of alleles, the frequency of the genes or alleles remains the same providing there is no mutation, selection, or differential mating. This implies genetic stability, gene frequenceis remaining constant.

hare. Any of a number of small to medium-sized, rodentlike mammals of the order Lagomorpha, family Leporidae, especially those of the genus *Lepas.* Term is commonly applied to the larger members of this family in contrast to the smaller members, called rabbits.

harelip. A congenital condition in which there is a cleft or clefts in the upper lip.

harem. A group of female animals which mate with and are looked after by a single male.

¹harrier. A hawk of the genus *Circus,* especially *C. cyaneus,* the marsh hawk.

²harrier. A hunting dog resembling the fox hound.

hart. The male of the red deer, especially after the fourth year.

harvestman. An arachnid with long, slender legs of the order Phalangida, commonly called daddy longlegs.

harvest mite. *See* chigger.

hastate. Shaped like an arrowhead but with basal lobes turned outward, said of leaves.

¹hatch. To bring forth young from an egg or eggs.

²hatch. That which is hatched.

haunch. The loin and hips taken together.

haustellum. A tubular structure in insects adapted for sucking blood or plant juices.

haustoria sing. **haustorium.** Rootlike structures of parasitic plants which penetrate the cells of the host plant.

haversian system. A unit of structure of compact bone consisting of concentric layers or lamella of bone surrounding a central canal; an osteon.

¹haw. The hawthorn or red haw (*Crataegus*) of the rose family, or the fruit it produces.

²haw. The nictitating membrane of various mammals, especially the horse.

hawk. Any of a number of diurnal birds of prey of the order Falconiformes, comprising the true hawks (accipiters), buzzard hawks (buteos), marsh hawks (harriers), fish hawks (ospreys), and falcons.

hawk moth. Any of a number of stout-bodied moths of the family Sphingidae which hover over flowers sucking nectar through a long proboscis, as the tobacco hornworm moth (*Protoparce sexta*).

hazelnut. The nut or fruit of the hazel, a shrub or small tree of the genus *Corylus;* a filbert.

head. 1. In vertebrates, the most anterior or uppermost portion of the body, containing the brain, sense organs, and openings to the digestive and respiratory tracts. **2.** In invertebrates, the corresponding portion of the body. **3.** The rounded end of a structure, especially when set off by a constricted region, as the *head* of a bone. **4.** *bot.* **a.** A compact mass of foliage leaves, as a *head* of cabbage. **b.** A compact mass of flowers or flower buds, as in cauliflower. **c.** A dense cluster of sessile flowers on a short axis, as the sunflower or clover flower. **5.** A capitulum.

head kidney. The pronephros when it persists in an adult, as in the hagfish.

head louse. A sucking louse (*Pediculus humanus capitis*).

hearing. Audition; the act of perceiving sound; the power of interpreting sensory impulses from auditory receptors.

heart. *zool.* **1.** A hollow, muscular organ or structure which, by rhythmical contractions propels the blood throughout the body. **2.** A contractile structure which propels fluid, as a branchial heart or lymph heart.

heart urchin. An irregular echinoderm of the class Echinoidea, as *Echinocardium.*

heartwood. The darker, denser wood in the central portion of the stem consisting of dead elements which have lost their conducting ability: DURAMEN.

heartworm. A large nematode worm (*Dirofilaria immitis*) found in the heart of carnivores. Females reach a length of 30 cm.

heat. ESTRUS.

heath. 1. An open, level area, especially in Great Britain, consisting of wasteland covered with a characteristic vegetation of plants of the heath family, Ericaceae. **2.** Any plant of the genus *Erica*, a low, evergreen shrub.

heath hen. A subspecies of grouse (*Tympanuchus cupido*) of northeastern United States, now extinct, the last individual dying on Martha's Vineyard in 1932.

hectocotylus. A modified arm of a squid or octopus which functions in the transfer of spermatophores to the female. The detached arm was originally thought to be a parasitic worm and described under the name *Hectocotylus*.

hedgehog. 1. An Old World, insectivorous mammal (*Erinaceus europaeus*) capable of rolling itself into a ball with its spines projecting outwardly. **2.** In America, the porcupine.

hedonic glands. Glands in amphibians and reptiles which secrete odoriferous substances which function as sexual attractants and excitants.

heel. 1. *anat.* The hind part of the foot in man or the corresponding part in other animals. **2.** *bot.* The base of a tuber, cutting, or some other part of a plant which has been removed for propagation.

heel bone. The calcaneus.

helad. A marsh plant.

heleoplankton. Pond plankton.

heliad. A heliophyte.

helical. In the form of a helix; spiral.

helicoid. Spirally coiled; in the form of a flattened coil.

heliophyte. A sun-loving plant.

heliotropic. Responding to sunlight as a stimulus.

heliozoan. A protozoan of the order Heliozoida with numerous, stiff, slender pseudopodia radiating from a central mass, as the sun animalcule.

helix. 1. A spiral structure. **2.** The outer rim of the auricle of the ear.

hellbender. A large salamander (*Cryptobranchus alleganiensis*) inhabiting streams of eastern United States.

hellgrammite. The large, aquatic larva of the dobsonfly (*Corydalis cornutus*).

helmet shell. A large, thick-shelled gastropód of the family Cassididae inhabiting tropical waters. Their shells are used in making cameos.

helminth. A worm or wormlike animal.

helminthology. The study of worms, especially parasitic worms.

hemal. 1. Of or pertaining to blood or the cardiovascular system. 2. Of or pertaining to the portion of the body lying ventral to the spinal axis, that is, the region containing the heart and major blood vessels.

hemal arch. An arch on the ventral side of the caudal vertebrae of fishes formed by chevron bones. Its tip forms a hemal spine.

hemal canal. A canal formed by successive hemal arches containing caudal blood vessels.

hemal system. An ill-defined circulatory system of echinoderms.

hematin. HEME.

hematochrome. A carotenoid astaxanthin present in red granules of euglenoids.

hematophagous. Feeding on blood, as a bloodsucking insect.

hematopoiesis. HEMOPOIESIS.

hematoxylin. A dye obtained from logwood which is widely used as a biological stain.

heme. An iron porphyrin compound which with a protein, globin, comprises hemoglobin.

hemelytron. The anterior wing of a hemipterous insect, the basal portion of which is thickened, the distal portion membranous.

hemeranthous. Day-flowering.

hemerythrin. An iron-containing respiratory pigment found in annelids, sipunculids, and other invertebrates; hemoerythrin.

hemibranch. 1. A half-gill. 2. A series of gill filaments or lamellae on one side of an interbranchial septum.

hemicarp. A half-carpel.

hemicellulose. One of a group of polysaccharides resembling cellulose but less complex, more soluble, and more readily hydrolyzed to simple sugars.

hemichordate. Any of a number of small, wormlike, marine

223

animals of the phylum Enteropneusta, which includes the acorn worms, thought to be related to the chordates.

hemimetabolous. Undergoing incomplete metamorphosis, with reference to insects.

hemipenis pl. **hemipenes.** One of two grooved copulatory structures present in the males of snakes and lizards.

hemipode. A quaillike bird of the families Turnicidae and Pedionomidae.

hemipteran. An insect of the order Hemiptera which includes the true bugs, characterized by sucking mouth parts and incomplete metamorphosis, as the squash bug, chinch bug.

hemixis. A process involving fragmentation and reorganization of the macronucleus, as occurs in certain ciliates.

hemizygoid. HAPLOID.

hemlock. 1. A poisonous herb (*Conium maculata*) of the family Umbelliferae, or the drink made from its fruit. 2. An evergreen, coniferous tree of the genus *Tsuga*.

hemocoel. A body cavity through which blood circulates, as in arthropods.

hemocyanin. A respiratory pigment containing copper instead of iron, present in the blood of various mollusks and arthropods.

hemocyte. 1. An ameboid, hemoglobin-containing coelomocyte of echinoderms. 2. Any blood cell.

hemocytozoon. A protozoan which lives as a parasite within red blood cells.

hemoflagellate. A flagellate protozoan which lives within the bloodstream.

hemoglobin. An iron-containing, respiratory pigment present in red blood cells of vertebrates and in the plasma of certain invertebrates. Abbreviated Hb.

hemolymph. The circulatory fluid of animals with an open circulatory system.

hemolysin. A substance which causes hemolysis.

hemolysis. The liberation of hemoglobin from red blood cells resulting from their destruction or by its diffusion through an altered cell membrane.

hemophilia. An hereditary, sex-linked disease occurring principally in males in which clotting time is excessively long and prolonged bleeding results.

hemopoiesis. The formation of blood cells.

hemorrhage. The escape of blood from a blood vessel or cavity.

hemostat. An agent or instrument which stops the flow of blood.

hemp. A stout, bushy plant (*Cannabis sativa*) or its fibers, extensively used for ropes, twine, and other products; also the source of cannabis or marijuana.

hen. The female of most birds, especially gallinaceous birds.

henequen. Yucatan sisal, a tough fiber used especially in binder twine, or the plant that produces it (*Agave fourcroydes*).

hen-feathered. Having plumage like that of a hen, said of castrated males.

heparin. An anticoagulant present in the liver.

[1]hepatic. Of or pertaining to the liver.

[2]hepatic. A liverwort.

hepatic portal vein. A large vein which drains the digestive organs and spleen and terminates in the liver.

hepatopancreas. A digestive organ which functions as a liver and pancreas, present in certain crustaceans and xiphosurans.

herb. **1.** A flowering plant in which persistent woody tissue does not develop. **2.** A plant used for medicinal purposes or for its scent or flavor.

herbaceous. Herblike; resembling or having the characteristics of an herb.

herbage. **1.** Vegetation consisting principally of herbs, especially that utilized in grazing. **2.** The succulent parts of herbaceous plants.

herbarium. **1.** A collection of dried plants systemmatically arranged and labeled. **2.** A building or room which houses such a collection.

herbicide. An agent which kills or inhibits the growth of herbs, especially weeds, as 2-4-D.

herbivore. A herbivorous animal.

herbivorous. Feeding or subsisting principally or entirely on plants or plant products: PHYTOPHAGOUS.

herd. A group of large animals which associate together, as cattle, sheep.

hereditary. Capable of being transmitted genetically from parents to offspring.

heredity. **1.** The capacity of an individual to develop traits present in parents or ancestors. **2.** All the characteristics, morphological or physiological, which are dependent upon genetic factors received from parents.

heritable. Capable of being inherited.

heritage. That which is inherited.

hermaphrodite. 1. *zool.* An individual which possesses both male and female sex organs or a combined ovotestis; a monoecious animal. **2.** *bot.* A flower which possesses both stamens and pistil; a monoclinous flower.

hermatypic. Possessing symbiotic, unicellular algae within their tissues, said of certain corals.

hermit crab. One of a number of decapod crustaceans of the genera *Pagurus, Pylopagurus,* and others which live in the empty shells of gastropods and other animals.

hernia. The abnormal protrusion of an organ or a part of it, through the wall of the cavity in which it normally lies, as inguinal *hernia.*

heron. A long-legged, long-necked, wading bird of the family Ardeidae, as the great blue heron (*Ardea herodias*).

herpetology. The study of amphibians and reptiles.

herring. One of a number of marine fishes of the family Clupeidae, as **a.** the common herring (*Clupea harengus*) of the North Atlantic, an important food fish occurring in schools numbering millions; **b.** the California herring (*C. pallasii*), a closely related species.

hesperidin. A bioflavonoid present in the rind of citrus fruits.

hesperidium. A syncarpous, polycarpellary berry with a separable rind, characteristic of citrus fruits, as the orange.

Hessian fly. A small dipterous insect (*Phytophaga destructor*), a serious pest of wheat and other cereals.

heterandrous. Having two sets of stamens, with reference to flowers with stamens of unequal size and length.

heterauxesis. ALLOMETRY.

heteroauxin. *See* auxin.

heterocarpous. Producing more than one kind of fruit.

heterocercal tail. One in which the vertebral column extends into the prominent upper lobe, as in elasmobranchs.

heterochromatin. The chromatin in a chromosome which stains

226

more densely than the euchromatin in adjacent regions. It is relatively devoid of genes.

heterochromous. Of unlike colors.

heterochrony. *embryol.* The unequal rate of development of parts.

heterochthonous. Foreign; not native or indigenous.

heterocoelous. With ends of the vertebrae saddle-shaped, as in the cervical vertebrae of birds.

heterocyst. A large, empty cell in the filaments of certain algae, as *Nostoc.*

heterodactyl. *ornith.* With toes I and IV directed backwards, as in trogons.

heterodont. 1. Having teeth which differ in shape and function. **2.** In bivalves, having both lateral and cardinal teeth fit into depressions in the opposite valve.

heteroecious. Having two unlike hosts, said of parasites, as a fluke.

heterogamete. Either of a pair of gametes which differ in size and structure, as an ovum and spermatozoon.

heterogamous. 1. Producing unlike gametes. **2.** Producing unlike flowers.

heterogeneous. Unlike, dissimilar, not uniform.

heterogenesis. ALTERNATION OF GENERATIONS.

heterogonous. Having two or more types of flowers.

heterogony. Alternation of a parthenogenetic and a zygogenetic generation, as in rotifers and aphids.

heterograft. A graft or transplant from one species to another; a xenograft.

heterogynism. Having taxonomic characters more pronounced in the female than in the male.

heterogynous. Having two or more kinds of females, as the fertile queens and infertile workers of honeybees.

heterokaryotic. Designating a hypha or mycelium which contains two or more genetically different nuclei, as may result from fusion of two hyphae.

heterokont. Having flagella of unequal length.

heteromerous. *bot.* Having one of a series of parts differing from the others, as a floral whorl with an unusual number.

heteromorphic. 1. Deviating from normal. **2.** Exhibiting a diversity of form during development, as flukes, insects. **3.** Exhibiting a dissimilarity, as individuals in a polymorphic colony.

heteromorphosis. 1. The development of an organ or structure in an unusual position. **2.** In regeneration, the development of a part that is different from the part injured or lost.

heteronomous. Having unlike segments.

heterophyllous. Having more than one type of foliage leaf borne on the same plant.

heterophyte. A plant which secures its food directly or indirectly from another organism or its products; a saprophyte or parasite.

heteroplastic. Pertaining to grafts in which the donor is of a different species than the host.

heteroploidy. A condition in which the chromosome number is not a multiple of the haploid number, brought about by a loss or gain in number of chromosomes.

heterosis. Hybrid vigor, as increased size, fecundity, or resistance to disease, resulting from a cross between individuals of different species, races, or varieties.

heterosporous. Producing more than one kind of spore.

heterostyly. Possessing styles differing in shape or length.

heterothallic. A condition in certain fungi and algae in which **a.** male and female gametes are not produced on the same filament or plant; **b.** morphologically similar sexual strains exist but conjugation occurs only between two inherently different sexual strains, usually designated plus (+) and minus (−), as in the Mucorales.

heterotopic. Occurring or developing in an abnormal position; displaced. *See* baculum.

heterotrichous. 1. *bot.* Having a body divided into a prostrate portion and an erect or upright portion, as in certain liverworts. **2.** *zool.* Having a body covered with cilia of uniform length but with longer cilia about the peristome, as in *Stentor*.

heterotroph. An organism which is incapable of manufacturing its own food; one which depends upon other organisms as sources of food, as all animals, saprophytes, and parasites.

heterotrophic. Designating a type of nutrition in which an organism requires preformed carbohydrates or other organic compounds as sources of energy.

heteroxenous. HETEROECIOUS.

heterozooid. One of a number of specialized zooids in ectoprocts.

heterozygote. 1. An organism in which one or more pairs of genes are unlike; a hybrid. **2.** An individual which does not breed true for a specific trait.

heterozygous. Having one or more dissimilar pairs of genes.

hexacanth. A six-hooked embryo of a tapeworm. ONCHOSPHERE.

hexapod. 1. A six-footed animal, as an insect. **2.** Any member of the class Insecta (Hexapoda).

hexose. A monosaccharide which contains six carbon atoms in its molecule, as glucose, fructose, galactose.

hiatus. A gap or opening.

hibernaculum. 1. *bot.* The wintering stage of a plant or a part of it, as a winter bud. **2.** *zool.* **a.** A case or covering which protects an organism from extreme cold. **b.** The winter quarters of a hibernating animal.

hibernate. To pass the winter in a dormant or lethargic state.

hibernating gland. A mass of brown fat present in the neck and shoulder region of certain rodents.

hickory. One of a number of trees of the genus *Carya*, family Juglandaceae, containing several species which produce valuable timber and nuts.

hiemal. Of or pertaining to winter.

hilum. 1. *bot.* **a.** A mark or scar on a seed marking the point of attachment to the funiculus. **b.** The denser region of a starch grain surrounded by concentric layers. **2.** *anat.* A hilus.

hilus. An indented region of an organ, as the kidney or spleen, where blood vessels and nerves enter and make their exit; a hilum.

¹hind. 1. The adult female of the red deer. **2.** A marine fish related to the groupers, as the red hind (*Epinephelus guttatus*), a food fish of the West Indies.

²hind. Of or pertaining to the posterior portion or the region that is behind.

hindbrain. The rhombencephalon, the posterior of the three primary vesicles of the embryonic, vertebrate brain. It gives rise to the cerebellum, pons, and medulla oblongata.

hindgut. The caudal portion of the embryonic digestive cavity; the portion extending into the tail fold.

hind limbs. The posterior pair of appendages of a tetrapod.

hind wings. The second or posterior pair of wings of an insect; the metathoracic wings.

hinge. The articulation or joint between the two valves of a clam or brachiopod.

hinge joint. A ginglymus, a joint which works like a hinge permitting movement in one plane only, as the elbow joint.

hinny. A hybrid resulting from mating a stallion to a female ass.

¹hip. *bot.* The fruit of a rose consisting of the fleshy floral tube surrounding the matured ovaries.

²hip. The upper portion of the thigh at its junction with the body; the lateral portion of the body at the region of the hip joint.

hipbone. The os coxae or innominate bone consisting of three fused bones, the pubis, ischium, and ilium, forming the lateral half of the pelvic girdle. In some mammals, it includes the cotyloid or acetabular bone.

hip joint. The articulation between the head of the femur and the hip bone, a ball and socket joint.

hippocampus. A portion of the rhinencephalon which projects into the floor of the lateral ventricle.

hippopotamus. The river horse, a large, herbivorous mammal (*Hippopotamus amphibius*) inhabiting rivers and lakes of Africa.

hirsute. Hairy, shaggy; covered with hair, bristles, or hairlike structures.

hirudin. An anticoagulant present in the secretion of the buccal glands of a leech.

hirudinean. A leech; of or pertaining to leeches.

hispid. Rough due to the presence of stiff hairs, bristles, or minute spines.

histamine. An amine, $C_5H_9N_3$, present in various plant and animal tissues, resulting from the decomposition of histidine, an amino acid. It is associated with various allergic reactions as it stimulates smooth muscle, dilates capillaries, and increases the secretion by various glands.

histiocyte. A macrophage of loose connective tissue.

histogen. Meristem tissue in plants from which definitive plant tissues develop.

histogenesis. The development of tissues from undifferentiated cells.

histology. Microscopic anatomy; the science which deals with tissues and their organization into organs.

histolysis. The disintegration or dissolution of tissues.

histone. One of a group of simple proteins which, upon hydrolysis, yield only amino acids, as globin.

hoary. 1. White or gray with age. 2. *bot.* Canescent; white or grayish-white from a white pubescence.

hoatzin. A pheasantlike bird (*Opisthocomus hoazin*) of the Amazon valley, the young of which have claws on the first and second digits of the wings.

hog. A pig, sow, or boar. *See* swine.

hognose snake. A moderate-sized, nonpoisonous snake with an upturned snout of the genus *Heterodon* which feigns death when disturbed. Also called puff adder or spreading viper.

holandric gene. One in the nonhomologous portion of the Y-chromosome.

holdfast. An organ of attachment or an anchoring structure, as that of various algae, protozoans, and parasitic worms.

holly. A tree or shrub of the family Aquifoliaceae with persistent red berries and thick, spiny-toothed leaves widely used for Christmas decorations, as the American holly (*Ilex opaca*).

holoblastic. Designating a type of cleavage which is total, resulting in blastomeres of approximately equal size, as occurs in isolecithal ova.

holobranch. A complete gill consisting of two hemibranchs and the interbranchial septum.

holocarpic. Condition in which the entire thallus is converted into one or more reproductive cells.

holocrine. Designating a type of gland in which the secretory product consists of the substance of dead and disintegrating cells, as a sebaceous gland.

holoenzyme. An entire or complete enzyme consisting of a protein portion or apoenzyme and the coenzyme.

hologynic. Entirely female, with reference to a mode of inheritance in which a trait is transmitted through successive generations through females only.

holometabolous. Designating insects which undergo complete metamorphosis.

holonephridium. A type of nephridium, as that characteristic of most oligochaete worms.

holophytic. Having a mode of nutrition like that of a green plant: AUTOTROPHIC.

holopneustic. Pertaining to the type of respiratory system in insects in which all spiracles are functional.

holoptic. Having the two compound eyes united or in contact as in certain flies.

holosericeous. Covered with fine, silky hairs.

holothurian. An echinoderm of the class Holothuroidea which includes the sea cucumbers.

holotrich. A protozoan of the class Holotricha with cilia of uniform length over entire body, as *Paramecium*.

holotype. A single specimen, the type specimen, upon which the description of a new species is based.

holozoic. Animallike with reference to nutrition, that is, ingesting complex organic compounds which require digestion before utilization.

homeostasis. A steady state or state of constancy of the internal environment of an organism.

homeothermic. Having a constant body temperature: WARM-BLOODED; HOMOIOTHERMOUS.

homing. Home-returning.

homing pigeon. A pigeon trained to return home, formerly used in the transmission of messages over long distances.

hominid. A member of the primate family Hominidae which includes the genus *Homo* to which modern man belongs.

Homo. The genus of the order Primates which includes all races of modern man of the species *Homo sapiens* and recently extinct species.

homocercal tail. A tail with two symmetrical or nearly symmetrical parts, as characterizes most present day teleost fishes.

homodont. Having teeth which are alike in shape and form.

homogametic. Producing only one type of gamete, with reference to the sex chromosomes, as XX females.

homogamous. Characterized by homogamy.

homogamy. **1.** Condition in which stamens and pistils of a flower mature at the same time. **2.** Having flowers of the same type.

homogeneous. 1. Composed of parts or consisting of individuals of the same kind. **2.** Of uniform composition.

homogenous. Consisting of homologous parts.

homograft. A graft between individuals of the same species.

homoiomerous. Unstratified or scattered throughout the thallus, with reference to algal cells in a lichen thallus.

homoiothermous. Having a more or less constant body temperature; warm-blooded, as birds and mammals: HOMEOTHERMIC. *Compare* poikilothermous.

homolecithal. ISOLECITHAL.

homologous. Similar in fundamental structure, position, and development but not necessarily in function; similarity of structures due to common ancestry.

homologous chromosomes. A pair of chromosomes which have identical genes or their alleles located at corresponding loci.

homology. State or condition of being homologous.

homomorphic. 1. Corresponding in external form or appearance but not in fundamental structure. **2.** Having perfect flowers of one type only.

homonomous. Having uniform or similar segments, as in the earthworm.

homonym. *taxonomy.* One of two or more identical but independently proposed names for the same or different taxa.

homoplastic. From another individual of the same species, with reference to grafts.

homopteran. A sucking insect of the order Homoptera, which contains the cicadas, leafhoppers, aphids, and scale insects.

homosporous. Having spores of one kind only; isosporous.

homostyly. A condition in which flowers have styles of equal length.

homothallic. Condition in which male and female gametes are produced on the same filament or plant, as in certain algae and fungi.

homothetogenic. Designating a type of fission in ciliate protozoans in which the fission plane cuts across the longitudinal rows of cilia or basal granules.

homotropous. Curved in one direction with reference to the embryo in an anatropous seed.

homozygote. A homozygous individual; an organism that breeds true for a specific trait.

homozygous. A condition in which two members of a pair or a series of pairs of genes are alike, as AA or aa, consequently all eggs or sperm produced by such an organism are genetically alike with respect to these particular genes.

honey. A thick, viscid substance prepared from nectar in the honey sac of various bees, especially the honeybee. It is stored in cells in the hive and used for food.

honey ant. A genus of ants (*Myrmecocystus*) of arid regions which has a special caste known as repletes which store honey brought to them by workers.

honeybee. One of a number of honey-producing hymenopterous insects of the genus *Apis*, especially *A. mellifera*, a social insect living in colonies.

honey buzzard. A soaring hawk (*Pernis apivorus*).

honeycomb. 1. A structure built of wax in the nest or hive of bees consisting of compartments in which pollen and honey are stored and the young develop. **2.** The reticulum or second portion of the stomach of a ruminant.

honeydew. 1. A substance passed from the anus of aphids which is relished by ants and other insects. **2.** A sticky exudate which appears on plants infected with ergot, wheat rust, or other fungi.

honey eater. A nectar-feeding, oscine bird of the family Meliphagidae, as *zanthomiza phrygia*, the regent honey eater of Australia.

honey guide. A parasitic bird of the family Indicatoridae which directs man and other animals to stores of wild honey upon which it feeds, as *Indicator indicator*, the greater honey guide of Southeast Asia and Africa.

honeypot. A replete. *See* honey ant.

honey stomach. In bees, the crop, a dilated portion of the esophagus in which nectar is transformed into honey; also called honey sac.

hood. 1. *zool.* **a.** A flexible covering for the head and neck or a structure resembling such, as that present in arrowworms, lancelets, and other invertebrates. **b.** The expanded neck of a cobra. **c.** The crest of a bird. **2.** *bot.* A hoodlike petal or sepal.

hooded. 1. Possessing a hood or hoodlike structure: CUCULLATE. **2.** *ornith.* Having head plumage different from body plumage.

hoof pl. **hooves** or **hoofs. 1.** A curved, horny structure which covers the end of the digit in various ungulates. **2.** The entire foot of the horse and related species.

hoofed mammals. Odd-toed mammals of the order Perissodactyla (horse, zebra, tapir, rhinoceros) and even-toed mammals of the order Artiodactyla (cattle, sheep, goats, deer); ungulates.

hooklet. A small hook, especially the barbicel of a feather.

hookworm. A small, parasitic nematode worm inhabiting the intestine of man and other mammals. Human species are *Ancylostoma duodenale* and *Necator americanus.*

hoolock. A gibbon (*Hylobates hoolock*).

hop. A twining, perennial vine (*Humulus lupulus*) of the hemp family.

hopper. Any of a number of leaping insects, especially those of the order Homoptera (leafhoppers, planthoppers, treehoppers) and Orthoptera (grasshoppers).

hops. The ripe, dried, pistillate catkins of the hop plant used in the brewing of malt liquors.

hormogonium pl. **hormogonia.** A multicellular section of a fragment following fragmentation, as in certain algae.

hormone. 1. *zool.* A substance secreted by an endocrine gland or tissue which is transported by the bloodstream to other parts of the body where it evokes a response, usually in a specific organ or tissue designated target organ. *See* endocrine gland, parahormone, pheromone, neurohumor. **2.** *bot.* A substance produced in one part of a plant which, in minute quantities, is capable of producing effects in other parts of the plant. *See* auxin, kinin, phytohormone.

horn. 1. *zool.* **a.** A stiff, pointed process borne on the head of various vertebrates, used principally for offense and defense. Types include the hollow horns of ungulates, the pronghorn of the antelope, the antlers of deer, and the keratin-fiber horn of the rhinoceros. **b.** Any projection on the head which resembles a horn. **c.** The tough substance composed of keratin which covers, or forms the horns of animals, hooves, claws, or nails. **2.** *bot.* An appendage resembling the horn of an animal.

hornbill. A nonpasserine, tropical bird of the family Bucerotidae characterized by possession of an enormous bill.

horned. Having a horn, horns, or hornlike processes.

horned toad. *Phrynosoma cornutum*, an insectivorous, iguanid lizard of western arid regions.

hornet. A social, hymenopterous insect of the family Vespidae which constructs a large, exposed, paper-covered nest, as *Vespula maculata*, the common bald-faced hornet, which stings viciously.

horn fly. A dipterous insect (*Siphona irritans*), a serious pest to cattle.

horntail. A large, hymenopterous insect of the family Siricidae, as *Tremex colomba*, the pigeon tremex. Its larvae are wood-boring, infesting shade trees.

hornworm. The larva of various sphinx moths which bears a spinelike process on the eighth abdominal segment.

hornwort. Any bryophyte of the class Anthocerotae, so named from their long, hornlike capsule.

horse. **1.** A large, solid-hoofed, herbivorous mammal (*Equus caballus*) of the family Equidae, domesticated since ancient times. **2.** A male horse or stallion. **3.** A castrated male or gelding as distinguished from a stallion or mare.

horsefly. A large, bloodsucking, dipterous insect of the genus *Tabanus*, family Tabanidae, a serious pest to horses. Also called gadfly.

horsehair worm. A long slender worm of the phylum (class) Nematomorpha, as *Gordius*. Adults are free-living; larvae are parasitic in arthropods. Also called hairworm or hair snake.

horseshoe crab. *See* king crab.

horsetail. A primitive vascular plant (*Equisetum*) of the order Equisetales related to the ferns; a scouring rush.

hospitator. A plant which houses ants.

host. **1.** An organism upon or in which another organism lives and for which it provides lodgment or nourishment or both. *See* symbiosis. **2.** The recipient of a graft in tissue or organ transplantation. **3.** In conjoined twins, the larger of the two.

housefly. A dipterous insect (*Musca domestica*) of the family Muscidae, a common household pest and carrier of disease organisms.

hover fly. A syrphid or flower fly, one of a number of dipterous insects of the family Syrphidae which hover about flowers feeding upon pollen.

howling monkey. A monkey of Central and South America of the genus *Alouatta*, noted for its remarkable howling sounds, also called howler.

huia. *See* wattlebird.

human. Of, pertaining to, or related to man (*Homo sapiens*).

humble-bee. The bumblebee.

humeral. 1. Of or pertaining to the humerus or region of the shoulder. 2. *entomol.* Pertaining to the anterior basal portion of the wing or the lateral angle of the pronotum.

humerus. In man, the brachium, the bone of the upper arm between shoulder and elbow; in other vertebrates, the corresponding bone of the forelimb.

humifuse. Spread over the ground.

hummingbird. One of a number of small, nonpasserine birds of the family Trochilidae, noted for their small size, rapid flight, long, needlelike bill, extensible tongue, and brilliant iridescent plumage, as *Archilochus colubris*, the ruby-throated hummingbird.

hummingbird moth. A hawkmoth.

humor. Any fluid or semifluid substance of the body.

humoral. Of or pertaining to a humor.

humoral agent. A substance transmitted by the blood or lymph, as a hormone.

humpback. 1. A humpbacked person. *See* kyphosis. 2. A whale of the genus *Megaptera*, a rorqual whale.

humus. Dark material in the soil consisting principally of organic matter.

husk. 1. The dry, outer covering of various seeds and fruits. 2. The bracts enclosing the ear of Indian corn or maize.

hutia. A rodent (*Geocapromys*) of the West Indies, related to the porcupine.

hutia-conga. A nocturnal, arboreal rodent (*Capromys pilorides*) of Cuba.

hyaline. Glassy; transparent or translucent, like glass.

hyaline cartilage. A type of cartilage with a clear, homogeneous matrix, found in the septum of the nose, the trachea, and covering the articular surfaces of bones.

hyaline cell. One of the numerous, large, empty cells in sphagnum moss which are responsible for its water-absorbing properties.

hyaloplasm. The clear, ground substance of the protoplasm of a cell in which organelles and inclusions are suspended.

hyaluronic acid. A viscous mucopolysaccharide present in the intercellular material of connective tissue which acts as a binding agent. Also present in the capsules of certain bacteria.

hyaluronidase. An enzyme produced by certain bacteria; also present in snake venom and semen. It depolymerizes hyaluronic acid, reducing its protective action in resisting the spread of an invading organism. Also called spreading factor, diffusing factor, invasin.

hybrid. 1. An individual resulting from the union of gametes differing in one or more genes; a heterozygous individual. **2.** The offspring of a cross between two different species, races, or varieties; a crossbred animal or plant.

hybridization. The act or process of producing a hybrid; cross-breeding; cross-pollination; interbreeding.

hybrid vigor. HETEROSIS.

hydathode. A minute pore or specialized structure through which water is extruded from a leaf in guttation.

hydatid. A fluid-filled cyst containing brood capsules and scolices resulting from development of the larva of a dog tapeworm (*Echinococcus granulosus*). *See* hydatid worm.

hydatid worm. *Echinococcus granulosus*, a minute tapeworm of four segments which infests dogs and cats and other carnivores, its larva developing in man and various domestic and wild animals, forming hydatid cysts.

hydra. Any of a number of sessile, freshwater hydrozoans of the genera *Hydra* and *Chlorohydra*.

hydranth. 1. A hydralike feeding polyp of a colonial coelenterate. **2.** The elongated, vase-shaped oral end of a hydrozoan polyp bearing the terminal mouth and surrounded by tentacles.

hydrarch succession. Pond succession or the successive changes which occur in plant and animal communities in the conversion of a pond or lake to a dry land community.

hydration. The act of incorporating or combining with water to form a hydrate.

hydratuba. A small polyp which develops from the planula in the life cycle of a scyphozoan jellyfish.

hydric. 1. Of, characterized by, or requiring considerable moisture. **2.** Of, pertaining to, or containing hydrogen.

hydrion. The hydrogen ion.

hydrobiology. The study of aquatic organisms.

hydrocarpic. Pollinated above, but developing below the surface of water, with reference to flowers of aquatic plants.

hydrocaulus. The main stalk of a colonial hydrozoan.

hydrocladium. A branch of a hydrocaulus.

hydroecium. A sheathlike extension of the bell in certain siphonophores.

hydrofuge hair-pile. In aquatic insects, an accumulation of dense, fine hairs which trap air to be utilized in respiration following submergence.

hydrogen. A chemical element, symbol H, a colorless, inflammable gas, the lightest element known; at. wt. 1.008. Three isotopes are: protium (H^1), deuterium (H^2), and tritium (H^3).

hydrogen ion concentration. *See* pH, reaction.

hydroid. 1. A coelenterate which has a polypoid form resembling that of a hydra. **2.** A polyp. **3.** Any member of the class Hydrozoa.

hydrolase. An enzyme which catalyzes a hydrolytic reaction.

hydrolysis. 1. A reaction in which water is a reacting agent. **2.** A decomposition reaction in which a compound on reacting with water is split into two parts, one reacting with the H of water, the other with the OH, as in the digestion of foods.

hydrolytic. Of, pertaining to, or inducing hydrolysis.

hydromedusa. A free-swimming, craspedote medusa of a hydrozoan coelenterate.

hydrophilic. Having a strong affinity for water; capable of combining with, attracting, or holding water.

hydrophilous. 1. Preferring moist places or water as a habitat. **2.** Pollinated by the agency of water, said of flowers.

239

hydrophyte. A plant which lives in water or a moist environment.

hydroponics. The growth of plants in aqueous solutions containing essential inorganic substances; soilless growth.

hydrorhiza. The anastomosing, rootlike structure by which a colonial hydrozoan is attached to the substratum.

hydrosere. A pond sere; a succession of plants originating in a water habitat and terminating in terrestrial vegetation, characteristic stages being submerged vegetation, floating vegetation, emergent or marsh vegetation, swamp shrubs, and finally swamp forest.

hydrosphere. The aqueous envelope of the earth comprising all water (oceans, lakes, streams, subterranean water, and vapor of the atmosphere).

hydrostatic organ. An organ or structure which functions in the maintenance of equilibrium between the body of an aquatic organism and the surrounding water, as the swim bladder of a fish.

hydrotaxis. A movement of an animal in response to moisture as a stimulus.

hydrotheca. A transparent, vase-shaped structure enclosing a nutrient polyp of a colonial coelenterate.

hydrotropism. A growth movement of a plant in response to water or moisture as a stimulus.

hydroxide. A compound formed by the union of a metal or a radical with one or more hydroxyl (OH) groups, as sodium hydroxide, NaOH.

hydroxyl radical. The univalent radical, OH^-, characteristic of hydroxides.

hydrozoan. Any coelenterate of the class Hydrozoa which includes solitary polyps as *Hydra*, colonial forms as *Obelia* and the Portugese man-of-war, and a few medusae as *Gonionemus*.

hyena. A large, nocturnal, carrion-feeding carnivore of the family Hyaenidae as *Hyaena hyaena*, the striped hyena of Africa and southern Asia.

hygroscopic. Readily absorbing or taking up moisture from the air.

hygroscopic movements. Movements resulting from imbibition or loss of water, as that exhibited by the elaters of the spores of *Equisetum*.

hylacolous. Dwelling in trees.

hylophagous. Feeding upon wood: XYLOPHAGOUS.

hylophilous. Inhabiting forests or wooded areas.

hylophyte. A plant which grows in forests or wooded areas.

hylotomous. Woodcutting, with reference to insects.

hymen. A membrane which partially blocks the orifice to the vagina.

hymenium. A layer of spore-forming tissue on the surface of the fruiting body of various fungi, especially ascomycetes and basidiomycetes.

hymenopteran. Any insect of the order Hymenoptera which includes the wasps, ants, and bees, characterized by possession of two pairs of membranous wings, complete metamorphosis, and chewing mouth parts sometimes modified for lapping. The order also includes ichneumon flies, braconids, gall wasps, sawflies, and horntails.

¹hyoid. Pertaining to the second visceral arch or structures derived from it.

²hyoid. The hyoid bone, a U-shaped bone in higher vertebrates to which the larynx is attached.

hypanthium. A floral tube formed by the growth and enlargement of the receptacle which bears on its rim stamens, petals, and sepals.

hypaxial. 1. Located below an axis. 2. Located ventral to the vertebral column. 3. Derived from the ventrolateral portions of myotomes, said of trunk muscles.

hyperbranchial. Above the gills: EPIBRANCHIAL.

hyperemia. Excessive blood flow to a tissue; congestion.

hyperglycemia. An excess quantity of sugar in the blood.

hypermastia. Possessing more than the normal number of mammary glands; polymastia.

hypermetamorphosis. In insects, a form of metamorphosis in which there are two or more distinctly different larval forms, as in blister beetles.

hypermetropia. Farsightedness, a condition in which the light rays come to a focus behind the retina: HYPEROPIA.

hyperopia. HYPERMETROPIA; FARSIGHTEDNESS.

hyperparasitism. Condition in which a parasite is the host of a parasite, which in turn may be parasitized.

hyperphagia. Excessive intake of food, associated with bulimia (excessive appetite).

hyperplasia. Increase in size of a structure resulting from multiplication of cells.

hypertonic. Designating a solution whose osmotic pressure is greater than an isotonic solution. Cells placed in such a solution lose water and shrinkage or plasmolysis occurs.

hypertrophy. Increase in size of an organ or structure, usually the result of increased functional activity. *Compare* hyperplasia.

hypha pl. **hyphae.** A filament of the vegetative body or mycelium of a fungus.

hyphodromous. Beneath the surface and invisible, with reference to veins of a leaf.

hyphopodium. A specialized hypha for attachment, as in certain ascomycetes.

hypnospore. A resting or quiescent spore.

hypoblast. Old term for endoderm.

hypobranchial. Below the gills or branchial arches.

hypobranchial groove. *See* endostyle.

hypocarpogenous. HYPOGEAL.

hypocone. 1. In dinoflagellates, the portion of the body posterior to the annulus. **2.** The inner, posterior (distolingual) cusp of an upper molar tooth.

hypoconid. The outer, posterior (distobuccal) cusp of a lower molar tooth.

hypocotyl. The portion of a plant embryo or seedling which lies below the cotyledons. At its tip is the radicle or embryonic root.

hypodermic. 1. Below or underneath the skin. **2.** Of or pertaining to the hypodermis.

hypodermis. 1. *zool.* The outermost layer of cells of the body wall of arthropods and other invertebrates. **2.** *bot.* A layer of supporting or protective cells lying immediately under the epidermis.

hypodynamics. Inactivity.

hypogastric. Pertaining to the lower median region of the abdomen.

hypogeal. Subterranean; pertaining to, located in, or occurring beneath the surface of the soil; remaining underground, said of cotyledons of seedlings.

hypogenous. Growing on a lower surface, as fungi on the underside of a leaf.

hypogeous. HYPOGEAL.

hypoglossal. At the base of or beneath the tongue.

hypoglossal nerve. The twelfth cranial nerve (XII) consisting principally of motor fibers innervating muscles of the tongue and present only in amniotes.

hypoglycemia. A condition in which blood sugar level is below normal.

hypognathous. 1. With lower jaw longer than the upper. 2. With the head vertical and the mouth parts located ventrally, in reference to the head of an insect.

hypogravics. Weightlessness.

hypogynous. Designating a flower in which sepals, petals, and stamens arise from the receptacle below and entirely free from the ovary, as in the lily. *Compare* epigynous, perigynous.

hypolimnion. A zone of water in a lake extending from thermocline to the bottom, with temperature fairly uniform and cold.

hypomere. The ventral portion of a myotome.

hyponasty. Condition in plants in which the ventral surface of a part grows more rapidly than the upper surface resulting in an upward curvature, as in leaves.

hyponome. The funnel of a cephalopod.

hypopharynx. In insects, a median process on the dorsal surface of the labium, elongated in sucking insects with main salivary opening on its tip.

hypophloeodal. Growing beneath bark, with reference to fungi.

hypophyllous. Growing beneath or on the under surface of a leaf.

hypophyseal. Of or pertaining to the hypophysis.

hypophysis. The pituitary gland.

hypopodium. The basal portion of the petiole of a leaf.

hypopus. The travelling stage of certain mites characterized by possession of special attachment structures, as in grain mites.

hypopygium. The male clasping organ in certain insects, as dipterans.

hypopyle. The posterior opening of the clasper tube of fishes. *Compare* apopyle.

hyporachis. The aftershaft of a feather.

hypostasis. 1. The settling of blood in a dependent part of the body. 2. *genetics.* Condition in which expression of a gene is suppressed by a nonallelic gene. The former gene is said to be hypostatic; the latter, epistatic.

hypostome. 1. In hydrozoans, a conical elevation bearing the mouth. 2. A median-ventral toothed structure forming part of the capitulum of ticks.

hypothalamus. The ventral portion of the diencephalon comprising the optic chiasma, mammillary bodies, tuber cinereum, infundibulum, and neurohypophysis. It forms the floor of the third ventricle.

hypothallus. A thin, shiny membrane at the base of the fruiting body of slime molds.

hypotheca. 1. The inner frustule of a diatom. 2. The portion of a dinoflagellate posterior to the annulus.

hypothecium. In an ascocarp, the layer of hyphae which supports the hymenium.

hypothermia. Body temperature below normal.

hypotonic. Below normal strength or tension.

hypotonic solution. A solution in which osmotic pressure is less than that of an isotonic solution. Cells in such a solution swell and become turgid.

hypotrich. A ciliate of the order Hypotrichida, mostly creeping protozoans with fused cilia or cirri on ventral surface, as *Stylonychia*.

hypoxia. Condition in which oxygen tension of the blood is below normal.

hypsodont. Having or designating high-crowned, prism-shaped teeth, characteristic of grazing animals.

hyraceum. A substance used in perfumery obtained from the dried deposits of hyrax dung.

hyrax. One of a number of small, herbivorous, hoofed mammals of the order Hyracoidea of the Mediterranean region, as *Procavia capensis*, the rock hyrax. Also called dassie, coney, cony, rock rabbit, klipdas.

hyssop. A European plant (*Hyssopus officinalis*) of the mint family.

hysteranthous. Developing leaves after the appearance of flowers, as the almond.

I

ibex. A wild goat of the Old World, best known of which is the Alpine ibex (*Capra ibex*).

ibis. A wading bird with a long, recurved bill of the family Threskiornithidae, as the white ibis (*Eudocimus albus*).

icefish. An antarctic fish (*Chaenocephalus*) whose blood contains no hemoglobin.

ichneumon. 1. *Herpestes ichneumon*, the Egyptian mongoose. **2.** An ichneumon fly.

ichneumon fly. One of a large number of hymenopterous insects of the family Ichneumonidae, whose larvae are parasitic living in or on the immature stages of other insects. *See* thalessa.

ichthyology. The science which deals with fishes.

ichthyornis. An extinct, toothed bird of the genus *Ichthyornis*.

ICSH. Interstitial cell-stimulating hormone, *q.v.*

icterus. JAUNDICE.

idiogram. A diagrammatic representation of a set of chromosomes.

iguana. Any of a number of New World lizards of the family Iguanidae, especially those of the genus *Iguana*, a large tropical form.

ileum. 1. In vertebrates, the third portion of the small intestine extending from the jejunum to the large intestine. **2.** In insects, the first portion of the hindgut extending from crop to colon.

iliac. Of, pertaining to, or located near the ilium.

ilium. The more dorsal of the three bones comprising the os coxae or hipbone.

imaginal. Of or pertaining to an imago.

imago. The adult, sexually mature stage of an insect.

imagochrysalis. The stage preceding the adult in mites of the family Trombiculidae.

imbibition. Absorption, especially the taking in of water by a solid as by hydrophilic colloids, such as dried agar or gelatin.

imbricate. Overlapping, said of scales when they overlap each other like tiles of a roof.

immature. Not fully developed; unripe.

immersed. **1.** Embedded in or sunk below the surface of another part or structure. **2.** *bot.* Entirely submerged or growing under water.

immiscible. Incapable of being mixed.

immobile. Fixed, immovable.

immobilize. To make immovable or render motionless.

immotile. Not motile; incapable of moving from place to place.

immune. Safe from disease; protected from the effects of disease or disease processes.

immunity. The state or quality of being immune, that is, having the ability to resist infection or to overcome the effects of infection.

immunology. The science which deals with immunity or resistance to disease.

impala. A graceful African antelope (*Aepyceros melampus*).

imperfect. Defective, incomplete; lacking a part.

imperfect flower. One which lacks either stamens or pistil: DICLINOUS.

imperfect fungi. *See* Fungi Imperfecti.

imperforate. Lacking openings or pores; nonporous.

impermeable. Not permeable; incapable of being penetrated by solids or fluids.

¹**implant.** To fix or set in securely.

²**implant.** **1.** A tissue graft placed within another tissue. **2.** A radioactive tube or needle placed within a tissue for therapeutic purposes.

implantation. **1.** The act of placing an implant within an organ or tissue. **2.** In mammals, the process by which a blastocyst attaches to and becomes embedded within the endometrium of the uterus.

implicated. Entangled, interwoven, twisted together.

impotence. Lack of strength, power, or vigor, especially the inability of the male to engage in sexual intercourse.

impregnate. 1. To make pregnant or cause to conceive; to fertilize; to fecundate. **2.** To infuse, charge, or saturate throughout.

impressed. Imprinted, bent inward, or marked as though pressure were applied.

imprint. A type of fossil in which the original substance has disappeared but the outline or form has been retained as an impression, as that of a leaf or animal track.

imprinting. The development of a behavior pattern in waterfowl and gallinaceous birds in which birds reared from eggs react to human keepers as their parents. Once established, the behavior is fixed and irreversible.

impulse. 1. An action or an impelling force which drives an individual or an object onward. **2.** A sudden mental activity which impels an individual to action, as the *impulse* to cry. **3.** A physicochemical change propagated along a nerve fiber or other conductile tissue which initiates a physiological activity in nerve, muscle, or gland cells.

inactivate. To render or make inactive or noneffective.

inanimate. Without life: DEAD.

inanition. Weakness or exhaustion resulting from lack of food.

inarticulate. 1. Not distinct, clear, or understandable, with reference to speech. **2.** *zool.* **a.** Not jointed or segmented. **b.** Lacking a hinge, as in brachiopods.

inborn. Present at birth, innate, inherent, natural.

inbreeding. The mating of closely related individuals, as self-pollination in plants or brother-sister matings in animals.

incarnate. Pink or flesh-colored.

inchworm. The larva of a moth of the family Geometridae, so called because of its looping method of locomotion. Also called looper, measuring worm, earth-measurer, geometer.

incineration. Cremation; combustion of organic material, leaving only ash.

incipient. Beginning to exist or manifest itself.

incised. 1. Cut into. **2.** *bot.* Having a deeply notched margin, as an *incised* leaf.

incision. A cut or gash; a notch in a margin.

incisor. A cutting tooth, as one of the two pairs of anterior teeth in the jaw of a mammal.

incisure. A slit, notch, or indentation.

included. Enclosed; not extending beyond the enclosing structure, as stamens and pistil not projecting beyond mouth of the corolla. *Compare* exserted.

inclusion. A discrete mass of nonliving material within the cytoplasm of a cell, as fat droplets, nutrient, pigment, yolk, or secretory granules; called paraplasm, metaplastic substance.

inclusion body. A mass of virus particles within the cytoplasm or nucleus of a cell in an animal infected with a virus, as negri bodies in the cytoplasm of Purkinje cells in rabies-infected animals.

incomplete. 1. Not complete; imperfect. 2. *bot.* Lacking one or more of the four floral parts, said of flowers.

incrassate. Thickened; swollen.

incrustation. A hard crust or coating upon a surface.

incubation. 1. The act of sitting upon and hatching eggs by the heat of the body, as by fowls. 2. Artificially maintaining conditions of warmth and moisture favorable for the development of eggs, embryos, bacterial cultures, etc. 3. Maintenance of normal body temperature for a premature baby.

incubation period. 1. The period of an infectious disease between infection and appearance of the first symptoms. 2. The period a virus must remain within a vector before it is capable of causing the disease in its host.

incubous. A type of leaf insertion in which the upper portion of a leaf covers the base of the leaf directly above it. *Compare* succubous.

incumbent. 1. Resting against or lying upon. 2. Lying with its back against the radicle, said of cotyledons.

incurrent. Pertaining to or conducting an inflowing current.

incurvation. Curving inward.

incus. The anvil, the middle of the three ear bones.

indeciduous. Not deciduous, with reference to leaves; evergreen, with reference to trees.

indefinite. 1. Not definite; indeterminate. 2. *bot.* Numerous; not easily counted.

indehiscent. 1. Remaining closed at maturity. 2. Not opening along a regular line, with reference to fruits, seed pods, and anthers.

independent. Not dependent; free from or not subject to control.

independent assortment. Mendel's second law or principle which accounts for the 9:3:3:1 ratio in dihybrid crosses. It is based on the fact that in the formation of gametes, unless linked, each pair of genes operates independently of all other pairs resulting in random combination with other genes in the formation of eggs and sperm and random combination of pairs at fertilization.

indeterminate. Indefinite, not precise.

index pl. **indices. 1.** A sign, token, or indicator. **2.** A ratio or formula expressing the ratio of one measurement or dimension to another, as cephalic *index*.

index fossil. A fossil which identifies the stratum in which it is found. Sometimes called indicator species.

Indian corn. A North American cereal grass (*Zea mays*) widely cultivated for its seeds borne on ears. Also called maize.

indicator. 1. *chem.* A substance which, by a change in color, indicates a change in the nature of the solution containing it, as litmus or phenolphthalein. **2.** *ecol.* A plant or animal species which is characteristic of a particular seral stage.

indifferent. Undifferentiated or not specialized, said of cells and tissues.

indigen. An indigenous animal or plant.

indigenous. Native; not imported or introduced; growing naturally in a country or climate.

induce. To bring about; to cause to occur.

induction. 1. The process of inducing. **2.** *embryol.* The development of a specific structure as a result of a chemical substance transmitted from one part of the organism to another: EVOCATION. *See* organizer.

inductive. 1. Capable of producing an internal change or response. **2.** *bacteriol.* Designating enzymes which are produced only when specific substrates are present in the medium.

indumentum. 1. *zool.* The plumage of a bird. **2.** *bot.* Any dense, hairy covering or pubescence.

indurate. Hardened.

indusium pl. **indusia.** A membrane which serves as a covering, as

249

a. *bot.* that which covers the sorus of a fern; **b.** *zool.* the double membrane which encloses the developing embryo of certain insects.

induviae. **1.** The persistent portion of a perianth. **2.** The withered remnants of leaves or perianth which remain attached to the petiole or pedicel.

inermous. Lacking spines or prickles; unarmed.

inert. Inactive; physiologically ineffective.

infantile paralysis. POLIOMYELITIS.

infantilism. Persistence of infantile characteristics into adolescence and adult life.

infauna. Benthic animals living in bottom material.

infect. To introduce or implant an infective or disease-producing organism.

infection. **1.** The introduction or transmission of a disease-producing agent or organism to a disease-free organism. **2.** The state or condition of being infected by a pathogenic organism.

infectious agent. Any virus or microorganism (bacterium, spirochaete, rickettsia, fungus, protozoan), helminth, or arthropod which is capable of causing an infection or infectious disease.

inferior. In a lower position; beneath or underneath.

infertile. Incapable of reproduction; barren, sterile.

infestation. **1.** The state or condition of being infested by numbers of a disturbing or annoying organism, as lice or fleas. **2.** The presence of multicellular organisms, especially helminths, which do not multiply within the body. *Compare* infection.

infiltrate. To pass into or through pores or spaces; to permeate.

inflammation. The reaction of tissues to injury characterized by pain or tenderness, redness, swelling, heat, and usually disordered function.

inflate. To distend with air or gas.

inflated. Swollen, distended, puffed out.

inflexed. Turned or bent abruptly inward or downward.

inflorescence. The arrangement of flowers on an axis, whether a single flower or a group; a floral axis and its appendages; a flower cluster. *See* cymose, racemose, spike, catkin, raceme, corymb, umbel, capitulum, panicle, spadix.

infraneuston. Aquatic organisms which live on the underside of a surface film.

infrared rays. Invisible heat rays beyond the red end of the visible spectrum with wavelengths ranging from 7700 to 500,000 AU.

infundibulate. Shaped like a funnel or cone.

infundibuliform. Funnel-shaped.

infundibulum. A funnel-shaped structure, as **a.** the stalk of the pituitary gland; **b.** the lateral distended portion of the uterine tube.

infusion. 1. The steeping or soaking of a substance in water in order to extract its active principles. 2. The slow introduction of a substance into a vein or subcutaneous tissues.

infusorian. Any of the miscellaneous microorganisms found in infusions of decomposing organic matter, especially any ciliated protozoan.

ingest. To take in a solid substance, as food into the body.

ingesta. That which is ingested. *Compare* egesta.

ingluvies. The crop of birds.

ingression. A method of gastrulation in which cells fill the blastocoel forming a stereogastrula, as in sponges and coelenterates.

inguinal. Of or pertaining to the region of the groin.

inguinal canal. An oblique canal in the lower abdominal wall which, in the male, transmits the spermatic cord, and in the female, the round ligament of the uterus.

inhabit. To dwell in or on or to occupy, as a place of habitation.

inhale. To draw in air, as in breathing.

inherent. Inborn, innate, natural.

inherit. To receive from one's parents, as genetic factors.

inheritance. All the characteristics of an individual which are dependent upon hereditary determiners or genes received from one's parents.

inhibition. The act of restraining, checking, slowing down, or stopping an activity.

inhibitor. An agent which checks or stops an activity.

initial. *bot.* An initiating cell; a self-perpetuating cell present in apical meristem and vascular cambium from which specialized tissues develop.

inject. To introduce or force a fluid into any of the vessels, cavities, or tissues of the body.

251

injection. The act of injecting or the substance injected.

ink sac. A rectal gland in cephalopods which produces and stores sepia ink, a melanin-containing fluid which is discharged when the animal is disturbed.

innate. 1. Inborn, native, natural; dependent upon hereditary factors. **2.** *bot.* Borne at the tip of a supporting structure, as certain anthers.

innervation. The distribution of nerves to a particular region or structure.

innocuous. Harmless; producing no injury or ill effects.

innominate bone. The hipbone or os coxae.

inoculation. 1. The introduction of a microorganism or virus into a suitable medium for growth. **2.** The introduction of a pathogenic organism into an animal for the purpose of inducing immunity.

inorganic. 1. *chem.* Not organic; not containing carbon or compounds of carbon (with a few exceptions). **2.** Not of plant or animal origin.

inquiline. An organism that lives within the nest or abode of another, as insects that live in the nests of termites or ants. *See* myrmecophile.

insect. Any member of the class Insecta (Hexapoda), air breathing arthropods with a body of three parts (head, thorax, abdomen), three pairs of legs, usually two pairs of wings. Head bears one pair of antennae and a pair of compound eyes. Comprises the most numerous group of animals, with over 700,000 described species.

insectary. A place where insects are kept and propagated.

insectivore. 1. Any insect-eating plant or animal. **2.** Any member of the order Insectivora, small, insect-eating, placental mammals, as moles, shrews, and hedgehogs.

insectivorous. Feeding principally upon insects.

inseminate. 1. To impregnate; to introduce semen into the genital tract of a female. **2.** To sow or distribute seed.

insensible. 1. Devoid of feeling or sensation; unconscious. **2.** Incapable of being perceived by the senses.

insert. To place, put, or set in.

insertion. 1. The act of inserting. **2.** The mode of attachment or the place at which a structure is attached, as **a.** *bot.* the *insertion* of parts of a flower; **b.** *anat.* the attachment of a muscle to a relatively more movable part, in contrast to the origin.

insessorial. Adapted for or pertaining to perching.

in situ. In the normal or natural position.

insoluble. Incapable of being dissolved.

inspiration. The act of drawing air into the lungs; inhalation.

inspissate. To thicken or increase the consistency of, as by evaporation.

instar. The stage in the development or the form assumed by an insect between successive molts. The first *instar* is between hatching and the first molt.

instillation. Introduction of a liquid, drop by drop, into a cavity.

instinct. A natural, involuntary, inherited tendency to perform a specific action or to follow a certain behavior pattern.

insufflation. The act of blowing or breathing into, especially the blowing of a gas or powder into a cavity of the body.

insulin. An antidiabetic hormone secreted by the beta cells of the islets of Langerhans of the pancreas essential for the proper utilization of blood sugar. It is used in the treatment of diabetes mellitus.

integument. **1.** *bot.* One of the two layers (inner and outer integuments) surrounding the mature ovule of a flowering plant. **2.** *zool.* **a.** The surface covering of the body in various invertebrates. **b.** The skin of a vertebrate.

interbreed. **1.** To breed by mating individuals of different races, varieties, or species; to crossbreed. **2.** To breed within a small, closely related population.

intercalated. Inserted or interposed among.

intercellular. Between or among cells.

intercostal. Between ribs.

interface. The common border or surface between two structures, phases, or systems.

interfascicular. Between fascicles or bundles.

interference. **1.** Coming into contact with; checking, obstructing, or stopping. **2.** *genetics.* The blocking of crossing-over at a certain locus by adjacent or nearby crossing-over.

interferon. A substance formed within a virus-infected cell which prevents the entrance of another virus.

interkinesis. The interphase of mitosis.

intermedin. The melanophore-stimulating hormone (MSH) secreted by the intermediate lobe of the pituitary gland.

intermolt. The fourth period in the molting cycle of crustaceans. *See* molting.

internal. On the inside; within, enclosed.

internode. The portion between two nodes, as in a plant stem.

internuncial neuron. An association or connector neuron; one which transmits the impulse between two neurons in a neural pathway. Also called an association or adjustor neuron or interneuron.

interoceptor. A sensory receptor located within a visceral organ.

interphase. The period between two successive mitotic divisions: INTERKINESIS.

interrenal bodies. Structures in lower vertebrates which correspond to the cortex of the adrenal gland.

inter se. Between or among themselves, with reference to stock breeding.

intersex. An individual whose sexual characters are intermediate between male and female.

interstice pl. **interstices.** A space or interval between closely placed structures.

interstitial. Of or pertaining to interspaces between the cells or the structural units of an organ or structure.

interstitial cells. 1. Cells in the epidermis of coelenterates which give rise to sex cells. **2.** In vertebrates, the cells of Leydig, located between the seminiferous tubules of the testis. They secrete the male hormone testosterone.

interstitial cell-stimulating hormone. ICSH, produced by the anterior lobe of the pituitary. *See* luteinizing hormone.

interstitial fauna. Microscopic animals which occupy interstices or spaces between mud and sand particles in lake and marine bottoms.

intestine. 1. The portion of the alimentary canal which extends from the stomach or crop to the cloaca or anus. In man it consists of the small intestine (duodenum, jejunum, ileum) and large intestine (cecum, colon, rectum, and anal canal). **2.** The digestive cavity of certain invertebrates, as turbellarians.

intima. The innermost layer of a blood vessel; the tunica intima.

intine. The innermost layer of the wall of a spore or pollen grain. *Compare* exine.

intolerance. Inability to endure or extreme sensitivity to a chemical agent or physical condition.

intoxication. Poisoning, or a state of being poisoned by alcohol, drugs, serum, or an excess of any substance; a state of drunkenness; inebriation.

intracellular. Within a cell or cells.

intramembranous. Occurring within a membrane.

intramembranous ossification. Formation of bone within a connective tissue membrane without a preceding cartilaginous stage, as cranial bones.

intramural. Within the substance of the wall of an organ or structure.

intravital. Within a live organism or living tissue.

intravital stain. A dye that will stain living cells without killing them, as neutral red.

intra vitam. During life or while still living.

intrinsic. Enclosed within, peculiar to, or an inherent part of an organ or structure, as *intrinsic* muscles of the tongue.

intromission. The act of inserting one structure into another.

intromittent organ. Any male copulatory organ, as a penis or cirrus.

introrse. Turning inward; facing toward the axis.

introvert. 1. a. To turn or bend inward. b. To invaginate a tubular structure into another. 2. A structure that has been introverted, as the anterior portion of certain invertebrates (sipunculids or ectoprocts).

intubation. Insertion of a tube into a hollow organ or structure.

intumescent. Enlarged, swollen, distended.

intussusception. 1. A method of growth in which new particles of matter are added or deposited within those already existing, as in living matter. 2. Invagination of or the slipping of one part of a tubular organ within an adjacent region.

inulin. A carbohydrate which upon hydrolysis yields levulose.

invagination. 1. The act or process of infolding so that a pocket-like cavity is formed. 2. The ensheathing of a tubular structure: INTUSSUSCEPTION. 3. A part invaginated.

invasin. HYALURONIDASE.

255

invasion. The process of entering into and spreading throughout.

inversion. 1. The act of or state of being inverted. 2. *genetics.* A 180° rotation of a block of genes within a chromosome so that a sequence as ABCDEF becomes ABEDCF, the block CDE reversing itself.

invert. To turn upside down or inside out.

¹**invertebrate.** 1. Lacking a backbone or spinal column. 2. Of or pertaining to invertebrate animals

²**invertebrate.** An invertebrate animal.

invest. To enclose or envelop.

in vitro. In glass, with reference to processes or reactions which take place in glass vessels, as test tubes, petri dishes.

in vivo. Within a living organism.

involucel. A small or secondary involucre.

involucre. 1. A protective or investing structure. 2. A group of bracts which surrounds and encloses the base of a flower or fruit.

involuntary. Not under the control of the will; not voluntary.

involute. 1. Rolled or turned in at the margins, said of leaves. 2. Curved inward or spirally. 3. Having whorls closely wound, said of shells.

involution. 1. The process of rolling or turning inward. 2. Retrograde evolution or development in which structures become less complex or are lost. 3. The return of an organ to normal size following enlargement due to normal demands, as the uterus after pregnancy. 4. Retrogressive changes in an organ following peak of functional activity, as *involution* of the thymus after puberty or ovary after menopause. 5. *bacteriol.* The appearance of unusual or abnormal forms in bacterial cultures grown under unfavorable conditions.

iodine. A nonmetallic halogen element occurring as bluish-black plates or scales. It is a component of the thyroid hormone and used in medicine as an antiseptic. Symbol I, at. wt. 126.91.

ipsilateral. Located on, or occurring on the same side. *Compare* contralateral.

iridescence. Displaying changeable, rainbowlike colors, as mother-of-pearl.

iris. 1. *anat.* A pigmented diaphragm in the eye of cephalopods and vertebrates containing an opening, the pupil, through which light enters. **2.** *bot.* A cultivated flowering plant of the genus *Iris*.

Irish dulse. A red alga (*Rhodymenia palmata*) used as food; sheep's weed.

Irish moss. A red alga (*Chondrus crispus*) of the North Atlantic, the source of carrageen, an emulsifying agent.

iron. A silvery-white or gray, hard, ductile, malleable element. Symbol Fe; at. wt. 55.85. It is an important constituent of various respiratory pigments and enzyme systems.

irradiation. 1. Exposure to radiant energy. **2.** Emission or radiation of heat waves. **3.** Spreading or diffusion from a common center, as nerve impulses.

irregular. Not uniform; not conforming to the usual pattern.

irregular flower. One in which one or more members of at least one whorl is different from other members of the same whorl, as a bean or pea flower.

irreversibility of evolution. A principle which holds that an organ once lost can never be regained and that a specialized form or structure never becomes generalized.

irritability. Excitability; the ability of living matter to react or respond to a stimulus.

isadelphous. Having an equal number of stamens in each group or phalanx.

ischium. The most posterior of the three bones comprising the hipbone.

isidium. A propagative structure on the upper surface of a lichen thallus.

island. A small isolated structure, especially a mass of cells differentiated from surrounding tissue.

islet. A small island.

islet of Langerhans. A minute mass of cells in the pancreas which is the source of the hormones insulin and glucagon.

isobilateral. Having leaves illuminated on both sides, their surfaces being structurally similar.

isocyclic. Having the same number of members in each of the whorls.

isodactylous. Having digits of equal length.

isodiametric. Having equal diameters.

isodont. HOMODONT.

isoenzyme. One of several forms of the same enzyme; an isozyme.

isogamete. A reproductive cell which is similar in shape and size to the cell with which it unites.

isogamous. Having undifferentiated gametes.

isogamy. Reproduction in which conjugation of isogametes occurs, as in certain algae and protozoans.

isogeneric. 1. From the same genus. 2. Of the same or similar origin: ISOGENOUS.

isogenic. Genetically identical; possessing the same genes.

isogenous. Of the same or similar origin.

isograft. A graft or transplant to a genetically identical individual.

isokont. Possessing flagella of equal length.

¹isolate. To separate or set apart.

²isolate. A group in a large population which does not interbreed with other members of the population, mating being principally between members of the group; a breeding unit.

isolation. The separation of a race or species from other closely related groups so that interbreeding does not occur. May result from geographic, anatomical, physiological, ecological, genetic, cultural, or other factors.

isolecithal. Having yolk uniformly distributed throughout the cytoplasm, as in mammalian ova: HOMOLECITHAL.

isomere. One of a number of homologous structures.

isomerous. 1. Having an equal number of parts, as grooves, ridges. 2. *bot.* Having the same number of parts in each whorl, with reference to flowers.

isometric contraction. A type of muscle contraction in which no shortening occurs.

isomorphic. Structurally alike.

isomyarian. Equal in size, said of the two adductor muscles of a bivalve.

isoplanogametes. Flagellated sex cells which are alike.

isopod. Any crustacean of the order Isopoda which includes pill bugs, sow bugs, wood lice, and related forms.

isopteran. An insect of the order Isoptera which includes the termites or white ants, wood-eating insects which live in colonies.

isorhiza. A form of nematocyst with tube of uniform diameter and open at the tip.

isosmotic. ISOTONIC.

isotonic. Having equal osmotic pressure: ISOSMOTIC.

isotonic contraction. A type of muscle contraction in which a muscle shortens, resulting in movement. *Compare* isometric contraction.

isotonic solution. A solution in which living cells will neither gain nor lose water, as a 0.85% NaCl solution for human red blood cells.

isotope. One of two or more elements, the atoms of which have the same atomic number but different atomic weights, as Cl^{35} and Cl^{37}, *isotopes* of chlorine.

isozyme. *See* isoenzyme.

isthmus. A narrow or constricted region which connects **a.** two larger portions of an organ, as the *isthmus* of the thyroid gland; **b.** two cavities, as the *isthmus* of the fauces.

itch mite. An arachnid, *Sarcoptes scabiei*, which burrows into the skin causing intense itching. *See* scabies.

iter. A passageway.

ivory. A hard, white, fine-grained substance, a modified type of dentine obtained from the tusks of elephants and other large mammals.

ixodid. A hard tick of the family Ixodidae containing the genera *Ixodes*, *Boophilus*, *Amblyomma*, and *Dermacentor*. *See* wood tick, cattle tick.

J

jabiru. A large stork (*Jabiru mycteria*) of tropical America.

jacamar. A graceful forest bird of the family Galbulidae inhabiting forests of Central and South America.

jacana. A lily-trotter of the family Jacanidae, a wading bird with extremely long legs and toes enabling it to walk on floating

JACKAL

vegetation, as the American jacana (*Jacana spinosa*), ranging from Texas to Argentina.

jackal. A wild dog (*Canis aureus*) of the Old World, a nocturnal carnivore feeding principally on carrion.

jackass. A male ass or donkey.

jackdaw. *Corvus monedula*, of the family Corvidae, a common bird of Europe.

jacket cells. Sterile cells surrounding developing spores and gametes in various plants.

jackrabbit. One of several large American hares with long ears and long hind legs, as *Lepas californicus*, the black-tailed jackrabbit.

Jacobson's organ. The vomeronasal organ.

jaeger. A predaceous, gull-like bird of the genus *Stercorarius* which harasses other birds causing them to drop their prey or disgorge their food.

jaguar. *Panthera* (*Felis*) *onca*, a large, feline carnivore inhabiting forests of South America ranging north as far as Texas.

Japanese beetle. A shining leaf chafer (*Popillia japonica*) introduced into the United States in 1916. It is a serious pest, as adults and larvae feed on a variety of plants.

jaundice. Icterus, a condition characterized by yellowness of skin and mucous membranes due to presence of bile pigments in the blood.

Java man. *Homo* (*Pithecanthropus*) *erectus*, a species of fossil man, one of the earliest of human or subhuman types estimated to have lived about 700,000 years ago.

javelina. The collared peccary (*Tayassu angulatus*).

jaw. 1. In vertebrates, **a.** a bony or cartilaginous structure, the mandible or maxilla, which bears teeth and is utilized in the mastication of food; **b.** in the plural, the bones and associated structures which surround the mouth and enclose the mouth cavity. **2.** In invertebrates, **a.** a structure comparable to a vertebrate jaw which performs a chewing function; **b.** one of the movable parts of a forcepslike structure.

jawbone. Either of the bones, maxilla or mandible, which support teeth, especially the mandible or bone of the lower jaw.

jay. One of a number of aggressive, noisy birds of the family

Corvidae, as the Canada jay (*Perisoreus canadensis*) or blue jay (*Cyanocitta cristata*).

jejunum. The second portion of the small intestine in mammals extending from duodenum to ileum.

jelly. A soft, gelatinous, semisolid substance, as gelatin which has imbibed water.

jellyfish. A medusa; any of a number of free-swimming coelenterates shaped like a saucer or umbrella and having a jellylike consistency, as *Aurelia* of the class Scyphozoa or *Gonionemus* of the class Hydrozoa.

jerboa. A small, nocturnal, jumping rodent of arid regions of the Old World, as *Jaculus jaculus*, the Egyptian jerboa of the family Dipodidae.

jewfish. One of a number of extremely large marine fishes of the family Serranidae.

jigger. The chigoe or sand flea.

jimsonweed. *Datura stramonium*, of the nightshade family Solanaceae, the source of a number of alkaloids, as hyoscine, hyoscyamine, and atropine, used medicinally. Also called Jamestown weed, thorn apple.

joey. In Australia, the young of an animal, especially a young kangaroo.

joint. 1. *zool.* **a.** An articulation. **b.** The region about or between two parts, segments, or divisions of an organism or between two parts of a limb or appendage. 2. *bot.* A node, as in grasses.

jointed. Possessing nodes or articulations.

jugal. 1. Connecting or uniting. 2. Pertaining to the zygoma or malar bone.

jugate. 1. Occurring in pairs. 2. Connected together.

jugular. 1. Of or pertaining to the neck or throat. 2. Of or pertaining to the jugular vein which drains the head region.

jugular fins. The pelvic fins of a fish when located in the throat region sometimes anterior to the pectoral fins, as in the cod.

jugum. 1. *bot.* **a.** A pair, as a pair of opposite leaflets. **b.** A ridge on certain fruits, especially those of the Umbelliferae. 2. *zool.* **a.** A small lobe on the posterior margin of the forewing of lepidopterous insects. **b.** A sclerite in the head of certain insects, as hemipterans and homopterans.

juice. 1. *bot.* The fluid contained in any part of a plant, especially that which can be expressed by squeezing, as orange *juice*. **2.** *zool.* The secretion of a gland, as pancreatic *juice*.

julaceous. Bearing catkins: AMENTACEOUS.

jumping bean. The seed of certain Mexican shrubs of the genera *Sebastiana* and *Sapium* which contains the larva of a moth (*Laspeyresia saltitans*). Violent movements of the larva cause movements of the bean.

junco. A finch of the genus *Junco*, family Fringillidae: SNOW-BIRD.

junction. A connection or point of union between two structures.

June bug. A June beetle or May beetle, one of several scarabaeid beetles of the genus *Phyllophaga* whose larvae are the familiar white grubs of the soil.

jungle fowl. One of several species of wild birds of Asia, especially *Gallus gallus*, the red jungle fowl, considered to be the ancestor of modern domestic fowl.

juniper. A coniferous tree of the genus *Juniperus*, as *J. virginianus*, the red cedar.

Jurassic period. The middle period of the Mesozoic era.

jute. 1. A coarse fiber obtained from an Asiatic plant (*Corchorus*), used in the manufacture of burlap, gunny sacks, cheap twine, and coarse cloth. **2.** The plant which produces the fiber, principally two species (*C. capsularis* and *C. olitorius*) cultivated for commercial production.

juvenal. JUVENILE.

juvenal plumage. Plumage (of a bird) that replaces down; also, in birds which lack the natal down, the first plumage.

juvenile. 1. Young, youthful. **2.** Pertaining to a young or immature individual. **3.** A young individual.

juvenile hormone. JH, a hormone produced by the corpora alata of insects which suppresses pupal and imaginal differentiation, thus inhibiting metamorphosis and promoting retention of larval characters.

juxtapose. To place adjacent to each other.

juxtaposition. State of being placed side by side or adjacent to each other.

K

kaka. A New Zealand parrot of the genus *Nestor*.

kakapo. The owl parrot (*Strigops habroptilus*) of New Zealand.

kaki. The Japanese persimmon.

kala azar. Visceral leishmaniasis, an infectious disease of the Far East and tropical countries, caused by *Leishmania donovani*, a protozoan transmitted by sand flies. Also called black fever or dumdum fever.

kale. *Brassica oleracea acephala*, a variety of cabbage with loose leaves not forming a head.

kangaroo. One of a number of herbivorous, leaping marsupials of the family Macropodidae found in Australia and neighboring islands, comprising some 20 species ranging in size from the small rat kangaroo to the great gray kangaroo or forester (*Macropus major*). They have large, powerful hind legs and a long, thick tail.

kangaroo rat. A small, nocturnal, burrowing rodent of the genus *Dipodomys* of the western United States.

kapok. The silky fibers which invest the seeds of the kapok tree (*Ceiba pentandra*) of tropical countries, also called Java cotton.

kappa particles. Cytoplasmic particles in certain paramecia (killers) which contain a substance (paramecin) which is toxic to other paramecia (sensitives).

karakul. A broad-tailed sheep of the Asian province Bokhara or the glossy black fur obtained from their newborn lambs.

karyallagy. The union of two swarm cells in slime molds.

karyogamy. The fusion of male and female nuclei following union of gametes.

karyokinesis. Nuclear division, the phase of mitosis during which chromosomes are divided into equal portions.

karyolymph. Nuclear sap; the fluid substance of a nucleus.

karyolysis. Dissolution of a nucleus of a cell.

karyophore. In protozoans, an apparatus consisting of a nucleus suspended by ectoplasmic fibrils, as in *Isotricha*.

karyoplasm. Nuclear substance.

karyorrhexis. Nuclear fragmentation.

karyosome. A dense mass of chromatin other than the nucleolus within the nucleus of a cell. Also called chromocenter or false nucleolus.

karyotype. The general appearance or structure of a set of chromosomes within a nucleus. *See* idiogram.

katabolism. CATABOLISM.

katharobic. Living in clear water, as that of mountain streams.

katydid. A large, green orthopteran insect (*Pteryophylla camellifolia*) noted for its stridulatory sounds produced by the male.

kea. A large New Zealand parrot (*Nestor notabilis*) which sometimes attacks live sheep seeking kidney fat.

ked. *See* sheep tick.

keel. 1. A ridge or ridgelike process; a carina. **2.** *bot.* **a.** The edge of two united petals in a bean or pea flower; **b.** a ridge on a blade of grass.

kelp. Any of a number of large seaweeds, brown algae of the order Laminariales, as the giant kelp (*Macrocystis*).

kentrogon. The nonmotile larval stage which follows the cypris stage in *Sacculina*, a parasitic crustacean.

keratin. A fibrous protein of high sulfur content present in epidermal structures as horn, nails, claws, and feathers.

kermes. The dried bodies of scale insects from which a reddish-purple dye is obtained.

kernel. 1. The entire grain or seed of a cereal, as a *kernel* of corn. **2.** The inner and usually the more edible portion of a seed contained within the integuments or hard outer wall, as the *kernel* of a nut.

kestral. 1. A common European falcon (*Falco tinnunculus*). **2.** The American sparrow hawk (*Falco sparverius*).

ketone. A compound containing the characteristic group -CO, formed by the oxidation of a secondary alcohol, as acetone, CH_3COCH_3.

ketone bodies. Ketone acid, acetoacetic acid, acetone, and beta-hyroxybutyric acid, substances formed from the incomplete combustion of fats.

key. 1. A key fruit or samara. **2.** *taxonomy.* A tabular arrangement of the diagnostic characters of a group of animals or plants so arranged as to facilitate identification of taxa, as a genus or species.

kiang. A wild ass (*Equus hemionus kiang*) of the highlands of Tibet and Mongolia. Also called kulan.

kidney. 1. An excretory organ in vertebrates which forms and excretes urine. **2.** A corresponding organ in various invertebrates. *See* nephridium, nephron, urinary system.

kidney worm. One of two species of nematode worms which infest the kidneys and urinary ducts of certain vertebrates, as *Dioctophyme renale*, a large worm found in carnivores, and *Stephanurus dentatus*, found in swine.

killdeer. A plover (*Charadrius vociferus*), a North American shore bird.

killer. An individual which kills, as **a.** a paramecium which contains kappa particles; **b.** a killer whale.

killer whale. 1. The grampus (*Orcinus orca*), a predaceous cetacean which preys on fishes, seals, walruses, and other whales **2.** The false killer whale (*Pseudorca crassidens*) which feeds upon cephalopods.

killifish. One of several species of small fishes of the genus *Fundulus* found in fresh, brackish, and salt waters of the east and gulf coasts.

kilocalorie. A large calorie (Cal.). *See* calorie.

kinase. 1. A substance which converts a zymogen into an active enzyme; an activator, as enterokinase. **2.** An enzyme which catalyzes the transfer of terminal phosphate of ATP to an acceptor, as hexokinase.

kind. A natural group, as a class, order, family, etc.

kinesiology. The science which deals with muscular movement, especially in man.

kinesthesia. The sense of movement or position; the proprioceptive or muscle sense.

kinetic. Of or pertaining to motion or activity.

kinetodesma. A fine fibril or a set of fibrils connected to a kinetosome.

kinetoplast. A structure in trypanosomes consisting of a parabasal body and a blepharoplast.

kinetosome. A basal body of a cilium or flagellum.

kinety. In ciliates, a ciliary row consisting of kinetosomes and kinetodesma.

kingbird. A tyrant flycatcher of the family Tyrannidae, as *Tyrannus tyrannus*, the Eastern kingbird. Also called *bee martin*.

king crab. 1. A crustacean (*Paralithodes*), the commercial king crab of the North Pacific. **2.** *Xiphosura* (*Limulus*) *polyphemus*, the horseshoe crab, a primitive arachnid of the class Merostomata, found along the Atlantic and Gulf coasts.

kingdom. One of the three major categories into which natural objects are classified, namely the animal, plant, and mineral kingdoms.

kingfish. One of a number of marine fishes, as *Menticirrhus*, of the the Atlantic coast; *Genyonemus*, of the California coast; *Seriola*, of the Australian region.

kingfisher. A crested, fish-eating bird of the family Alcedinidae, as *Megaceryle alcyon*, the belted kingfisher.

kinglet. One of several small, crested birds of the family Sylviidae, as *Regulus satrapa*, the Eastern golden-crowned kinglet.

king snake. A moderately sized, nonvenomous snake of the family Colubridae, as *Lampropeltis getulus* of the southeastern United States.

kinin. One of a group of plant hormones which promote growth by stimulating cell division, cell enlargement, and initiation of development in buds and roots, as kinetin.

kinkajou. A nocturnal, arboreal mammal (*Potos caudivolvulus*) of the family Procyonidae, inhabiting Mexico, Central and South America.

kinorhynch. A member of the Kinorhyncha, a small group of marine pseudocoelomates, considered by some as a phylum, by others as a class of the Aschelminthes, as *Echinoderes*.

kipper. 1. A male salmon, especially during the spawning season. **2.** In commerce, a smoked herring or salmon.

kissing bug. One of several species of bloodsucking hemipterans, as *Melanolestes picipes*, the black corsair, which inflicts painful bites on the lips.

kite. A small, hawklike bird of prey of the family Accipitridae, as *Elanoides forficatus*, the swallow-tailed kite of the southern states.

kitten. The young of various members of the cat family.

kittiwake. A North Atlantic gull (*Rissa tridactyla*) of the family Laridae.

kiwi. A flightless, terrestrial, ratite bird of the order Apterygiformes, as *Apteryx australis* of New Zealand.

kiyi. A small whitefish (*Coregonus (Leucichthys) kiyi*) of the Great Lakes.

klendusity. The ability of a susceptible plant to avoid infection as a result of a particular growth characteristic, as an early-maturing variety escaping a late summer or fall disease.

klinotaxis. A group of reactions in which an organism moves its head from side to side as it is moving forward.

klipspringer. A small African antelope (*Oreotragus oreotragus*); klipbokke.

knee. 1. *zool.* A joint or articulation, as that **a.** in most tetrapods, between femur and tibia; **b.** in birds, the tarsal joint; **c.** in quadrupeds, the carpal or wrist joint of the forelimb. 2. *bot.* A sharp bend in a stem or branch.

kneecap. The patella.

¹**knot.** 1. A rounded swelling, knob, excrescence, or protuberance, as in a muscle, stem, or root. 2. *bot.* **a.** A hardened mass of wood formed at the base of a branch of a tree. **b.** A plant disease characterized by formation of excrescences.

²**knot.** A large sandpiper (*Calidris canutus*).

koala. An arboreal marsupial (*Phascolarctus cinereus*) of Australia.

kob. An African antelope (*Kobus kob* or *K. megaceros*).

Kodiak bear. *Ursus middendorffi*, the giant Alaskan brown bear.

kohlrabi. A variety of cabbage (*Brassica oleracea* var. *gongylodes*) with a greatly enlarged, edible stem.

koodoo. *See* kudu.

kookaburra. *See* laughing jackass.

krait. An extremely venomous snake (*Bungarus coeruleus*) of Southeast Asia.

Kreb's cycle. Tricarboxylic acid cycle or citric acid cycle, a complex series of reactions occurring in cells in which pyruvic acid, a product of carbohydrate metabolism, is broken down under aerobic conditions to CO_2 and H_2O with the release of a large amount of energy which is utilized principally in the synthesis of ATP. *See* citric acid cycle.

krill. Shrimplike crustaceans of the genus *Euphausia* existing in enormous numbers and comprising the principal food of certain whales.

kryoplankton. *See* cryoplankton.

kudu. A large, spiral-horned, African antelope (*Strepsiceros*). Also called koodoo.

kumquat. A citrus fruit, a native of China, produced by a tree of the genus *Fortunella.*

Kupffer's cells. Fixed macrophages present in sinusoids of the liver.

kymograph. An apparatus widely used in physiology by which changes brought about by various physiological activities are recorded graphically on a rotating drum.

kyphosis. Humpback, hunchback; excessive curvature of the spine with convexity of the curve being directed posterior, usually in the thoracic region.

L

labellum. 1. *bot.* The enlarged third petal in the corolla of an orchid. 2. *zool.* The expanded tip of the labium in certain insects, especially dipterans.

labia. Plural of LABIUM.

labial. Of or pertaining to lips or labia.

labiate. Having a lipped structure, as flowers of the mint family, Labiatae, in which petals of the corolla form a two-lipped structure.

labile. Capable of adapting, being changed, or modified; unstable.

labium. 1. A lip or liplike structure. 2. *zool.* **a.** The lower lip of an insect or crustacean. **b.** One of two pairs of folds, labia majora and labia minora, lying on each side of the vaginal orifice. **3.** *bot.* The lower lip of a labiate flower.

labor. 1. Childbirth: PARTURITION. **2.** Functional activity. *See* division of labor.

labrum. 1. The upper lip of an insect, usually attached to and lying below the clypeus. **2.** The upper lip of a crustacean. **3.** The outer margin of the aperture of a gastropod shell. **4.** A rim of fibrocartilage on the margin of the acetabulum or glenoid cavity.

labyrinth. 1. A maze of canals, cavities, or passageways. **2.** *anat.* The internal ear, consisting of the osseous or bony labyrinth (semicircular canals, vestibule, cochlea) within which lies the membranous labyrinth (semicircular ducts, utricle, saccule, and cochlear duct).

labyrinthine. Pertaining to or of the nature of a labyrinth.

labyrinthine receptors. The maculae and cristae.

labyrinthine sense. The sense of balance or equilibrium.

labyrinthodont. A fossil amphibian of the subclass Stegocephalia.

lac. One of a number of natural resins secreted by various species of scale insects used in the manufacture of waxes, varnishes, and lacquers. *See* lac insect.

lace bug. A hemipterous insect of the family Tingididae, with a sculptured, lacelike pattern on the dorsal side of body.

lacerated. 1. Torn; mangled. **2.** Irregularly cleft at the edge, as though torn, said of leaves.

laceration. 1. The act of tearing. **2.** In sea anemones, a method of asexual reproduction in which a portion of the pedal disc is left behind when the animal moves; pedal laceration.

lacewing. A neuropterous insect of the family Chrysopidae, with large, delicate, finely-veined wings, as *Chrysopa oculata*.

lachrymal. *See* lacrimal.

lacinate. Jagged, fringed, deeply incised.

lacinia. 1. *bot.* A narrow, incised portion of a leaf. **2.** *zool.* The innermost of two processes on the stipes of the maxilla of an insect.

lac insect. One of a number of scale insects of the family Coccidae, especially *Laccifer lacca*, of Southeast Asia, the source of lac used in the manufacture of shellac.

lacrimal. Of or pertaining to tears; lachrymal.

lacrimal apparatus. The organs involved in the production and flow of tears. Includes the lacrimal gland and excretory ducts, lacrimal canaliculi, lacrimal sac, and nasolacrimal duct leading to the nasal cavity.

lactase. An enzyme in intestinal juice which hydrolyzes lactose to dextrose and galactose.

lactation. 1. The secretion of milk by the mammary gland. **2.** Suckling, or the period during which young are nourished from the breast.

¹lacteal. Pertaining to or resembling milk.

²lacteal. *anat.* **1.** A minute lymph vessel in a villus of the intestine. **2.** One of the larger lymph vessels of the intestine, containing chyle.

lactescent. 1. Resembling milk. **2.** Producing or yielding milk or a milklike fluid, as latex.

lactic acid. $CH_3CHOHCOOH$, an organic acid formed by the action of bacteria on lactose and responsible for the souring of milk. It is also formed in carbohydrate metabolism by the action of lactic dehydrogenase on pyruvic acid.

lactiferous. Secreting or conveying milk or a milklike substance.

lactogenic. Stimulating the secretion of milk.

lactogenic hormone. Prolactin secreted by the anterior pituitary.

lactose. Milk sugar, $C_{12}H_{22}O_{11}$, a disaccharide present in milk.

lacuna. 1. *anat.* A small space or cavity, especially one in cartilage or bone, containing a cell; also a pit or depression. **2.** *bot.* An air space or cavity in the stems of certain plants.

lacunar. Consisting of or containing spaces or cavities.

lacunose. Containing cavities or lacunae; bearing pits, depressions, or perforations.

lacustrine. Of, pertaining to, or inhabiting lakes.

ladybird beetle. A small, brightly-colored, predaceous beetle of the family Coccinellidae, the larvae and adults feeding principally upon aphids.

ladybug. A ladybird beetle.

lady's slipper. The moccasin flower, an orchid of the genus *Cypripedium* and related genera with a large, showy, flower with a pouch-shaped labellum.

lagena. A small diverticulum on the sacculus of lower vertebrates corresponding to the cochlear duct of mammals.

lagomorph. Any mammal of the order Lagomorpha, principally terrestrial, short-tailed, gnawing herbivores as the pikas, rabbits, and hares; distinguished from rodents by presence of two pairs of incisors in upper jaw.

laking. *See* hemolysis.

Lamarckian theory. A theory of evolution proposed by Jean Baptiste Lamarck (1744–1829) that evolutionary changes result from changes in environment, use and disuse of organs, and that such changes (acquired characters) are inherited.

lamb. A young sheep.

lamella. 1. A thin, platelike structure or layer. **2.** *zool.* **a.** A layer of a gill in lamellibranchs. **b.** A layer of bone surrounding a haversian canal. **3.** *bot.* **a.** A gill of a mushroom. **b.** A submicroscopic layer of a chloroplast. **c.** The original thin wall (middle lamella) separating two protoplasts.

lamellate. Composed of thin plates or lamella.

lamellibranch. A bivalve mollusk of the class Pelecypoda.

lamellicorn. Designating antennae with large, flattened terminal segments.

lamellose. LAMELLATE.

lamina. 1. A thin layer or plate. **2.** *anat.* A thin plate of bone, as that which forms the roof of a neural arch. **3.** *bot.* **a.** The blade of a leaf. **b.** The flattened expanded part of a thallus, as in kelps.

laminarin. A polysaccharide present in brown algae.

laminate. Consisting of or bearing laminae.

lammergeier. The bearded vulture (*Gypaetus barbatus*) of Europe. *See* bonebreaker.

lamprey. A jawless, eellike vertebrate of the class Marsipobranchii, as *Petromyzon marinus*, the sea lamprey, now established in the Great Lakes, where it has become a serious pest preying upon fishes, sucking their blood.

lamp shell. A brachiopod.

lanate. Woolly; covered with long, fine hair.

lancelet. A marine chordate of the subphylum Cephalochordata; an amphioxus.

lanceolate. Pointed; lance-shaped.

land bridge. A connection which formerly existed between two land masses now separated by water.

271

landlocked. Confined to fresh water by being cut off from the sea by a barrier, said of certain fishes.

languet. One of numerous filiform processes which extend from the hyperbranchial band into the branchial chamber in tunicates.

langur. A slender-bodied, long-tailed monkey of the genus *Presbytis* of Southeast Asia; the sacred monkey of India.

lanolin. WOOL FAT.

lanose. Woolly: LANATE.

lanuginous. Covered with down; cottony: WOOLLY.

lanugo. A downy coat of hair on a fetus, shed shortly before or after birth.

lapillus. The utricular otolith in fishes.

lappet. **1.** A small fold or flap, especially one hanging pendant, as those on the margin of a scyphozoan jellyfish. **2.** In crinoids, a movable plate bordering an ambulacral groove. **3.** *ornith.* A wattle.

larch. A coniferous tree of the genus *Larix*, family *Pinaceae*.

lark. A small ground bird of the family Alaudidae, as *Eremophila alpestris*, the horned lark.

larkspur. A plant (*Delphinium*) of the crowfoot family or its dried, ripe seeds used in a tincture for the treatment of pediculosis.

larva. **1.** The young and immature form of an organism which is unlike the adult. **2.** *entomol.* The wormlike, wingless, immature, feeding form which hatches from the egg in insects which undergo complete metamorphosis, variously called caterpillar, maggot, grub, wiggler.

larva migrans. A condition of the skin resulting from invasion by a foreign species of hookworms, usually *Ancylostoma braziliense;* creeping eruption.

larviparous. A form of viparity in which the young produced are larvae, as in blowflies.

laryngeal. Of, pertaining to, or produced by the larynx.

larynx. A sound-producing organ in amphibians, reptiles, and mammals located at upper end of the trachea, a cartilaginous structure containing vocal folds (cords); the voice box.

lasso cell. A colloblast.

latent. Concealed, hidden, not manifest.

latent period. The interval between **a.** application of a stimulus and the response; **b.** exposure to radiation and appearance of effects. **c.** exposure to infection and appearance of symptoms.

laterad. Toward the side.

lateral. 1. *zool.* Of, pertaining to, or directed to the side; away from the midline or median plane. **2.** *bot.* Pertaining to, towards, or borne on the side of a structure; away from the central axis.

lateral line. A pale line extending from head to the tail along the sides of the body of most aquatic vertebrates. It contains minute pores which open into the lateral line canal which lies underneath it.

lateral line system. A system of sense organs in cyclostomes, fishes, and aquatic amphibians located in the lateral line canal. It contains neuromasts which are sensitive to vibrations and pressure.

laterigrade. Running or walking sideways, as certain crabs.

latex. The milky juice of a number of plants from which rubber is obtained.

latex tubes. A system of branching and anastomosing tubes in plants which produce latex.

laughing jackass. A large Australian kingfisher (*Dacelo novaeguineae* or *D. gigas*) noted for its raucous cry.

Laurer's canal. In flukes, a tube extending dorsally from the oviduct, usually ending blindly. When it opens to the outside, it may serve as a vagina.

laver. A dried edible food prepared from algae, green laver from *Ulva*, purple or red laver from *Porphyra*.

laxative. An agent which induces activity of the bowel; a mild cathartic.

layer. 1. A stratum; one thickness of a substance, as a *layer* of cells. **2.** A branch of a plant propagated by layerage. **3.** A bird which lays eggs.

layerage. A method of plant propagation in which a branch is bent downward, covered with soil at which point roots develop, then the branch is separated from parent plant.

leader. The terminal root of a tree; the primary root.

leaf pl. **leaves.** A green, lateral outgrowth of a stem of a plant in which the functions of photosynthesis, transpiration, and

273

respiration are concentrated, consisting typically of a blade, petiole, and stipules. Other functions may include protection, food and water storage, attachment, insect capture, and reproduction.

leaf arrangement. PHYLLOTAXY.

leaf beetle. A beetle of the family Chrysomelidae, as the Colorado potato beetle, striped cucumber beetle.

leaf bug. A sucking hemipterous insect of the family Miridae, as the tarnished plant bug, apple red bug, and garden fleahopper; also called plant bug.

leaf chafer. A beetle of the family Scarabaeidae which feeds on the roots, flowers, and foliage of various plants, as the rose chafer, Japanese beetle.

leaf curl. A virus disease of plants in which leaves curl.

leaf-cushion. A rhomboidal or circular area on the stem of a fossil lycopod.

leaf-cutter bee. A bee of the family Megachilidae which lines its nest with circular pieces of leaves.

leaf-cutting ant. A tropical ant of the genus *Atta* which cuts pieces of leaves used in the cultivation of fungus gardens.

leaf disease. Any disease in which leaves are the principal parts affected, as leaf blotch, mildew, curl, mold, rust, scald, scorch, smut, spot, streak, and others.

leaf gap. A gap in the fibrovascular ring at a node due to a part of the ring passing outward into the petiole of a leaf.

leafhopper. Any of a number of small, homopterous insects of the family Cicadellidae which suck plant juices. Many are serious pests.

leaflet. One of the divisions of a compound leaf.

leaf miner. One of a number of insects whose larva feeds on the tissues between upper and lower surfaces of leaves, causing blotchlike or serpentine mines, as moths of the family Gracilariidae; also certain dipterans and hymenopterans.

leaf mosaic. A rosette arrangement of first-formed leaves with a minimum of leaf overlapping, as in mullein.

leaf roller. A lepidopterous insect of the families Olethreutidae and Tortricidae whose larvae roll or tie leaves and pupate within them.

leaf scar. A scar on a stem marking former attachment of a leaf.

leaf sheath. The basal portion of the blade of a leaf which surrounds the stem, as in leaves of the corn plant.

leaf stalk. A petiole.

leaf tier. An insect which ties leaves together and feeds on the inner surface of the folded leaves, as the greenhouse (celery) leaf tier.

leaf trace. A vascular bundle which extends from the vascular tissue of the stem into the petiole of a leaf; foliar trace.

leafy. 1. Consisting principally of or having many leaves. 2. Of the nature of or resembling a leaf. 3. LAMINATE.

¹lean. Thin, not fat; having little flesh; deficient in quantity or quality.

²lean. To bend or incline from a vertical position, usually necessitating support.

learned. Acquired; not inherited, as *learned* behavior.

leather. The skin of an animal, tanned and prepared for use.

leatherback turtle. The largest marine turtle (*Dermochelys coriacea*), with a leathery skin containing dermal ossicles. May reach a length of 8 feet and weight of 1000 lbs.

leatherjacket. 1. The larva of a crane fly. 2. A carangid fish (*Oligoplites saurus*) of tropical waters.

lecithin. A choline-containing phospholipid universally present in animal and plant tissues, especially abundant in egg yolk and nervous tissue.

leech. A predatory or parasitic bloodsucking annelid worm of the class Hirudinea, as *Hirudo medicinalis*, the medicinal leech, formerly used for bloodletting.

leek. A liliaceous plant (*Allium porrum*) which resembles the onion.

leg. A limb or an appendage of the body used for support or for locomotion (walking or running); in tetrapods, the forelimb or the hindlimb; in bipeds, the hindlimb; in man, the portion of the lower extremity between the knee and the ankle.

legume. 1. A dry, dehiscent fruit which develops from a single carpel or simple ovary, when ripe splitting along both sutures, as that of the bean or pea plant; a pod. 2. Any member of the Leguminosae, the pulse family, a large group of dicots which produce legumes as the characteristic fruit, as beans, peas, peanuts, alfalfa, clover, vetch.

leguminous. 1. Of, pertaining to, or producing legumes. **2.** Belonging to the family Leguminosae, legume-producing plants.

leishmaniasis. A pathological condition or disease caused by a flagellate protozoan of the genus *Leishmania*, as kala azar and Oriental sore.

lelwel. The hartebeest (*Alcelaphus lelwel*) of Africa.

lemma. The lower of two bracts enclosing the flower of grasses.

lemming. One of several species of Arctic, mouselike rodents of the genera *Lemmus* and *Dicrostonyx*, noted for migrating in large numbers at irregular times.

lemniscus. 1. A band of nerve fibers in the brain. **2.** One of a pair of glandular structures in acanthocephalans.

lemon. The yellow, acid fruit of the citrus tree (*Citrus limon*) or the tree which produces it.

lemur. One of a number of primitive primates of the suborder Lemuroidea, solitary, arboreal, nocturnal mammals of Madagascar and neighboring islands. Most belong to the genus *Lemur*.

lenitic. LENTIC.

lens. 1. A piece of glass or other transparent material with one or both surfaces curved for bringing light rays to a focus. **2.** *zool.* **a.** A transparent, biconvex structure present in an eye which focuses light rays on the retina or other light-sensitive structure, as the crystalline *lens* of the vertebrate eye. **b.** A comparable structure in the eye of a cephalopod.

lentic. Of or pertaining to standing, inland waters, as lakes, ponds, swamps; lenitic.

lenticel. An opening on the root or stem of a woody plant through which air is admitted to underlying tissues.

lenticellate. Having lenticels.

lenticular. 1. *bot.* Resembling in shape the seed of a lentil. **2.** *anat.* Having the shape of a double convex lens.

lentil. 1. A leguminous plant (*Lens culinaris*) widely grown throughout Europe and Asia, whose seeds are used for food, the plant for fodder. **2.** The lens-shaped seeds of this plant.

leopard. A large, spotted feline carnivore (*Felis (Panthera) pardus*) of Southern Asia and Africa: PANTHER.

lepidoid. Scalelike, said of certain leaves.

lepidopteran. An insect of the order Lepidoptera which comprises the moths, butterflies, and skippers characterized by two pairs of wings covered by overlapping scales, sucking mouth parts, and complete metamorphosis, the larvae called caterpillars.

lepidopterous. Adapted for pollination by moths or butterflies, said of flowers.

lepidote. Covered by small scales.

lepidotrichia. Flexible, jointed fin rays in the fins of bony fishes.

leporid. A mammal of the family Leporidae which includes the hares and rabbits.

leptocephalus. The larva of freshwater eels of the genus *Anguilla*, which develop from eggs laid in deep waters of the Atlantic.

leptom or **leptome.** Phloem or bast.

leptome. The region in the exine of a pollen grain where the pollen tube makes its exit.

leptoscope. An apparatus for measuring the thickness of thin membranes, as the plasma membrane of a cell.

leptosporangiate. Originating from a single cell, as in most ferns, with reference to sporangium formation. *Compare* eusporangiate.

leptotene. A stage in prophase of meiosis in which the chromosomes are long, slender filaments.

lesion. A structural alteration of an organ resulting from injury or disease processes.

lespedeza. A bush clover (*Lespedeza*), an important forage legume.

lethal. Deadly, fatal; pertaining to or causing death.

lethal gene. A gene which, when it expresses itself, brings about the death of the individual. Also called lethal factor.

leucine. An essential amino acid, $C_6H_{13}O_2N$.

leuco-. For words not found here, see LEUKO-.

leuconoid. Pertaining to a type of canal system in sponges consisting of a complicated branching series of canals with choanocytes limited to small cavities. *See* rhagon.

leucoplast. A colorless plastid which functions in the synthesis and storage of starch; a leucoplastid.

leucosin. Chrysolaminarin, a storage carbohydrate present in certain algae.

leukemia. A neoplastic disorder of blood-forming tissues characterized by formation of an excessive number of white blood cells.

leukoblast. A cell which gives rise to a leukocyte.

leukocyte. A colorless or white blood cell, nucleated, ameboid cells of the following types: granulocytes (neutrophils, eosinophils, basophils) and agranulocytes (lymphocytes, monocytes). Also leucocyte.

leukocytosis. An increase in the number of white blood cells above normal numbers, as occurs in infections.

levator. A muscle which raises or elevates a part.

leveret. A hare or rabbit in its first year.

levorotatory. Pertaining to rotation or turning to the left (counterclockwise), with reference to a plane of polarized light.

levulose. Fructose or fruit sugar, $C_6H_{12}O_6$, a levorotatory hexose present in fruit juices and honey.

Leydig cells. *See* interstitial cells, 2.

liana. A climbing plant or vine, especially a perennial, woody, climbing plant of the tropics, characteristic of the Amazon forests.

lianous. Vinelike, creeping.

lice. Plural of LOUSE.

lichen. 1. An association of two plants consisting of a fungus, usually an ascomycete, living symbiotically with an alga, occurring on solid surfaces as rocks, trees. **2.** *med.* One of a number of skin disorders.

life. The sum of the properties of protoplasm, namely metabolism, growth, irritability, movement, and reproduction as manifested by a cell, group of cells, or an organism by which such is distinguished from inorganic or nonliving matter; the functional activities of an organism.

life cycle. The complete life history of an organism from any one stage to the recurrence of that stage.

life expectancy. 1. The average length of life of all individuals born at a certain time. **2.** The probable number of years that a person of a particular age group will survive.

life-form. The general term applied to the shape and appearance of an organism, as, for plants: annual, biennial, perennial, herbaceous, woody; for animals: aquatic, fossorial, cursorial, saltatorial, aerial.

life history. The complete series of changes in an organism beginning with its origin from an ovum, zygote, spore, or other primary stage and continuing to its natural death.

lifeless. Without life; devoid of vital activity.

life plant. A tropical plant (*Bryophyllum*) which reproduces by the development of plantlets on the margins of its leaves.

life span. The extreme limit of life for an individual of a given species.

ligament. 1. A band of dense, fibrous connective tissue which connects two bones at a joint. **2.** A fold of a peritoneum which connects two organs or supports an organ, as the *broad ligament* of the uterus. **3.** A fibrous cord representing the remains of a fetal structure, as the *round ligament* of the liver.

ligate. To tie off a vessel, as an artery, vein, or duct.

ligature. 1. The act of tying off a vessel or duct. **2.** A cord or band used in tying off a vessel. **3.** Any structure which binds.

light. A form of energy consisting of electromagnetic waves, which, when acting on a visual receptor, gives rise to the sensation of vision. Visible light consists of waves ranging from 4000 to 7800 AU in length.

lightning bug. A firefly.

ligneous. Of the nature of or resembling wood; woody.

lignicolous. Growing upon or in wood, as various fungi.

lignification. The impregnation of cellulose with lignin, resulting in formation of wood.

lignify. To convert into wood.

lignin. A substance present in the cell walls of woody plants which, with cellulose, forms wood. Also present in the shells of nuts and coats of certain seeds.

lignite. Wood coal or brown coal, a variety of coal intermediate between peat and soft coal.

ligula. 1. *zool.* **a.** The terminal, lobed portion of the labium of an insect. **b.** The thickened line of attachment of the tela choroidea to the medulla oblongata. **c.** A lobe of a parapodium of certain polychaetes. **2.** *bot.* A ligule.

ligulate. Straplike; possessing a ligula or ligule.

ligule. *bot.* **1.** A strap-shaped corolla of a ray flower of a

279

composite. **2.** A membranous projection from the summit of the leaf sheath in grasses. **3.** A scalelike structure on the leaves and sporophylls of club mosses.

lilac. A common garden shrub (*Syringa vulgaris*) of the family Oleaceae.

lily. A plant of the genus *Lilium* or the flower or bulb produced by it, of the family Liliaceae.

limacine. Of or pertaining to slugs.

limb. 1. *zool.* **a.** One of a pair of appendages as an arm, leg, or wing, used for locomotion or grasping. **b.** A limblike structure, as one of the parts of the internal capsule of the brain or of the loop of Henle of a nephron. **2.** *bot.* **a.** A large branch of a tree. **b.** The expanded portion of a gamopetalous corolla. **c.** The broad expanded portion of a leaf or petal.

limbate. Bordered, as when the margin of a petal or leaf differs in color from the central portion.

limbus. 1. An edge or border, especially one with a distinct color or structure. **2.** In anthozoans, a constricted region near the base.

¹**lime.** A tropical tree or shrub (*Citrus aurantifolia*) or its small, yellowish-green, acid fruit.

²**lime.** *chem.* Calcium oxide (CaO) or quicklime, prepared by heating (calcining) calcium carbonate thus driving off carbon dioxide.

limen. A boundary line or threshold.

limestone. Natural calcium carbonate.

limicoline. 1. Shore-inhabiting. **2.** Of or pertaining to shore-birds.

limicolous. Inhabiting mud.

liminal. Of or pertaining to a limen or threshold.

limivorous. Swallowing mud containing food.

limnetic. Pertaining to organisms inhabiting open waters of ponds, lakes, and inland seas.

limnobiology. LIMNOLOGY.

limnology. The science which deals with life in inland waters and all factors which influence it.

limnoplankton. Lake plankton.

limpet. A sessile, marine gastropod with a low conical shell.

limpid. Clear, transparent, translucent.

limpkin. A large tropical, marsh bird (*Aramus guarauna*) of the family Aramidae; courlan.

linden. Any tree of the genus *Tilia*, especially *T. americana*, the American basswood.

linea. A line or linelike mark or structure.

lineage. A line of descent with reference to **a.** the offspring or progeny of a single ancestor; **b.** the succession of ancestors from which an individual has descended.

lineal. **1.** In a straight line. **2.** In the direct line of ancestry: HEREDITARY.

linear. **1.** Of or pertaining to a line: LINEAL. **2.** Directly or in a straight direction. **3.** *bot.* Long, narrow, and of uniform width, with reference to leaves.

lineate. Bearing parallel lines or stripes.

linen. Thread or fabric made from the fibers of flax.

lineolate. Possessing or marked by fine, parallel lines; minutely lineate.

ling. **1.** A marine food fish (*Molva molva*) of the North Atlantic. **2.** BURBOT.

lingua. A tongue or tonguelike structure.

lingual. **1.** Of or pertaining to the tongue. **2.** Produced by or resulting from action of the tongue, as certain sounds.

linguatulid. A tongue worm.

linguiform. Tongue-shaped.

lingulate. Shaped like a tongue.

linin. Term formerly applied to a light-staining reticulum connecting the chromatin particles of a nucleus.

linkage. The tendency of certain traits (or their genes) to remain together and be inherited as a group rather than to assort independently, due to the genes being located on the same chromosome. *See* crossing-over.

linkage map. A chromosome map, a graphic representation of a chromosome showing the position of the genes and their relative distances from each other.

Linnaean. Pertaining to Carolus Linnaeus (Carl von Linné), a Swedish naturalist, the author of *Systema Natura*, the 10th edition (1758) of which became the basis of the binomial system of nomenclature employed in the classification of animals and plants.

linnet. An Old World finch (*Carduelis cannabina*).

linsang. A long-tailed, catlike carnivore of the genus *Prionodon* of Southeast Asia.

linseed. Flaxseed. *See* flax.

lion. A large, carnivorous mammal (*Panthera* (*Felis*) *leo*) of the family Felidae found principally in Africa and Southeast Asia.

lioness. A female lion.

lip. 1. One of two fleshy structures which surround the mouth opening in man and other vertebrates. **2.** A corresponding structure in other animals, as the labium or labrum of an insect. **3.** The rim or a projecting edge of a tube or cavity, as the glenoid lip of the acetabulum. **4.** The edge of a valve in a bivalve mollusk. **5.** *bot.* **a.** The labellum of an orchid. **b.** One of the two divisions of a bilabiate calyx or corolla. *See* labium.

lipase. An enzyme which catalyzes the hydrolysis of fats.

lip cells. In ferns, two thin-walled cells in the wall of a sporangium which mark the point of rupture.

lipid. Any of a group of organic compounds which are insoluble in water but soluble in organic solvents (alcohol, benzene, ether). Most upon hydrolysis yield fatty acids. The group includes true fats, oils, waxes, phospholipids, glycolipids, sterols, and some hydrocarbons, as carotene.

lipoblast. A cell which is capable of developing into a fat cell.

lipochrome. Any of a number of fat-soluble pigments as carotene, xanthophyll, or lutein present in yellow vegetables, egg yolk, and corpora lutea.

lipoid. Resembling fat or oil.

lipophore. A chromatophore containing red, yellow, or orange pigment.

lipoprotein. A conjugated protein consisting of a simple protein and a lipid.

lirella. A long, narrow apothecium of a lichen.

litchi. A tree (*Litchi chinensis*) of Southeast Asia or the fruit it produces.

lithocyst. A statocyst.

lithocyte. A cell containing a statolith; present in the statocysts of various invertebrates.

lithosere. A sere which develops upon a rock surface.

lithostyle. A club-shaped static organ in certain medusae; sense club: TENTACULOCYST.

litmus. A purple dyestuff obtained from lichens used as an indicator, with a pH range of 4.5 to 8.3., that turns red in acid solutions and blue in alkaline.

litter. Collectively, the young of a multiparous animal at one birth.

littoral. 1. Of or pertaining to the seashore, especially the region between tide lines. 2. In lakes, pertaining to the region between the shoreline and the outer limit of rooted plants.

live. Alive; viable, not dead.

liver. 1. In vertebrates, a large, glandular organ which secretes bile and performs a number of other functions as glycogenesis, glycogenolysis, deamination of amino acids and formation of urea, storage of vitamins and minerals, detoxication, denaturization of hormones, hemopoietic and phagocytic functions. 2. In various invertebrates, any of a number of glandular, digestive organs which perform comparable functions.

liver fluke. A parasitic trematode inhabiting the liver and bile ducts of vertebrates, as *Fasciola hepatica*, infesting sheep, and cattle, and *Opisthorchis* (*Clonorchis*) *sinensis*, the Chinese liver fluke, infesting cats and dogs. Both may infest man.

liverwort. A simple plant of the phylum Bryophyta, class Hepaticae, with a flattened gametophyte, as *Marchantia*.

livid. 1. Of a grayish-blue or lead color. 2. Black and blue, as bruised flesh.

living. Alive, not dead; possessing the properties of life; animate.

lizard. 1. In common usage, any small to moderate-sized, four-legged, long-tailed reptile. 2. Specifically, any reptile of the suborder Sauria (Lacertilia), which includes the geckos, chameleons, skinks, monitors, and some limbless forms.

llama. 1. A South American ruminant of the camel family, as the guanaco and vicuna. 2. A domesticated form of the guanaco used in Peru as a pack animal.

loach. A small Old World, freshwater fish of the family Cobitidae, some of which breathe by intestinal respiration.

loam. Soil containing considerable organic material.

loa worm. The African eye worm (*Loa loa*). *See* eye worm.

lobar. Of or pertaining to a lobe or lobes.

lobate. Composed of or possessing lobes or divisions resembling lobes.

lobe. A rounded division or projection of an organ.

lobo. In the western United States, the timber wolf.

lobopodium. A blunt, fingerlike pseudopodium; lobopod.

lobose. Having lobopodia.

lobster. A large malacostracan crustacean of the order Decapoda, especially those of the genus *Homarus* used for food, as *H. americanus*, the American lobster, and *H. gammarus*, the European lobster. Also those of the genus *Nephrops*, as *N. norvegicus*, the Norwegian lobster. *See* spiny lobster.

lobule. A subdivision of a lobe; a small lobe.

lobulus. A lobule.

local. Limited to or occupying a particular place or location; not general or widespread.

localization. The limitation of function or activity to a specific location or area.

localized. Confined to or restricted to a limited area; not general or widespread.

lockjaw. TETANUS.

loco disease. A disease of livestock resulting from eating locoweeds.

locomotion. The act of moving from place to place.

locoweed. One of a number of plants of the genera *Astragulus* and *Oxytropis* which contain unknown poisons or selenium taken up from the soil. Eating the plants causes poisoning of sheep, cattle, and horses.

locular. LOCULATE.

loculate. Possessing compartments or cells (loculi).

locule. A loculus.

loculocidal. A form of dehiscence in which the splitting occurs along a dorsal suture directly into each locule of the ovary.

loculus pl. **loculi.** **1.** A compartment, cavity, or chamber. **2.** *bot.* A cavity of a compound ovary, pollen sac, or certain ascocarps. **3.** *zool.* **a.** A chamber in the shell of a foraminiferan protozoan **b.** A pocket between the septa at the base of an anthozoan polyp.

locus. A place, locality, or position.

locust. **1.** *zool.* **a.** A short-horned grasshopper of the family

Acrididae, as *Melanoplus mexicanus*, the Rocky Mountain locust, a migrating species. **b.** Common name for the cicada. **2.** *bot.* A tree, *Robinia pseudoacacia*, the black locust, or *Gleditsia triacanthus*, the honey locust, of the family Leguminosae.

lodicule. In the flower of grasses, one of a pair of small scales located at the base of the ovary.

loggerhead sponge. A massive sponge (*Spheciospongia*) sometimes reaching a diameter of several meters.

loggerhead turtle. A large marine turtle (*Caretta caretta*) of the Atlantic coast, sometimes reaching a length of 4 feet and weight of 400 lbs.

log perch. A darter (*Percina caprodes*) of Central North America.

logwood. The heartwood of a tree (*Hematoxylon campechianum*) from which a dye, hematoxylin, is obtained. A native of Central America, it is now grown throughout the West Indies.

loin. The region of the body between the pelvis and false ribs.

loment. An indehiscent legume with constrictions between the seeds which, at maturity, breaks into one-seeded portions.

lomentum. A loment.

long-day plants. Plants which flower only when subjected to twelve hours or more of daily illumination, as most summer flowering annuals. *See* short-day plant.

longitudinal. Lengthwise; extending along the long axis.

longspur. A sparrowlike bird of the genera *Rhynchophanes* and *Calcarius*, of the family Fringillidae.

loon. A large, fish-eating, waterbird of the family Gavidae, as the common loon (*Gavus immer*).

loop. A bend or fold of a cordlike structure.

looper. A moth of the family Phalaenidae (Noctuidae) whose larvae have a reduced number of prolegs and move like measuring worms.

loop of Henle. A U-shaped portion of a renal tubule connecting proximal and distal convoluted portions.

lophodont. Designating a type of dentition in which molar teeth bear prominent transverse ridges and are adapted for grinding, as in ungulates and rodents.

[1]**lophophorate.** Possessing a lophophore, as animals of the phyla

285

Phoronida, Ectoprocta (Bryozoa), Entoprocta, and Brachiopoda.

²**lophophorate.** Any animal which possesses a lophophore.

lophophore. A circular or U-shaped, food-getting structure bearing ciliated tentacles present in certain invertebrates. *See* lophophorate.

lophotrichous. Having a tuft of flagella at one pole, as in certain bacteria.

loquat. A small evergreen tree (*Eriobotrya japonica*) widely grown in the Orient, or the fruit it produces.

lorate. Strap-shaped.

lore. In birds, the space between the eye and the bill or the corresponding space in other vertebrates, as reptiles.

lorica. 1. A protective case or covering. 2. *zool.* **a.** A case or shell enclosing ceratin ciliate protozoans. **b.** The thin, chitinlike covering of rotifers. 3. *bot.* That covering the protoplast of certain algae.

loricate. Possessing a lorica.

loris. A small nocturnal lemur of Southeast Asia, as *Nycticebus coucang*, the slow loris, and *Loris tardigradus*, the slender loris.

lory. A brilliantly colored Australian parrot.

lotic. Pertaining to running water (brooks, streams, creeks, rivers).

lotus. 1. A flowering water plant of the Far East, of religious significance, as the sacred lotus of India (*Nelumba nucifera*). 2. A leguminous plant of the genus *Lotus*, the bird's-foot trefoil.

louse pl. **lice.** 1. A wingless, sucking, parasitic insect of the order Anoplura comprising the true lice (head, body, and crab lice). 2. Any insect of the order Mallophaga, comprising the biting lice. 3. Any of a number of small anthropods resembling lice, as plant lice, book or bark lice, wood lice, fish lice, whale lice.

louse fly. A parasitic dipterous insect of the family Hippoboscidae, *Melophagus ovinus*, the sheep tick or ked.

lovebird. A small tropical parrot, widely sold as a cage bird, as *Agapornis* and *Psittacula* which show great affection for their mates.

loxodont. With shallow grooves between the ridges of molar teeth, as in those of the African elephant.

lubricous. Smooth, slippery, slimy.

luce. A full-grown pike.

lucid. 1. Bright, shining. **2.** Clear.

luciferase. An enzyme which catalyzes the oxidation of luciferin.

luciferin. A substance present in light-producing organs of bioluminescent organisms which, when oxidized, yields light energy, as in fireflies.

luciferous. Producing or emitting light.

lucifugous. Avoiding or moving away from light.

lugworm. A burrowing marine polychaete (*Arenicola marina*).

lumbar. Of or pertaining to the region of the loin.

lumen. 1. *zool.* A passageway within a tube, vessel, or duct. **2.** *bot.* A cavity within the cell wall after the protoplast has disappeared, as in tracheids. **3.** *physics.* A unit of luminous flux, that is, the quantity of visible radiation per unit of time.

luminescence. The emission of light without appreciable production of heat or incandescence, as in bioluminescent organisms.

luminous. Bright or shining; reflecting, emitting, or producing light; phosphorescent.

luminous organ. An organ which emits light; a photophore.

luna moth. A large, pale green, saturnid moth (*Tropaea* (*Actias*) *luna*) called the "pale empress of the night."

lunar. Of, pertaining to, or caused by the moon.

¹lunate. Crescent-shaped.

²lunate. The semilunar bone, a carpal bone.

lung. 1. One of a pair of saccular or spongy organs of respiration present in air-breathing vertebrates. **2.** One of a number of comparable organs, as **a.** the swim bladder of lungfishes; **b.** the modified mantle cavity of a pulmonate snail; **c.** the book lung of an arachnid.

lung book. *See* book lung.

lungfish. One of three genera of dipnoian fishes in which the swim bladder functions as a lung, as *Neoceratodus* of Australia, *Lepidosiren* of South America, and *Protopterus* of Africa.

lung fluke. A fluke which inhabits the lungs, as *Paragonimus westermanii* in man and other mammals.

lungworm. One of a number of parasitic nematodes living within the lungs, especially *Metastrongylus elongatus*, the lungworm of swine.

lungwort. The Virginia cowslip or bluebell (*Mertensia virginica*).

lunula. A narrow, crescent-shaped area or structure, as **a.** the clear area at the base of a nail; **b.** a slit above the base of the antenna in certain insects, especially dipterans; **c.** a notch or opening in the test of certain sand dollars.

lustrous. Glossy, radiant, shiny.

luteal. Of, pertaining to, or produced by the corpus luteum.

lutein. **1.** Xanthophyll, a yellow pigment in plants. **2.** A yellow pigment present in cells of the corpus luteum, egg yolk, and other lipochromes. **3.** A dried powdered preparation of corpora lutea from hogs.

luteinizing hormone. A gonadotrophic hormone (LH) secreted by the anterior lobe of the hypophysis. With FSH, it stimulates secretion of estrogens by the follicle, ovulation, and development of the corpus luteum; with prolactin, it stimulates secretion of progesterone. In the male, it stimulates production of testosterone by the interstitial cells of the testis, hence is called interstitial cell-stimulating hormone (ICSH).

luteotrophic hormone. *See* prolactin.

lutescent. Turning or becoming yellow.

luxation. A dislocation, as at a joint.

luxuriant. Rank, exuberant, profuse in growth.

lycopene. A carotene pigment present in the chromoplasts of ripe fruits.

lycopod. A club moss.

lymph. The clear fluid occupying lymphatic vessels and ducts, originating from the interstitial fluid of tissue spaces.

lymphatic. **1.** Of, pertaining to, or conveying lymph. **2.** A small lymph vessel.

lymphatic organ. An organ composed principally of lymphatic tissue, as Peyer's patch, lymph node, hemolymph node, tonsil, thymus, spleen.

lymphatic system. A system of vessels and associated organs consisting of lymphatic vessels and ducts and lymphatic organs which function in the return of tissue fluid to the blood stream, filtration of lymph, manufacture of leukocytes, and formation of antibodies.

lymphatic tissue. The tissue which comprises lymphatic organs, consisting of a stroma of reticular tissue and free cells (macrophages and lymphocytes) within the meshes of the fibers. It functions in filtration of lymph and formation of lymphocytes.

lymph heart. A dilatation of a lymph vessel with contractile tissue in its wall, as in amphibians.

lymph node. An encapsulated body consisting of several lymph nodules interposed in the course of a lymph vessel. It functions as a lymph filtering and blood-forming organ.

lymphocyte. A type of leukocyte originating in lymphatic tissue. Most are small, agranular cells with a large spherical nucleus and little cytoplasm. They are nonphagocytic but actively motile; they function in repair of tissue and formation of antibodies.

lymphoid. Resembling or of the nature of lymph. Term is sometimes used as a synonym for lymphatic.

lynx. A wildcat of the genus *Lynx* characterized by short stubby tail and usually tufted ears, as *Lynx canadensis*, the Canada lynx and *L. rufus*, the wildcat or bobcat.

lyophilization. Rapid freezing of a substance followed by dehydration under high vacuum. Also called freeze-drying or gelsiccation.

lyrebird. An Australian bird of the family Menuridae, the males possessing a beautiful, lyre-shaped tail when spread. Common genera are *Menura* and *Harriwhitea*.

lyriform organs. Lyre-shaped organs in certain spiders thought to have an olfactory function.

lysigenous. Originating by the dissolution of cells, with reference to secretory cavities or glands of plants, as oil cavities in the rinds of citrus fruits.

lysin. An antibody or other agent which brings about the dissolution of cells, as a cytolysin, hemolysin, bacteriolysin.

lysis. The dissolution or destruction of cells by lysins or other lytic agents.

lysogenic. 1. Inducing the production of lysins. **2.** Inducing lysis.

lysosome. A cytoplasmic body rich in hydrolytic enzymes, present in certain cells, especially phagocytes.

lysozyme. An antibacterial enzyme present in egg white, tears, nasal secretions, and saliva.

lytic. Inducing or bringing about lysis.

M

macaco. One of several lemurs, especially *Lemur macaco*, the black lemur.

macaque. The common rhesus monkey (*Macaca mulata*), a short-tailed monkey of Southeast Asia. Also called bandar.

macaw. A noisy, brightly-colored parrot of Central and South America, most species belonging to the genus *Ara*.

mace. An aromatic spice obtained from the shell of the nutmeg.

macerate. 1. To soften, as by boiling, to bring about the separation of parts. 2. In microtechnique, to treat tissues so as to cause dissociation of cells.

mackerel. One of a number of important food fishes of the family Scombridae, as *Scomber scombrus*, the Atlantic mackerel, and *Scomberomorus maculatus*, the Spanish mackerel.

macrandrous. Having large male plants or structures; producing antheridia on large filaments.

macrocarpous. Bearing or producing large fruit.

macroclimate. The general overall climate of a large area.

macrocyst. A reproductive body formed in certain slime bacteria within which develop microcysts.

macrodont. Possessing abnormally large teeth.

macroevolution. Evolution in which marked alterations occur in the genetic makeup of an organism, thus accounting for the origin of larger groups as classes and phyla. *Compare* microevolution.

macrogamete. The larger of two gametes which undergo conjugation, as in certain algae and protozoans, considered the female gamete as it is usually passive and nonmotile. *Compare* microgamete.

macromere. One of the four large blastomeres at the vegetal pole of a telolecithal ovum undergoing cleavage, as in the eggs of amphibians.

macromolecule. A large molecule composed of smaller units or monomers linked together, as a protein composed of amino acids.

macromutation. A pronounced, abrupt change which results in an individual differing markedly from its parents, in contrast to a small or point mutation.

macronucleus. The vegetative nucleus, the larger of two nuclei present in certain protozoans, as *Paramecium*.

macronutrient. A nutrient which is required by a plant in relatively large amounts, as nitrogen, phosphorus, potassium, calcium, magnesium, sulfur, and iron. *Compare* micronutrient.

macrophage. A phagocytic cell of the reticuloendothelial system which has the ability to engulf particulate matter and to store vital dyes, as a histiocyte of connective tissue or Kupffer cell of the liver.

macrophagous. Feeding on large particles of food.

macrophyll. A megaphyll.

macrophyllous. Having large leaflets or leaves.

macroplankton. Organisms in plankton which can be recognized by the unaided eye.

macropterous. Having large wings or winglike structures; having large fins.

macroscopic. Visible to the unaided eye.

macrosmatic. Having a well-developed olfactory sense.

macrospore. A megaspore.

macrostylous. With long styles.

macroszoospore. A large, asexual spore, as in *Ulothrix*.

macrurous. Having a long tail, with reference to crustaceans, as the lobster.

macula. A spot or blotch, especially a discolored spot on the skin.

macula lutea. A yellow spot on the retina of the eye in the center of which is the fovea centralis upon which light rays are focused. It is the point of keenest vision.

macula sacculi. An area of sensory epithelium in the saccule of the ear containing hair cells which are stimulated by the movement of otoliths in the otolithic membrane on its surface. It is concerned with static equilibrium.

macula utriculi. A sensory area similar to the macula sacculi but located in the utriculus.

madrepore. A coelenterate of the order Madreporaria

(Scleractinia) which includes the stony corals, reef-building corals of tropical waters.

madreporite. A porous, calcareous plate on the aboral surface of various echinoderms through which water enters the water-vascular system.

madtom. A small catfish of the genus *Noturus*.

maggot. **1.** The soft-bodied, legless larva of various dipterous insects, commonly found in decaying matter. **2.** A parasitic copepod. *See* gill maggot.

magnesium. A metallic element, symbol Mg, at. no. 12, widely distributed in the earth's crust, an essential constituent of chlorophyll and many enzyme systems.

magnification. The apparent enlargement of an object when seen through a series of lenses, as a microscope.

magnify. To increase the apparent size of an object; to enlarge.

magnolia. A tree of the genus *Magnolia*, noted for its large, beautiful, fragrant flowers. Also called cucumber tree.

magpie. A long-tailed corvine bird (*Pica pica*) of the Northern Hemisphere.

mahogany. **1.** A tree (*Swietenia mahogani*) in the Meliaceae family, of tropical America noted for its hard, reddish wood used in the manufacture of furniture. **2.** One of a number of trees producing similar wood, especially *S. macrophylla* of Mexico and Central America.

maidenhair fern. *Adiantum pedatum*, with finely divided, delicate fronds.

maidenhair tree. The ginkgo.

maidenhead. The hymen.

mail-cheeked. Designating fishes of the order Scleroparei (Scleropareida) with a bony process across the cheek, as sculpins, gurnards, sticklebacks, ocean perch.

maize. Indian corn (*Zea mays*), a cereal grain.

malaceous. Of or pertaining to apples.

malacia. Softening of an organ or tissue.

malacology. The branch of biology which deals with mollusks.

malacostracan. Of or pertaining to crustaceans of the subclass Malacostraca, which includes the lobsters, crayfishes, crabs, shrimps, and related forms.

malaise. A general feeling of not being well, accompanied by restlessness, loss of appetite, and decreased activity.

malar. 1. Of or pertaining to the cheek. **2.** The zygomatic bone.

malaria. An acute and sometimes chronic, febrile disease of man and other vertebrates, caused by sporozoans of the genus *Plasmodium* transmitted by anopheline mosquitoes.

¹male. 1. *zool.* Of, pertaining to, or designating an individual of the sex which produces spermatozoa. **2.** *bot.* **a.** Of, pertaining to, or designating an individual, organ, or structure which produces the cell by which fertilization is accomplished. **b.** In seed plants, staminate. **3.** With reference to tools or instruments, designating a part which fits into another part.

²male. An individual of the male sex.

male fern. Aspidium, the rhizome and stipes of *Dryopteris filix-mas*, which contains filicin, formerly used for expelling tapeworms.

male-sterile. Producing no visible pollen, said of plants.

malignant. Virulent; inducing harm or injury; threatening life.

mallard. A common wild duck (*Anas platyrhynchos*), a surface-feeding, freshwater duck, considered to be the ancestor of all domestic varieties of ducks.

¹malleate. Shaped like a hammer.

²malleate. A type of mastax in rotifers adapted for grinding, grasping, and pumping.

malleolus. One of the two processes on either side of the ankle joint, one on the tibia, the other on the fibula.

malleus. 1. The hammer, the first of the three ear ossicles. **2.** A portion of the trophi of a rotifer.

mallophagan. An insect of the order Mallophaga which includes the biting or chewing lice, external parasites of birds and mammals.

Malpighian body. 1. A renal corpuscle. **2.** A splenic corpuscle.

Malpighian layer. The stratum germinativum of the epidermis of the skin.

Malpighian tubule. One of the numerous, blind, threadlike, excretory tubules opening into the anterior end of the hindgut of insects.

malt. Grain, usually barley, which has softened by soaking in water, then allowed to germinate; used in the brewing and distilling industry.

maltase. An enzyme which converts maltose to dextrose, present in germinating grains, saliva, and intestinal juice.

malting. The production of malt.

maltose. A disaccharide, $C_{12}H_{22}O_{11}$, formed in the hydrolysis of starch.

malvaceous. Of or pertaining to plants of the mallow family, Malvaceae, including cotton, hemp, jute, okra, hibiscus, and hollyhock.

mamba. The tree cobra (*Dendraspis*), a poisonous snake of Africa.

mamma pl. **mammae. 1.** The mammary gland which secretes milk, characteristic of the class Mammalia. In males it is a rudimentary structure. **2.** The breast including the mammary gland and associated structures.

mammal. Any vertebrate of the class Mammalia, characterized by possession of hair and mammary glands. They are air-breathing, possessing lungs, have a four-chambered heart and are warm-blooded. All are viviparous except the egg-laying monotremes.

mammalian. Of, pertaining to, characteristic of, or belonging to the class Mammalia.

mammalogy. The study of mammals.

mammary gland. A compound alveolar gland present in female mammals, an apocrine gland which secretes milk which is discharged through lactiferous ducts opening on a nipple or teat.

mammiferous. Possessing mammary glands.

mammiform. Shaped like a breast; conical with a nipplelike or rounded tip.

mammilla. 1. *zool.* A small prominence or papilla; a nipple or teat. **2.** *bot.* A rounded protuberance, as that on a spore wall.

mammillate. Having nipples or nipplelike protuberances.

mammoth. An extinct elephantlike mammal (*Elephas* (*Mammuthus*) *primigenius*) with long, recurved tusks and long, thick hair, the remains of which have been found in the frozen tundra of Siberia and Alaska.

mammotrophic. Inducing the development of the mammary gland.

man. 1. A human being, especially an adult male of the human race. **2.** Any member of the genus *Homo*, all present-day individuals considered as belonging to a single species (*Homo* *sapiens*). **3.** The human race or mankind; all human beings considered collectively.

manakin. A tropical bird of the family Pipridae, brilliantly colored birds with a peculiar wing plumage, as the fandango bird (*Chiroxiphia linearis*).

manatee. A sea cow, a large, herbivorous, aquatic mammal of the order Sirenia, inhabiting warm waters, as *Trichetus latirostris*, the Florida manatee.

mandible. 1. In a general sense, a jaw, either upper or lower. **2.** In vertebrates, the skeleton of the lower jaw, whether cartilaginous or bony; the mandibular bone. **3.** In arthropods, one of a pair of mouth parts for cutting, crushing, or grinding, or a part homologous to the same.

mandibular. Of or pertaining to the mandible.

mandibular arch. The first visceral arch.

mandibulate. Any member of the Mandibulata, a subphylum of the Arthropoda characterized by possession of antennae and the first pair of jaws as mandibles. Includes the classes Crustacea, Insecta, Chilopoda, Diplopoda, Symphyla, and Pauropoda.

mandrake. 1. A European herb (*Mandragora officinarum*), the source of many superstitious beliefs pertaining to induction of conception. **2.** The mayapple (*Podophyllum peltatum*).

mandrill. A fierce, tailless baboon (*Mandrillus sphinx*) of West Africa, with brilliantly colored face and buttocks.

mane. The long, thick hair growing along the sides and back of the neck, as in lions, or along the back of the neck, as in horses.

manganese. A mineral element, symbol Mn, at. no. 25, present in minute amounts in water and in animal and plant tissues; a constituent of certain enzyme systems.

mange. A skin condition characterized by inflammation, exudation, and scab formation resulting from infestation by parasitic mites, as *Demodex*, the follicle mite and *Sarcoptes*, the itch or mange mite.

mangel. A form of beet grown extensively in Europe and Canada, used for feeding cattle.

mangel-wurzel. *See* mangel.

mango. A well-known tropical tree (*Mangifera indica*) or the fruit it produces, used extensively for food. Its seeds and bark are used medicinally.

mangrove. A tropical tree or shrub of the genus *Rhizophora*, with leaves and bark which are rich in tannin. Its prop roots form dense interlacing masses.

manikin also **mannikin** or **manakin. 1.** A model of the human body showing various organs in their normal positions. **2.** A sheet or series of sheets of drawings of the human body showing organs in their relative positions. **3.** A model of a full-term fetus used for instruction in obstetrics. **4.** A man of small stature; a dwarf.

manioc. CASSAVA.

manna. A sweet exudate of the manna ash (*Fraxinus ornus*) of Europe, used as a mild laxative.

mannitol. Mannite or manna sugar, a carbohydrate widely distributed in nature, present in manna, brown algae, and certain fungi.

mannose. A simple sugar obtained by hydrolysis of complex carbohydrates or by oxidizing mannitol. It does not occur free in nature.

manometer. A device for measuring the pressure exerted by gases or liquids. In its simplest form, it is a U-shaped tube containing a fluid (water or mercury) against which the pressure of an unknown is balanced.

manta ray. A large ray (*Manta biostris*) of the suborder Batoidea. Also called devilfish or sea devil.

mantid or **mantis.** A carnivorous, orthopteran insect of the family Mantidae with greatly enlarged front legs adapted for grasping and holding prey. Also called the praying mantid or soothsayer.

mantid fly. A neuropterous insect (*Mantispa cincticornis*) which resembles a mantid in appearance.

mantle. 1. A thin fold or pair of folds of the integument which covers the body or a portion of it in brachiopods and mollusks; in shelled forms, it lines the shell which it secretes; in cephalopods, it forms the body wall. **2.** The carapace of a barnacle. **3.** The body wall or tunic of a tunicate. **4.** *ornith.* The feathers of the back and folded wings; pallium, stragulum.

manubrium. 1. *zool.* A handlelike process or structure, as that seen in medusae or polyps, the mastax of a rotifer, the furcula of a springtail, the sternum of mammals, the malleus of the ear. **2.** *bot.* **a.** An inward-projecting cell in the antheridium of certain algae. **b.** The elongated cylindrical base of certain cymbas and spathes.

manure. **1.** Any fertilizing material, especially the excreta of animals which, when applied to the soil, yields nutrient materials. **2.** Crops plowed under (green manure) for enrichment of the soil.

manus. The distal portion of the forelimb in vertebrates. In man it comprises the hand; in tetrapods, the forefoot.

manyplies. The omasum or psalterium, the third portion of the stomach of a ruminant.

maple. A tree of the genus *Acer*, important as a source of wood, maple syrup and sugar, and as an ornamental tree.

map turtle. A freshwater turtle of the genus *Graptemys*.

mara. A large South American cavy (*Dolichotis patagona*) called Patagonian hare because of its long hind legs.

marabou. A large, ungainly looking stork (*Leptotilus crumeniferus*) of Africa, or *L. dubius* of India.

marbled. Marked by irregular streaks, marks, or veins.

marcescent. Withering, but persistent, that is, not falling off, as leaves.

mare. The mature female of the horse and related equines.

margay. A spotted cat (*Felis tigrina*) of Central and South America.

margin. A border or an edge.

marginal. Of, pertaining to, or located on or near a margin.

marginate. Possessing a margin, especially one of a distinctive nature.

mariculture. Sea farming, as the raising of marine animals, as shrimps or oysters for commercial purposes.

marigold. A yellow- or orange-flowered plant of the genus *Tagetes*.

marihuana or **marijuana.** Cannabis, the dried flowering tops of pistillate plants of hemp (*Cannabis sativa*).

marine. Of, pertaining to, living in, or related to the seas or oceans.

marita. A sexually mature fluke.

markhor. A wild goat (*Capra falconeri*) of the Himalayas; markhoor.

marlin. A large, marine, game fish of the family Istiophoridae, as the blue marlin (*Makaira ampla*).

marmoset. A small, squirrellike monkey of the family Callithricidae of Central and South America, as *Callithrix jacchus*, the common marmoset.

marmot. A stout-bodied burrowing rodent of the genus *Marmota*. *See* whistler, woodchuck.

marrow. The soft, myeloid tissue filling the spaces in bone, *yellow marrow*, consisting principally of fat, filling the medullary cavities of long bones; *red marrow*, composed of hemopoietic tissue, filling the spaces of cancellous bone.

marsh. A swamp or an area of soft, wet land; a morass.

marsh grass. Any of a number of coarse grasses which grow in marshes, especially those of the genus *Spartina*.

marsh hawk. *Circus cyaneus hudsonius*, a harrier which frequents marshy areas.

marsipobranch. Any member of the Marsipobranchii, which includes the cyclostomes.

marsupial. Any member of the Marsupialia, an order of mammals which includes the kangaroos, wallabies, wombats, and opossum. Females bear immature young which are nourished in a marsupium which encloses the nipples.

marsupial frog. A South American frog (*Gastrotheca*), the female of which carries the eggs in a pouch on her back.

marsupium. **1.** *zool.* A pouch or cavity in which young develop, as **a.** that in marsupials and certain monotremes; **b.** the gill of a bivalve mollusk when containing embryos or larvae; **c.** the brood pouch of various crustaceans or the sea horse. **2.** *bot.* The pendant perigynium of various bryophytes.

marten. A carnivorous, fur-bearing mammal (*Martes americana*) of the family Mustelidae.

martin. A swallow (*Progne subis*) of the family Hirundinidae, variously called purple, house, gourd, or black martin.

masculine. **1.** *zool.* Of the male sex; belonging to or having qualities of a male. **2.** *bot.* STAMINATE.

¹**mask.** To cover, hide, conceal, or prevent from showing.

²**mask.** A structure which conceals, as the distal end of the labium of a dragonfly nymph.

mason wasp. A potter wasp of the genus *Eumenes*, which constructs a juglike nest provisioned with caterpillars.

massula pl. **massulae.** A portion of a periplasmodium.

mast. The fruit of various forest trees, especially beechnuts, which is utilized as food by hogs and other animals.

mastax. An enlarged portion of the pharynx of rotifers which contains chitinous jaws used for grinding food; dental mill.

mastic. A resin which exudes from the mastic tree (*Pistacia lentiscus*), used medicinally and as an ingredient of varnishes.

masticate. To grind, chew, and crush the food in preparation for swallowing.

¹**masticatory.** Adapted for or pertaining to the chewing of food.

²**masticatory.** A substance that is chewed but not swallowed, as chewing gum.

mastigate. Having one or more flagella.

mastigium. A protrusible, whiplike anal organ in certain caterpillars.

mastigobranch. A slender, fringed extension of the epipodite of the maxillipeds of decapod crustaceans.

mastigoneme. A flimmer, one of the numerous minute lateral processes on the sides of the tinsel type of flagellum.

mastigophoran. Any protozoan of the class or superclass Mastigophora characterized by possession of one or more flagella, as *Euglena, Trypanosoma;* a flagellate.

mastigote. MASTIGATE.

mastigure. A lizard of the genus *Uromastix* of Southern Asia and Africa.

mastitis. Inflammation of the breast.

mastodon. An extinct proboscidean of the genera *Mastodon* and *Mammut* differing from the mammoth in having molar teeth with high conical cusps. Fossil remains found throughout Europe and United States.

mastoid. Nipplelike.

mastoid cells. Air cells of the mastoid portion of the temporal bone which communicate with the tympanic cavity.

mastoid process. A process of the temporal bone located behind the ear.

mat. 1. A thick, dense, tangled mass of fibers, hair, weeds, or other filamentous structures. **2.** A thin crust formed by the stolons of stoloniferous corals.

¹mate. One of a pair of animals which associate together for mating.

²mate. To join together for breeding.

maté. 1. A South American holly (*Ilex paraguariensis*). **2.** A beverage, Paraguay tea, prepared from the leaves of this holly.

maternal. Pertaining to, produced by, or derived from the female parent or mother.

mating. The joining of individuals of different sexes for reproductive purposes.

matrix. 1. The intercellular substance of cartilage or bone. **2.** A mass of cells (germinal matrix) from which a hair develops. **3.** The posterior portion of the nail bed from which a nail develops.

matroclinous. Resembling the female parent more than the male, with reference to hereditary traits.

maturation. 1. The process of coming to a full state of maturity or development. **2.** The processes involved in the formation of functional gametes, as in spermatogenesis and oogenesis.

mature. Fully grown or developed; ripe.

maturity. The state or quality of being mature.

matutinal. Of, pertaining to, or occurring in the morning.

maw. The portion of the digestive tract into which food is taken, as **a.** the mouth and throat of a carnivore; **b.** the crop of a bird; **c.** the stomach.

maw seed. The seed of the opium poppy, used for bird feed.

maxilla. 1. The bone of the upper jaw. **2.** One of a pair of appendages located posterior to the mandibles in most arthropods.

maxilliped. 1. One of a pair or several pairs of appendages located posterior to the maxillae in crustaceans. **2.** The poison claw of a centipede.

maxillulla. One of the first pair of maxillae in arthropods having two pair.

maximum. The highest, greatest, or largest amount, degree or extent; the upper limit.

mayapple. *See* mandrake.

May beetle. *See* June bug.

mayfly. A soft-bodied insect of the order Ephemeroptera; adults are short-lived, nymphs are aquatic living 1 to 3 years.

maze. A confusing, interconnected series of lines or pathways including blind alleys or compartments used extensively in intelligence tests.

mazzard. A wild cherry (*Prunus avium*) used extensively as a rootstock upon which commercial varieties of cherries are grafted.

meadowlark. A blackbird of the genus *Sturnella*, family Icteridae.

meadow mouse. A small rodent of the genus *Microtus*. Also called field mouse or field vole.

mealworm. The larva of a small, black beetle (*Tenebrio*) which feeds on stored grain or grain products. Commonly reared in laboratories as food for insect-eating animals.

mealybug. A homopterous insect of the family Pseudococcidae which produces a mealy or waxy secretion which covers their bodies, as those of the genus *Pseudococcus*, a serious fruit pest.

mean. Intermediate or occupying a median position; average; midway between extremes.

measles. 1. Rubeola, an acute, infectious disease caused by a virus. **2.** Rubella (German measles, three-day measles) resembling rubeola but milder. In pregnant women, it may cause birth defects. **3.** A disease of domestic animals in which muscles are infected with the larvae (cysticerci) of tapeworms of the genus *Taenia*.

measly meat. Meat infested with the larvae (bladderworms or cysticerci) of tapeworms.

measuring worm. The larva or caterpillar of a moth of the family Geometridae, which moves in a characteristic, looping fashion; a looper or inchworm; a geometer.

meat. 1. The flesh of an animal which is used for food. **2.** The edible part of anything, as the *meat* of a nut.

meatus. A natural opening, passageway, or canal.

mechanical. 1. Of or pertaining to a mechanism or a machine. **2.** Of or pertaining to the action of physical force on matter as distinguished from chemical, vital, mental, or spiritual forces.

mechanism. The doctrine that all phenomena in nature are the result of natural processes and explicable by the laws of chemistry and physics. *Compare* vitalism.

mechanoreceptor. A receptor or sense organ which is stimulated by physical factors as contact with solid objects, air and water currents, etc. It mediates the senses of touch, pressure, equilibrium, and position.

meconium. 1. The intestinal contents of a fetus which accumulates before birth. **2.** The intestinal contents of an organism which accumulates during its dormant state, as that of the pupa of an insect.

mecopteran. An insect of the order Mecoptera which includes the scorpion flies.

¹media. Plural of MEDIUM.

²media. 1. The middle coat of a blood or lymph vessel, as the *tunica media*. **2.** A vein in the wing of an insect.

mediad. Toward the median plane or midline of the body.

medial. Located in or toward the median plane.

median. 1. Located in the middle of the body or a part of it, as the *median* plane. **2.** In statistics, in a series of numerical values, the point above which the number of individuals in the series equals the number below it.

mediastinum. A partition which separates two adjacent parts, especially the *thoracic mediastinum* or *mediastinal septum* which separates the two pleural cavities.

mediate. To serve as an intermediary or a medium by which an effect is produced, as a sensory receptor which *mediates* a sensation.

medicinal. 1. Having therapeutic or healing properties. **2.** Pertaining to or caused by a medicine, as a *medicinal* rash.

medicine. 1. Any substance used in the diagnosis, treatment, or prevention of disease. **2.** Medical science; the science and art of diagnosing, treating, and preventing disease, especially internal diseases requiring the services of a physician.

medina worm. The guinea worm.

medium pl. **media. 1.** The substance surrounding an organism or in which it lives. **2.** *bacteriol.* Food or materials prepared for the growth and culture of bacteria or other microorganisms, commonly called culture *medium*.

medulla. 1. *anat.* **a.** The marrow of a bone. **b.** The inner, central portion of an organ in contrast to the outer cortex, as in the kidney. **c.** Shortened form for medulla oblongata. **2.** *bot.* **a.** The pith of a stem. **b.** The central portion of a thallus in certain algae. **c.** The central layer of a crustose lichen consisting principally of fungal hyphae.

medulla oblongata. The lowermost or most posterior portion of the vertebrate brain continuous with the spinal cord.

medullary. 1. *anat.* Of or pertaining to a medulla, marrow, or myelin. **2.** *bot.* Of or pertaining to the medulla or pith.

medullary ray. 1. *anat.* A raylike extension of medullary substance which projects into the cortex of a metanephric kidney. **2.** *bot.* **a.** A sheet of vascular tissue extending radially in the xylem of a stem of a woody plant; xylem ray, wood ray, pith ray. **b.** A sheet of a parenchymatous tissue extending radially from the central pith in the stem of certain vines and herbs.

medullary sheath. 1. *anat.* The myelin sheath. **2.** *bot.* A layer of cells surrounding the pith in the stem of certain plants.

medullated. Myelinated; possessing a myelin sheath.

medusa. A jellyfish; a free-swimming sexual form in coelenterates.

medusoid. 1. Resembling a medusa. **2.** In colonial coelenterates, designating a type of zooid which gives rise to medusae.

meerkat. A suricate, a small mongoose (*Suricata suricata*) of South Africa.

megacephalic. Possessing an abnormally large head.

megagamete. A macrogamete.

megagametophyte. The female gametophyte; the plant which develops from a megaspore.

megakaryocyte. A large, multinucleated cell of red bone marrow, the source of blood platelets.

megalocephalic. MEGACEPHALIC.

megalops larva. The free-swimming, larval stage of a crab following the zoea stage and preceding the adult stage.

megalopteran. Any insect of the order Megaloptera which includes the alderflies and dobsonflies.

megalospheric. Having a large first chamber, with reference to the shell of a foraminiferan.

megaphyll. A macrophyll, a large leaf with a complex venation, as that of a fern or higher plant.

megasporangium. A sporangium which produces macro- or megaspores.

megaspore. In heterosporous plants, the larger of two kinds of spores; a spore which gives rise to a female gametophyte; a macrospore.

megasporocyte. A cell from which a megaspore develops.

megasporophyll. A leaf or modified leaf which produces or bears sporangia; a macrosporophyll; in higher plants, a carpel.

Mehlis' gland. A group of unicellular glands surrounding the ootype in trematodes and cestodes. Formerly called shell gland.

meiocyte. A cell which gives rise to a meiospore.

meiosis. A form of nuclear division in which the chromosome number is reduced from the diploid (2n) to the haploid (n) number, occurring in animals in gametogenesis, in plants in sporogenesis.

meiospore. A haploid spore produced by meiosis.

Meissner's corpuscle. A sensory receptor which mediates the sense of light touch; a tactile corpuscle.

melanin. One of a number of dark brown or black pigments present in animal and plant tissues.

melanism. The excessive accumulation of pigment (melanin) in the skin or other tissues.

melanoblast. A cell in the skin of neural crest origin which produces melanin.

melanocyte. A branched, melanin-containing cell present in the basal layer of the epidermis.

melanophore. A branched, pigment cell (chromatophore) which contains granules of melanin present in many animals; responsible for pigmentation and color changes.

melanosome. A pigment mass in the ocellus of a dinoflagellate.

melatonin. A substance isolated from the pineal body which causes the concentration of melanin in melanophores.

melon. A muskmelon or watermelon.

membranaceous. 1. MEMBRANOUS. 2. *bot.* Like a thin, soft, delicate, and usually transparent or translucent membrane.

membrane. 1. A thin layer of soft, pliable tissue. 2. A thin sheet or layer which covers a part, lines a tube or cavity, or connects organs or structures.

membrane bone. A dermal bone; one formed by intramembranous ossification.

membranelle. A platelike, vibratile structure composed of fused cilia found in certain protozoans.

membranous. Of, pertaining to, consisting of, originating in, or resembling a membrane.

menarche. The onset of menses at puberty.

Mendelian. Of or pertaining to Gregor Johann Mendel (1822-84) or his laws of heredity. *See* Mendel's laws.

Mendel's laws. A series of laws or principles governing the inheritance of certain traits discovered by Gregor Mendel, published in 1866, but their significance was not realized until their rediscovery in 1900. They include the principles of unit characters, dominance, segregation, and independent assortment.

menhaden. A marine fish (*Brevoortia tyrannis*) of the family Clupeidae; also called mossbunker.

meninges sing. **meninx.** The three membranes, dura mater, arachnoid, and pia mater, which enclose the brain and spinal cord.

meniscus. 1. The curved (concave or covex) surface of a column of a liquid. 2. A crescent-shaped body, especially one of the two semilunar cartilages of the knee joint.

menopause. The time of cessation of menstrual cycles and ovarian function occurring normally in women between the ages of 45 and 50; climacteric, change of life.

menotaxis. A light-compass reaction in which an animal moves at a fixed angle to the source of light.

menses. Menstruation; the cyclic discharge of blood or bloody fluid from the vagina.

menstrual cycle. The periodic, recurring series of changes in the endometrium of the uterus culminating in menstruation. The cycle averages 28 days in length and includes the following phases: repair, proliferation, secretion, menstruation.

menstruate. To undergo menstruation.

menstruation. The discharge of menstrual fluid consisting of blood, necrotic tissue, and secretions of glands, occurring monthly from puberty to menopause.

menstruum. A solvent, especially one used in crude drug extraction.

mensuration. The act of measuring.

¹**mental.** Of or pertaining to the mind.

²**mental.** Of or pertaining to the chin.

mentum. **1.** *zool.* **a.** The chin. **b.** The distal portion of the post-mentum of the labium of an insect. **c.** The basal portion of the gnathochilarium of a diplopod. **2.** *bot.* An extension of the column in the flower of an orchid; a perula.

meraspis. The second larval stage of a trilobite.

mercuric chloride. Bichloride of mercury, $HgCl_2$, used in biology as an antiseptic and fixing agent; corrosive sublimate.

mercury. A silvery-colored metal, symbol Hg, at. no. 80, commonly called quicksilver.

merganser. A fish-eating, diving duck of the genera *Mergus* and *Lophodytes*, family Anatidae, subfamily Merginae.

mericarp. A one-seeded portion of a schizocarp.

meristele. A vascular bundle (dictyostele) of a fern, consisting of pericycle, xylem, and phloem surrounded by an endodermis.

meristem. Embryonic tissue of a plant from which definitive tissues arise by cell multiplication and differentiation.

meristematic. Of, pertaining to, or arising from meristem.

meristic. Divided into segments or metameres.

meristoderm. The epidermis of certain brown algae.

merlin. A small falcon (*Falco columbarius*), the Eastern pigeon hawk.

mermaid's purse. The horny egg capsule of certain elasmobranchs.

meroblastic. Undergoing partial cleavage in which cell divisions are limited to region of the animal pole, as in telolecithal ova.

merocrine. Designating a type of gland in which the gland cells remain intact during the process of secretion, as the parotid.

merogon. An embryo or larva resulting from merogony.

merogony. **1.** The development of a part of an egg or an egg fragment. **2.** The splitting of a schizont into merozoites, as occurs in sporozoans.

meromixis. Sexual phenomena in bacteria which involve the unidirectional transfer of genetic material as in transformation and transduction.

meromyarian. Having only two or a few muscle cells in each quadrant, as in certain nematodes.

306

meronephridium. A multiple nephridium arising from a single embryonic cord of cells, as in oligochaetes.

meroplankton. Plankton organisms which exhibit daily vertical movements.

merosome. A segment, somite, or metamere.

merostomate. An arachnid of the subclass Merostomata which contains the king or horseshoe crab (*Xiphosura* or *Limulus*).

merozoite. One of the divisions or segments of a schizont seen in asexual reproduction of certain sporozoans, as *Plasmodium*.

mesarch. Designating the type of xylem in which the protoxylem is surrounded by metaxylem, as in certain ferns.

mesaxonic. Designating a type of foot in which the midaxis passes through the middle digit, as in the horse.

mescal. An intoxicating drink of Mexico distilled from the leaves of maguey (*Agave*).

mescal buttons. The flowering heads of a cactus (*Lophophora williamsii*), the source of mescaline. *See* peyote.

mescaline. An alkaloid with hallucinogenic properties obtained from mescal buttons, used by American Indians during religious ceremonies.

mesectoderm. Mesenchyme derived from neural crest cells.

mesencephalon. The midbrain, one of the three primary vesicles of a vertebrate brain located between forebrain and hindbrain.

mesenchyme. A form of embryonic connective tissue derived principally from mesoderm consisting of a diffuse network of stellate cells which gives rise to connective tissue, cartilage, bone, blood and blood vessels, cardiac and smooth muscle.

mesenteron. Old term for midgut.

mesentery. 1. In vertebrates, a membrane consisting of two layers of peritoneum and enclosed structures (nerves, blood and lymph vessels), which attaches a visceral organ to the body wall. **2.** In invertebrates, a comparable structure which connects the gut or a part of it to the body wall.

mesial. Of, in, or toward the middle.

mesic. Characterized by moderately moist conditions; neither too moist nor too dry.

mesoblast. Old term for mesoderm.

mesocarp. The middle layer of the pericarp.

mesocercaria. A stage in the development of certain flukes interposed between the cercaria and metacercaria, as in strigeids.

mesocolon. The mesentery which supports the colon.

mesoderm. In a triploblastic animal, the middle of the three germ layers from which connective tissues, muscles, blood, circulatory system, urinogenital organs, serous cavities and their linings develop.

mesodont. 1. With medium-sized teeth. 2. With medium-sized mandibles, with reference to insects.

mesoglea. The jellylike substance lying between the epidermis and gastrodermis of coelenterates.

mesome or **mesom.** The region of a plant axis between successive branches.

mesomere. The intermediate mesoderm; a nephrotome.

mesonephric duct. The duct of a mesonephros terminating caudally at the cloaca: WOLFFIAN DUCT.

mesonephros. A type of embryonic kidney in amniotes consisting of a number of paired tubules which open into a mesonephric duct. It is transitory, following the pronephros and preceding the metanephros. Also called Wolffian body, middle kidney.

mesophil. An organism which grows best at moderate temperatures (23°–40°C.)

mesophyll. The chlorophyll-bearing tissue (chlorenchyma) between the upper and lower epidermal layers of a leaf, usually in two distinct layers, palisade and spongy parenchyma.

mesophyte. A plant whose water requirement is intermediate between that of xerophytes and hydrophytes, as most plants growing in watered land of the temperate zone. *Compare* hydrophyte, xerophyte.

mesorchium. The mesentery which supports the testis.

mesosaprobic. Living in waters in which active oxidation and decomposition of organic matter is occurring, said of protozoans.

mesosoma. 1. The preabdomen of eurypterids. 2. The first portion of the abdomen (opisthosoma) of an arthropod.

mesosome. The middle portion of the body of a deuterostome, as in hemichordates.

mesothelium. Simple squamous epithelium of mesodermal origin which lines the serous cavities of vertebrates.

mesothorax. In insects, the middle region of the thorax usually bearing the first pair of wings and second pair of legs.

mesovarium. The mesentery which supports the ovary.

mesozoan. One of a group (phylum?) of animals of uncertain affinities intermediate between protozoans and true multicellular animals, as *Dicyema, Rhopalura*.

Mesozoic era. A division of geological time between the Paleozoic and Cenozoic eras; the Age of Reptiles. Includes the Triassic, Jurassic, and Cretaceous periods.

mesquite. **1.** A spiny shrub, of the genus *Prosopis*, especially *P. glandulosa* and *P. juliflora* of arid regions of Southwest United States and Mexico, its pulpy pods a source of food. **2.** The algaroba or keawe (*Prosopis chilensis*), a leguminous tree of the West Indies, Mexico, and Central America; an important food source.

mestome sheath. The sheath enclosing a vascular bundle in the leaf of grasses; the inner bundle sheath.

metabolic. Of or pertaining to metabolism.

metabolic rate. The rate of metabolism as determined by the amount of food consumed, heat produced, or oxygen utilized. *Basal metabolic rate* (BMR) is the energy output as determined under special conditions.

metabolism. The chemical or energy changes which occur within a living organism or a part of it which are involved in various life activities. *Basal metabolism* refers to the minimum energy requirements for the maintenance of normal functions; *intermediary metabolism* includes all changes which occur in the utilization of a foodstuff after its absorption into the blood.

metabolite. **1.** A product of metabolic activity, especially one formed in intermediary metabolism. **2.** A substance utilized in a metabolic process, as certain vitamins.

metaboly. Squirming, undulating movements, as in *Euglena*.

metacarpal. Of or pertaining to the metacarpus; a metacarpal bone.

metacarpus. The region of the forefoot or hand which contains the metacarpal bones located between the carpus and phalanges.

metacercaria. A stage in development of a digenetic fluke between cercaria and adult during which the tail is lost and encystment usually occurs.

metachromatic. Assuming a color different from that usually resulting from a specific stain or dye.

metachromatic granules. Babes-Ernst bodies; inclusion bodies present in certain bacteria, yeasts, and algae. *See* volutin.

metachronism. The orderly succession in the initiation of beat of cilia, resulting in a wavelike action on a ciliated surface.

metachrosis. The ability to change color, as seen in fishes, amphibians, reptiles, and other animals.

metacneme. A secondary or incomplete septum in an anthozoan.

metacoel. The cavity of a metasome.

metagenesis. ALTERNATION OF GENERATIONS.

metagnathous. With tips of the mandibles crossed, as in crossbills.

metamere. A somite or segment; one of a linear series of more or less similar divisions or parts which make up the body of a segmented animal.

metameric. Consisting of segments or parts repeated one after another.

metamerism. The state or condition or being composed of segments or metameres.

metamorphosis. 1. *zool.* A change in shape or form which an animal undergoes in its development from egg to adult as seen in annelids (trochophore to adult worm); insects (egg, nymph, adult; egg, larva, pupa, adult); crustaceans (egg, nauplius, zoea, mysis, adult); amphibians (egg, tadpole, adult). **2.** *med.* A retrogressive change of one type of tissue to another, as fatty degeneration of the liver.

metanauplius. In crustaceans, a larval stage preceding the nauplius, as in the shrimp.

metandrous. Maturing before male flowers, said of female flowers.

metanephric. Of or pertaining to a metanephros.

metanephridium. A nephridium with an internal opening or nephrostome opening into the coelom and a nephridiopore opening to the outside.

metanephros. An embryonic type of kidney in amniotes which develops posterior to the mesonephros. It becomes the functional kidney in reptiles, birds, and mammals.

metaphase. The phase or stage of mitosis during which the chromosomes, each consisting of a pair of chromatids, arrange themselves with their centromeres in an equatorial plane.

metaphloem. The portion of the primary phloem which matures after growth of the surrounding tissues is complete.

metaphysis. In the development of a bone, the region between diaphysis and epiphysis; the epiphyseal plate.

metaphyte. Any plant of the Metaphyta, a major category of living things comprising the bryophytes and tracheophytes, complex multicellular plants.

metaplasia. The transformation of one type of tissue into another. May be progressive (prosoplastic) or retrogressive.

metaplasm. The nonliving constituents of protoplasm; cell inclusions.

metapleural folds. Two ventrolateral folds in lancelets which meet in the region of the atriopore and continue posteriorly as the caudal fin.

metapterygium. *See* basalia.

metarteriole. A minute blood vessel which lies between an arteriole and a true capillary.

metasoma. 1. The posterior portion of the abdomen of an arthropod: OPISTHOSOMA. **2.** The postabdomen of a eurypterid.

metasome. 1. The midbody of a copepod. **2.** The third portion of the body of a deuterostome.

metastasis. The spread from one part of the body to another of a pathological state or condition resulting from the dispersal by way of blood or lymphatic channels of the causative agents or cells from the primary focus of infection, as the spread of a malignant neoplasm.

metasternum. 1. In insects, the sternum of the metathorax. **2.** In vertebrates, the posterior portion of the sternum; the xiphisternum.

metastoma. 1. The lower lip of certain crustaceans, usually divided into two paragnatha. **2.** A pregenital plate in eurypterids.

[1]metatarsal. Of or pertaining to the metatarsus.

[2]metatarsal. A bone of the metatarsus.

metatarsus. 1. The portion of the foot (in tetrapods, the hind foot) between the tarsus and phalanges. **2.** In the leg of a spider, the segment between the tibia and tarsus.

311

metatherian. Any mammal of the Metatheria which includes the marsupials.

metathorax. The third and most posterior division of the thorax in insects, usually bearing the second pair of wings and the third pair of legs.

metatroch. A girdle of cilia located just below the mouth in a trochophore.

metatrochophore. A larval stage between trochophore and nectochaeta in certain polychaetes.

metatrophic. Saprophytic or saprozoic.

metaxenic. *bot.* Pertaining to the influence of the male parent on maternal tissue outside the embryo and endosperm.

metazoan. **1.** A multicellular animal. **2.** Any member of the Metazoa, a major category of the animal kingdom comprising all animals except the Protozoa.

metazoea. In certain decapod crustaceans, a stage in development between the zoea and megalops stages.

metencephalon. The anterior portion of the hindbrain which gives rise to the pons and cerebellum.

meter. A unit of linear measurement, the basic unit of the metric system, equivalent to 39.37 inches. Abbreviated m.

metestrus. The stage in the estrous cycle which follows estrus during which corpora lutea develop; in the vagina the cornified lining is shed and leukocytes appear in its lumen.

methanol. Methyl or wood alcohol.

methyl alcohol. Wood alcohol, CH_3OH.

mice. Plural of MOUSE.

micelle. A submicroscopic unit of protoplasm; a molecular aggregate, as that of a colloid. Term now falling into disuse.

microaerophilic. Growing in minute quantities of free oxygen.

microbe. Any minute living thing; a microorganism, especially a disease-causing bacterium.

microbiology. The study of microorganisms, including their culture, economic importance, pathogenicity, etc. Organisms studied include viruses, rickettsias, algae, yeasts, molds, bacteria, protozoans, and microzoa. *See* bacteriology.

microbiostasis. A condition in which microorganisms, although alive or viable, do not multiply.

microcyst. One of the numerous resting cells in the fruiting body of a myxobacter.

microbiota. The microscopic fauna and flora of a region or locality.

microcercous. Having a minute tail, said of cercaria.

microclimate. The climate immediately about an organism in its microhabitat.

microcneme. A microseptum.

microconjugant. The smaller of two conjugating protozoans.

microdissection. Dissection of minute structures requiring the use of a microscope and usually with the aid of a micromanipulator.

microevolution. Evolution resulting from the accumulation of many small mutations resulting in the development of subspecies within a species. *Compare* macroevolution.

microfauna. The microscopic animals present in a particular habitat.

microfilariae. The minute larvae of parasitic nematodes of the order Filarioidea found in the bloodstream or tissues.

microgamete. In a heterogamous organism, the smaller of the two types of gametes, usually the male gamete.

microgametocyte. A cell which gives rise to a microgamete.

microgametophyte. *bot.* A male gametophyte; the plant which develops from a microspore and produces microgametes.

microglia. Neuroglial cells of the central nervous system of mesodermal origin. They are ameboid and capable of phagocytic activity.

micrograph. A drawing or photograph of an object as it appears magnified by a microscope.

microhabitat. The local conditions which immediately surround an organism.

microincineration. The subjection of a glass slide bearing a thin slice of tissue to a high temperature resulting in combustion of all organic materials. The remaining ash gives information as to quantity and distribution of minerals before incineration. *See* spodogram.

micromanipulator. A micropositioner, an apparatus by which microtools (needles, pipettes, electrodes, etc.) can be moved about with extreme precision.

313

micromere. One of the smaller blastomeres in an egg undergoing unequal cleavage.

micrometazoa. Microscopic multicellular animals.

micrometer. A device used with optical instruments for measuring minute distances.

micron. One-thousandth of a millimeter or approximately one twenty-five thousandth (1/25,000) of an inch. Symbol μ.

microniscus. A larval stage of an isopod of the suborder Epicaridea parasitic upon copepods.

micronucleus. A small nucleus, especially the smaller of two nuclei present in certain protozoans, as *Paramecium*. It functions in the processes of reproduction and rejuvenation. *Compare* macronucleus.

micronutrient. A microelement, a trace element or substance present in minute amounts but essential for normal growth and development, as **a.** for plants, boron, copper, zinc, manganese, and molybdenum; **b.** for animals, most of the minerals and certain vitamins.

microorganism. A small, microscopic animal or plant.

microphyll. A very small leaf, especially one with a single vein, as in most lycopods.

microphyllous. Having small leaves (microphylls).

micropterous. Possessing small wings.

micropyle. 1. *zool.* A small opening, as **a.** that in the membrane of certain eggs through which the sperm enters; **b.** that in the shell of the gemmule of a sponge through which cells make their exit. 2. *bot.* A small opening between the integuments of an ovule through which a pollen tube gains access to the embryo sac.

microsclere. A minute spicule in sponges.

microscope. A device or apparatus which produces enlarged images, thus enabling minute objects to be seen. Types include compound (monocular, binocular), dark-field, electron, fluorescent, phase-contrast, polarizing, and ultraviolet.

microscopy. The use of the microscope; the examination of objects by means of a microscope.

microseptum. An incomplete septum in an anthozoan.

microsere. A successional series in a microhabitat.

314

microsome. A minute, ultramicroscopic body in the proto-plasm of a cell. *See* ribosome.

microspheric. Designating a foraminiferan in which the first chamber of the shell is small.

microsporangium. A sporangium which produces micro- or meiospores.

microspore. 1. *bot.* **a.** The smaller of two kinds of spores produced by a heterosporous plant; a spore which develops into a male gametophyte. **b.** In seed plants, a pollen grain. **2.** *zool.* The smaller of two kinds of spores produced by certain protozoans.

microsporophyll. A modified leaf which bears microsporangia, as **a.** the scalelike leaves of the carpellate cone of a gymnosperm; **b.** the stamen of a flower.

microsurgery. Micrurgy or surgery on extremely small organisms or individual cells, usually performed under a microscope or by means of a micromanipulator; includes microdissection, microinjection, microisolation, and various other procedures.

microtome. An instrument for cutting extremely thin sections of tissues for mounting on slides for microscopic examination.

microvilli. Extremely minute cylindrical processes which form the striated border of columnar epithelium or brush border of cuboidal epithelium.

microzoa. Microscopic multicellular animals.

micrurgy. MICROSURGERY.

mictic. Of or pertaining to the haploid egg of rotifers which, if fertilized, forms a thick-shelled, dormant egg which after a time hatches into a female; if not fertilized, it develops parthenogenetically into a male.

micturition. Urination or the act of passing urine; voiding.

midbrain. The mesencephalon or the portion of the adult brain which develops from it, in man comprising the corpora quadrigemina and cerebral peduncles.

midge. One of a number of small, dipterous insects, as **a.** chironomids of the family Chironomidae whose aquatic larvae are called bloodworms; **b.** biting midges, bloodsucking insects of the family Ceratopogonidae, commonly called punkies or no-see-ums; **c.** phantom midges of the family Culicidae; **d.**

315

gall midges or gnats of the family Cecidomyiidae, some of which form galls, others live as inquilines in galls produced by other insects.

midget. A dwarf or undersized human; an adult who has not attained full growth.

midgut. 1. The midintestine or mesenteron, the middle portion of the digestive tract in a vertebrate embryo. **2.** In arthropods, the portion of the digestive tract between stomach and intestine.

midrib. The central or principal vein of a leaf.

midriff. 1. The diaphragm. **2.** The upper portion of the abdomen.

migration. Movement from one place to another, as **a.** the movement of leukocytes through capillary walls or through tissues; **b.** movement of macrophages through tissues; **c.** movement of germ cells through reproductive ducts; **d.** the mass movement of animals to and from feeding or reproductive and nesting areas; **e.** the movement of ions through cell membranes.

migratory. Having a tendency to migrate; migrating periodically.

mildew. 1. A white, powdery growth which appears on plants or organic matter produced by phycomycete fungi. **2.** A fungus which causes mildew to develop. *See* downy mildew, powdery mildew. **3.** A plant disease caused by a fungus whose mycelial growth forms a white coating on infected parts.

¹**milk. 1.** The whitish fluid secreted by the mammary gland of a mammal and used for the nourishment of the young. **2.** Any fluid resembling milk, as the fluid of a coconut or latex.

²**milk.** To draw milk from the breast or udder, or to extract any fluid by a similar action, as to *milk* venom from a snake.

milk snake. A harmless snake (*Lampropeltis doliata*), so called because of the erroneous belief that they milk cows; also called king snake.

milk sugar. LACTOSE.

milk teeth. The temporary or deciduous teeth of a mammal; baby teeth.

milkweed. A common plant (*Asclepius syriaca*), a source of latex.

milky disease. A bacterial disease of insects caused by *Bacillus popilliae*, used in the control of grubs of the Japanese beetle.

millepore. A colonial, hydrozoan coelenterate of the genus *Millepora* which produces a massive, calcareous skeleton of importance in reef formation.

miller. Any of a number of moths of the suborder Frenatae whose wings appear to be colored by a fine dust or powder.

miller's thumb. *See* sculpin.

millet. One of a number of small-grained, rapidly-growing grasses, especially *Panicum miliaceum*, grown extensively for hay and forage. Its seed is also used extensively for human, poultry, and bird feed.

milliequivalent. The weight (number of grams) of a substance in one milliliter of a one normal (1N) solution.

milliliter. One thousandth of a liter, approximately one cubic centimeter. Abbreviated ml.

millimeter. One thousandth of a meter or one tenth of a centimeter, approximately 1/25 of an inch. Abbreviated mm.

millimicron. One thousandth of a micron. Abbreviated mμ.

millipede. Any arthropod of the class Diplopoda which includes the thousand-legged worms, terrestrial animals with many segments each bearing two pairs of appendages, as *Julus terrestris*.

milt. 1. The testes of fishes or the seminal fluid produced by them. 2. Breeding males of fishes, as *milt* salmon.

mimesis. MIMICRY.

mimetic. Mimicking, imitating.

mimic. 1. An organism which imitates or mimics. 2. A bird of the family Mimidae which mimics the songs of other birds, as a mockingbird.

mimicry. The superficial resemblance of an organism to one of another species by which it gains an advantage presumably of survival value, as the palatable species of butterfly, the viceroy, mimicking the unpalatable monarch.

mine. To form a burrow or excavate a tunnel.

miner. *See* leaf miner.

mineral. An inorganic compound, usually a solid, having a definite chemical composition and usually formed as a result of inorganic processes. When in crystalline form, each has a characteristic molecular structure. Water and mercury are minerals which are liquid at ordinary temperatures.

mineralocorticoid. One of a group of adrenal cortical hormones which regulate salt and water balance, as 11-deoxycorticosterone.

mineral salts. Inorganic constituents of the body which, upon combustion, yield ash, as salts of sodium, calcium, and potassium.

minimum. The least, smallest, or lowest in strength, effectiveness, or intensity.

mink. A weasellike, fur-bearing mammal (*Mustela vison*) of the family Mustelidae.

minnow. A small fish, especially any of the small, freshwater fishes of the carp family, Cyprinidae.

minor. Lesser or smaller; of little consequence or importance.

mint. Any plant of the mint family, Labiatae, especially those of the genus *Mentha*, as peppermint.

Miocene epoch. An epoch of the Tertiary period preceding the Pliocene and following the Oligocene.

miosis. *See* meiosis.

miracidium. The first larval stage of a fluke, a ciliated, free-swimming form which, upon hatching from the egg, enters a snail or bivalve in which it transforms into a sporocyst.

mirror. 1. A glass or polished surface which reflects light. 2. A tense, shining membrane in the sound-producing apparatus of a cicada. 3. A resonating structure in the stridulatory apparatus of a male cricket. 4. A speculum.

miscarriage. Expulsion of the fetus before the age of viability. *See* abort.

miscegenation. The mixing of races, especially the intermarriage of individuals of white and colored races.

miscible. Capable of being mixed or dissolved.

mistletoe. A semiparasitic seed plant of the family Loranthaceae which grows on various woody plants, as *Viscum album* of Europe and *Phoradendron flavescens* of America.

mite. Any of a large number of arachnids of the order Acarina, including free-living and parasitic forms, as the itch mite, follicle mite, harvest mite.

mitochondrion pl. **mitochondria.** One of the minute, spherical, rod-shaped or filamentous organelles present in all cells. They contain many enzymes of Kreb's citric acid cycle and the electron transport systems, hence are of primary importance in the metabolic activities of cells.

mitosis. In cell division, karyokinesis or division of the nucleus usually occurring in a series of stages designated prophase, metaphase, anaphase, and telophase followed by an interphase. In the process there is duplication of the chromosomes and an equal division of the duplicates to the daughter cells. *See* cytokinesis, meiosis.

mitosporangium. A sporangium which produces spores by mitosis. Spores may be haploid or diploid.

mitospore. A spore produced by mitosis.

mitotic. Of or pertaining to mitosis.

mitral valve. The left atrioventricular or bicuspid valve of the heart.

mitraria larva. A peculiar trochophore larva of certain polychaetes.

mixed. Of more than one type or kind; of different or dissimilar parts or components, as a *mixed* gland or a *mixed* bud.

mixonephridium. A combined nephridium and coelomoduct.

mixotrophic. Nourished by more than one method, with reference to nutrition in protozoans.

moa. A recently extinct bird of the family Diornithidae, remains of which have been found in New Zealand. They were large, flightless, terrestrial birds sometimes attaining a height of 8 to 10 feet.

moccasin. The common name for two venomous pit vipers of the genus *Agkistrodon* (*Ancistrodon*), as *A. piscivorus*, the water moccasin or cottonmouth and *A. contortrix*, the highland moccasin or copperhead.

mockingbird. One of several birds of the family Mimidae which have the ability to imitate the songs of other birds, as the Eastern mockingbird (*Mimus polyglottos*).

mode. 1. In statistics, the value which occurs most frequently in a series of observations. 2. A manner or method of doing or being. 3. A particular form or kind.

model. In mimicry, the organism or the structure mimicked.

modifier. A modifying gene.

modifying gene. A gene which influences the expression of another nonallelic gene, as that which affects the degree and extent of spotting.

modiolus. The central axis or pillar of the cochlea of the ear.

mohr. A West African gazelle (*Gazella dama mohrr*).

mol. A mole or gram-molecule.

mola. A large ocean sunfish (*Mola mola*).

molar. 1. A molar tooth. 2. Of or pertaining to molar teeth. 3. *chem.* Pertaining to or expressed is gram-molecules.

molar solution. *chem.* A solution that contains one mole of solute per liter of solution.

molasses. A thick, viscid syrup obtained from sugar in the process of refining.

mold. 1. A slimy, cottony, or filamentous growth which develops on moist organic material as the result of a growth of saprophytic fungi. 2. Any mold-producing fungus, as a black mold, blue-green mold.

[1]mole. *zool.* A small, burrowing, insectivorous mammal of the family Talpidae, as *Scalopus aquaticus*, the common North American mole.

[2]mole. *med.* 1. A small growth on the skin, usually pigmented and sometimes hairy. *See* nevus. 2. A benign tumor of the uterus which develops from a fetus or fetal membranes following cessation of growth or degeneration.

[3]mole. *chem.* The molecular weight of a substance expressed in grams. Also called gram-molecule or mol.

mole cricket. A burrowing cricket (*Scapteriscus acletus*).

molecular. Of, pertaining to, or composed of molecules.

molecular biology. The quantitative and qualitative study of molecular structures within cells and changes which occur in or among them in relation to biological activities, as the analysis of genes and their relationship to the synthesis of enzymes and other proteins.

molecule. The smallest particle of a substance which can exist separately and still retain the characteristic properties of the substance; the smallest combination of atoms which forms a specific chemical compound.

mollusc. A mollusk.

mollusk. Any invertebrate of the phylum Mollusca, unsegmented animals with a body consisting of a ventral foot and a dorsal visceral mass. Most possess a mantle which secretes a calcareous shell. Common representatives are chitons, tooth shells, snails, slugs, mussels, clams, oysters, squids, octopuses, and nautili.

molly. A small, freshwater fish of the genus *Mollienisia*, commonly kept in aquaria.

molt. To shed and develop anew the outer covering of the body, as the cuticle, exoskeleton, skin, hair, or feathers.

molt cycle. A cycle in crustaceans which includes four periods: premolt, molt, postmolt, and intermolt.

molting. The act or process of shedding the outer covering of the body or a part of it: ECDYSIS.

molting gland. The prothoracic gland of insects and the Y-organ of crustaceans, both of which secrete hormones which control molting.

molting hormone. A hormone, MH, which promotes growth and differentiation to the adult stage, produced by the prothoracic glands of insects. Also called growth and differentiation hormone: ECDYSONE.

molybdenum. A metallic element of the chromium group required by plants in minute amounts. Symbol Mo; at. no. 42.

monad. A simple, one-celled organism, especially a flagellate of the genus *Monas*.

monadelphous. *bot*. With stamens united to form a single tube or column. *Compare* didelphous.

monandrous. Having one stamen.

monanthus. Having a single flower.

monarch. 1. *bot*. Having a single protoxylem. 2. *zool*. The milkweed butterfly (*Danaus plexippus*), a migratory butterfly of the family Danaidae. It is mimicked by the viceroy butterfly.

monaxon. A single, needlelike spicule of sponges.

moneran. Any organism of the Monera, a major taxon which includes the bacteria and blue-green algae, characterized by the absence of true nuclei and chromosomes.

mongoose. One of several species of ferretlike, carnivorous mammals of the family Viverridae distributed throughout Asia and Africa. Most belong to the genus *Herpestes;* their food, snakes and rodents.

mongrel. An individual resulting from the interbreeding of two different strains or races; a hybrid; an animal of unknown ancestry.

moniliform. Constricted at intervals, resembling a string of beads.

monitor. A large lizard of the genus *Varanus* found in Southeast Asia, Africa, and Australia, some reaching a length of eight feet.

monkey. 1. In a general sense, any mammal of the order Primates except the lemurs, tarsiers, and man. **2.** More specifically, one of the smaller, long-tailed primates as distinguished from the apes, as the macaques, langurs, marmosets.

monocarpic. Bearing fruit once and then dying.

monocarpous. Having a single carpel.

monocephalous. Bearing a single head or capitulum, said of flowers.

monochasium. A cyme consisting of a single branch.

monoclinous. Hermaphroditic; having stamens and pistil in one flower.

monocolpate. Having a single furrow or colpas, with reference to pollen grains.

monocot. A monocotyledon.

monocotyledon. Any plant of the Monocotyledoneae, a class of the Angiospermae which includes vascular plants characterized by an embryo with a single cotyledon, flower parts usually in threes or sixes, not in fives; leaves with parallel veins, and a stem with scattered vascular bundles. Comprises over 75,000 described species, as the grasses, palms, arums, lilies, irises, and orchids.

monocular. 1. Having a single eye. **2.** In optics, having a single ocular or eyepiece.

monocyclic. Arranged in one whorl or circle, said of floral parts.

monocyte. A large, agranular leukocyte with a large, oval or slightly indented nucleus.

monodactylous. Having a single digit or claw.

monodisc. Designating one ephyra formed at a time. *See* strobilization.

monodont. Having a single tooth, as the narwhal.

monodynamous. Having one stamen longer than the others.

monoecious. **1.** *zool.* Having both testes and ovaries in the same individual: HERMAPHRODITIC. **2.** *bot.* **a.** Having both antheridia and archegonia on the same plant, as in mosses and ferns. **b.** Having both staminate and carpellate cones or staminate and pistillate flowers on the same plant.

monogamous. Having only one mate.

monogeneid. Any trematode of the subclass Monogenea containing flukes in which development is direct, there being no intermediate host and no asexual generations. Most are ectoparasites on fishes and amphibians, as *Gyrodactylus*; a few are endoparasites, as *Polystoma* in the urinary bladder of frogs.

monogenetic. **1.** Of or pertaining to direct development in which there is no alternation of generations. **2.** Of or pertaining to flukes of the subclass Monogenea.

monogonoporous. Having a single opening for both male and female sex cells.

monogynous. **1.** *bot.* Having a single pistil. **2.** *zool.* Mating with one female only.

monohybrid. The offspring of a cross between parents which differ in only one character.

monokaryotic. Having a single haploid nucleus, with reference to the hyphae of certain fungi.

monolete. Possessing a single, longitudinal suture, with reference to the spores of certain plants, as *Psilotum*.

monolocular. Unilocular or one-celled, with reference to the ovaries of flowers.

monomerous. Designating flowers with only one member in each whorl.

monomorphic. Having only one form. *Compare* dimorphic, polymorphic.

monopetalous. Having a single petal: GAMOPETALOUS.

monophagous. Subsisting on a single type of food, with special reference to insects.

monophyletic. Derived from a single ancestral stock or type.

monophyllous. Possessing or composed of a single leaf.

monophyodont. Having only one set of nonreplaceable teeth.

monoploid. HAPLOID.

monopodial. Having a single main axis of growth from which lateral branches arise. *Compare* sympodial.

monorchid. Having a single testis.

monosaccharide. A monose or simple sugar; one that cannot be decomposed by hydrolysis, as glucose, fructose.

monosepalous. Having one sepal.

monosomic. Having one less than the normal diploid number of chromosomes. Somatic number is 2n-1.

monospermous. Producing only one seed.

monospermy. Fertilization of an egg by only one sperm.

monospondyly. The condition in which a vertebra possesses a single centrum, as in amniotes. *Compare* diplospondyly.

monosporangium. A vegetative cell that produces a single spore, as in certain red algae.

monospore. A single spore produced by metamorphosis of a vegetative cell.

monostachous. Having a single spike.

monostichous. 1. *bot.* Arranged along one side of an axis. **2.** *zool.* Arranged in a single row.

¹monostome. A fluke with a single sucker.

²monostome. Possessing a single sucker.

monostylous. Having a single style.

monotocous. Producing one young at birth.

monotreme. An egg-laying mammal of the order Monotremata, as the duckbill platypus and the spiny anteater.

monotrichous. Having a single flagellum.

monotrophic. MONOPHAGOUS.

monotypic. In taxonomy, having only one subordinate unit, as a genus with a single species.

monoxenous. Having a single species as a host, said of parasites.

monozoic. Lacking segmentation, with reference to tapeworms.

monozygotic. Derived from a single fertilized ovum or zygote, as identical twins.

mons. *anat.* An eminence, as the *mons pubis*, an elevated area over the pubic symphysis covered with pubic hair; in the female, sometimes designated *mons veneris*.

monster. 1. A huge living thing. **2.** *embryol.* A fetus which, as a result of maldevelopment differs markedly from the normal, usually having a grotesque appearance and with reduced viability. A double monster is one consisting of conjoined twins.

¹**montane.** Pertaining to the fauna and flora of the mountains.

²**montane.** In mountainous regions, a zone extending downward from the timberline, a vertical distance of about 1500 feet.

moose. A large American deer of the family Cervidae, the males possessing broad, heavy antlers, as *Alces americanus*, the common moose, and *A. gigas*, the Alaskan moose.

morass. A swamp or marsh.

moray. A voracious, predaceous marine eel of the family Muraenidae, order Apodes, inhabiting coral reefs in tropical waters.

mordant. In microtechnique, a substance which combines with a dye or stain to form an insoluble compound, as alum.

morel. An edible ascomycete fungus (*Morchella esculenta*).

moribund. In a dying state or condition; near death.

morphallaxis. Regeneration of a structure through the remodeling or transformation of parts of the parent, with only limited production of new tissue.

morphogenesis. The development of specific organs and structures which result in an organism having its characteristic size, form, and structure.

morphology. 1. The study of structure and form. **2.** The structure and form of an organism.

mortality. The death of individuals of a population, especially death in large numbers.

mortality rate. The death rate, especially the specific-cause-of-death rate, the number of deaths from a certain cause in a unit of population in a certain time unit, as the number of deaths from typhoid fever per 100,000 per annum.

morula. *embryol.* A solid mass of cells resulting from cleavage of the zygote preceding the formation of the blastula.

mosaic. 1. An individual having patches of somatic tissue differing genetically, thought to be the result of mutations which occur during development. **2.** A type of leaf arrangement in which there is a minimum of leaf overlap, as in the rosette of mullein.

mosaic development. Determinate development in which the fate of the blastomeres is determined at an early stage, destruction of a single blastomere resulting in a defective embryo. *Compare* regulative development.

mosaic disease. Any of a number of virus-caused plant diseases characterized by irregularly distributed light and dark spots on the leaves, as tobacco mosaic.

mosaic image. An apposition image or that formed from the functioning of the numerous ommatidia of a compound eye.

mosasaur. A carnivorous, marine lizard of the Cretaceous period.

mosquito. One of several species of small dipterous, blood-sucking insects of the family Culicidae. Important genera are *Culex*, the house mosquito; *Anopheles*, malaria mosquito; *Aedes*, yellow-fever mosquito.

moss. 1. A small, leafy, nonvascular land plant of the class Musci, division Bryophyta, with sex organs borne at the tip of the gametophyte plant, as *Polytrichum*, the hairy-cap moss. *See* sphagnum moss. **2.** A clump or mass of these plants. **3.** Any of a number of plants which resemble moss in appearance, as club moss, Irish moss, reindeer moss, Spanish moss.

moss animal. An ectoproct.

mossbunker. The menhadden.

moth. A lepidopterous insect of the suborder Frenatae. Most are nocturnal insects with a siphoning proboscis, blunt abdomen, antennae not clubbed, and wings held together when in flight. Their larvae (caterpillars) are often serious pests, as those of the codling moth, clothes moth, gypsy moth.

¹**mother. 1.** The female parent, especially of a human. **2.** The source from which something arises or is formed. **3.** Designating a cell from which other cells arise, as a spore mother-cell.

²**mother.** A thick, gelatinous or slimy membrane which forms on the surface of a liquid undergoing fermentation, as mother-of-vinegar.

Mother Carey's chicken. Wilson's storm petrel (*Oceanites oceanicus*).

mother-of-pearl. The inner, nacreous layer of the shell of many mollusks; nacre.

motile. Capable of spontaneous movement; exhibiting motility.

motility. Having the ability to move or to change position spontaneously.

motion. Movement or activity; change in place or position.

motmot. A jaylike tropical bird of the family Momotidae.

motoneuron. A motor neuron.

motor. *physiol.* Of or pertaining to muscles or muscular activity.

motor area. An area in the cerebral cortex from which impulses initiating voluntary movement arise.

motor cells. *bot.* Large, thin-walled cells in the upper epidermis of the leaves of grasses and other plants which, when turgid, keep the leaf blade flat, but in dry weather, on loosing their turgidity, cause the leaf margins to turn inward. *See* rolling.

motor end plate. The termination of the axon of a motor neuron in a striated muscle fiber; myoneural junction, neuromuscular junction.

motor nerve. A nerve composed entirely of the axons of motor neurons.

motor neuron. A neuron which conveys impulses which result in muscle contraction. The cell bodies of upper motor neurons lie in the cerebral cortex, their axons passing downward in the spinal cord; the cell bodies of lower motor neurons lie in ventral horn of the gray matter of the spinal cord, their axons passing outward in spinal nerves and terminating in striated muscle fibers.

mouflon. A wild sheep (*Ovis musimon*) of the mountains of Sardinia and Corsica.

moult. MOLT.

mountain goat. *See* Rocky Mountain goat.

mountain lion. *See* cougar.

mountain sheep. One of a number of species of wild sheep inhabiting mountainous regions, especially the bighorn.

mourning dove. The turtle dove (*Zenaidura macroura*), a wild dove of North America.

mouse pl. **mice. 1.** Any of a large number of species of small rodents of the family Muridae, as the house mouse (*Mus musculus*). **2.** Any of a number of mouselike mammals of other families. *See* vole. **3.** One of a number of animals which superficially resembles a mouse, as a mousefish, sea mouse.

mousebird. A small bird of the family Coliidae, of South Africa.

mouse deer. A chevrotain.

mouse hare. *See* pika.

mouth. 1. The opening through which food is ingested. 2. In vertebrates, the first portion of the alimentary canal in which food is masticated; the oral or buccal cavity. 3. In a restricted sense, the opening between the lips. 4. In general, the opening or orifice of any tube, canal, or cavity.

mouthparts. Structures which surround the mouth and function in securing, manipulation, and mastication of food, as **a.** in insects, labrum, mandibles, maxillae, labium; **b.** in arachnids, chelicerae; **c.** in crustaceans, mandibles, maxillae, maxillipeds; **d.** in mollusks, radula.

movement. Motion or any activity which results in change of shape, form, place, or position. Various types include ameboid, Brownian, ciliary, cytoplasmic.

mucigenic bodies. Organelles in ciliates which produce a mucoid material utilized in the formation of cysts and protective coverings.

mucilage. 1. A thick, viscid, adhesive substance produced by mixing various vegetable gums and water. 2. A gelatinous substance produced by various plants which absorbs water readily, increasing in bulk and becoming a hydrogel.

mucilage canal. An elongated cell in certain brown algae and cycads which conducts mucilaginous substances.

mucilage cell. A cell which secretes mucilage, as the cells in the gemmae of certain bryophytes.

mucilaginous. Of the nature of or pertaining to mucilage; slimy; sticky.

mucin. A glycoprotein or a mixture of glycoproteins secreted by mucous cells or glands which is responsible for the high viscosity of mucus.

mucket. A freshwater mussel, especially one of the genus *Actinonaias* which possesses a thick shell used in the manufacture of pearl buttons.

¹mucoid. Resembling mucus.

²mucoid. One of a group of glycoproteins resembling mucin present in connective tissue, egg white, and other substances.

mucorales. Phycomycete fungi of the order Mucorales which includes the bread mold and other common molds.

mucosa. A mucous membrane.

mucous. Of, pertaining to, containing, or secreting mucus.

mucous cell. Any cell which secretes mucin, as a goblet cell of the intestine or the mucous cells of mixed glands.

mucous membrane. An epithelial membrane whose surface is moistened by mucus, as that lining the gastrointestinal, respiratory, and urinogenital passageways, tubes which open onto an external surface.

mucro. A sharp pointed tip or process.

mucron. An anchoring device possessed by certain acephaline gregarines.

mucronate. Possessing a sharp, pointed tip.

mucus. A viscous, slimy substance present on the surface of mucous membranes which serves to moisten and lubricate. It is formed from a mixture of mucin and water.

mud cat. A catfish.

mud dauber. One of several species of wasps of the family Sphecidae which construct nests, often many-celled, of mud, as *Sceliphron.*

mud eel. An amphibian (*Siren lacertina*) of southeastern United States without hindlimbs and reaching a length of 30 inches.

mud hen. The coot (*Fulica americana*).

mudminnow. A small, freshwater fish of the genus *Umbra,* as *U. limi,* which frequents muddy or stagnant waters.

mud puppy. A large, aquatic salamander, of the genus *Necturus,* especially *N. maculosus* of eastern United States. Also called water dog.

mud turtle. A turtle of the genus *Kinosteron.*

mugger. A broad-nosed crocodile (*Crocodylus palustris*) of southern Asia. Also muggar, muggur.

mule. **1.** The hybrid and usually sterile offspring of a cross between a male ass or jack and a mare, noted for its strength, endurance, and stubbornness. **2.** Any sterile hybrid.

mule deer. A large deer (*Odocoileus hemionus*) of western United States.

mullein. A common plant (*Verbascum thapsus*) with a rosette of coarse, woolly leaves; also called flannelleaf or velvet plant.

Mullerian duct. One of a pair of ducts present in the embryos of vertebrates which gives rise to the uterine tubes, uterus, and a portion of the vagina.

Muller's larva. The free-swimming larva of certain polyclad turbellarian worms, with eight ciliated arms or lobes directed posteriorly.

mullet. A marine food fish as *Mugil cephalus*, the gray mullet of the family Mugilidae, and *Pseudopeneus maculatus*, the red mullet or goat fish of the family Mullidae.

multicellular. Consisting of more than one cell; many-celled.

multifid. Divided into many lobes or divisions.

multifoliate. Having many leaves.

multilocular. 1. Containing many cells or compartments. **2.** Containing many cysts, as a hydatid cyst; polycystic.

multipartite. Divided into many parts.

multiple. 1. Consisting of or involving more than one. **2.** Having several or many units or parts. **3.** Affecting many parts simultaneously.

multiple alleles. A series of three or more alternative forms of a gene which occupy the same locus in a chromosome.

multiple genes. A condition in which two or more sets of genes produce more or less equal and cumulative effects on the same character. Such genes are primarily involved in the inheritance of quantitative characters as weight, size, yield, etc.

multipolar. 1. Having more than two poles. **2.** Having more than two processes, said of neurons.

multivoltine. Having more than two broods or generations a year.

mummified. Dried, shrunken, withered; converted into a mummy.

mummy. 1. *zool.* The embalmed, preserved body of a human or other animal. **2.** *bot.* A dried, shrunken fruit, as that resulting from fungus infections, as brown rot.

muntjac. The rib-faced or barking deer (*Muntiacus*) of southeast Asia.

mural. Of, pertaining to, or resembling a wall.

mural plates. Those forming the carina and rostrum of a sessile barnacle.

murex. A carnivorus, marine snail of the genus *Murex*.

muricate. Covered with sharp, pointed projections.

muriculate. Finely muricate.

murid. Any rodent of the family Muridae which includes the rats and mice.

muriform. Having bricklike markings; arranged in a bricklike fashion.

murine. 1. Of or pertaining to the family Muridae. **2.** Of, pertaining to, or resembling a mouse or rat.

murre. An open sea bird of the family Alcidae, as the common murre or guillemot (*Uria aalge*).

musang. The common Indian palm civet or toddy cat (*Paradoxurus hermaphroditus*) of southeast Asia.

muscarian. Having a putrid odor which attracts flies, with reference to flowers.

muscarine. A poisonous alkaloid present in certain mushrooms, as *Amanita muscaria*.

¹muscid. 1. *bot.* A moss plant. *See* moss. **2.** *zool.* A dipterous insect of the family Muscidae which includes the flies (housefly, stable fly, horn fly, tsetse fly, and others).

²muscid. Of, pertaining to, or resembling flies.

musciform. Resembling a fly in appearance or structure.

muscle. 1. A specialized organ or tissue whose special function is contraction by means of which change of shape or movements of parts is accomplished. **2.** A specific contractile structure, as the biceps *muscle* of the arm or an adductor *muscle* of a bivalve. **3.** A contractile layer or structure, as the circular *muscle* of an earthworm or a sphincter *muscle*.

muscle sense. The kinesthetic or proprioceptive sense; the perception of movement or position of parts of the body.

muscle tissue. A specialized type of contractile tissue composed of elongated cells or fibers. In vertebrates, there are three types: striated (striped, voluntary, skeletal), smooth (nonstriated, involuntary, visceral), and cardiac.

muscular. 1. Of, pertaining to, or composed of muscles or muscle tissue. **2.** Due to or resulting from the activity of muscles.

muscularis. A layer of muscle in the wall of a visceral organ, as the intestine.

musculature. The muscles of the body or a portion of it.

musculocutaneous. Pertaining to the muscles and the skin, with special reference to nerves and blood vessels which supply both structures.

mushroom. 1. A conspicuous, fleshy fungus, especially a gill fungus of the Basidiomycetes. **2.** Any of the edible fungi of the order Agaricales, especially *Agaricus campestris*, the mushroom of commerce. *Compare* toadstool.

musk. 1. An odiferous substance obtained from the musk sac or gland of the male musk deer; used in the manufacture of perfumes. **2.** A substance secreted by the musk glands of a number of animals, as turtles and crocodiles.

musk bag. A sac containing musk located under the skin of the abdomen of the male musk deer.

musk deer. A small deer (*Moschus moschiferus*) of the highlands of Asia, the source of musk.

muskeg. In the northern portion of North America, a bog characterized by extensive growth of sphagnum moss.

muskellunge. A large game fish (*Esox masquinongy*) of the pike family, Esocidae, in lakes and streams of northern United States and Canada. Also called muskie.

musk gland. A musk-secreting gland, as **a.** that of certain mammals, especially the musk deer; **b.** that of various reptiles, as the musk turtle and alligator.

musk hog. The collared peccary. *See* peccary.

muskmelon. An oval melon, a fruit of the vine, *Cucumis melo*, or the plant that produces it; a cantaloupe.

musk-ox. A small, wild ox (*Ovibos moschatus*), a hollow-horned ungulate of the arctic regions of North America.

muskrat. An aquatic rodent (*Ondatra zibethica*) of the United States and Canada, a valuable fur-bearing animal; musquash, musk beaver.

musquash. A muskrat.

mussel. Any bivalve mollusk of the class Pelecypoda (Bivalvia), especially the freshwater forms, *Unio* and *Anodonta*, of central Unites States, and marine or sea mussels, *Mytilus* and *Modiolus* (horse mussel), both used for food.

must. The expressed juice of grapes or other fruits before fermentation occurs. For red wines, it contains the skins and seeds.

mustard. 1. A powder prepared from the ground seeds of the mustard plant used as a condiment; in paste form it is used medicinally in poultices. **2.** A plant (*Brassica*) of the mustard family, Cruciferae.

mutable. Capable of mutation; having the capacity of undergoing change.

mutagen. An agent which is capable of inducing a mutation, as radium, mustard gas.

mutagenic. Capable of causing or inducing a mutation.

mutant. An organism or a gene which has undergone mutation.

mutation. 1. A sudden change in the characteristics of an organism which is capable of being transmitted to offspring, as that which results from a change in the structure of a gene (gene mutation) or changes in the number or structure of chromosomes (chromosomal mutations). **2.** An individual or a species which has undergone mutation.

muticous. Blunt; lacking a point or sharp tip; awnless.

muton. A unit of mutation; the smallest portion of a chromosome or gene which, by mutation, can give rise to a new trait.

mutualism. A form of symbiosis in which two organisms of different species live in intimate association with each other to the advantage of both; eusymbiosis.

muzzle. 1. The projecting portion of the head of an animal, especially the mouth, jaws, and nose, as the *muzzle* of a cow. **2.** A restraining device covering the jaws to prevent them from being opened.

mycelioid. Resembling a mycelium; having the structure of a mold, said of bacterial colonies.

mycelium. A tangled mass of hyphal filaments which make up the vegetative body or thallus of a fungus.

mycetangium. A tube- or baglike structure which functions in fungus storage and transmission, as in the ambrosia beetle.

mycetocyte. A hypertrophied cell in the gut of insects which houses symbiotic fungi.

mycetome. A structure in larval or adult insects which houses symbionts.

mycetozoan. 1. A myxomycete or slime mold. **2.** Any member of the Mycetozoa (Mycetozoida), regarded by some as an order of the phylum Protozoa.

mycobiont. The fungal component in a symbiotic relationship, as in a lichen.

mycocecidium. A gall caused by a fungus.

mycoid. Funguslike.

mycology. The division of botany which deals with fungi.

mycoplasma. A bacterium lacking a cell wall of the genus *Mycoplasma*, the smallest known free-living organisms. They cause bovine pleuropneumonia and related diseases.

mycorrhiza. A symbiotic relationship of fungi with seed plants in which the mycelium forms a thick web over the surface of roots (exotrophic mycorrhiza) or their hyphae invade the roots (endotrophic mycorrhiza).

mycosis. A pathological condition resulting from infection by a fungus, as tinea capitis, ringworm of the scalp.

mycostatic. Inhibiting the growth of molds.

myelencephalon. The caudal portion of the hindbrain from which the medulla oblongata develops.

myelin. A complex mixture of lipids present in the myelin sheath of nerve fibers.

myelinated. Possessing a myelin sheath; medullated.

myelin sheath. A white, glistening sheath investing certain nerve fibers (axons) consisting of concentric layers of myelin laid down by the neurilemmal cells of Schwann.

myelocyte. A cell in red bone marrow which gives rise to a granular leukocyte.

myeloid. Of or pertaining to bone marrow; marrowlike.

myeloid elements. Blood cells (erythrocytes and granular leukocytes) formed in red bone marrow.

myelon. The spinal cord.

myenteric. Of or pertaining to the muscle layers of the intestine.

myiasis. Infestation by maggots or the larvae of flies, as by screwworms, wool maggots, bots, or warbles. The larvae may localize in the skin (cutaneous myiasis) or invade cavities or passageways.

myna also **mynah** or **mina.** A starlinglike bird (*Gracula religiosa*), the talking myna or Indian grackle, commonly caged and taught to talk.

myoblast. A cell which develops into a muscle cell or fiber.

myocardium. The muscular layer of the heart located between the outer epicardium and the inner endocardium.

myochordotonal organ. A sensory receptor on the walking legs of decapod crustaceans.

myocoel. The cavity within a myotome.

myocomma. A myoseptum.

myocyte. 1. A contractile cell, especially one in sponges. 2. A muscle cell.

myoepithelial cell. A cell of ectodermal origin which possesses contractile processes, present in the secreting portions and ducts of various glands. Also called basal or basket cell.

myofibril. A minute, contractile element in the cytoplasm of a contractile cell, as a muscle cell.

myofilament. A minute element of a myofibril.

myogenic. Originating within muscle.

myoglobin. A form of hemoglobin present in muscle tissue, abundant in red muscle fibers.

myology. The study of muscles.

myomere. 1. A V-shaped muscle segment in the body wall of lancelets and fishes. 2. A myotome.

myometrium. In mammals, the muscular layer of the uterus.

myoneme. A contractile fibril present in certain ciliates and in the epithelial-muscular cells of coelenterates.

myoneural. Of or pertaining to muscles and nerves.

myoneural junction. A motor end plate.

myopia. A visual defect in which the visual image is formed in front of the retina; nearsightedness.

myoseptum. A sheet of connective tissue between two myotomes.

myosin. A protein of high molecular weight present in striated muscle tissue. *See* actin.

myotome. 1. *embryol.* The portion of a somite which gives rise to striated or skeletal muscles. 2. A myomere.

myriapod. 1. An arthropod with an elongated, segmented body, each segment bearing one or two pairs of jointed legs, as a centipede, millipede, pauropod, or symphylan. 2. Any member of the Myriapoda, formerly a class of the Arthropoda which included the centipedes and millipedes.

myriapodous. Having many pairs of legs.

myrmecology. The study of ants.

myrmecophagous. Feeding upon ants.

myrmecophile. An ant guest; an organism, especially an insect which regularly inhabits an ant's nest. *See* inquiline, symphile.

myrmecophilous. Ant-loving, with reference to plants which are regularly inhabited by or pollinated by ants.

myrrh. A gum resin occurring in two forms: **a.** herabol myrrh, from *Commiphora myrrha*, a shrub of the Middle East, used for medicinal purposes; **b.** bisabol myrrh, from *C. erythraea*, common in Arabia, used in perfumes, incense, and embalming fluids.

mysis larva. A schizopod larva of various crustaceans, so called because of its resemblance to the adult crustacean, *Mysis*.

myxedema. A pathological condition resulting from inadequate secretion of the thyroid hormone.

myxobacter. A bacterium of the order Myxobacteriales which comprise the slime bacteria.

myxoid. Of the nature of or resembling mucus.

myxoma. A tumor of connective tissue consisting of stellate cells embedded in a matrix of mucoid material.

myxomycete. A slime mold.

myxosporidian. A sporozoan of the order Myxosporidia which includes a number of parasites of fishes, amphibians, reptiles, and some invertebrates.

myxovirus. One of a group of viruses which have an affinity for mucoproteins. Includes the causative agents for human and swine influenza, mumps, and measles.

myzostome. An invertebrate of the taxon Myzostomida generally regarded as an order of polychaete annelids, members of which live as commensals or parasites in echinoderms, especially crinoids, as *Myzostoma*.

N

Na. Symbol for sodium.

nacre. MOTHER-OF-PEARL.

nacreous. Composed of mother-of-pearl.

nagana. An infectious disease of horses and cattle in East Africa caused by *Trypanosoma brucei* transmitted by the tsetse fly.

naiad. An aquatic, gill-bearing nymph of an insect.

nail. **1.** A flattened, horny, epidermal structure on the tips of the digits of primates. **2.** A horny structure or scale on the upper mandible (maxilla) of ducks, geese, and swans.

naked. **1.** Nude, bare, uncovered. **2.** *zool.* Lacking hair, feathers, or a shell. **3.** *bot.* **a.** Lacking pubescence, said of leaves. **b.** Lacking enveloping structures, as a *naked* bud.

namaycush. A large lake trout (*Cristivomer namaycush*) of North America.

name. *biol.* The term applied to an individual or the group to which it belongs. *See* binomial nomenclature.

nanander. *bot.* A dwarf male.

nanandrous. Produced on dwarf-sized filaments, as antheridia of algae.

nanism. Abnormal smallness; dwarfism.

nannoplankton. Microplankton; organisms in plankton less than 40 microns in length or diameter.

nape. The back part of the neck.

napiform. Turnip-shaped.

narcosis. A state of stupor or unconsciousness induced by a drug.

narcotic. A drug which, in moderate doses, relieves pain, acts as a sedative, and induces profound sleep but, in excess, produces stupor and coma, as opium and its derivatives, belladonna and its derivatives, and alcohol.

narcotize. **1.** To subject to the influence of a narcotic. **2.** In microtechnique, to render immobile or incapable of contracting preparatory to examination, fixing, or preserving.

nares sing. **naris.** The openings of the nasal cavity. *See* nostril, choana.

narwhal. A cetacean of northern waters (*Monodon monoceros*), males possessing a single, long, twisted tusk projecting forward from the upper jaw.

¹**nasal.** Of or pertaining to the nose or nasal cavity.

²**nasal.** The nasal bone.

nasal cavity. In vertebrates, a cavity containing olfactory receptors opening to the outside through the external nostrils or nares. In air-breathing vertebrates, it functions as a respiratory passageway opening posteriorly into the mouth or pharynx through the internal nares or choanae.

nascent. Being formed or coming into existence.

nastic movements. Growth movements in plants, especially in parts with upper and lower surfaces, as in a leaf, usually resulting from unequal concentrations of growth substances. *See* hyponasty, epinasty.

nasties. Paratonic or stimulus movements in plants. *See* photonastic, thermonasty, nyctinasty.

¹**nasute. 1.** Having a large, prominent nose. **2.** *entomol.* With a large beaklike snout.

²**nasute.** A soldier in a termite colony, with an elongated snout from which a sticky secretion is exuded which repels invaders.

natal. Of or pertaining to birth.

natatory. Of, pertaining to, or adapted for swimming.

nates. The buttocks.

¹**native. 1.** Occurring in a region at the time of its discovery: INDIGENOUS. **2** Occurring or produced naturally, as a *native* protein; not denatured.

²**native.** A plant or animal raised or produced in a particular region.

natural. 1. Innate or inborn, as a *natural* instinct. **2.** Of or pertaining to nature, as a *natural* science. **3.** Occurring or produced in a state of nature; not artificial, as a *natural* color. **4.** In keeping with or like that which occurs in nature, as a *natural* food.

natural classification. One based on phylogenetic relationships.

natural history. Formerly, the study of animals, plants, and minerals in a general and more or less unsystematized manner.

338

Now, generally restricted to the general study of a limited group, as the *natural history* of mammals, in which various aspects of mammalian life, as the classification, ecology, geographic distribution, life cycles, and economic importance, are considered.

naturalist. A student of or one versed in natural history.

naturalize. To introduce and cause to grow in a region to which an organism is not native.

natural selection. A principle called "survival of the fittest" proposed by Charles Darwin in 1859 to account for the origin of species. Basic principles are: all species tend to overproduce, hence there is a struggle for existence. Variations occur and individuals with favorable variations survive, those with less favorable are eliminated. Through successive generations, gradual and continuous changes occur, resulting in the production of new types and species.

nature. 1. The entire physical universe. **2.** All the processes and activities which take place in or occur within the universe. **3.** In a limited sense, the outdoors, especially the plants and animals which comprise the natural scenery. **4.** The physical and physiological makeup of an organism, its mode of life, and its manner of responding to stimuli.

nauplius. A free-swimming larva, the first stage in the development of certain crustaceans, as shrimps and prawns.

nautilus. A cephalopod mollusk of the genus *Nautilus. See* chambered nautilus.

navel. The umbilicus.

¹**navicular.** Boat-shaped.

²**navicular. 1.** A tarsal bone. **2.** The scaphoid bone of the wrist.

Neanderthal man. A species of fossil man (*Homo neanderthalensis*) which inhabited Europe during the third and final interglacial period, becoming extinct about 25,000 years ago.

Nearctic realm. A biogeographical realm which includes Greenland and all of North America from Mexico northward. Characteristic animals are the mountain goat, caribou, muskrat, raccoon, and pronghorned antelope.

neck. 1. *zool.* **a.** A constricted region which connects the head with the main portion of the body; the cervical region. **b.** A constricted region which resembles a neck, as that of a tooth or bone. **2.** *bot.* **a.** An elongated portion of an archegonium through which sperm gain access to the venter. **b.** The junction between the stem and root of a plant.

necrobiosis. The natural death of cells. *See* necrosis.

necrophagous. Feeding upon decaying flesh.

necrosis. The pathologic death of cells or tissues, especially those in contact with or associated with living cells.

nectar. A sweet substance secreted by nectaries or specialized cells in the flower of a plant, especially insect-pollinated plants.

nectariferous. Producing or possessing nectar, with reference to flowers.

nectarine. A smooth-skinned variety of peach.

nectary. A gland or a group of specialized cells which secrete nectar, located on any of the floral parts or upon bracts outside a flower.

nectocalyx. A nectophore.

nectochaeta. A swimming-crawling larva of certain polychaetes, as *Nereis.*

nectophore. A swimming bell, a medusa of certain siphonophores.

nectosome. The region of a siphonophore from which swimming bells arise by budding.

needle. A stiff, linear leaf, as in pines.

negroid. Of or pertaining to the Negroid division of the human species comprising several dark-skinned races as the Forest, Nilotic, and Oceanic Negroes inhabiting western, central, and southern Africa.

nekton. The larger, free-swimming animals of the sea.

nema. **1.** A nematode. **2.** A narrow filament or thread.

nemathelminth. Any member of the Nemathelminthes, formerly a phylum which included the threadworms (Nematoda) and several other groups now comprising the Aschelminthes.

nemathybome. A sac containing bundles of nematocysts, usually occurring in rows on various anthozoans.

nematocyst. A stinging or nettle cell characteristic of coelenterates but also occurring in ctenophores; each consisting of an oval or spherical capsule containing a coiled tube or thread capable of being everted or discharged. They function as protective, food-getting, and adhesive structures.

nematode. Any member of the Nematoda, a phylum or class of pseudocoelomate animals which includes elongated, cylindrical, unsegmented worms commonly called roundworms, threadworms, or eelworms, as *Ascaris*.

nematogen. A mesozoan which bears vermiform embryos. *Compare* rhombogen.

nematomorph. Any member of the Nematomorpha, a class of the Aschelminthes which includes the hairworms.

nematophore. In coelenterates, a dactylozooid or sarcostyle.

nemertean. Any member of the phylum Rhynchocoela (Nemertinea) which includes the proboscis worms or ribbon worms, as *Prostoma, Micrura*.

nemertine. A nemertean.

neolithic. Of or pertaining to the New Stone Age culture originating in the Near East, characterized by the beginnings of agriculture and animal industry.

neonatal. Newly born or recently hatched.

neopallium. The cerebral cortex, excluding the rhinencephalon.

neoplasm. A tumor; any abnormal new growth.

neopteran. Any insect of the Neoptera, a taxon which includes all winged insects with wings folded over the body when at rest.

neoptile. An immature type of feather; a down or nestling feather. *Compare* teleoptile.

neornithine. Pertaining to the Neornithes, a subclass of Aves which includes all living birds.

neotenous. Neotenic; exhibiting neoteny.

neoteny. The condition in which larval characteristics persist into the adult stage, as in certain salamanders which do not metamorphose but retain their gills and other larval characters. *See* axolotl, paedogenesis.

Neotropical realm. A biogeographic realm which includes South America, Central America, the lowlands of Mexico, and the West Indies.

neotype. In taxonomy, a specimen selected as a type specimen subsequent to the original description when it is known that the original type specimen has been lost or destroyed.

nephric. Of or pertaining to a nephridium or kidney: RENAL.

nephridiopore. The external opening of a nephridium.

nephridium. In various invertebrates, a simple or branched, tubular structure which functions as an excretory organ, opening to the outside through a nephridiopore. *See* protonephridium, metanephridium.

nephrocyte. A phagocytic cell present in arachnids, crustaceans, and insects.

nephroid. Kidney-shaped: RENIFORM.

nephromixium. A combined coelomoduct and nephridium functioning as an excretory organ and genital duct, as in certain polychaetes.

nephron. A renal tubule, the structural and functional unit of the kidney consisting of a glomerular capsule and its subjoined tubule; a uriniferous tubule.

nephrostome. 1. The ciliated, opening of a nephridium into the coelom. **2.** *embryol.* The opening of a pronephric tubule into the coelom.

nephrotome. In vertebrate embryos, a narrow longitudinal band of mesoderm connecting a somite to the lateral mesoderm from which the urinary tubules develop; intermediate cell mass: MESOMERE.

nepionic. Postembryonic; infantile.

nereid. A marine polychaete annelid of the family Nereidae, as *Nereis*.

neritic. Inhabiting waters of the continental shelf, roughly the region from low tide line to a depth of about 600 feet.

nervate. *bot.* Having veins, said of leaves.

nervation. Venation; the manner in which the nerves or veins of a leaf or wing of an insect are arranged. Also called neuration.

nerve. 1. *zool.* **a.** An elongated structure lying outside the central nervous system composed of a bundle (fasciculus) or several bundles of nerve fibers enclosed within a connective tissue sheath or epineurium. **b.** A vein in the wing of an insect. **2.** *bot.* A vein in a leaf, especially a simple, unbranched vein.

nerve cell. A neuron.

nerve center. A group of neurons within the central nervous system which controls a specific activity, as the respiratory center. *See* nucleus, 2b.

nerve cord. An elongated structure composed principally of nerve fibers extending posteriorly from the brain or cerebral ganglion, as **a.** the ventral, ganglionated structure in most invertebrates; **b.** the dorsal, tubular, spinal cord in vertebrates; **c.** the paired, interganglionic connecting cords in mollusks.

nerve ending. *See* receptor, motor end plate.

nerve fiber. A process of a neuron, especially an axon or the axonlike peripheral process of a sensory neuron.

nerve impulse. A wave of excitation initiated by a stimulus which passes along a nerve fiber. *See* action potential.

nerve net. A primitive type of nervous system consisting of a network of neurons which connects various parts of the body; impulses pass in any or all directions giving a generalized response, as in coelenterates.

nerve ring. A nerve cord which passes around a structure, as that encircling the esophagus in echinoderms.

nerve root. The connection of a spinal nerve with the spinal cord comprising a dorsal sensory root composed of afferent fibers and a ventral motor root composed of efferent fibers.

nervous. 1. Of, composed of, or related to nerves or nerve cells. **2.** Excessively excitable, as a *nervous* individual. **3.** Caused by or brought about by nervous impulses, as *nervous* control of the heart.

nervous system. A system composed of nervous tissue by which, through nerve impulses, regulation and coordination of bodily activities is accomplished; a stimulus-response mechanism involving conducting pathways connecting receptors which respond to stimuli to effector organs (muscles or glands) which respond. In vertebrates, it consists of the *central nervous system*, CNS, (brain and spinal cord) and *peripheral nervous system*, PNS, (nerves and ganglia).

nervous tissue. Tissue composed of neurons and neuroglia which comprise the organs of a nervous system (brain, ganglia, nerves, nerve cords, and sensory portion of sense organs).

nervure. 1. *bot.* A vein in a leaf. **2.** *zool.* A vein in the wing of an insect.

nest. 1. A structure or a place prepared by a bird for receiving eggs, for their incubation and hatching, and for the rearing of the young. **2.** In general, any place where an animal deposits its eggs and rears its young, as a turtle *nest*. **3.** A place for concealment or retreat or for rearing young, as a rabbit *nest*. **4.** *embryol.* and *histol.* An isolated mass of cells.

nestling. A young bird that has not left the nest.

netted. Reticulated.

nettle. 1. A plant, especially one of the genus *Urtica* which possesses glandular stinging hairs. 2. A scyphozoan medusa which inflicts painful stings.

nettle cell. A nematocyst.

network. A reticulum; an interlacing arrangement of vessels or fibers.

neural. Of or pertaining to a nervous system or nervous tissue.

neural arch. The portion of a vertebra which forms the roof and sides of a vertebral foramen, consisting of two pedicles, two lamina, and their processes.

neural canal. 1. *embryol.* The neurocoel or cavity of the neural tube. 2. The vertebral canal formed by the arches of successive vertebrae.

neural crest. A band of cells of ectodermal origin located along each side of the neural tube following its formation.

neural fold. One of two folds which develop from the neural plate and, upon fusion, form the neural tube.

neural gland. *See* subneural gland.

neural groove. A median, longitudinal groove which lies between the neural folds prior to their closing.

neuralization. The formation of the neural tube.

neural lobe. A portion of the posterior lobe of the pituitary gland.

neural plate. 1. A median, dorsal sheet of ectoderm from which the neural tube and neural crest cells develop; medullary plate. 2. In turtles, one of a set of plates overlying, or in some cases, fused with the vertebrae.

neural spine. The spine of a neural arch; the spinous process of a vertebra.

neural tube. A longitudinal ectodermal tube which forms in the middorsal region of a vertebrate embryo from which the brain and spinal cord develop.

neuration. NERVATION.

neurenteric canal. In vertebrate embryos, the connection between **a.** the neurocoel and the tail gut, as in the frog; **b.** the amniotic cavity and the yolk sac, as in mammals.

neurilemma. A delicate sheath composed of Schwann's cells

344

which forms the outermost covering of a nerve fiber. Also neurolemma.

neurobiotaxis. *embryol.* The migration of nerve cell bodies in the brain in the direction from which they receive their impulses.

neuroblast. An embryonic cell which gives rise to a neuron.

neurocoel. The cavity of the neural tube.

neurocranium. The portion of the skull which encloses and protects the brain; the cranium.

neuroendocrine. Pertaining to both nervous and endocrine systems.

neuroepithelial cells. Highly specialized receptor cells, as olfactory, gustatory, rod and cone cells, and hair cells of the internal ear.

neurofibrils. Minute fibrils in the cell body and processes of a neuron.

neurogenic. 1. Of nervous origin. **2.** Under nervous control.

neuroglia. The non-nervous, interstitial tissue of the nervous system which includes ependyma, neuroglia proper (astrocytes, oligodendrocytes, microglia), satellite cells, and cells of Schwann.

neurohemal organ. An organ which stores and liberates neurohormones, as **a.** in invertebrates, the corpus cardiacum, sinus gland, pericardial and postcommissural organs; **b.** in vertebrates, the pars nervosa of the pituitary gland.

neurohormone. A hormone secreted by a neurosecretory cell, as vasopressin.

neurohumor. A transmitter substance produced at the termination of an axon, as acetylcholine and norepinephrine; a local hormone.

neurohypophysis. The posterior lobe of the pituitary gland which includes the neural stalk (infundibulum) and neural lobe.

neurolemma. NEURILEMMA.

neurology. The science which deals with the nervous system.

neuromast. 1. A sensory papilla of the lateral line system of lower vertebrates. **2.** A crista of the internal ear. **3.** In general, any neuroepithelial cell, as a gustatory cell.

neuromotor apparatus. A complicated fibrillar system in ciliates through which the movement of cilia in various parts of the organism is coordinated.

neuromuscular junction. *See* motor end plate.

neuromuscular spindle. A proprioceptor which is stimulated by changes in length and tension of muscle fibers.

neuron. A nerve cell, the structural and functional unit of the nervous system, typically consisting of a cell body (perikaryon) and its processes (axon and dendrites).

neuropil. A network of nerve fibers, as **a.** that between terminations of axons and dendrites in an axo-dendritic synapse; **b.** that in a ganglion of an invertebrate consisting of fibers of sensory, association, and motor neurons.

neuroplasm. The interfibrillar substance in the cell body of a neuron.

neuropodium. The ventral portion of a parapodium.

neuropore. The anterior or posterior aperture of the neurocoel before complete closure of the neural tube.

neuropteran. Any insect of the order Neuroptera, which comprises the nerve-winged insects as the lacewings, ant lions, and dobsonfly.

neurosecretion. The elaboration and discharge of neurohormones or neurohumors by neurosecretory cells; the product produced.

neurosecretory cell. A neuron or nerve cell which produces and discharges a neurohormone or neurohumor.

neurotransmitter. A substance which functions in the transmission of nerve impulses at synapses or neuroeffector junctions, as acetylcholine, epinephrine, and norepinephrine; a local or diffusion hormone.

neurotrophic. Pertaining to the influence of nerves or impulses conducted by them upon the normal integrity of tissues.

neurotropic. Having an affinity for, turning toward, or affecting nervous tissue, as *neurotropic* virus.

neurula. An embryo during the period in which it possesses a neural plate.

neurulation. *embryol.* The formation of a neural plate with the subsequent formation of the neural groove and its closure to form a neural tube.

neuston. All organisms present in a surface film of water, those in the upper layer comprising the supraneuston, those in the lower layer the infraneuston.

neuter. 1. *zool.* Sexless; lacking reproductive organs; having imperfectly developed or functionless reproductive organs. **2.** *bot.* Lacking stamens and pistil, said of flowers.

neutral. 1. Lacking definite characteristics of one type or the other. **2.** *biol.* Sexless; neither male nor female: NEUTER. **3.** *chem.* Neither acid nor alkaline; with a pH of 7.0.

neutron. An atomic particle comprising a part of the nucleus consisting of an electron closely bound to a proton and bearing no electric charge.

neutrophil. 1. A cell or part of a cell which stains readily with neutral dyes. **2.** A polymorphonuclear leukocyte with cytoplasmic granules which stain with neutral dyes; a heterophil leukocyte.

nevus. A small, congenital growth or pigmentary area on the skin. *See* mole.

newt. One of a number of small, tailed salamanders especially those of the genus *Triturus* (*Diemictylus*).

nexus. 1. An interconnection or bond between members of a group or series. **2.** A connected group or series.

niacin. A vitamin of the B complex which serves as an essential component of certain coenzymes concerned with hydrogen transport in cellular respiration; nicotinic acid, niacinamide.

nib. A pointed structure, as a bill or beak; a prong.

niche. *ecol.* A place or position occupied by or a type of activity engaged in by an organism with reference to other organisms.

nicotine. A poisonous alkaloid, $C_{10}H_{14}N_2$, obtained principally from tobacco, widely used as an agricultural insecticide, usually as a 40% solution of nicotine sulfate.

nicotinic acid. Niacin; niacinamide.

nictitating membrane. A thin, transparent membrane which functions as a third eyelid in many vertebrates; vestigial in man. *See* haw.

nidamental glands. Four accessory reproductive organs in a squid which function in the formation of the eggshell.

nidation. IMPLANTATION.

nidicole. A nest parasite, especially a mite which parasitizes nest- or burrow-dwelling vertebrates.

nidicolous. 1. Remaining in the nest for some time: ALTRICIAL. **2.** Living in or inhabiting a nest. *See* myrmecophile.

347

nidification. Nest construction or the manner of building a nest.

nidifugous. Leaving the nest soon after hatching: PRECOCIAL.

nidus. **1.** A nest. **2.** A breeding place or place of origin. **3.** A source or focus of infection.

night blindness. Reduced ability to see in dim light, usually resulting from a deficiency in vitamin A.

night-blooming cereus. An American cactus of the genera *Hyalocereus* and *Selenicereus* whose flowers open at night.

night crawler. A large earthworm, as *Lumbricus terrestris*.

nighthawk. A night-flying, insectivorous bird of the goatsucker family, Caprimulgidae, as the eastern nighthawk (*Chordeiles minor*).

nightingale. **1.** An Old World thrush of the genus *Luscinia*. **2.** Common name applied to the ovenbird or hermit thrush.

nightshade. A plant of the family Solanaceae, as *Atropa belladonna*, the deadly nightshade, the source of atropine and belladonna.

nilgai. A large antelope of India (*Boselaphus tragocamelus*), commonly called the blue cow or blue bull.

nipple. **1.** A conical protuberance of the mammary gland which bears the openings of the lactiferous ducts. **2.** Any protuberance which resembles the nipple of the breast.

Nissl bodies. Clumps or granules of chromophil substance present in the cell body and dendrites of a neuron.

nit. **1.** The egg of the human body louse usually attached to a hair or to fibers of clothing. **2.** The newly hatched young from these eggs.

nitrate. A salt or ester of nitric acid, as sodium nitrate, $NaNO_3$.

nitric acid. A monobasic acid, HNO_3.

nitrification. The oxidative conversion of ammonia or ammonium salts to nitrates, accomplished by nitrifying bacteria.

nitrifying bacteria. Soil bacteria which oxidize **a.** ammonia to nitrite, as *Nitrosomonas;* **b.** nitrite to nitrate, as *Nitrobacter*.

nitrite. A salt of nitrous acid.

nitrogen. A gaseous element, symbol N, comprising the major portion (78% by volume) of the atmosphere; at. no. 7; an essential constituent of all plant and animal tissues.

nitrogen cycle. The cycle through which nitrogen atoms pass from the atmosphere into nitrogen compounds of the soil which, in turn, are utilized by plants and animals, and the eventual return of nitrogen to the atmosphere.

nitrogen fixation. The process by which free, atmospheric nitrogen is converted into compounds such as ammonia and nitrates, accomplished naturally by nitrogen-fixing bacteria, certain photosynthetic bacteria, and some blue-green algae.

nitrogen-fixing bacteria. Bacteria which are able to utilize atmospheric nitrogen in the synthesis of nitrogen-containing compounds, as **a.** free-living forms as *Azotobacter* and *Clostridium*; **b.** symbiotic forms, as *Rhizobium*, which form nodules on the roots of legumes.

nitrous. Pertaining to a compound in which nitrogen has a lower valence than in a nitric compound, as nitrous acid, HNO_2.

nitta. A leguminous tree (*Parkia biglobosa*) with large, edible pods, of West Africa.

nociceptive. Pertaining to the sense of pain; injurious, as *nociceptive* stimuli.

noctilucous. Shining at night; phosphorescent.

nocturnal. 1. Of, pertaining to, or occurring during the night. **2.** Active at night, as animals which move about and seek their food principally at night.

nodal. Of or pertaining to a node or nodes.

node. 1. A rounded mass, protuberance, or swelling. **2.** *bot.* A joint or a region on a stem at which a leaf or leaves are attached. **3.** *anat.* **a.** A rounded structure, as a lymph *node*. **b.** A constricted region, as a *node* of Ranvier.

node of Ranvier. One of a series of constrictions in the myelin sheath of a myelinated nerve fiber.

nodiform. Resembling a knob or knot.

nodose. Possessing nodes; swollen at intervals.

nodular. Pertaining to, possessing, or composed of nodes or nodules.

nodule. A small rounded mass, node, or lump; a tubercle.

nodulus. A swelling in the middle of a seta, as in oligochaetes.

nodus. 1. A node or knot. **2.** *entomol.* A notch marking the position of a prominent crossvein near the middle of the anterior margin of the front wings of odonatans.

nomenclature. A system of names used in a particular field of knowledge, as the binomial system of *nomenclature* employed in naming animals and plants.

nondisjunction. *genetics.* The failure of a pair of homologous chromosomes to separate during the reduction division.

nonmedullated. Unmyelinated; lacking a myelin sheath.

nonmotile. Lacking the power of spontaneous movement.

nonstriated. Lacking striations or cross markings, with reference to smooth muscle.

noolbender. The honey mouse (*Tarsipes spencerae*), an Australian marsupial; also called honey possum.

noradrenaline. NOREPINEPHRINE.

norepinephrine. A hormone secreted by the adrenal medulla; also produced at the terminations of sympathetic nerve fibers; a heart stimulant and general vasoconstrictor. Also called noradrenaline; sympathin E.

nori. Purple laver, a food product prepared from a red alga (*Porphyra*).

norm. A standard, usually the average or median.

normal. 1. *biol.* Typical, regular, standard, conforming to type. 2. *chem.* Containing one gram equivalent of the reagent or one gram of replaceable hydrogen ions per liter, with reference to solutions.

nose. A structure in the center of the face which contains the olfactory receptors and whose cavities form the first portion of the respiratory passageways. It includes the *external nose* which surrounds the nostrils and the *internal nose* or nasal cavity with the associated nasal sinuses.

no-see-um. A biting midge or punkie.

nose leaf. An elaborate skin fold on the nose of certain bats which functions in ultrasonic echolocation.

nostril. One of the external openings of the olfactory sac or nasal cavity of a vertebrate; an external naris.

notochord. In chordates, an elongated structure composed of vacuolated cells located ventral to the neural tube or spinal cord. It forms the axial skeleton of lancelets and larval tunicates and is present in the embryos of all vertebrates. In adult vertebrates, it is partially or completely replaced by the centra of the vertebrae.

notonectal. Swimming upside-down or with ventral side up.

notopodium. The dorsal portion of a parapodium of a poly-chaete worm.

notum. *entomol.* The dorsal portion of a body segment, especially in the thoracic region.

noxious. Harmful, injurious, deleterious.

nucellus. In gymnosperms, the sporophyte tissue of an ovule which lies adjacent to the female gametophyte; in angiosperms, the megasporangium or the tissue which encloses an embryo sac in the ovule.

nuchal. Of or pertaining to the nape of the neck.

nuchal organs. A pair of ciliated, sensory pits on the head of most polychaetes.

nuciferous. Bearing nuts.

nuclear. Of, pertaining to, within, or forming a part of a nucleus.

nuclear body. One or more bodies composed of a single strand of DNA present in a bacterium.

nuclear division. KARYOKINESIS. *See* mitosis.

nuclear membrane. A double-layered membrane forming the surface of the nucleus of a cell, the outer layer containing pores (annuli) which communicate with the endoplasmic reticulum

nuclear sap. KARYOLYMPH.

nuclease. One of a large group of enzymes which catalyze the hydrolysis of nucleic acids or their products (nucleotides, nucleosides).

nucleated. Possessing a nucleus.

nucleic acid. One of a group of organic compounds present in the nuclei and cytoplasm of cells, as ribonucleic acid (RNA) and deoxyribonucleic acid (DNA).

nucleocaspid. A caspid and its enclosed nucleic acid.

nucleolar. Of or pertaining to a nucleolus.

nucleolus. One or more densely staining, basophilic bodies present in the nucleus of a cell in the interphase; composed principally of RNA.

nucleoplasm. The protoplasm of a nucleus.

nucleoprotein. One of a group of conjugated proteins each consisting of a protein combined with a nucleic acid. They are present in the nuclei and cytoplasm of all living cells, being concentrated in the chromosomes.

351

nucleoside. A substance resulting from the removal of phosphoric acid from a nucleotide, consisting of a purine or pyrimidine base linked to a sugar (ribose or 2-deoxyribose), as uridine, adenosine.

nucleotide. A nucleoside phosphate; a nucleoside in which the phosphoric acid group is attached to a sugar.

nucleus. **1.** The central portion of a mass about which associated material is gathered. **2.** *biol.* **a.** A structure present in most animal and plant cells, typically a spheroid mass enclosed in a membrane which is essential for the life of a cell. It is the center for the synthesis of nucleic acids, DNA and RNA, through which the synthesis of specific cellular proteins and the transmission of hereditary traits is accomplished. **b.** *anat.* A mass of gray matter in the brain or spinal cord consisting principally of the cell bodies of neurons, as motor and sensory nuclei of the brain. **c.** *zool.* The initial whorl in the shell of a gastropod. **3.** *chem.* and *physics.* **a.** The central portion of an atom about which extranuclear electrons revolve; composed of protons and neutrons. **b.** A complex of atoms about which other atoms are attached. **c.** A point at which crystals begin to accumulate.

nucule. The female fructification (oogonium) of certain green algae, as *Chara.*

nude. Naked, bare, uncovered.

nudibranch. A sea slug, a gastropod mollusk of the order Nudibranchia characterized by lack of a mantle and shell, as *Aeolis, Doris. See* cerata.

numbat. A small, ant-eating marsupial of the genus *Myrmecobius,* of Australia.

nuptial. *biol.* Pertaining to mating or the breeding season, as *nuptial* plumage.

nurse. **1.** An individual that nurses or takes care of another. **2.** To nourish the young at the breast. **3.** *zool.* **a.** In social insects, a worker which takes care of the young. **b.** In ascidians, an asexual zooid which acts as a nurse to other zooids, as in *Doliolum.*

nurse cell. A cell which provides nourishment or protection to other cells, especially developing germ cells, as archeocytes of sponges or nurse cells in insect ovarioles.

nurse tree. *forestry.* A tree planted to protect young and growing trees.

nurture. That which encompasses the breeding, nourishment, care, education, and training of an individual.

nut. 1. An indehiscent, one-seeded, dry fruit usually formed from a compound ovary and typically possessing a hard outer wall or pericarp, as an acorn, hazelnut, chestnut. **2.** One of a number of structures which are not true nuts, as the Brazil nut, a seed, and certain dry drupes after the removal of the hard parts, as an almond, coconut, pecan, and walnut.

nutant. Nodding, drooping; with head bent forward.

nutation. 1. *anat.* A nodding movement, as that of the head. **2.** *bot.* A bending of the apical portion of the stem as a result of differential growth rate, as in circumnutation.

nutgall. A nutlike excrescence or gall which develops on the twigs of oaks.

nuthatch. A small, square-tailed passerine bird of the family Sittidae which creeps over the trunks of trees with head pointed downward, as *Sitta carolinensis*, the white-tailed nuthatch.

nutlet. A small, nutlike fruit or seed, as those produced by heliotrope.

nutmeg. A tree (*Myristica fragrans*) cultivated in the East Indies for its seed sold commercially as a culinary spice.

nutria. The coypu or its fur.

nutrient. A nutritious or nourishing substance; any substance which promotes growth or provides energy for physiological processes.

nutrition. 1. *zool.* The total of the processes involved in the absorption and utilization of foods and food accessories. **2.** *bot.* The taking in of mineral substances and their utilization in plant metabolism, excluding the processes of photosynthesis and fat and protein synthesis.

nutritive. Providing food or nourishment.

nutshell. The hard external covering (pericarp) enclosing the meat of a nut.

nut shell. A small, marine bivalve of the genus *Nucula*.

nux vomica. The dried, ripe seed of an Asiatic tree (*Strychnos* *nux-vomica*), the source of strychnine, brucine, and other poisonous alkaloids.

nuzzle. 1. To work with the nose, as to dig, push, rub, or poke with the nose or snout. **2.** To lie close together; to snuggle up to.

nyala. A South African antelope of the genus *Tragelaphus.*

nyctalopia. NIGHT BLINDNESS.

nyctinasty. A movement of a plant in response to day and night periods or changes in the intensity of light, as sleep movements or the opening or closing of flowers. *See* photonastic, thermonasty.

nyctitropism. A turning in response to darkness; a sleep movement.

nymph. 1. The young of an insect which undergoes simple metamorphosis, having a body form resembling the parent but with undeveloped structures, as wings and reproductive organs, as in the grasshopper. **2.** In the Acarina, the eight-legged immature form which follows the six-legged larva and precedes the adult stage.

nymphochrysalis. In water mites, the larval exoskeleton within which transformation of the parasitic larva to the nymph takes place.

nystagmus. Involuntary oscillatory movements of the eyeballs.

O

oak. A tree or shrub of the genus *Quercus*, family Fagaceae, the fruit of which is the acorn; a valuable lumber tree.

oak apple. An applelike gall which develops on oak leaves caused by a gall insect of the genus *Amphibolips.*

oarfish. The ribbon fish, a long, slender, marine fish of the genus *Regalecus*, sometimes reaching a length of 25 feet.

oat. 1. The seed or grain of a cereal grass (*Avena sativa*). **2.** The oat plant.

oat coleoptile test. A test for determining the relative quantities of auxins present in an unknown material by affixing a block of agar containing the auxin unilaterally to a detipped oat coleoptile. The amount of curvature resulting from asymmetric growth is proportional to the amount of auxin in the block.

obclavate. Inversely clavate; attached at club-shaped end.

obcompressed. Compressed dorsoventrally.

obconical. Inversely conical; attached at apex instead of base.

obcordate. Inversely cordate or heart-shaped; attached at apex.

obese. Extremely fat or corpulent.

oblanceolate. Inversely lanceolate; with broadest portion towards the apex.

oblate. Flattened at the poles.

obligate. Limited; bound to a restricted environment, as an *obligate* parasite. *Compare* facultative.

oblong. Elongated, with sides nearly parallel and corners rounded, as an *oblong* leaf.

obovate. Inversely ovate, with distal end broader than basal end, with reference to leaves.

obsolete. Indistinct, not evident, rudimentary, vestigial.

obstetrical. Of or pertaining to labor or childbirth.

obstetrical toad. A toad (*Alytes obstetricans*) of southern Europe, the male of which carries about eggs attached to his hind legs.

obtect. With appendages adfixed to the body surface, with reference to the pupae of insects, as in lepidopterans.

obturator. Closing an opening, as the *obturator* membrane of the pelvic girdle.

obtuse. Blunt or rounded at the end; not pointed.

occipital. Of or pertaining to the occiput.

occipital bone. The bone forming the posterior portion of the skull bearing one or two *occipital* condyles which articulate with the atlas. It contains the foramen magnum.

occiput. 1. In vertebrates, the back or posterior region of the head. **2.** *entomol.* The dorsal, posterior region of the head.

occlude. To block, obstruct, or close a passageway.

occlusion. 1. *anat.* **a.** The state of being closed or blocked, said of vessels. **b.** The coming into contact of the teeth of the upper and lower jaws. **2.** *bot.* The closing of a wound in a tree.

oceanic. Of or pertaining to the seas or oceans.

oceanic islands. Those formed by volcanic action in waters of great depth, as the Hawaiian and Galapagos Islands.

oceanic zone. The open seas beyond the continental shelf, the surface layers comprising the euphotic zone; deeper layers the bathyal and abyssal zones.

ocellar. Of or pertaining to an ocellus.

ocellar bulb. An enlargement at the base of a tentacle in a medusa: TENTACULAR BULB.

ocellated. 1. Bearing an ocellus or ocelli. **2.** Bearing an eyelike marking. **3.** Resembling an ocellus in appearance.

ocellus pl. **ocelli. 1.** A simple eye or pigment spot which functions as a light receptor in various invertebrates. **2.** The stigma or eyespot of a protozoan, as that of a dinoflagellate. **3.** An eyelike spot, as that on certain feathers and leaves.

ocelot. A spotted, medium-sized cat (*Felis pardalis*) found from Texas through South America. Also called tiger cat, leopard cat, spotted cat.

ocrea. 1. *bot.* A tubular structure consisting of one or a pair of stipules which ensheaths the base of a petiole. **2.** *zool.* A sheath, as that in the booted tarsus of a bird.

ocreated. Possessing a sheath or ocrea.

octamerous. Having parts which are eight in number or occur in multiples of eight.

octandrous. Having eight stamens.

octopus. A dibranchiate, cephalopod mollusk of the order Octopoda, as *Octopus vulgaris* with a globular body bearing eight long, sucker-bearing arms encircling a parrotlike beak; commonly called devilfish.

¹ocular. Pertaining to or of the nature of an eye.

²ocular. The eyepiece of an optical instrument, as a microscope.

ocular plate. One of a series of ossicles in an echinoid.

oculomotor. Of or pertaining to eye movements.

oculomotor nerve. The third cranial nerve (III), fibers of which innervate all eye muscles except the superior oblique and lateral rectus.

oculus. 1. *anat.* An eye. **2.** *bot.* An eye or bud of a tuber.

odd-pinnate. Pinnate but terminating in a single leaflet, said of leaves.

odonate. An insect of the order Odonata which includes the dragonflies and damselflies.

odontoblast. 1. A cell in the outer layer of dental pulp which is involved in the formation of dentin. **2.** A tooth-forming cell in the buccal mass of a mollusk.

odontoid. Resembling a tooth; toothlike.

odontoid process. A toothlike process on the axis about which the atlas rotates: DENS.

odontophore. A cartilagelike mass which underlies and supports the radula of various mollusks, especially gastropods.

odor. A scent, fragrance, or smell; a sensation resulting from the stimulation of olfactory receptors.

odoriferous. Having a pronounced odor or smell.

odorous. Having a distinctive odor or smell: ODORIFEROUS.

oe-. For words beginning with OE- not found here, see E-.

oenocyte. A large cell found in clusters about the tracheae and fat body of most insects.

offset. A short branch of a stem which bears leaves, usually arising from underground axils of leaves; a tiller.

offshoot. A lateral branch of the main stem.

oidiophore. A hyphal branch which produces oidia.

oidium pl. **oidia.** A thin-walled spore of certain basidiomycetes.

oil. A fluid lipid, a viscous liquid soluble in ether or other organic solvents but insoluble in water.

oil gland. A gland which secretes oil, as the uropygial gland of birds or the sebaceous gland of mammals.

okapi. A ruminant (*Okapia johnstoni*) of Africa, similar to a giraffe but smaller in size and with a shorter neck.

okra. A garden plant (*Hibiscus esculentus*), native of Africa.

old-squaw. A sea duck (*Clangula hyemalis*), a diving duck of the Arctic coasts and eastern North America; oldwife.

oleaceous. Of or belonging to the olive family, Oleaceae.

oleaginous. Fatty, oily, unctuous; producing oil.

olecranon. A bony process at the elbow forming proximal end of the ulna.

olfaction. Smelling; the sense of smell.

olfactory. Of or pertaining to the sense of smell.

olfactory brain. The rhinencephalon.

olfactory bulb. The expanded anterior portion of the olfactory tract.

olfactory cells. Bipolar receptor cells present in olfactory epithelium which respond to olfactory stimuli, their axons forming the olfactory nerves.

olfactory epithelium. The epithelium covering the olfactory areas of the nasal cavity, containing olfactory, sustentacular, and basal cells.

olfactory lobe. The anterior portion of the telencephalon: RHINENCEPHALON.

olfactory nerve. The first cranial nerve (I), a sensory nerve composed of the unmyelinated axons of olfactory cells.

olfactory pit. 1. A depression which contains olfactory receptors, as that of a lancelet. 2. *embryol.* In vertebrate embryos, a depression which develops from an olfactory placode which gives rise to the olfactory sac or nasal cavity.

olfactory sac. A blind pouch in lower vertebrates which contains olfactory receptors. Its opening is the nasal aperture.

olfactory tract. A bundle of nerve fibers which connects the olfactory bulb with the cerebral hemisphere.

oligandrous. Possessing few stamens.

oligocarpous. With few carpels or fruits.

Oligocene epoch. An epoch of the Tertiary period characterized by the rise of flowering plants and archaic mammals. It follows the Eocene epoch.

oligochaete. An annelid of the class Oligochaeta which includes the earthworms, as *Lumbricus.*

oligodendroglia. Small neuroglia cells with few branches, usually located close to the cell bodies of neurons.

oligolectic. Visiting few food plants or flowers, said of insects.

oligophagous. Having food restricted to a single food plant, with reference to insects.

oligopneustic. Having few functional spiracles, said of insects.

oligopod larva. An insect larva which lacks abdominal appendages, as the larva of scarab beetles.

oligosaprobic. Pertaining to an aquatic habitat rich in mineral matter and high in oxygen content with decomposition and purification processes reduced to a minimum.

oligospermia. Deficient production of sperm or semen.

oligospermous. Producing few seeds.

oligotrophic lake. A deep lake with clear water containing little organic matter. *Compare* dystrophic lake, euthrophic lake.

oligoxenous. Having a limited host range, with reference to parasites.

olive. 1. *bot.* A small, slow-growing tree (*Olea europaea*) of the Mediterranean region or its fruit, the latter a source of olive oil. **2.** *anat.* The oliva or olivary body, a rounded eminence on the medulla oblongata.

olive shell. A cylindrical-shaped, marine gastropod with a smooth, highly polished shell, as *Oliva sayana*, the lettered olive.

olm. A salamander (*Proteus anguinus*) of southeastern Europe.

olynthus. A developmental stage in certain sponges in which the body is vase-shaped with a sycon type of canal system.

omasum. The third compartment in the stomach of a ruminant; manyplies; psalterium. *See* ruminant.

omental bursa. The lesser peritoneal sac, a cavity within the greater omentum which communicates with the peritoneal cavity via vestibule and epiploic foramen.

omentum pl. **omenta.** A fold of peritoneum which connects and protects certain visceral organs consisting of the *greater omentum*, which hangs like an apron over the intestines and the *lesser omentum* which lies between the stomach and the liver.

ommachrome. A brown pigment present in ommatidia.

ommatidium. One of the numerous elongated units which make up a compound eye of an arthropod. *See* mosaic image.

omnivore. An omnivorous animal.

omnivorous. Feeding on both animal and plant food.

omosternum. A bone in the sternum of an amphibian which lies anterior to the junction of the clavicles and the epicoracoid cartilages.

omphalos. The umbilicus or navel.

onager. A wild ass (*Equus hemionus onager*) of southwest Asia. *See* kiang.

onchomiracidium. The ciliated larva of a monogenetic fluke; also oncomiracidium.

onchosphere. A hexacanth, a six-hooked embryo which develops within the egg of various tapeworms, especially those of the genus *Taenia*. Also oncosphere.

oncology. The study of tumors.

onion. A liliaceous plant (*Allium cepa*) or its succulent bulb, with a characteristic taste and odor due to a volatile oil, allyl sulfide.

onisciform. With a flattened body; platyform.

ontogeny. The development of an individual; ontogenesis. *Compare* phylogeny.

onychium. A pulvillus.

onychophoran. Any member of the Onychophora, a phylum or a subphylum or class of the Arthropoda, which contains a small number of wormlike forms mostly of the genus *Peripatus* which possess characteristics of both annelids and arthropods.

oocyst. 1. A cyst containing sporozoites of *Plasmodium* which develops in the wall of the stomach of a mosquito. **2.** The encysted zygote of various protozoans, as coccidia.

oocyte. In oogenesis, a cell which develops from an oogonium and, following meiosis, becomes a mature ovum. *See* oogenesis.

oogamy. Sexual reproduction involving gametes of unequal size, as in all metazoans.

oogenesis. The processes involved in the origin, growth, and development of a mature, haploid ovum; gametogenesis in the female.

oogonium. 1. *bot.* A single-celled organ or structure in algae or fungi which produces the female gamete. **2.** *zool.* A cell which in oogenesis gives rise to a primary oocyte.

ookinete. The active, motile zygote of certain parasitic protozoans, as *Plasmodium;* vermicule. In a mosquito, it develops into an oocyst.

oology. The study of eggs, especially the eggs of birds.

oosperm. A fertilized ovum; a zygote.

oosphere. A large, nonmotile, female gamete formed within an oogonium; an egg cell.

oospore. A resting zygote formed by the fusion of heterogametes, as in various algae and fungi.

oostegite. One of several flattened structures borne on the coxae of thoracic appendages of certain crustaceans which form a brood pouch for the eggs and young, characteristic of the Pericarida.

ootheca. An egg case, as that produced by certain insects, mollusks, and elasmobranchs.

ootid. A mature ovum following the second maturation division.

ootype. A dilated region of the oviduct in flatworms in which yolk from the vitellaria is added to the fertilized ovum and a shell formed before passage of the ovum to the uterus.

oozooid. In tunicates, an individual which develops from an ovum.

opah. A large, brilliantly colored, marine fish (*Lampris regius*). Also called moonfish or Jerusalem haddock.

open cell. A cell in the wing of an insect which extends to the margin; one not entirely surrounded by veins.

operator. A controlling gene located close to or forming a part of the structural genes it regulates.

operculate. Having a lid, covering, or operculum.

operculum. 1. A lid or covering. 2. *bot.* **a.** The lid covering the discharge pore in the sporangium of certain fungi. **b.** The lid of a spore capsule in mosses. **c.** The lid of a pyxis. 3. *anat.* A cerebral fold which conceals the insula. 4. *zool.* A flap or plate which **a.** closes the opening of a hydrotheca in coelenterates; **b.** closes the aperture of the shell in gastropods; **c.** closes the opening of an egg through which an embryo or larva makes its exit, as in trematodes and insects; **d.** closes a burrow when occupant is retracted, as in serpulid polychaetes; **e.** covers the cavity containing sound-producing structures in cicadas; **f.** covers the book gills and genital openings of the king crab; **g.** forms a respiratory structure in terrestrial isopods; **h.** covers the gill chamber of fishes; **i.** covers the gills and developing legs in amphibian larvae; **j.** is the second ear ossicle in urodeles and some anurans.

operon. An operator and the gene or group of genes it controls.

[1]ophidian. Of or pertaining to snakes.

[2]ophidian. A snake.

ophiuroid. An echinoderm of the class Ophiuroidea which includes the brittle stars, basket stars, and serpent stars.

ophthalmic. Of or pertaining to the eyes; ocular.

opisthaptor. *See* opisthohaptor.

opisthobranchiate. A gastropod of the order Opisthobranchiata in which gills and mantle cavity lie posterior to the heart, as in bubble shells, sea hares, pteropods, and sea slugs.

opisthocoelous. Concave behind, with reference to vertebrae in which the posterior end of centrum is concave, anterior end convex, as in the garpike.

opisthoglyph. Having one or more pairs of rear teeth or fangs attached to the posterior portion of the maxillary bones of the upper jaw.

opisthognathous. Having a receding lower jaw.

opisthohaptor. A haptor located at the posterior end, as in monogenetic flukes.

opisthonephros. The functional kidney of lampreys, most fishes, and amphibians.

opisthosoma. The abdomen or posterior portion of the body, as in chelicerates.

opium. A narcotic drug prepared from the dried juice of the unripe capsules of the opium poppy (*Papaver somniferum*). It contains a number of alkaloids, including morphine.

opossum. 1. A marsupial (*Didelphis virginiana*) of the United States, noted for feigning death (playing possum). 2. One of a number of related marsupials of Australia.

opossum shrimp. A mysid, a crustacean of the order Mysidacea, females of which possess a brood pouch or marsupium.

opsiblastic. Undergoing a period of dormancy before developing, with reference to eggs.

opsonin. An antibody or antibodylike substance which renders bacteria more susceptible to phagocytosis.

optic. Of or pertaining to the eyes or vision.

optic chiasma. The crossing and partial decussation of fibers of the optic nerves located on ventral surface of the brain.

optic lobes. The corpora bigemina, two rounded bodies forming the roof of the midbrain and concerned with vision; well developed in amphibians, reptiles, and birds; absent in mammals.

optic nerve. The second cranial nerve (II), a sensory nerve formed of fibers which arise from ganglion cells in the retina of the eye.

optic tract. A band of fibers which extends from the optic chiasma to the lateral geniculate body of the thalamus.

optic vesicle. *embryol.* A lateral evagination of the forebrain which gives rise to the retina of the eye and optic nerve.

oral. **1.** Of or pertaining to the mouth or region of the mouth. **2.** Spoken, in contrast to written. **3.** Taken into, or used within the mouth.

oral arm. One of four extensions of the manubrium which surround the mouth in scyphozoan medusae.

oral cone. The hypostome of a hydroid.

oral disc. In anthozoans, the flattened, circular area surrounding the mouth.

oral funnel. *See* buccal funnel.

oral groove. In ciliates, a surface groove which leads to the cytostome.

oral hood. A hoodlike fold bearing cirri which surrounds the vestibule of a lancelet.

oral spear. A pointed structure or stylet in soil nematodes used in penetrating roots, or once within a root, in piercing cells.

oral sucker. In flukes, the sucker which surrounds the mouth.

oral valves. In fishes, two folds within the mouth opening which prevent the backflow of water during respiratory movements.

orange. A spherical, globose fruit, a berry of a small evergreen tree of the genus *Citrus* or the tree that produces it.

orangutan. A large, arboreal, anthropoid ape (*Pongo pygmaeus*) of the swamps of Sumatra and Borneo.

orbicular. Having the shape of an orb; spherical; globular.

orbiculate. Circular or nearly circular in shape, said of leaves; disc-shaped.

orbit. **1.** In crustaceans, the cavity which contains the eyestalk. **2.** In insects, the region along the inner border of the eye, as in dipterans. **3.** In vertebrates, the eye socket, a cavity of the skull which contains the eye.

orbital. Of, pertaining to, or located in or near an orbit.

orbital glands. Large mucous glands located in the orbit whose ducts open into the mouth cavity, as in members of the dog family.

orchid. Any monocot plant of the family Orchidaceae, noted for the beauty of their flowers and high degree of specialization, especially with respect to pollination. Most are tropical plants; many are epiphytic.

order. A taxon or taxonomic group made up of families and forming a subdivision of a class or subclass.

Ordovician period. The second period of the Paleozoic era following the Cambrian and preceding the Silurian characterized by the rise of the first vertebrates, the ostracoderms.

ordure. Excrement, dung, feces. *See* scat.

organ. A part of an organism having a definite form and structure which performs one or more specific functions, as a leaf in a plant or a gland in an animal.

organelle. A specialized structure in a cell which performs a definite function, as the Golgi apparatus, a cilium, flagellum, mitochondrion, or plastid.

organic. **1.** Pertaining to, consisting of, or related to organs. **2.** Pertaining to, originating in, or derived from living organisms. **3.** *chem.* Pertaining to compounds which contain carbon.

organic evolution. The evolution of life and living things; the origin of species.

organism. **1.** A living individual. **2.** An individual capable of carrying on metabolic activities and reproducing, as a plant, animal, or protist.

organizer. **1.** Something that brings about the development of a specific structure. **2.** *embryol.* A specific part or region of an embryo which, through a chemical substance produced by it, influences the development of another part; an inductor. *See* induction, competence.

organ of Bojanus. **1.** A nephridium. **2.** The excretory organ of a bivalve.

organ of Corti. The end organ of hearing, an elongated spiral structure resting on the basilar membrane of the cochlea. It contains hair cells, receptors for auditory stimuli.

organogenesis. The development of specific organs or structures from undifferentiated tissue.

[1]**organoid.** Organlike.

[2]**organoid.** An organelle.

organ-pipe coral. An alcyonarian coral of the genus *Tubipora* with long, slender polyps encased in brick-red skeletal tubes.

organ system. *See* system.

Oriental realm. A major biogeographic realm comprising Asia south of the Himalayas. It includes India, Malay Peninsula, Sumatra, Borneo, Java, Celebes, and the Philippine Islands.

orifice. An opening or aperture; a mouth or mouthlike opening.

origin. 1. The act of coming into existence. **2.** That from which anything originates. **3.** The relatively fixed end of attachment of a muscle.

oriole. 1. One of a number of brightly-colored passerine birds of the family Icteridae, as the Baltimore oriole (*Icterus galbula*), which builds a long, purselike, hanging nest; fire hangbird, firebird. **2.** An Old World passerine bird (*Oriolus*) of the family Oriolidae.

ormer. An abalone, especially *Haliotis tuberculata*, of the English channel.

ornate. Elaborately marked or embellished, as the ornate scallop (*Pecten ornatus*) or ornate ticks, as *Dermacentor* and *Amblyoma*.

ornithine. An amino acid, $C_5H_{12}O_2N_2$, present in the excreta of birds.

ornithine cycle. A cyclic series of reactions occurring in the liver resulting in the formation of urea; Krebs-Henseleit or urea cycle.

ornithischian. Any fossil reptile of the Ornithischia, comprising birdlike, bipedal, herbivorous dinosaurs; the beaked and duck-billed dinosaurs.

ornithology. The study of birds and bird life.

ornithophily. Pollination by birds, as hummingbirds.

ornithosis. A virus disease of birds, especially pigeons and domestic fowl, transmissible to man; resembles psittacosis.

oronasal groove. An external groove connecting the olfactory pit to the mouth, as in elasmobranchs.

orphan virus. A virus which has not been shown to cause a disease.

orthogenesis. Determinate or straight-line evolution; a theory that variations and resulting evolutionary changes occur along definite lines or follow a specific trend rather than occurring haphazardly in all directions. Such is thought to account for parallelism and overspecialization.

orthognathous. With jaws evenly aligned so that neither protrudes.

365

orthopedics. The branch of medicine which deals with diseases of the bones and joints, especially the correction of skeletal deformities.

orthopnea. A condition in which breathing is difficult unless subject is sitting in an upright position.

orthopteran. Any insect of the order Orthoptera which includes the grasshoppers, locusts, katydids, crickets, cockroaches, mantids, and walking sticks. Front wings are straight and narrow and cover the hind wings which are folded like a fan; mouth parts biting; metamorphosis incomplete.

orthostichous. In phyllotaxy, arranged in straight, vertical rows or ranks.

orthotropous. Erect with micropyle tip pointing away from point of attachment to the placenta, with reference to ovules.

oryx. A large African antelope of the genus *Oryx*, with long, straight horns.

¹os. A mouth or opening.

²os. A bone.

oscillate. To move back and forth; to swing like a pendulum.

oscine. Of or pertaining to passerine birds of the suborder Oscines which includes most of the common songbirds, characterized by possession of a syrinx.

os cordis. A heterotopic bone present in the heart of certain mammals, as the deer.

os cornu. A mass on the frontal bone of a mammal from which a horn develops.

os coxae. The innominate or hipbone.

osculum. The excurrent opening of a sponge.

os falciforme. A sickle-shaped heterotopic bone in the forelimb of a mole.

osmeterium. A repugnatorial gland in insects.

osmic acid. Osmium tetroxide, OsO_4, a crystalline substance used in microtechnique as an immobilizing and fixing agent.

osmoconformer. An organism which is able to change its body fluid osmotically to adjust to the osmotic pressure of the medium in which it lives. *Compare* osmoregulator.

osmophilic. Adapted for living in a medium with a high osmotic pressure. *Compare* halophile.

osmoreceptor. A receptor which is stimulated by changes in osmotic pressure.

osmoregulation. The maintenance of proper water and electrolyte balance.

osmoregulator. An organism which maintains a constant osmotic concentration in its body fluid regardless of the medium in which it lives.

osmoregulatory. Pertaining to or involved in the regulation or control of osmotic pressure within an organism, accomplished by maintenance of proper water and electrolyte balance.

osmosis. The passage of water through a semipermeable membrane as a result of differences in the concentrations of the solutions.

osmotic pressure. The pressure needed to prevent water from passing into a solution from which it is separated by a membrane permeable only to water.

os palpebrae. A heterotopic bony plate embedded in the eyelid of a crocodile.

os penis. A heterotopic bone of the penis. *See* baculum.

osphradium. A sensory receptor in gastropods and bivalves which functions as a chemoreceptor and for detection of sediment in water passing over the gills.

osprey. A large fish hawk (*Pandion haliaetus*).

os priapi. *See* os penis.

ossein. A protein present in the organic matrix of bone.

osseous. Bony; composed of, resembling, or of the nature of bone.

ossicle. A small bone or bonelike structure, as **a.** one of the bones of the middle ear; *see* ear ossicle; **b.** a small, calcareous plate in the skeleton of echinoderms; **c.** a toothlike structure in the gastric mill of a crustacean; **d.** a small bony plate embedded in the sclera of certain birds and reptiles.

ossification. The formation of bone. *See* endochondral and intramembranous ossification.

ossify. **1.** *anat.* To convert into bone or bonelike material. **2.** *bot.* To develop a hard, bonelike structure, as the stony layer of a drupe.

osteichthyan. Any vertebrate of the class Osteichthyes which includes the bony fishes, comprising the subclasses Choanichthyes (lung- and lobe-finned fishes) and Actinopterygii (ray-finned fishes).

osteoblast. A bone-forming cell.

osteoclast. A large, multinuclear cell found where resorption of bone is taking place and thought to be the active agent in the process.

osteocranium. The bony portion of the cranium. *Compare* chondrocranium.

osteocyte. A mature bone cell lodged within a lacuna in the matrix of bone tissue.

osteoderm. A minute bone in the dermis of certain reptiles.

osteogenesis. Bone formation; ossification.

osteogenic. Of or pertaining to bone formation.

osteoid. The organic matrix of developing bone within which calcium salts are deposited.

osteology. The study of bones, especially their formation and organization into the skeletal system.

osteomalacia. Adult rickets; the softening of bones due to inadequate intake of calcium, phosphorus, or vitamin D.

osteone. A haversian system.

ostiole. A small aperture, opening, or pore.

ostium pl. **ostia.** An opening into a tube or cavity, as **a.** in sponges, an opening through which water enters a radial canal or spongocoel; **b.** in anthozoans, an opening in a septum; **c.** in arthropods, an opening in the heart through which blood enters; **d.** in bivalves, an opening through which water enters a water tube of a gill; **e.** in vertebrates, the opening to an oviduct or uterine tube.

ostracod. A crustacean of the Ostracoda which includes mostly minute freshwater or marine forms with a bivalve carapace, as *Cypris*.

ostracoderm. Any member of the Ostracoderma, an order of primitive, jawless, fishlike vertebrates with a heavy armor of dermal bone encasing head and anterior part of the trunk. All are fossil forms.

ostrich. A large, flightless bird (*Struthio camelus*) inhabiting arid regions of Arabia, Syria, and North Africa; the largest of living birds.

otic. Of or pertaining to the ear; auditory, auricular.

otic vesicle. An otocyst.

otoconia. Ear dust or particles of calcium carbonate present in the otolithic membrane of the macula of the inner ear.

otocyst. 1. A statocyst. **2.** *embryol.* The auditory or otic vesicle from which the membranous labyrinth of the inner ear develops.

otolith. A calcareous body present in the statocyst of invertebrates or the saccule of vertebrates. *See* ear stone, otoconia.

otolithic membrane. A gelatinous substance on the surface of a macula into which the hairs of hair cells project. It contains otoliths (otoconia), movement of which stimulates the hair cells.

otology. The science which deals with the ear.

otoporpa. In hydrozoan medusae, a tract containing nematocysts which extends upward from a lithostyle. *See* peronium.

otter. An aquatic, fur-bearing mammal of the family Mustelidae, as *Lutra lutra*, the Old World otter; *L. canadensis*, the North American or river otter, and *Enhydra lutris*, the sea otter.

ounce. The snow leopard (*Uncia uncia*) of central Asia.

outbreeding. The mating of unrelated individuals employed to introduce new genes and increase heterozygosity. *Compare* inbreeding. *See* heterosis.

outcrossing. OUTBREEDING.

ouzel also **ousel.** The ring ouzel, a European bird (*Turdus torquatus*) related to the thrushes.

oval window. The fenestra ovalis, an opening in the vestibule of the inner ear which receives the footplate of the stapes.

ovarian. Of or pertaining to the ovary.

ovariectomy. Surgical removal of an ovary.

ovariole. An egg-forming tubule in the ovary of an insect.

ovariotomy. Surgical incision into an ovary: OVARIECTOMY.

ovary. 1. *zool.* The female gonad or reproductive organ; the organ which produces ova or eggs; also the source of hormones (estrogens, progesterones, androgens, and relaxin). **2.** *bot.* In seed plants, the enlarged basal portion of the pistil composed of one or more carpels; the part of the flower which develops into a fruit.

ovate. Oval; shaped like an egg with basal end slightly wider.

369

ovenbird. An American warbler (*Seiurus aurocapillus*); teacher bird.

overturn. In limnology, the process occurring in lakes when, in the spring, surface water on warming becomes heavier and sinks to the bottom (*spring overturn*) or in the fall when surface water becomes relatively cooler and sinks to the bottom (*fall overturn*).

ovicell. In ectoprocts, a modified zooecium which serves as a brood chamber or pouch.

oviducal. Of or pertaining to an oviduct.

oviduct. A duct which conveys ova or eggs; in mammals, the fallopian tube or uterine tube.

ovigerous. Egg-bearing, as an *ovigerous* mantle in barnacles or *ovigerous* legs, in certain spiders.

oviparous. Designating a type of reproduction in which females lay eggs which hatch outside the body, as in birds and insects. *Compare* ovoviviparous, viviparous.

oviposit. To discharge eggs from the body.

ovipositor. 1. A specialized organ or structure for the deposition of eggs, as that possessed by many insects. 2. A similar structure in teleost fishes consisting of a tubular extension of the genital papilla.

ovisac. A structure in which eggs are stored or, in some cases, developed, as **a.** the uterus in elasmobranchs and amphibians; **b.** the egg sac in the fairy shrimp.

ovotestis. An organ which produces both spermatozoa and ova, present in certain hermaphroditic animals, as snails.

ovoviviparous. Producing shelled eggs with considerable yolk which develop within the body of the female but the young do not receive nourishment from the mother through a placenta or similar structure, as in sharks.

ovulation. The discharge of an ovum or ova from the ovary.

ovule. An oval body in the ovary of a flower which develops into a seed. Before fertilization, it consists of a nucellus, its enclosed embryo sac (megagametophyte), and the integuments.

ovuliferous. Bearing ovules.

ovum pl. **ova.** 1. A mature female gamete or germ cell. 2. An egg cell or egg. 3. A specialized cell capable of developing into a new individual of the same species. It is usually a large, immobile cell with a haploid nucleus and cytoplasm containing yolk or deutoplasm and is produced by the female gonad or ovary. *See* egg.

owl. A nocturnal bird of prey of the order Strigiformes characterized by a large head, short neck, large eyes surrounded by facial discs, short, hooked bill, and long, sharp, curved talons. They feed on small animals, principally rodents and insects.

owlet. A small owl; a young owl.

ox pl. **oxen.** A domestic bovine (*Bos taurus*), especially the castrated male, formerly used extensively as a draft animal.

oxidase. An enzyme which catalyzes an oxidation reaction.

oxidation. 1. A reaction in which oxygen combines with another substance with the liberation of heat; combustion; burning. **2.** The removal of hydrogen from a compound. **3.** The loss of electrons with an increase in positive valence of an element. *Compare* reduction.

oxidize. To bring about oxidation.

oxpecker. An African starling of the genus *Buphagus* which perches on the backs of large game animals and cattle and feeds on ectoparasites. Also called tickbird.

oxygen. An element, symbol O, at. no. 8, an odorless, tasteless, colorless gas, the most abundant element in nature comprising 21% of the atmosphere, 90% of water, and 50% of the earth's crust. It combines readily with most elements and supports combustion. It is essential for life in animals, being utilized in respiration.

oxygenation. 1. A combination with oxygen: OXIDATION. **2.** The loose combination of oxygen with hemoglobin to form oxyhemoglobin.

oxyhemoglobin. Oxygenated hemoglobin (Hb_4O_8), a compound formed in red blood cells on their passage through the gills or lungs.

oxytocin. A neurohormone produced in the hypothalamus and stored and released by the neural lobe of the pituitary. It induces uterine contractions and stimulates milk ejection by the mammary gland. Formerly called pitocin.

oyster. A sessile, marine bivalve of the genus *Ostrea*, widely sought for and cultivated as a source of food. Adults have a rough, asymmetrical shell with a single adductor muscle; foot is lacking. Adults function alternately as males and females. Larva is a veliger; young are called spat. Common species is *Ostrea virginica*.

oyster catcher. A sandpiperlike bird (*Haematopus palliatus*), also called clam bird or oysterbird.

oyster crab. A small crab (*Pinnotheres ostreum*) which lives as a commensal in the mantle cavity of an oyster.

oyster drill. A carnivorous, boring gastropod (*Urosalpinx cinerea*), a serious pest in oyster beds. Also a related form, *Eupleura caudata*.

oyster leech. A polyclad turbellarian worm (*Stylochus*) which feeds on oysters.

oystershell scale. A scale insect (*Lepidosaphes ulmi*), a serious pest to fruit and ornamental trees and shrubs.

oyt. The least tern (*Sterna antillarum*).

P

P$_1$. In genetics, the first parental generation; the parents in a particular cross, considered to be homozygous unless otherwise stated.

paca. A large, tailless, South American rodent (*Cuniculus paca*).

pacarana. A South American rodent (*Dinomys branickii*) similar to a paca.

pacemaker. Something that sets a pace, especially the sinoatrial node of the heart.

pachyderm. A large, thick-skinned, hoofed mammal, as an elephant, rhinoceros, or hippopotamus.

pachytene stage. A stage in the early prophase of mitosis during which pairing of the chromosomes is completed and chromosomes become shorter and thicker. It follows the zygotene stage and precedes the diplotene stage.

Pacinian corpuscle. A capsulated sensory end organ for deep pressure.

pack rat. A rodent (*Neotoma floridans*) of southern and western United States. Also called .wood rat or trade rat.

paddle. *See* skipping blades.

paddlefish. The spoonbill (*Polyodon spathula*), a large ganoid fish of the lower Mississippi, with a spatulalike snout and heterocercal tail.

pae-. For words beginning with PAE not found here, see PE-.

paedogenesis. Reproduction by immature or larval forms or by individuals which have not acquired adult characteristics, as certain gall insects, trematodes. *See* neoteny.

pagina. *bot.* The blade of a leaf.

pain. An acutely uncomfortable feeling or sensation of discomfort or distress, usually the result of stimuli which are injurious to the body, which may result from dysfunction, excessive stretching or contraction of distensible structures, inadequate or excessive blood supply, bacterial infection, presence of foreign proteins, mechanical injury, or trauma. It is usually associated with impaired function.

painted. With colored streaks or blotches of varying intensity.

painter. *See* cougar.

palaceous. Spade-shaped, said of leaves.

palama. The web in a web-footed bird.

palatal folds. In reptiles and birds, two folds which partially separate the oral and nasal cavities.

palate. 1. *anat.* The roof of the mouth consisting of an anterior bony portion (hard palate) and a posterior membranous portion (soft palate). 2. *bot.* A rounded projection of the lower lip of a personate corolla which closes or nearly closes the throat, as in orchids.

¹palatine. Of or pertaining to the palate.

²palatine. The palatine bone.

palea. The upper scale or bract which, with the lemma, encloses a single flower of a grass spikelet.

Palearctic realm. A biogeographical realm which includes the continents of Europe and Asia excluding Southeast Asia and the tip of Arabia but including northern Africa and Iceland.

Sometimes combined with the Nearctic realm to form the Holarctic realm.

paleobotany. The study of fossil plants.

Paleocene epoch. The first epoch of the Tertiary period marked by the rise of archaic placental mammals and birds.

Paleolithic period. A period in early human culture characterized by the use of stone instruments which, with the Neolithic period, comprises the Stone Age.

paleontology. The study of animals and plants of former geological periods as represented by their fossil remains.

Paleozoic era. An era of geological history following the Proterozoic and preceding the Mesozoic, commonly called the Age of Ancient Life. It was marked by the rise of land animals and plants.

palingenetic. Pertaining to characteristics developed during the life of an individual that are inherited from ancestors. *Compare* cenogenetic. *See* recapitulation theory.

palisade parenchyma. The photosynthetic tissue of a leaf forming one or more layers of chlorophyll-containing cells (palisade cells) which, with the spongy parenchyma, comprises the mesophyll of a leaf.

pallet. One of a pair of calcareous valves which close the tube of the shipworm (*Teredo*) when siphons are retracted.

pallial. Of or pertaining to a mantle, especially that of a mollusk.

pallial groove. In chitons, a groove between the foot and the mantle.

pallial line. A linear scar on the inner surface of a bivalve shell marking attachment of the mantle.

pallial sinus. An indentation in a pallial line, marking point of attachment of the siphon retractor muscle.

pallium. 1. *zool.* The mantle of a mollusk, brachiopod, tunicate, or bird. 2. *embryol.* The thin wall of the lateral vesicles of the telencephalon. 3. *anat.* The cerebral cortex.

¹**palm.** A monocotyledonous plant of the order Palmales, most possessing a large, unbranched trunk bearing at its tip a cluster of large leaves, as the coconut palm.

²**palm.** 1. The inner surface of the hand from wrist to proximal ends of fingers. 2. The corresponding part in the forefoot of a tetrapod.

374

palmar. 1. Of or pertaining to the palm of the hand or manus. 2. Of or pertaining to the ventral surface of the wing of a bird.

palmate. 1. Having the general shape of a hand. 2. *ornithol.* Having the three front toes fully webbed; palmiped. 3. *bot.* Designating leaves with lobes radiating from a common point.

palmella formation. A stage in the development of certain flagellates in which they lose their flagella, undergo division, and form a multicellular, gelatinous mass.

palmelloid. Designating a growth form in which individual cells are embedded in a gelatinous mass.

palmiped. Web-footed.

palmitic acid. A white, crystalline fatty acid $C_{16}H_{32}O_2$, common in animal and plant fats.

palola worm. 1. A burrowing polychaete (*Eunice viridis*) of the South Pacific which appears periodically in enormous numbers during certain phases of the moon. *See* atoke, epitoky. 2. A polychaete (*Eunice schemacephala*) of the West Indies, which swarms in June or July.

palp. A palpus; a small and usually blunt appendage present in various invertebrates, usually sensory in function but sometimes performing other functions.

palpal organ. An intromittent organ on the pedipalp of an arachnid.

palpate. 1. Possessing a palp or palpi. 2. To touch or feel for purposes of diagnosis.

palpation. The act of palpating.

palpebral. Of or pertaining to the eyelids.

palpifer. A palp-bearing sclerite on the stipes of an insect maxilla.

palpiger. A palp-bearing sclerite on the mentum of an insect labium.

palpitation. A rapid pulsation, throbbing, or beating, as that of the heart.

palpon. A tactile zooid in certain siphonophores; a dactylozooid.

palpus pl. **palpi.** A palp.

palsy. PARALYSIS.

paludal. Of or pertaining to marshes.

palus. A thickened ridge between the columella and sclerosepta in the skeleton of a coral.

palustrine. Growing in or inhabiting marshes.

palynology. The study of spores and pollen grains, especially their presence in air and their relationship to respiratory disease (aeropalynology), their presence in food, as honey (melitopalynology), and their presence in sediments of lakes and bogs (pollen stratigraphy).

pampas. The treeless plains of South America, especially those of Argentina.

¹pan. A shallow area or depression containing standing water which usually dries up during the dry season.

²pan. The leaf of the betel pepper (*Piper betle*) or a masticatory made of these leaves smeared with lime to which slices of betel nuts and flavoring substances are added.

pancreas. 1. In vertebrates, a large, compound tubuloacinar gland which secretes digestive enzymes (trypsin, chymotrypsin, amylase, lipase) which enter the duodenum through the pancreatic duct. It also secretes the hormones, insulin and glucagon. *See* islet of Langerhans. 2. A digestive gland in certain invertebrates, as the squid. *See* hepatopancreas.

pancreozymin. A hormone secreted by the duodenal mucosa which stimulates the secretion of enzymes by the pancreas.

panda. A carnivore of the family Procyonidae, as **a.** the lesser panda (*Ailurus fulgens*) of the Eastern Himalayas, also called red cat bear; **b.** the giant panda (*Ailuropoda melanoleuca*), a large bearlike mammal of eastern Asia.

pandemic. Of continental or worldwide distribution, said of a disease.

pandurate. Fiddle-shaped, as certain leaves.

panfish. Any small, edible fish, especially those prepared whole, as perch, sunfish, and other nongame fish.

pangolin. A toothless, anteating mammal of the order Pholidota, as *Manis temmincki*, the scaly anteater, with broad, overlapping scales covering dorsal surface of body and tail.

panicle. A compound raceme, an inflorescence in which the main axis bears several branches each of which bears one or more flowers, as the lilac or grape.

panmictic. Characterized by random mating and crossbreeding, said of populations.

panniculus. *anat.* A layer or membrane.

pansy. A garden flower (*Viola tricolor*) of the violet family, Violaceae.

panther. **1.** In the Americas, the cougar (puma, mountain lion). **2.** In India, any of the large leopards, especially *Felis pardus*.

pantothenic acid. A vitamin of the B complex essential for normal growth of microorganisms and laboratory animals; its role in human nutrition has not been established.

pantropic. Affecting more than one type of tissue, said of viruses.

panzootic. Affecting many species of animals, said of a disease.

papain. *See* papaya.

papaw. An American shrub or tree (*Asimina triloba*) or the large cylindric, pulpy fruit it produces. Also pawpaw.

papaya. The oblong, yellow, edible fruit of a tropical tree (*Carica papaya*), the source of papain, a proteolytic enzyme. Also called mamao.

paper wasp. A social vespid which constructs a nest out of papery materials, as the bald-faced hornet and the yellow jacket.

papery. Resembling paper in appearance or texture; chartaceous.

papilionaceous. Having a pealike flower, as the pea and other legumes.

papilla. A nipple; a small nipplelike eminence or projection.

papillate. Possessing papillae.

papilloma. A neoplastic growth on a surface epithelium, as a wart.

papillose. Covered with or bearing numerous papillae.

papovirus. A tumor-causing virus, as that causing warts.

pappose. Resembling a pappus; possessing a pappus; downy.

pappus. **1.** *zool.* A tuft of hairs, setae, or bristles. **2.** *anat.* The first downy hair that covers the cheeks and chin. **3.** *bot.* A tuft of delicate hairs or a circle of fine hairs or bristles on the achenes of various fruits, as the dandelion.

paprika. The fruit of certain European varieties of *Capsicum*, especially when dried and sold as a condiment.

papula pl. **papulae.** One of the numerous ciliated projections on the body surface of echinoderms which function in respiration and excretion. Also called dermal branchia.

papule. A small, solid elevation of the skin; a pimple.

parabasal body. A minute body of unknown function lying close to the basal granule in certain protozoans, as trypanosomes.

parabiosis. The experimental joining together of individuals so as to form parabiotic twins.

parachordal. Alongside the cephalic portion of the notochord, as *parachordal* cartilages of the embryonic skull.

parachuting. 1. Ballooning. 2. Wind dispersal of seeds which bear a parachutelike tuft of hairs, as the dandelion seed.

paracone. The exterior, anterior cusp of an upper molar tooth.

paraconid. The interior, anterior cusp of a lower molar tooth.

paraffin method. A widely-used procedure employed in the preparation of tissues for microscopical examination. Steps include killing and fixing, dehydration, clearing, embedding in paraffin, sectioning, and mounting of sections. Tissues may be stained before or after sectioning.

paraganglia. Small masses of chromaffin tissue located between the kidneys in lower vertebrates, as fishes.

paragaster. The spongocoel, the central cavity of a sponge.

paraglossa. The outermost of a pair of lobes on the labium of certain insects.

paraglycogen. A polymerized product of glycogen which forms a reserve food supply occurring as refractile granules in various protozoans.

paragnath. 1. One of a pair of chitinized processes bearing jaws in the lining of the eversible pharynx of certain polychaetes, as *Nereis*. 2. A lobe on the lower lip or metastoma of certain crustaceans.

paragnathous. Having mandibles of equal length, with reference to birds.

parahormone. A substance resembling a hormone in many respects but differing in that it is not a secretory product, as carbon dioxide.

parakeet. One of a number of small-sized parrots, popular as cage birds.

parallelism. 1. A similar course of evolution followed by two unrelated groups of organisms. 2. The occurrence of superficially similar organs in unrelated groups, as the eye in a squid and man.

paralysis. Loss of sensation or loss of muscular function usually due to an injury to a nerve or a lesion within the central nervous system.

paramere. 1. One of the halves of a bilaterally symmetrical animal or of a metamere. **2.** In radially symmetrical animals, a perradius and half of an interradius on each side. **3.** One of a pair of copulatory structures lying lateral to the penis, in certain insects.

paramitosis. Mitosis in which chromosomal behavior is atypical, as in protozoans.

paramylon. A polysaccharide occurring in the form of ovoid granules (paramylon bodies) in various protozoans; paramylum.

parapet. The collar of a sea anemone.

paraphysis pl. **paraphyses. 1.** *bot.* One of numerous slender, sterile filamentous structures present in spore-producing or gamete-producing structures in fungi and bryophytes. **2.** *zool.* In lower vertebrates, a non-nervous portion of the roof of the telencephalon.

parapineal body. *See* parietal body.

parapodium. 1. A fleshy appendage extending laterally from the body wall of a polychaete which functions as a locomotor and respiratory organ. **2.** A finlike projection from the foot in certain mollusks, as in heteropods and pteropods.

parapophysis. A process on the body of a vertebra with which the capitulum of a rib articulates.

parasitemia. The presence of parasites in the blood, especially the malarial parasite.

paraproct. A podical plate.

parapsid. Designating a type of skull in which a single opening, the upper supratemporal fossa, exists, as in certain extinct reptiles.

parapsis pl. **parapsides.** A lateral sclerite on the thorax of an insect. Sometimes called scapula.

parasite. 1. An organism which lives in or on another organism from which it derives its nourishment. **2.** The smaller of a pair of conjoined twins.

parasite-mix. A parasitocoenosis.

parasitism. A form of symbiosis in which two organisms live in close association with each other, the one, a parasite, depending upon the other, the host, for some essential food factor. *See* social parasitism, symbiosis.

parasitocoenosis. The combined populations of organisms, plant and animal, which live together in a host organ or in the entire host, as all the organisms (bacteria, yeasts, molds, protozoans, worms) which live in the intestine of a vertebrate, or all the organisms (ecto- and endoparasites) which live in or on a single host.

parasitoid. 1. A parasite which lives within its host only during its larval development, eventually killing the host. **2.** A mite of the superfamily Parasitoidea, as the red mite or roost mite of poultry.

parasitology. The study of parasites and parasitism.

parasitosis. A pathological state or condition caused by parasites.

parasymbiosis. MUTUALISM.

parasympathetic. Pertaining to that division of the autonomic nervous system which comprises the *craniosacral outflow*. Includes cranial nerves III, VII, IX, and X and sacral nerves 2, 3, and 4 containing preganglionic fibers which synapse in terminal ganglia with postganglionic neurons.

paratenic host. In parasitology, a transport host. *See* phoresis.

parathormone. A hormone, PTH, secreted by the parathyroid glands, which regulates calcium and phosphorus metabolism.

parathyroid gland. In vertebrates above fishes, an endocrine gland usually four in number, located in the neck region near or embedded within the thyroid. It secretes parathormone (PTH).

paratonic. Not automatic; due to external causes, with reference to plant movements.

paratype. One of a group of specimens other than the holotype from which the description of a new species was prepared.

paraxial. Lying near or alongside the main axis of the body.

paraxonic. Having the axis of symmetry between the third and fourth digits, as in even-toed ungulates. *Compare* mesaxonic.

parazoan. Any member of the Parazoa, a division of the Metazoa which includes the Porifera (sponges), characterized by lack of organized tissues and absence of a digestive cavity.

parenchyma. 1. *zool.* The loose cellular tissue filling the spaces between organs in various invertebrates, as the flatworms. **2.**

anat. The essential functional tissue of an organ as distinquished from the stroma. **3.** *bot.* The tissue in plants which comprises the soft parts, as the pith and cortex of stems, spongy tissue of leaves, and the major portion of most fruits. Cells are thin-walled, vacuolated, and retain their protoplasm indefinitely.

parenchymatous. Resembling, of the nature of, or composed of parenchyma.

parenchymula larva. The stereogastrula of a sponge.

parent. 1. A father or mother. **2.** Something which brings forth offspring, as a *parent* cell.

parenteral. Not by way of the intestine or digestive tract, with reference to substances being taken into the body.

paresis. General paralysis.

¹**parietal. 1.** Pertaining to the wall of an organ or of the body. *Compare* visceral. **2.** *bot.* Attached to the wall of an ovary, with reference to ovules or placenta.

²**parietal.** A paired membrane bone forming a part of the roof and sides of the cranial cavity.

parietal body. The parapineal body, the anterior portion of the epiphyseal apparatus located anterior to the pineal body; present in lower vertebrates, as in *Sphenodon* where it forms a third eye.

parietal cell. A cell of a gastric gland which secretes hydro-chloric acid.

parotid gland. A salivary gland located anterior to the ear, its duct opening into the mouth cavity; a compound tubuloacinous gland composed of serous cells.

parotoid gland. A poison gland in the skin of an amphibian.

paroxysm. A sudden, violent, uncontrollable action.

parr. The young of salmon and other marine fishes, especially those with short vertical marks (parr marks) along sides of body.

parrot. Any of a mumber of brilliantly colored birds of the order Psittaciformes with hooked beaks and loud raucous voices, including the parakeets, macaws, cockatoos, lories, lorikeets, lovebirds, and allied species.

parrot fever. PSITTACOSIS.

parrot fish. A brightly colored tropical fish of the family Scaridae with parrotlike jaws, as *Sparisoma, Scarus.*

parsley. A garden herb (*Petroselinum*) of the parsley family, Umbelliferae.

parsnip. A biennial herb (*Pastinaca sativa*) of the parsley family. When cultivated its root is edible; in the wild state, poisonous.

parson bird. *See* tui.

parted. Cleft or divided, but not entirely to the base, with reference to leaves.

parthenitae. The sporocysts and redia of flukes. *See* adolescariae, marita.

parthenocarpy. The development of fruits without fertilization, resulting in the production of fruits lacking seeds. May occur naturally or be induced by the application of chemicals, dead pollen, or pollen extracts.

parthenogenesis. The development of an ovum without fertilization. May occur naturally, as in rotifers or aphids or be induced artifically by subjecting eggs to mechanical or chemical stimuli.

parthenogone. A parthenogenetically produced individual.

parthenogonic. Reproducing by parthenogenesis.

parthenogonidium. In colonial protozoans, an asexual reproductive zooid, as in *Volvox*.

parthenote. A haploid individual produced by parthenogenesis.

partridge. **1.** One of several Old World gallinaceous birds, as *Perdix perdix*, the gray or Hungarian partridge, and *Alectoris graeca*, the chukar partridge, both introduced into the United States as game birds. **2.** Local name applied to the ruffed grouse and bobwhite quail.

parturition. Giving birth to young; childbirth, labor.

PAS reaction. The Periodic Acid Schiff reaction, a cytological technique by which the localization of polysaccharides within a cell can be accomplished. It is dependent upon the Feulgen reaction.

PAS reagent. The Schiff reagent, basic fuchsin which has been bleached with sulfurous acid.

passenger pigeon. A wild pigeon (*Ectopistes migratorius*) formerly abundant in the United States but now extinct.

passeriform. PASSERINE.

[1]**passerine.** Of or pertaining to the order Passeriformes, the largest order of the class Aves containing the perching birds, num-

bering over 5100 species. Includes the suborder Passeres (Oscines) which comprises the songbirds.

²**passerine.** Any bird of the order Passeriformes.

passive. 1. Not active. **2.** Resulting from or brought about by outside forces.

passive transport. The movement of substances across a plasma membrane which does not involve the expenditure of energy by the cell. *Compare* active transport.

Pasteur, Louis. A French scientist, 1822–1895, chemist, biologist, bacteriologist, noted for his studies on fermentation, germ theory of disease, immunology, pasteurization, and who laid the foundations of modern bacteriology.

pasteurization. The subjection of milk, fruit juices, and alcoholic beverages (beer and wine) to heat to destroy certain undesirable microorganisms. Milk is pasteurized by subjecting it to 62° C. for 30 min., or 72° C. for 15 sec.

pasture. Grass or other plants used for feeding grazing animals, or land bearing such that is used for grazing.

patagium. 1. *ornith.* The patagial membrane, a web of skin which extends from body to wing. **2.** In mammals, a fold of skin between fore and hind limbs as in flying squirrels and flying lemurs. **3.** The wing membrane of a bat. **4.** A lateral process on the thorax of certain insects, as in lepidopterans.

patella. 1. The kneecap, a sesamoid bone located in the quadriceps tendon at the knee. **2.** In arachnids, a segment of the leg between femur and tibia.

patelliform. 1. *zool.* Shaped like a limpet of the genus *Patella.* **2.** *bot.* Disc-shaped with a narrow rim.

patent. 1. *anat.* Exposed, open, said of tubes or ducts. **2.** *bot.* Patulous, spreading.

paternal. Of or pertaining to the father or male parent; inherited from the male side.

pathogen. A disease-causing organism.

pathogenesis. The sequence of changes occurring in the development of a disease.

pathogenic. Causing or capable of causing a disease.

pathology. The scientific study of disease, including causes, symptoms, signs, and various structural and functional alterations which may occur in its course.

patronymic. In nomenclature, a name derived from a person or persons.

patulous. Expanded, open, dilated.

paunch. 1. The rumen. **2.** The abdominal cavity and its contents.

paurometabolous. Pertaining to the group Paurometabola, which comprises insects which undergo gradual metamorphosis, as orthopterans, hemipterans.

pauropod. An arthropod of the class Pauropoda, comprising mostly minute, terrestrial forms living in damp places, as *Pauropus*.

paw. The foot of an animal, especially the clawed foot of a quadruped.

pawpaw. *See* papaw.

pea. A round, edible seed produced in dehiscent pods by the vine, *Pisum sativum*, a legume, or the plant which produces it, of the pea family, Leguminosae.

peach. 1. A tree (*Prunus persica*) of the rose family, Rosaceae. **2.** The fruit, a drupe with a fleshy pericarp, of the peach tree.

peacock. The male of the peafowl, with elongated, erectile, ocellated tail feathers of striking iridescent colors. *See* peafowl, peahen.

pea crab. One of a number of small crabs of the family Pinnotheridae, most of which are commensals inhabiting the tubes and burrows, and the mantle, cloacal, and pharyngeal cavities of various invertebrates.

peafowl. One of a number of gallinaceous birds of the genus *Pavo*, especially *P. cristatus*, a domesticated species, a native of India. *See* peacock, peahen.

peahen. The female of the peafowl.

peanut. A bushy, creeping annual (*Arachis hypogaea*), a legume, or its fruit which ripens underground.

peanut worm. A sipunculid.

pear. The fleshy fruit, a pome, of a tree (*Pyrus communis*) or the tree which produces it, a member of the rose family, Rosaceae.

pearl. A dense, lustrous body which develops as an abnormal growth in the inner layer of the shell of certain mollusks due to a foreign object or parasite lodging between the mantle and the shell and nacre being deposited about it in concentric layers.

pearl organ. A cornified area of the epidermis which appears in certain teleost fishes during the breeding season.

pearl oyster. One of several large, marine bivalves of the family Pteridae, a source of pearls, found principally in the Persian Gulf, and off the coasts of Australia and Lower California, as *Avicula* (*Pteria*) and *Margaritifera.*

peat. Partially decomposed vegetation, especially that of swamps and bogs which has become compacted and carbonized; sometimes dried and used for fuel.

pebrine. SILKWORM DISEASE.

pecan. A hickory tree (*Carya illinoensis*) or the thin-shelled nut it produces.

peccary. A piglike mammal of the Americas ranging from Texas south to Patagonia, as the collared peccary (*Tayassu angulatus*), also called javelina or musk hog.

pecking order. A social hierarchy observed in chickens in which there is a clear-cut dominance-subordinance relationship between individuals of a group in which one individual at the top dominates all others, that is, can peck all others but is not pecked by any, while at the bottom is an individual pecked by all; in between there are all degrees of subjugation.

pecten. 1. pl. **pectines.** A comblike structure, as **a.** a serrated, fan-shaped structure in the eye of a bird; **b.** the pollen rake, a part of the pollen packer of a honeybee; **c.** a sensory structure attached to the second mesosomatic segment of a scorpion. **2.** pl. **pectens.** A scallop, a bivalve of the genus *Pecten.*

pectin. Any of a group of water-soluble substances present in plant tissues which, in the presence of sugar (65–70%) and proper acid concentation (pH 3.2–3.5) causes the formation of jelly.

pectinate. Shaped like or having teeth resembling a comb; comblike.

pectoral. Of, pertaining to, or located in the region of the breast or chest.

pectoral fin. One of the paired, anterior lateral fins of a fish.

pectoral girdle. The cartilaginous or bony structure supporting the pectoral fins of a fish or the forelimb of a tetrapod. In man, it comprises the scapulae and clavicles. Also called shoulder girdle.

pedal. Of or pertaining to the foot or feet.

pedal cyst. A podocyst.

pedal disk. The basal disk of a hydrozoan.

pedal ganglion. A ganglion in the foot of various mollusks.

pedal gland. A gland in the foot of a rotifer or a mollusk.

pedalium. The flattened, basal portion of a tentacle in certain coelenterates, as the cubomedusae.

pedal laceration. *See* laceration.

pedate. 1. *bot.* Designating a palmately lobed leaf in which the two side lobes are also divided. 2. *zool.* **a.** Having podia or feet, with reference to holothuroids. **b.** Having a foot or feet.

peddler. The larva of the tortoise beetle, which possesses a forked caudal process to which cast skins and excrement are attached, the whole forming a shield over the body.

pedicel. 1. A slender stalk or process which serves for support or attachment; a pedicle. 2. *bot.* The stalk of a single flower in an inflorescence. 3. *zool.* **a.** The second segment of an insect antenna. **b.** The stalk of a halter. **c.** The petiole of a hymenopterous insect or an arachnid. **d.** The stalk of a brachiopod. **e.** The lateral wall of a neural arch.

pedicellaria. A small pincherlike organ on the body surface of various echinoderms.

pedicellate. Borne on a pedicel.

pedicle. A pedicel or peduncle.

pediculosis. Infestation with lice. *See* louse.

pediform. Shaped like a foot.

pedigerous. Bearing feet or limbs.

pedipalp. One of the second pair of appendages in arachnids, resembling a leg in appearance and structure; in scorpions they are well developed chelate structures.

peduncle. 1. A pedicel; a stalk or stem. 2. *bot.* The stalk of a flower cluster or a strobilus. 3. *zool.* **a.** The stalk of a barnacle or a brachiopod. **b.** The pedicel of a spider. **c.** A band of nerve fibers in the brain.

pedunculate. Possessing a peduncle.

pekan. A fisher.

Peking man. *Pithecanthropus* (*Sinanthropus*) *pekinensis,* a species of prehistoric man related to the Java man.

pelage. The hairy coat of a mammal; wool, fur, hair.

pelagic. Of or pertaining to the open sea.

pelagic life. All the life of open waters, as plankton, nekton, neuston.

pelecypod. Any mollusk of the class Pelecypoda (Lamellibranchia, Bivalvia) comprising bivalves with laminate gills, as clams, mussels, scallops, oysters, cockles, and shipworms.

pelican. A large, fish-eating bird (*Pelecanus erythrorhynchos*) with a long, flat bill bearing a large, distensible gular pouch.

pellagra. A deficiency disease resulting from inadequate intake or absorption of niacin or niacinamide.

pellet. 1. A small, spherical mass of material, as food or medicine. **2.** *ornith.* A mass consisting of the undigestible remains of animals regurgitated by predatory birds, as herons and owls which swallow their prey whole.

pellicle. A thin membrane or covering, as **a.** the thin, elastic, gelated, surface layer of protoplasm in various protozoans; **b.** a periplast.

pellucid. Clear; transparent.

pelt. The hide or skin of a fur-bearing mammal, especially when undressed.

pelta. A bract attached in its midregion, as in peppers.

peltate. Shield-shaped, especially with reference to a leaf whose petiole is attached to its lower surface instead of at the margin or base.

pelvic. Of or pertaining to the pelvis or pelvic girdle.

pelvic girdle. The portion of the appendicular skeleton which supports the pelvic fins of fishes or the hind limbs of a tetrapod, in the latter consisting of three paired bones, the ilium, ischium, and pubis. *See* innominate bone.

pelvis. A basinlike structure or cavity, as **a.** the *bony pelvis,* in man, a ring of bones formed by the two innominate bones, sacrum, and coccyx; **b.** the cavity within the bony pelvis; **c.** the cavity of the kidney consisting of the expanded end of the ureter.

pen. 1. *bot.* The midrib of a leaf. **2.** *zool.* **a.** The horny, internal shell of a squid. **b.** A female swan. *Compare* cob.

pencilled. Marked by fine, distinct lines, especially parallel, radiating lines.

pendant. Hanging downward, suspended.

pendulous. PENDANT.

penetrance. *genetics.* The degree to which a gene or genes express themselves: EXPRESSIVITY.

penetrant. A type of nematocyst with a tube which, when discharged, penetrates prey producing a toxic effect.

penguin. A short-legged, aquatic, flightless bird of the southern hemisphere with wings adapted for swimming, as *Aptenodytes patagonica*, the king penguin of the family Spheniscidae.

penial. Of or pertaining to the penis.

penial setae. Copulatory setae in a male nematode.

penicillate. Ending in a tuft of hairs, like that of a brush.

penicillin. One of several antibiotics produced by ascomycete molds of the genus *Penicillium*, especially *P. notatum*. Its action is principally bacteriostatic, inhibiting the growth of gram-positive bacteria and other microorganisms.

penis pl. **penes** or **penises.** The male copulatory or intromittent organ. *See* aedaegus, cirrus, verge.

penna pl. **pennae.** A contour feather or pluma as distinguished from a down feather (plumula). *See* remex.

pennaceous. Penniform; like a feather.

pennate. PINNATE.

penniculus. A curved band of cilia in the cytopharynx of certain ciliates.

penniform. Pennate; resembling a feather, said of certain muscles.

Pennsylvanian period. The Upper Carbonaceous, a period of the Paleozoic era marked by warm and humid climate and the formation of great coal forests and swamps.

pensile. Hanging or pendant, with reference to the nest of certain birds.

pentacula larva. The final larval stage in the development of a sea cucumber.

pentadactyl. Having five digits in the hand or foot.

pentadelphous. In groups of five, with reference to the filaments of stamens.

pentamerous. 1. Consisting of or divided into five parts. 2. *bot.*

Designating a floral whorl consisting of five or a multiple of five parts.

pentandrous. With five stamens.

pentapetalous. With five petals.

pentasepalous. With five sepals.

pentastomid. A tongue worm, a parasitic, wormlike animal of the class Pentastomida (Linguatulida), phylum Arthropoda, as *Linguatula serrata*, inhabiting the paranasal sinuses of carnivores.

pentose. One of a class of carbohydrates with a basic formula, $C_5H_{10}O_5$.

peony. A flower (*Paeonia*) of the buttercup family, Ranunculaceae.

pepo. 1. The characteristic fruit of the gourd family, Cucurbitaceae, a fleshy, many-seeded berry, as the cucumber, melon, pumpkin, and squash. **2.** The seed of the cucumber (*Cucurbito pepo*), formerly used as a taeniafuge.

pepper. 1. A pungent spice obtained from plants of the genus *Piper*, black pepper being obtained from the dried ripe fruit of *P. nigrum*. **2.** Any of a number of plants of the genus *Capsicum* or their fruits, the source of cayenne or red pepper and paprika.

peppermint. The dried leaf and flowering top of *Mentha piperita*, the source of peppermint oil, used as a flavoring and a carminative.

pepsin. 1. An enzyme present in gastric juice which catalyzes the hydrolysis of proteins and polypeptides. **2.** A substance containing this enzyme obtained from the mucosa of the stomach of a hog.

pepsinogen. The precursor (zymogen) of pepsin produced by the chief cells of gastric glands.

peptidase. A proteinase which hydrolyzes peptides.

peptide. A compound formed by the union of two or more amino acids, as a dipeptide or polypeptide.

peptide bond. One that holds two amino acids together.

peptide linkage. The union of an amino group (NH_2) with a carboxyl group (COOH), as occurs in the union of two or more amino acids.

peptone. A product of the hydrolysis of a protein, intermediate between a proteose and a peptide.

¹perch. A horizontal bar, pole, or limb of a tree upon which birds sit or roost.

²perch. 1. A small, freshwater percomorph fish of the family Percidae, especially *Perca flavescens*, the yellow perch. **2.** Any of a number of freshwater or marine spiny-finned fishes, as the climbing perch (*Anabas testudineus*), the pike perch or walleye (*Stizostedion vitreum*), or white perch (*Roccus americanus*).

percomorph. Of or pertaining to the Percomorphi (Percomorphida), an order of ray-finned fishes which includes the perches, basses, sunfishes, mackerels, and tunas.

peregrine. Foreign, exotic; wandering, migratory.

pereiopod. A thoracic appendage of a crustacean which functions as a walking leg.

perennation. Surviving from year to year; the act of being perennial.

perennial. *bot.* A plant which lives for two years or longer. *Herbaceous perennials* have aerial stems which die, the underground parts remaining alive, as asparagus; *woody perennials* have aerial stems which may live for many years, as shrubs and trees.

perennibranch. An amphibian which retains its gills throughout life, as the mud puppy.

perfect. *bot.* Having both stamens and pistil, with reference to flowers.

perfoliate. Having a stem which apparently passes through it, with reference to a leaf.

perforated. Possessing holes, pores, or openings.

perforation. An opening, especially one of the numerous openings in vessel members of xylem.

perforation plate. The region in the lateral wall of a vessel member which bears one or many perforations.

perfusion. 1. The spreading of a liquid over or through a substance. **2.** The introduction of a fluid into an organ or tissue by injection into the blood vessel supplying the structure.

perianth. 1. A floral envelope consisting of the calyx and corolla. **2.** The enveloping structure which surrounds the base of the archegonium and developing sporophyte in certain bryophytes.

periblast. In a highly telolecithal ovum, the rapidly dividing cells about the margin of the blastodisc.

periblem. A plant histogen which gives rise to the cortex.

peribranchial. Surrounding the gills.

peribranchial cavity. The atrium of tunicates and lancelets.

pericardial cavity. The cavity within the pericardium which surrounds the heart.

pericardial sinus. The portion of the hemocoel in invertebrates which surrounds the heart.

pericardium. 1. In vertebrates, the closed membranous sac enclosing the heart consisting of outer *parietal* and inner *visceral* layers. **2.** In invertebrates, the pericardial sinus.

pericarp. 1. In seed plants, the wall of the ripened ovary or fruit, usually consisting of three layers, exocarp, mesocarp, and endocarp. **2.** In red algae, an urn-shaped structure surrounding the carposporophyte.

pericentric. Pericentral; located near or about a center.

perichaetium. A group of terminal leaves surrounding the archegonium in mosses.

perichondrium. A fibrous membrane which forms an investing membrane about cartilage except at articular surfaces.

perichordal. Surrounding the notochord, as the *perichordal* sheath.

periclinal. Parallel to the circumference.

pericycle. A layer of parenchymous cells located between the endodermis and phloem, or in some cases, xylem and phloem, in the roots and stems of vascular plants.

periderm. 1. *bot.* The outer, protective layer or bark of older roots and stems consisting of cork cambium and cork. **2.** *zool.* **a.** The perisarc of a hydroid. **b.** *embryol.* The outermost layer of the developing skin of a vertebrate.

peridiole. A spore-producing structure in certain basidiomycetes.

peridium. 1. The hardened envelope of a sporangium in myxomycetes. **2.** The outer wall of the fruiting body of various fungi, as **a.** the parchmentlike wall of a puffball; **b.** the outer layer of an ascocarp composed of haploid vegetative hyphae.

perigonium. 1. In bryophytes, an antheridium and its surrounding leaves. **2.** In angiosperms, a perianth.

391

perigynium. 1. In liverworts, a tubular extension of the stem or thallus which encloses the archegonium. **2.** A saclike structure (utricle) which encloses the ovary in sedges. **3.** The reduced perianth in various sedges.

perigynous. Designating flowers in which the sepals, petals, and stamens are attached above the ovary to the margin of a cuplike receptacle.

perigyny. The state of being perigynous.

perihemal. Around a blood vessel, as *perihemal canals* of a sea star or the *perihemal coelom* of an enteropneuston.

perikaryon. The cell body of a neuron.

perilymph. The fluid which separates the membranous labyrinth from the osseous labyrinth of the inner ear.

perimysium. The fibrous membrane which surrounds a bundle (fasciculus) of muscle fibers.

perineum. 1. The region of the outlet of the pelvis including the pelvic floor and associated structures. **2.** In a restricted sense, **a.** in a male, the region between the anus and the scrotum; **b.** in a female, the region between the anus and vulva.

perineurium. The fibrous sheath which surrounds a bundle (fasciculus) of nerve fibers.

period. A time interval during which a certain series of events occurs, as **a.** *geol.* a subdivision of an era; **b.** *physiol.* the menses.

Periodic Acid Schiff reaction. *See* PAS reaction.

periodical. Occurring at fixed intervals.

periodontal. Surrounding a tooth.

periosteum. A fibrous connective tissue membrane which covers the surface of bone, its innermost layer containing osteoblasts.

periostracum. The thin, horny, outermost layer of the shell of a mollusk.

periotic. About or around the ear.

peripheral. Of, towards, or constituting a periphery; away from a center; towards the surface.

periphery. An external surface or boundary; the outer regions of a body or structure.

periphysis pl. **periphyses.** Sterile hyphae in an ascocarp, especially those lining the osteolar canal.

periphyte. A bacterium which grows attached to a solid surface.

periphyton. The organisms adhering to submerged plants.

periplasmodium. A multinucleate structure which develops from the wall of a sporangium in ferns of the order Salvinales.

periplast. The surface membrane or pellicle of certain algae.

peripneustic. With spiracles arranged in a row along the side of the body.

periproct. The structure surrounding the anus in a sea urchin.

periproctal. About or surrounding the anus or rectum.

perisarc. The thin, transparent, chitinous layer surrounding the coenosarc in various colonial hydrozoans, as *Obelia:* PERIDERM.

perisperm. Nutritive tissue derived from the nucellus present in some seeds, as the pine, water lily.

perissodactyl. Any mammal of the Perissodactyla, an order which includes the odd-toed, hoofed quadrupeds, as the horse, zebra, tapir, rhinoceros.

peristalsis. Progressive wavelike movements occurring in the intestines or other tubular structures in which a wave of contraction moves along the tube preceded by a wave of relaxation.

peristome. 1. *zool.* The region about the mouth, as **a.** in ciliates, the buccal cavity which leads to the cytostome; **b.** in anthozoans and echinoderms, a smooth membranous structure about the mouth; **c.** in gastropods, the edge of the aperture of the shell; **d.** *entomol.* the ventral margin of the head surrounding and bordering the mouth. 2. *bot.* A ring of hygroscopic teeth surrounding the opening of the spore capsule of a moss.

peristomium. The segment or segments behind the head and surrounding the mouth, as in various polychaetes.

perithecium. A spherical or urn-shaped ascocarp, usually with an opening, the osteole.

peritoneum. A serous membrane which lines the abdominal cavity and covers the enclosed visceral organs.

peritreme. 1. The edge of an opening to a shell. 2. A sclerite surrounding the spiracle of an insect. 3. A sclerite surrounding, or a chitinous tube leading to, the stigma of an arachnid.

peritrich. A ciliate protozoan of the order Peritrichida, as *Vorticella.*

peritrichous. 1. Having an enlarged, disklike anterior region bearing cilia, with reference to protozoans. 2. Having flagella. which cover the entire surface, with reference to bacteria.

peritrophic. Surrounding food.

peritrophic membrane. 1. A chitinous tube free within the cavity of the midgut of insects which separates food from the lining epithelium. **2.** A thin, mucoid sheath surrounding the feces of certain crustaceans.

perivisceral. About or around the viscera.

perivisceral cavity. The principal body cavity or coelom.

periwinkle. 1. *zool.* A small marine gastropod of the genus *Littorina*. **2.** *bot.* A plant (*Vinca*) of the dogbane family, Apocynaceae; myrtle.

permanent. Enduring, lasting, as *permanent* teeth.

permeability. The sum of the factors associated with the transfer of substances across a plasma membrane.

permeant. An animal which can move readily from one area to another.

permeate. 1. To pass into or through a substance. **2.** To pass through the pores of a membrane.

Permian period. The last period of the Paleozoic era; the period preceding the Mesozoic era.

peronium. In certain medusa, an epidermal tract containing nematocysts, muscle and nerve fibers, extending from the inner surface of a tentacle onto the exumbrella.

peroxidase. An enzyme, common in plants, which catalyzes the removal of oxygen from hydrogen peroxide.

persimmon. A tree (*Driospyrus virginiana*) of the ebony family, Ebenaceae, or its fruit, edible when ripe but extremely bitter and astringent when unripe.

persistent. 1. *bot.* Long-continuing or enduring, as the corolla remaining attached until maturation; or leaves remaining attached throughout winter. **2.** *zool.* Remaining present in the adult, with reference to embryonic or larval structures which usually disappear, as a *persistent* notochord or *persistent* gills.

persister. An organism, as a bacterium, which has been exposed to an antibiotic, as penicillin, and survived, or a culture derived from such an organism.

personate. Designating a bilabiate corolla, the throat of which is closed by a projection, the palate.

perspiration. Sweat; the secretion of sweat glands.

perula. 1. The scale of a leaf bud. **2.** The mentum of an orchid.

perulate. Bearing scales, with reference to a leaf bud.

pervious. Open, perforate; capable of admitting passage.

pes pl. **pedes.** The foot or hindlimb of a vertebrate, comprising the tarsus, metatarsus, and digits.

pessimum. The least favorable conditions under which an organism can survive.

pessulus. A skeletal structure in birds which crosses the lower end of the trachea.

pest. Any organism which is troublesome, disturbing, noxious, or destructive.

pesticide. A substance or a mixture of substances used to destroy or control any undesirable form of animal or plant life, as a fungicide, herbicide, rodenticide, or insecticide.

pestilence. An infectious disease which is especially virulent and devastating in its effects, as the bubonic plague.

petal. A modified leaf, usually colored, forming a part of the corolla of a flower.

¹petaloid. Like or resembling a petal.

²petaloid. In echinoids, an ambulacral area shaped like a petal, radiating from the center of the aboral surface.

petalous. Possessing petals.

petiolar. Pertaining to, borne on, or arising from a petiole.

petiolate. Possessing a stalk or petiole.

petiole. 1. *bot.* The slender stalk which supports the blade of a leaf. **2.** *zool.* The slender stalk connecting the thorax and abdomen in certain insects (ants, bees, wasps, dipterans) and spiders; a pedicel or peduncle.

petrel. One of a number of oceanic birds of the family Procellariidae, commonly called stormy petrels or Mother Carey's chickens.

petricolous. Living in, on, or among rocks.

petrifaction. The conversion of organic matter into stone by the interstitial addition or gradual replacement, molecule by molecule, of the original substance, as trees of the petrified forest in Arizona.

petrophilous. Inhabiting rocks or frequenting rocky areas.

petrous. Hard, like stone, as the *petrous* portion of the temporal bone.

petunia. A flowering herb (*Petunia*) of the nightshade family.

pewee. A small bird of the flycatcher family, Tyrannidae, as the wood pewee (*Contopus virens*).

Peyer's patch. An aggregated mass of lymph nodules in the wall of the small intestine, usually the ileum.

peyote. Mescal buttons, the dried flowering tops of the cactus (*Lophophora williamsii*).

PGA. Pteroylglutamic acid (folic acid), one of the B- complex vitamins, effective in treatment of anemia.

pH. A symbol for expressing the concentration of hydrogen (H^+) ions or hydroxyol (OH^-) ions. *See* pH scale.

phacella. A row of gastric filaments (digitelli) on the sub-umbrellar surface of a septum in scyphozoan jellyfishes.

phaeophyte. Any plant of the division Phaeophyta which includes the brown algae, mostly marine plants some reaching enormous size, as the kelps and rockweeds. Some are the source of algin.

phage. *See* bacteriophage.

phagocyte. A cell which has the ability to engulf particulate matter, as a leukocyte or macrophage.

phagocytosis. The engulfing of substances, especially foreign bodies (bacteria, dust particles, colloidal dyes) by phagocytes. *Compare* pinocytosis.

phagotroph. An organism that ingests or engulfs solid particles of food.

phalanger. Any marsupial of the family Phalangeridae, arboreal mammals with long tails.

phalanx pl. **phalanges. 1.** *zool.* A bone of a finger or toe. **2.** *bot.* A bundle of stamens united by their filaments.

phalarope. A small shorebird resembling a sandpiper; of the family Phalaropodidae.

phallus. 1. The penis. **2.** *embryol.* The embryonic structure from which the penis or clitoris develops.

phanerogam. Any plant of the Phanerogamia, a group in older classifications which included the flowering or seed plants. *Compare* cryptogam.

¹**pharmaceutical.** Of or pertaining to pharmacy.

²**pharmaceutical.** A drug.

pharmacy. 1. The science which deals with drugs, especially their preparation, properties, and actions. **2.** The profession which deals with the preparation and distribution of medicinal products. **3.** A place where medicinal products are sold.

pharyngeal. Of or pertaining to the pharynx.

pharynx. 1. In vertebrates, the portion of the digestive tract which connects the mouth cavity and esophagus. In tetrapods, it also connects the nasal cavities and the larynx. **2.** In invertebrates, the region of the alimentary canal which connects the mouth or buccal cavity with the next succeeding portion, usually the esophagus.

phasmid. A small, pouch-shaped sensory receptor located post-anally in certain nematodes.

pheasant. 1. A long-tailed, gallinaceous bird of the family Phasianidae common throughout Asia. **2.** The ring-necked pheasant (*Phasianus colchicus torquatus*) introduced into the United States as a game bird and now well established.

phellem. Cork, the middle layer of the periderm.

phelloderm. The innermost layer of the periderm.

phellogen. The cork cambium, a layer of meristematic tissue forming the middle layer of the periderm.

phenocopy. A condition in which an acquired or artificially induced characteristic is identical with that produced by a gene, as pigmentation induced by ultraviolet rays.

phenogenetics. Developmental genetics, the study of gene action in relation to developmental processes.

phenology. The science which deals with relationships between weather and cyclic biological phenomena, as migrations of animals, flowering of plants.

phenotype. *genetics* **1.** The physical makeup or appearance of an individual in contrast to its genetic constitution or genotype. **2.** A group of individuals which exhibit the same physical traits.

phenylthiocarbamide. PTC, a substance which is bitter to some individuals but tasteless to others.

pheromone. An ectohormone; a substance produced and discharged by an organism which induces a physiological response in another of the same species, as the sexual attractants of insects.

phlegm. Mucus, especially thick, viscid mucus discharged through the mouth.

phloem. Food-conducting tissue in vascular plants consisting principally of sieve tubes and companion cells; also called bast, leptom. *Compare* xylem.

phlox. A flowering herb of the genus *Phlox*.

phoebe. A flycatcher of the family Tyrannidae, as *Sayornis phoebe*, the Eastern flycatcher.

pholidosis. The arrangement of scales on an animal or a part.

phoresis. A form of symbiosis in which the host transports the symbiont, as barnacles on whales or sessile ciliates on arthropods; phoresy.

phoronid. Any invertebrate of the phylum Phoronidea which contains a small number of lophophorate, coelomate marine animals of uncertain relationships, as *Phoronis*.

phoront. The symbiont or transported individual in phoresis.

phorozooid. A specialized zooid in certain pelagic tunicates which acts as a nurse bearing gonozooids.

phosphagen. Creatine phosphate or arginine phosphate.

phosphatase. One of a number of enzymes which catalyze the hydrolysis of esters of phosphoric acid, widely distributed in animal and plant kingdoms and of importance in cellular metabolism.

phosphate. A salt of phosphoric acid.

phosphate bonds. High-energy bonds, designated as $P \frown P$ bonds, especially those in the molecules of adenosine triphosphate (ATP) which are the principal sources of energy for cellular activity.

phosphatide. A phospholipid.

phosphocreatine. Creatine phosphate, $C_4H_{10}N_3O_5$, a source of energy for cellular activity.

phospholipid. A phosphatide; a compound formed of a lipid and an ester of phosphoric acid, as lecithin, cephalin, sphingomyelin.

phosphoprotein. A protein combined with a phosphorus-containing compound.

phosphorescence. 1. The emission of light without heat, as occurs in certain bacteria, fungi, protozoans, insects, and fishes. *See* bioluminescence. 2. The reemission of light waves of a different wavelength and for a considerable time following subjection to certain types of radiation.

phosphoric acid. H_3PO_4, an important compound in cellular metabolism.

phosphorolysis. The process by which glucosidic bonds of glycogen are broken at the 1–4 linkage by reaction with inorganic phosphates resulting in the formation of glucose-1-phosphate or Cori ester, a reaction catalyzed by phosphorylase.

phosphorus. A nonmetallic element, symbol P, at. no. 15, occurring in two allotrophic forms, yellow and red. It is an essential component of every living cell being a constituent of hexose phosphates, nucleic acids, phospholipids, phosphoproteins, and other compounds involved in metabolic processes, especially those involved in the transfer and release of energy.

phosphorylase. An enzyme which catalyzes the phosphorolysis of glycogen.

phosphorylation. The union of a substance with phosphorus or the phosphate radical, PO_4^{3-}.

photic. Pertaining to light.

photic zone. The surface layer of oceans or lakes which light penetrates. *See* euphotic zone.

photoautotrophic. Using light energy to synthesize organic compounds from inorganic substances, as in all green plants.

photolysis. Decomposition resulting from the action of light, as in the dissociation of water in photosynthesis.

photon. A quantum, or the unit of visible light; the unit of all radiant energy.

photonastic. Pertaining to nastic movements resulting from changes in light intensity, as the opening and closing of flowers at night.

photoorganotroph. An organism which utilizes light energy in the manufacture of food from organic materials.

photoperiodism. The responses of an organism to changes in light intensity or in length of days, as **a.** in animals, seasonal and cyclic events as migrations, reproductive cycles, changes in pelage or plumage; **b.** in plants, the growth and development of flowers and fruits, the shedding of leaves, or onset of winter dormancy.

photophore. A light-producing structure; a luminescent organ.

photopsin. A component of iodopsin, a visual pigment present in cone cells of the retina.

photoreceptor. **1.** *zool.* A receptor which is capable of being stimulated by light rays, as an eye or ocellus. **2.** *bot.* A structure containing a light-sensitive pigment, as an eyespot or stigma.

photoreversibility. The capacity of opposite responses occurring through the action of a single photoreceptor.

photosensitive. Sensitive to or capable of being stimulated by light.

photosynthesis. The process which occurs in the chloroplasts of green plants in which simple sugars are formed from carbon dioxide and water in the presence of light and chlorophyll. The basic reaction is $6CO_2 + 6H_2O \rightarrow C_6H_{12}O_6 = 6O_2$ by which light or radiant energy is converted to chemical energy and stored in the molecules of carbohydrates.

phototaxis. The movement of an animal in response to light.

phototroph. A phototrophic organism.

phototrophic. Designating organisms which can utilize radiant energy, as green plants.

phototropic. Responding to light as a stimulus.

phototropism. The bending or turning of a sessile organism in response to light.

phragma pl. **phragmata.** A dorsal apodeme in insects and other arthropods which serves for attachment of muscle fibers; endotergite.

phragmoplast. In plant mitosis, all the structures in the region of the cell equator during the division of the cytoplasm.

phragmosis. Using a part of the body to close a burrow.

phragmosome. In plant mitosis, one of the many small bodies present in the equatorial plane which contribute to the formation of the middle lamella of the cell wall.

phreatophyte. A plant with extremely long roots reaching to the water table.

phrenic. **1.** Of or pertaining to the diaphragm. **2.** Of or pertaining to the mind.

pH scale. A scale ranging from 0 to 14 which represents the acidity or alkalinity of a solution. A neutral solution, as water, has a pH of 7; acid solutions less than 7; alkaline solutions more than 7.

phycobilin. A metal-free pigment associated with phycocyanin or phycoerythrin.

phycobiont. The algal member of a symbiotic relationship, as in a lichen.

phycocolloid. One of a number of complex, colloidal substances present in the cell walls of various brown algae, as algin, fucoidin.

phycocyanin. A blue, light-absorbing pigment present in blue-green and other algae.

phycoerythrin. A red, light-absorbing pigment present in red and other algae.

phycology. The study of algae; algology.

phycomycete. Any member of the class Phycomycetes, which includes the algalike fungi, as the chytrids, water molds, downy mildews, and the true and black molds.

phylactocarp. A protective structure, usually a modified hydrocladium, which protects a gonangium in various hydrozoans. *See* basket, corbula.

phyllid. A flattened lateral leaf of a leafy liverwort.

phyllidium. A bothridium.

phyllobranchia pl. **phyllobranchiae.** A flat, leaflike gill composed of thin plates or lamella, as in certain crustaceans.

phylloclad. A cladode.

phyllode. 1. A flattened, leaflike petiole which functions as a leaf, as in *Acacia*. 2. One of the oral, ambulacral areas in certain echinoids.

phyllodium pl. **phyllodia.** A phyllode.

phyllopodium. 1. *bot.* The axis of a leaf. 2. *zool.* A broad, flat appendage in various crustaceans, as the maxilla of a crayfish.

phyllosoma larva. The long-legged zoea larva of the spiny lobster.

phyllotaxy. The arrangement of leaves on a plant; phyllotaxis.

phyllozooid. A leaflike or helmet-shaped bract in a siphonophore; hydrophyllium.

phylogenetic. Of or pertaining to ancestral development.

phylogeny. 1. The racial history of an animal or plant. 2. The evolution of a race or species; phylogenesis. *Compare* ontogeny.

phylum pl. **phyla.** A taxon comprising the largest division of the animal or plant kingdom, consisting of one or more classes. In some plant classifications, the term division is generally used instead of phylum.

physa. In burrowing anthozoans, the end bulb or base, a constricted portion of the column.

physical. 1. Of or pertaining to natural or material things. **2.** Of or pertaining to matter, especially its form and structure. **3.** Of or pertaining to the science of physics.

physiological. 1. Of or pertaining to physiology. **2.** Pertaining to normal processes or activities in contrast to those that are abnormal or pathologic.

physiology. 1. The branch of biology which deals with the various processes and activities which occur within a living organism or one of its parts; the study of function. **2.** The functioning of an organism or any of its parts.

physoclistous. Not connected to the alimentary canal by a duct, with reference to swim bladders in fishes.

physogastric. Having an enlarged abdomen.

physostomous. Connected to the alimentary canal by a duct, with reference to swim bladders.

phytochrome. A photoreversible pigment present in certain plants, extractable from corn seedlings.

phytoflagellate. A plantlike flagellate of the class Phytomastigophorea, as *Euglena* and related forms.

phytogeography. Plant geography, the science which deals with the distribution of plants over the surface of the earth and the factors accounting for such.

phytohormone. A plant hormone. *See* hormone, 2.

phytol. An alcohol containing a chain of 20 carbon atoms which forms the tail of a chlorophyll molecule.

phytology. The study of plants: BOTANY.

phytophagous. Plant-eating or herbivorous, said of animals, especially insects.

phytoplankton. The plant organisms of plankton.

phytotoxin. A toxin produced by a plant, as ricin, croton.

phytozoan. A biological unit consisting of an animal and plant living in symbiotic mutualism, as a coral polyp and their contained zooxanthellae.

pia mater. The innermost of the three meninges of the brain and spinal cord.

pickerel. A small, carnivorous, freshwater fish of the pike family, as the chain pickerel (*Esox niger*).

picornavirus. One of a group of viruses comprising the smallest viruses known, including most of the enteroviruses of man and lower animals, as the Coxsackie, ECHO, and polio viruses.

piddock. A burrowing bivalve of the family Pholadidae, as *Pholas*.

piebald. Spotted; mottled.

pig. A swine of any age; a boar, sow, hog, shoat.

pigeon. A dovelike bird of the family Columbidae, as the common rock pigeon or rock dove (*Columba livia*) from which many domesticated varieties as the carrier, fantail, homing, pouter, turbut, and others originated.

pigeon milk. A nutritious, cheeselike substance produced by the crop glands of pigeons which, when regurgitated, serves as food for the young.

pigeon pea. The cajan pea (*Cajanus cajan*), a food and forage legume.

pigment. 1. Any coloring matter. 2. *biol.* A biochrome, an organic coloring matter present in tissues, body fluids, or secretions, as a bile, blood, respiratory, visual, or plant pigment.

pigmentation. The amount and quality of the coloring in animals or plants.

pigment cell. A pigment-containing cell, especially a chromatophore. *See* erythrophore, guanophore, lipophore, melanophore, xanthophore. *Also see* melanoblast.

pignut. The bitter nut of a hickory (*Carya glabra*) or the tree that produces it.

pika. A small mammal of the order Lagomorpha, as *Ocotona princeps*, the Rocky Mountain pika of western United States. Also called mouse hare, calling hare.

pike. 1. One of several large, carnivorous fishes of the family Esocidae, as *Esox lucius*, the Northern pike, a food and game fish. 2. One of a number of similar fishes, as the garpike, walleyed pike.

pike perch. The walleyed pike. *See* walleye.

pilchard. The California herring or sardine (*Sardinops caerulea*).

pile. A dense growth of short, fine hairs.

pileate. Bearing a cap or caplike structure.

pileated. Having the entire pileum crested, said of birds.

pileous. Hairy: PILOSE.

pileum. In birds, the entire top of the head; pileus.

pileus. 1. *zool.* A pileum. 2. *bot.* The umbrella-shaped cap of the fruiting body of a mushroom.

pileworm. *See* shipworm.

pilidium. The free-swimming, helmet-shaped larva of certain nemerteans.

pilifer. A lateral projection on the labrum of certain insects, as lepidopterans.

piliferous. Bearing hairs or hairlike processes.

pill. 1. A small, rounded mass of medicinal substances which is swallowed whole. 2. One of several oral contraceptive pills containing synthetic hormones which suppress the development of Graafian follicles and the release of ova from the ovary.

pill bug. A small, terrestrial, isopod crustacean (*Armadillidium vulgare*) which rolls itself into a ball when disturbed.

pilose. Covered with fine, soft hairs or down.

pilot fish. A marine fish (*Naucrates ductor*) which accompanies sharks.

pilus pl. **pili.** 1. A hair or hairlike structure. 2. One of the minute filamentous appendages extending from the cell wall of a bacterium. Also called fimbria.

pimento. The dried, unripe fruits of *Pimenta officinalis,* or the tree which produces them, of the American tropics. Also known as allspice or Jamaica pepper.

pimiento. Spanish paprika (*Capsicum frutescens* or *C. annuum*), a garden vegetable.

pinacocyte. A flattened, contractile cell in the epidermis of a sponge.

pincers. A forcipulate organ, as a chela, forceps, or pedicellaria.

pine. A coniferous tree of the genus *Pinus* with long needles in groups of two to five; a valuable timber tree, also the source of turpentine, rosin, tar, pitch, and other products.

pineal body. The epiphysis cerebri, a small, cone-shaped structure forming a part of the epithalamus. *See* parietal body.

pineal eye. A well-developed parapineal organ which contains photoreceptors, considered a third eye; present in lampreys, tadpoles, lizards.

pinfeather. A growing feather just emerging from the skin and still enclosed within its sheath.

pinion. In birds, the hand bone or the portion of the wing beyond the wrist, exclusive of the feathers.

pinna. 1. *zool.* **a.** Any protruding structure, as a fin, feather, or flipper. **b.** The auricle of the external ear. **2.** *bot.* A primary division of a pinnate leaf; a leaflet.

pinnate. 1. Shaped like a feather. **2.** *bot.* Bearing leaflets or pinnae along each side of a central axis or rachis, as in a pinnately compound leaf.

pinnated. With winglike tufts of feathers along the sides of the neck.

pinnatifid. Cleft or divided in a pinnate manner, said of leaves with narrow clefts extending almost to the midrib.

pinnatisect. Lobed almost to the midrib, said of leaves.

pinniform. Like a fin, as the wing of a penguin.

pinniped. An aquatic mammal of the Pinnipedia, a suborder of the Carnivora, which includes the seals, sea lions, and walruses, with limbs modified as flippers.

pinnule. 1. *zool.* A small branch of a feather-shaped structure. **2.** *bot.* One of the primary divisions of a pinna of a double compound (decompound) frond or leaf.

pinocytosis. The taking in or engulfing of globules of a fluid by a cell. *Compare* phagocytosis.

piñon. The nut pine (*Pinus parryana*) and related species, or its edible seed.

pintail. A surface-feeding duck (*Anas acuta*).

pin-tailed. With central tail feathers long and narrow.

pinule. A ray of a sponge spicule, when covered with spines.

pinworm. A small nematode (*Enterobius vermicularis*) parasitic in the intestine of man, especially children. Also called seat worm.

pipefish. A long, slender, marine fish of the family Syngnathidae.

pipit. A small, tail-wagging ground bird of the family Motacillidae, as *Anthus spinoletta*, the American pipit.

pirana. A voracious, sharp-toothed fish (*Serrasalmus*) which operates in schools in South American rivers, noted for its ferocity; also called piranha, caribe.

405

pirarucu. A large freshwater fish (*Arapaima gigas*) of South America.

piscine. Of or pertaining to fishes.

pistil. A gynoecium or a unit of a gynoecium, the female structure of a flower consisting of an ovary, style, and stigma; a female organ composed of one or more carpels.

pistillate. 1. Having a pistil or pistils but no stamens. **2.** Female, with reference to flowers.

pit. 1. A depression or an open cavity. **2.** *bot.* **a.** A thin, depressed area in a cell wall. **b.** The stone of a drupe, as a cherry. **3.** *zool.* A depression, as an armpit, facial, or loreal pit, or gastric pit. *See* pit organ.

pit canal. A minute opening in the wall of a sclerid.

pitch. 1. *bot.* **a.** A resinous exudate of certain conifers, especially the hemlock. **b.** The dark residue resulting from the distillation of coal tar, wood tar, and other substances. **2.** *physiol.* The quality of sound which depends upon the frequency of vibrations.

pitcher plant. An insect-eating bog plant (*Sarracenia purpurea*) with a leaf or leaves forming a pitcher-shaped structure which traps insects.

pit field. A depression in a primary cell wall in plants.

pith. A soft, spongy, parenchymatous tissue which occupies the central portion of the stem in most angiosperms.

pithecanthropoid. Of or pertaining to *Pithecanthropus*. *See* Java man.

pit organ. 1. A pit containing sensory cells sensitive to heat located between the eye and the nose in pit vipers and along the margins of the jaws in pythons and boas. **2.** A pit in the lateral line system of fishes containing cells sensitive to pressure changes and electrical stimuli.

pit-pair. A pair of pits of adjacent cells.

pitted. Bearing or marked by minute depressions or pits.

pitting. The arrangement and distribution of pits in the cell walls of cells in plant stems.

pituitary. The pituitary gland or hypophysis cerebri, an endocrine gland attached to the ventral surface of the diencephalon, consisting of an anterior lobe which secretes a number of trophic hormones (somatotrophin, adrenocorticotrophin, thyrotrophin, and several gonadotrophins), an intermediate lobe which secretes a melanocyte-stimulating hormone, and a posterior lobe which stores and releases vasopressin and oxytocin secreted by the hypothalamus. *See* infundibulum, Rathke's pouch.

pit viper. One of several venomous snakes of the family Crotalidae characterized by possession of a pit organ, as the rattlesnake, copperhead, water moccasin, fer-de-lance, and bushmaster.

placenta. 1. *anat.* In mammals, a structure attached to the inner surface of the uterus through which the embryo or fetus obtains its nourishment and discharges its wastes. 2. *zool.* A comparable structure in invertebrates, as in certain scorpions and tunicates. 3. *bot.* The structure by which an ovule is attached to the wall of the ovary.

placental. Of, pertaining to, or having young nourished through a placenta.

placentate. Having a placenta.

placentation. 1. *anat.* The processes involved in implantation, the establishment of the embryo, and the development of a placenta. 2. *bot.* The mode of attachment of an ovule to the ovarian wall, as axile, parietal, marginal, free-central, basal, and lamellate.

placode. A localized thickening of an epithelium in an embryo which comprises the anlage of a structure, as a lens *placode*.

placoid scale. A type of scale found in elasmobranchs consisting of a bony plate embedded in the dermis. From the plate a dentine-covered spine projects through the epidermis.

plagiotropism. The tendency to grow in an oblique or horizontal direction.

plague. Any infectious, highly contagious, epidemic disease, especially that caused by a bacterium (*Pasteurella pestis*) existing in three forms: bubonic septicemic, and pneumonic.

plaice. A flatfish, as **a.** the European flounder (*Pleuronectes platessa*); **b.** a North Atlantic flounder, *Hippoglossoides*.

plaited. Folded: PLICATE.

plakea. A stage of development of certain algae and protozoans in which the cells form a hollow ball with flagella directed inward, as in *Volvox*.

planarian. Any free-swimming, triclad, turbellarian flatworm, especially those of the genera *Planaria*, *Euplanaria*, and *Dugesia*.

planidium. The first in two or more larval stages in an insect undergoing hypermetamorphosis.

plankter. An individual organism or species of plankton.

plankton. Aquatic organisms of fresh, brackish, or sea water which float passively or exhibit limited locomotor activity. *See* nannoplankton, phytoplankton, zooplankton.

plant. 1. Any member of the plant kingdom, Planta. 2. A living organism distinguished in general from an animal by the following characteristics: capacity for unlimited growth by persistence of embryonic tissue throughout life; lack of locomotor structures; ability to carry on photosynthesis; slow responses to stimuli; cells with cell walls containing cellulose; reproduction involving alternation of an asexual with a sexual generation. *Compare* animal.

planta. The sole of the foot.

plantain. 1. The stemless herb *Plantago*, a common weed; ribwort. 2. A tropical food plant (*Musa paradisiaca*) of southern Asia.

plantar. Of or pertaining to the sole or the under surface of the foot.

plantar-flexion. Extension of or straightening the foot.

plant bug. A hemipterous insect of the family Miridae, also called leaf bug.

plantigrade. Walking on the sole of the foot with heel touching the ground, as man, bears, and certain insectivores.

plant lice. *See* aphid.

plantula. A pulvilluslike structure on the underneath side of the tarsal segments in various insects, as the cockroach.

planula. The ciliated, free-swimming larva of certain coelenterates, especially hydrozoans and one genus of ctenophorans, *Gastrodes*, a parasite.

plaque. 1. A small plate or platelike structure. 2. A clear area in a bacterial culture resulting from the action of a bacteriophage.

plasma. 1. The clear fluid portion of blood or lymph. 2. In a general sense, protoplasm.

plasmagel. The more solid, jellylike, outermost portion of the protoplasm of a rhizopod protozoan. *Compare* a plasmasol.

plasmagene. A self-duplicating hereditary determiner present in the cytoplasm of a cell. *See* kappa particles.

plasmalemma. The plasma membrane; cell membrane.

plasma membrane. The cell membrane.

plasmasol. The more fluid central portion of the protoplasm of a rhizopod protozoan. *Compare* plasmagel.

plasmid. A term proposed to include plasmagenes and plastogenes.

plasmodesma pl. **plasmodesmata.** Minute protoplasmic threads which pass through cellulose cell walls of plants connecting adjoining cells.

plasmodiocarp. A sessile sporangium which develops from a branching plasmodium in myxomycetes.

plasmodium. 1. A multinucleate, ameboid mass of naked protoplasm which forms the vegetative body of a slime mold (myxomycete). 2. A protozoan of the genus *Plasmodium*, which includes the causative agents of malaria.

plasmogamy. Condition in which there is a fusion of the cytoplasm of two haploid cells without fusion of the nuclei, as in certain fungi.

plasmolysis. The shrinking of protoplasm from the cell wall with loss of turgidity as a result of exosmosis. *Compare* deplasmolysis.

plasmoptysis. The escape of protoplasm from a cell following rupture of the cell membrane.

plasmotomy. The division of a multinucleate cell into portions which are still multinucleate, as occurs in certain myxosporidians.

plastic. 1. Capable of being altered or modified. 2. Capable of undergoing change. 3. Concerned with the restoration or building up of tissues.

plastid. Any of a number of self-duplicating bodies present in the cytoplasm of cells, especially plant cells, varying in structure, pigment content, and function. *See* chloroplast, chromoplast, leucoplast, amyloplast, elaioplast.

plastid inheritance. *See* plastome.

plastogene. 1. A plasmagene which gives rise to a plastid. 2. A cytoplasmic unit of heredity associated with the development of plastids and responsible for their characteristics.

plastome. The phenomena by which the primordia of plastids pass through the cytoplasm of the egg and give rise to plastids of the next generation.

plastron. 1. The ventral portion of the shell of a turtle. 2. A thin film of air held by hydrofuge hairs or scales of an insect.

plate. 1. *zool.* A thin, flat structure; a lamina, a scute. 2. *microbiol.* A thin, flat, culture dish, as a petri dish, which contains a solid medium upon which microorganisms can be grown.

plate count. A method employed for estimating the number of bacteria in a substance as milk or water based on introducing a measured amount of the material into a petri dish containing a suitable culture medium, and then, after a period of incubation, counting the number of colonies which develop.

platelet. *See* blood platelets.

plating. The cultivation of microorganisms in a petri dish containing a solid nutrient medium, usually nutrient agar.

platyhelminth. A flatworm, an invertebrate of the phylum Platyhelminthes which includes free-living turbellarians and the parasitic trematodes (flukes) and cestodes (tapeworms).

platypus. *See* duckbill.

platyrrhine. With a broad, flat nose.

plecopteran. An insect of the order Plecoptera comprising the stoneflies.

plectenchyma. The hyphae of a fungus when organized into tissues.

plectrum. 1. The stapes in anurans, reptiles, and birds. 2. The rasping part of a stridulatory organ.

pleiomerous. Having more than the normal number of floral parts, as petals or sepals.

pleiotropism. A condition in which there are multiple effects produced by a single gene.

Pleistocene epoch. The first epoch of the Quaternary period characterized by repeated glaciation and the first indications of social life in man.

pleomorphic. Having more than one form during a life cycle, as in certain bacteria and fungi.

pleon. 1. The abdomen of a crustacean. 2. The telson of a king crab.

pleopod. A slender, biramous abdominal appendage of a malacostracan crustacean.

plerocercoid. The wormlike, second larval stage of a fish tapeworm which develops in a fish following ingestion of a procercoid.

plerocercus. The larval stage of a tapeworm consisting of a solid, globular body containing the invaginated head or scolex.

plerome. Meristem tissue comprising the central portion of a growing root or stem.

plesiosaur. An extinct, marine reptile of the order Plesiosauria with a long neck and paddlelike limbs; common in the Mesozoic era.

pleura. 1. A pleuron. 2. In mammals, a serous membrane which lines the thoracic cavity and covers the surface of the lungs.

pleural. Of or pertaining to the pleura or a pleuron.

pleurapophysis. A lateral process of a vertebra with which a rib fuses.

pleurite. A sclerite of a pleuron.

pleurobranch. A gill borne on the side of a thorax.

pleurocarpous. Designating a growth form in mosses in which the gametophyte is much branched and creeping with sporophytes borne on the tips of lateral branches. *Compare* acrocarpic.

pleurocentrum. A lateral element in the centrum of a vertebra of certain fishes and amphibians.

pleuron pl. **pleura.** The lateral portion of the body wall of a segment of an arthropod.

pleuston. A mass of free-floating organisms, especially a mat of algae.

plexus. An interlacing network of vessels or fibers.

plica pl. **plicae.** A fold.

plicae circulares. Circular folds on the inner surface of the intestine.

plicate. 1. Folded like a fan. 2. Possessing a surface with parallel ridges.

Pliocene epoch. The last epoch of the Tertiary period during which man and most species of modern mammals came into existence.

ploidy. *genetics.* A condition in which the chromosome number is a multiple of the haploid number. *See* polyploidy.

411

plover. A shorebird of the family Charadriidae, as the golden plover (*Pluvialis dominica*).

pluck. The heart, liver, and lungs of an animal.

plum. The fruit, a drupe, from the plan tree (*Prunus*).

plumage. The entire covering of feathers of a bird.

plumbeous. Resembling lead; lead-colored.

plume. **1.** A feather, especially one that is long and conspicuously colored. **2.** An ornamental tuft of feathers on the head of a bird. **3.** A tuft of cilia or hairs which resembles a feather.

plumelet. A small plume.

plumose. **1.** Like a plume or feather. **2.** Covered with fine hairs, as the stigma of many grasses. **3.** Having fine hairs or bristles resembling a plume.

plumule. **1.** *bot.* The primary bud or epicotyl of a plant embryo. **2.** *zool.* A down feather.

plurilocular. Having more than one loculus or compartment, with reference to **a.** the ovary of a flower; **b.** a sporangium or gametangium of an alga.

pluteus. The ciliated, bilaterally symmetrical larva of certain echinoderms, as the sea urchin and brittle star.

PMS. Pregnant mare's serum gonadotrophin, with effects similar to FSH and LH.

pneumatic. **1.** Pertaining to air or gas. **2.** Operated by air pressure. **3.** Conducting or containing air.

pneumatic duct. The duct which connects the swim bladder of a fish with the pharynx or esophagus.

pneumatic sinus. An air sinus in a cranial bone; a paranasal sinus.

pneumatocyst. An air-containing cavity, as **a.** the swim bladder of a fish; **b.** the air bladder or float of various algae, as rockweeds.

pneumatophore. In coelenterates, a modified zooid which forms a gas-filled float, as in the Portugese man-of-war and various siphonophores.

pneumatopore. The opening of an air sac or pneumatophore.

pneumonia. A disease of the lungs characterized by inflammation and consolidation, usually due to a bacterial or virus infection.

pneumostome. The external opening of the respiratory chamber in arachnids and pulmonate snails.

pneumothorax. Air or gas in the pleural cavity.

poacher. A marine fish of the family Agonidae.

poad. A meadow plant.

pocket. A pouch or diverticulum.

pod. 1. *bot.* A dry, dehiscent fruit, splitting along two sutures, as that of a legume, as a bean or pea. 2. *zool.* **a.** A number of animals clustered together, as a school of fishes, flock of birds, or a group of whales or porpoises. *See* gam. **b.** A mass of insect eggs cemented together. *See* egg pod.

podetium pl. **podetia.** Loosely, a stalklike structure as, in lichens, the stalk which supports an apothecium.

podex. The rump or anal region.

podical plate. One of two platelike structures representing the eleventh segment of the abdomen in certain insects, as the grasshopper.

podium pl. **podia.** The tube foot of an echinoderm.

podobranchium pl. **podobranchia.** A gill borne on a thoracic appendage of a decapod crustacean.

podocyst. A chitinous cyst containing a ciliated larva present in the basal disk of certain coelenterates; pedal cyst.

podocyte. A cell with numerous cytoplasmic processes which come in contact with the capillaries of a renal corpuscle.

pogonophoran. Any of a small group of coelomate, marine worms of the phylum Pogonophora; a beard worm.

poikilothermous. 1. Having a body temperature which varies with the environment. 2. Cold-blooded, said of animals as amphibians and reptiles.

poison. 1. A substance which, upon contact with or being introduced into an organism, impairs or prevents normal metabolic processes from taking place, thus altering the normal functioning of organs or tissues. 2. An agent with an injurious or deadly effect. *See* toxin, venom.

poison claw. 1. A clawlike fang on the tip of a chelicera of a spider. 2. The maxilliped of a centipede.

poison gland. A gland which secretes a poisonous substance.

poison hemlock. A poisonous plant (*Conium maculatum*) of the family Umbelliferae.

poison ivy. A climbing, three-leafed vine (*Rhus radicans*) which produces a volatile oil capable of causing a severe dermatitis.

poison oak. A shrub (*Rhus toxicodendron*) which causes a contact dermatitis similar to poison ivy.

poisonous. 1. Having the properties of or producing the effects of a poison. **2.** Containing a poison, as a *poisonous* plant or animal. **3.** Venomous, toxic.

poisonous snake. A venom-producing snake, including those of the following families: Elapidae (cobra, krait, coral snake); Hydrophiidae (sea snakes); Viperidae (Old World Vipers); Crotalidae (pit vipers).

poison sac. A sac for storing poison, as that of the sting of a honeybee.

poison sumac. A poisonous shrub (*Rhus vernix*), also called poison dogwood or poison elder.

polar bear. A large white bear (*Thalarctos maritimus*) of the polar regions.

polar body. A polocyte, a minute cell produced during oogenesis.

polar capsule. A special cell which develops within sporoblasts of various sporozoans, as myxosporidians.

polarity. A state or condition of having poles or possessing parts or regions of opposite or contrasting effects.

polarization. 1. The state or condition of being polarized. **2.** *physiol.* The existence of opposite charges on the two sides of a cell membrane, as in a nerve fiber.

polar nuclei. In angiosperms, the two nuclei in the midregion of the embryo sac, which, when fertilized, develop into the triploid endosperm.

pole. Either of the two extemities of a structure which exhibits polarity, as an ovum. *See* animal pole, vegetal pole.

polecat. A European weasel (*Putorius* (*Mustela*) *putorius*) which ejects a foul-smelling fluid from anal glands when disturbed. Also called fitchew, foumart, or foul-marten.

Polian vesicle. One of several stalked sacs attached to the ring canal of the water-vascular system in echinoderms.

poliomyelitis. An acute, febrile disease caused by a virus in which motor neurons in the spinal cord are affected; also called infantile paralysis.

pollack also **pollock.** A marine food fish of the family Gadidae, as *Pollachius virens*, the American pollack or bluefish.

pollard. 1. A hornless ruminant. **2.** A stag which has shed its antlers. **3.** *bot.* A tree that has had its crown removed in order to promote growth of numerous branches.

polled. Lacking horns, with reference to cattle; lacking antlers, with reference to deer.

pollen. The pollen grains of seed plants, considered collectively.

pollen analysis. PALYNOLOGY.

pollen basket. A concave depression on the outer surface of the tibia of the third leg of a honeybee, its anterior margin bearing recurved hairs.

pollen brush. A group of stiff hairs on the first tarsal segment of the first and second pairs of legs in the honeybee or on the abdomen of certain bees; also called scopa.

pollen chamber. In gymnosperms and cycads, a small cavity in which pollen grains germinate located at tip of the nucellus under the micropyle.

pollen comb. A series of rows of stiff bristles on the inner surface of the metatarsus of the third leg of a honeybee.

pollen flower. One which produces pollen but no nectar.

pollen grain. 1. A microspore produced in the pollen sac of angiosperms or the microsporangium of gymnosperms. **2.** A haploid cell containing two nuclei, a tube nucleus and a generative nucleus.

pollen packer. A structure in the honeybee for compressing pollen into a compact mass consisting of a pecten and auricle located at the joint between tibia and metatarsus of the hind leg.

pollen sac. An elongated sac in the anther of a stamen in which pollen grains (microspores) are produced.

pollen stratigraphy. The study of pollen present in successive layers of sediment deposited in bogs or lakes. *See* palynology.

pollen tube. A slender tube which develops upon germination of a pollen grain through which two sperm nuclei gain access to the female gametophyte and its contained ovum.

pollex. The first digit of the forelimb of a tetrapod; the thumb.

pollinate. To perform the act of pollination.

pollination. The transfer of pollen from an anther to a stigma, accomplished by the wind, water, insects, birds, bats, or artificially.

polliniferous. Bearing or producing pollen.

pollinium pl. **pollinia.** A mass of coherent pollen grains, as that formed in orchids.

polliwog. A tadpole.

polocyte. A polar body.

polyadelphous. The condition in which the filaments of stamens are united into three or more groups or fascicles.

polyandrous. 1. *bot.* Having an indefinite number of stamens. **2.** *zool.* Having more than one mate, said of females.

polychaete. Any annelid of the class Polychaeta, mostly marine forms with well developed parapodia, as the wormclam, *Nereis.*

polycythemia. A condition in which there is an abnormally large number of red blood cells.

polydactyl. Having extra fingers or toes or both.

polydisc. Designating a multiple disc scyphistoma consisting of several ephyrae.

polyembryony. The development of more than one individual from a fertilized ovum.

polyestrous. Having more than one estrous cycle per year, as in rats, rabbits.

polygamous. 1. *bot.* Bearing staminate, pistillate, and hermaphroditic flowers at the same time. **2.** *zool.* Having more than one mate at one time.

polygene. A multiple gene.

polygenetic. POLYPHYLETIC.

polygenic. Due to many genes, said of hereditary traits of a quantitative character. *See* multiple genes.

polymer. A substance resulting from the combination of two or more molecules of the same type.

polymorph. A polymorphonuclear leukocyte.

polymorphic. Polymorphous; exhibiting polymorphism.

¹polymorphism. Existing in more than one form.

²polymorphism. The occurrence in the same habitat of two or more distinct forms of a species.

polymorphonuclear. Having a nucleus consisting of two or more lobes joined together by slender strands, as in granular leukocytes.

polymyarian. Having numerous muscle cells in each quadrant of the body wall, as in certain nematodes. *Compare* meromyarian.

polyp. 1. A sessile form of coelenterate with a cylindrical body attached at its basal end, its free end bearing a mouth surrounded by tentacles; a hydranth, as *Hydra*. **2.** *med.* A pedunculated, neoplastic structure which develops on a mucous membrane, as in the nasal cavity.

polypeptide. *See* peptide.

polypetalous. Having many distinct petals.

polyphagous. Subsisting on many kinds of foods.

polyphyletic. Having more than one ancestral line, with reference to species and blood cells.

polypide. The visceral mass of a bryozoan capable of being retracted into the zooecium.

polyploid. An individual or species whose chromosome number is a multiple other than two of the haploid number.

polyploidy. A condition in which an individual possesses one or more sets of homologous chromosomes in excess of the normal diploid sets, as triploidy (3n), tetraploidy (4n), hexaploidy (6n), octoploidy (8n).

polypod. Having many legs or feet, with reference to insect larvae.

polypoid. Polyplike in appearance or structure.

polysaccharide. One of a group of complex carbohydrates which, upon hydrolysis, yields more than two molecules of a monosaccharide, as cellulose, starch, glycogen.

polysaprobic. Capable of living in waters of low oxygen content containing much decomposing organic material.

polysepalous. Possessing many sepals.

polyspermia. Excessive production of spermatozoa or semen.

polyspermy. The entrance of more than one sperm into an egg.

polystachyous. Bearing many spikes.

polystichous. Arranged in many ranks or rows.

polystomatous. Having many or multiple pores, mouths, openings, or suckers.

polytocous. Producing more than one at birth.

polytrophic. 1. Utilizing various kinds of foods. **2.** Having nutritive cells alternating with oocytes, with reference to insect ovarioles.

polytropic. Affecting several kinds of tissues, said of viruses.

polytypic. Consisting of more than one subspecies.

polyuria. Production of an excessive quantity of urine.

polyvoltine. Having many broods in a single season, as the silkworm.

polyxenous. Having many possible hosts, with reference to parasites.

polyzoan. A bryozoan.

polyzoic. 1. Consisting of many zooids. **2.** Having many proglottids, with reference to tapeworms.

pomace fly. A fruit fly, especially *Drosophila melanogaster*.

pomaceous. 1. Of or pertaining to a pome. **2.** Of or pertaining to apples.

pome. An inferior, indehiscent, many-seeded fruit in which the receptacle forms the outer, fleshy portion, as an apple, pear.

pomegranate. A bush or low tree (*Punica granatum*) or its fruit, a large, thick-rinded berry.

pomiferous. Bearing pomes.

pompano. A marine food fish (*Trachinotus carolinus*) of the Atlantic and Gulf coasts.

pond. A small body of standing water.

pondweed. Any of a number of aquatic plants of the family Zosteraceae, especially those of the genus *Potamogeton*.

pongid. Any anthropoid ape of the family Pongidae.

pons. A bridge or bridgelike structure, especially the *pons Varolii*, a part of the brain with nerve fibers connecting the midbrain, cerebellum, and medulla oblongata.

poplar. A rapid-growing tree of the willow family, Salicaceae, especially those of the genus *Populus*.

poppy. A plant of the genus *Papaver*, family Papaveraceae, with showy, regular flowers, capsular fruits, and a milky juice, especially *P. somniferum*, the opium poppy.

population. All the individuals belonging to a single species or several species which are closely associated and occupy a particular area or space.

porate. Having pores, with reference to the exine of a pollen grain.

porbeagle. A medium-sized, voracious shark (*Lamnia nasus*) of northern waters.

porcine. Of or pertaining to swine.

porcupine. One of a number of large, stout-bodied rodents with barbed, erectile quills, comprising two families, Erethizontidae (New World porcupines), as *Erethizon dorsatum*, the Canadian porcupine, and Hystricidae (Old World porcupines), as *Hystrix cristata*, the crested porcupine.

porcupine fish. A marine fish of the genus *Diodon*, family Diodontidae, which is capable of inflating itself and becoming globose bringing about the erection of its spiny scales.

poricidal. Having pores through which the contents are shed, with reference to seed pods or anthers.

poriferan. Any invertebrate of the phylum Porifera which includes the sponges, asymmetrical or radially symmetrical, sessile metazoans lacking organs or organ systems; body wall permeated with pores; possess an internal skeleton of calcareous spicules or spongin or both.

porocyte. A tubular cell found in the wall of simple sponges.

porogamy. The condition in plants in which the pollen tube gains entrance to the ovule through the micropyle. *Compare* aporogamy.

porose. POROUS.

porous. Possessing openings or pores: POROSE.

porpoise. 1. One of a number of cetaceans of the genus *Phocaena*, with a blunt head and spade-shaped teeth, as *P. phocaena*, the harbor porpoise. 2. One of a number of dolphins.

porrect. Extended outward and forward.

portal. An entryway, especially the hilus of an organ.

portal heart. A contractile portion of the portal vein in cyclostomes.

portal of entry. The avenue through which an infectious organism gains entrance to the body.

portal system. A vein with a capillary network at each end, as the hepatic, hypophyseal, or renal *portal systems*.

portal vein. The principal vein of a portal system. When unqualified, it usually means the *hepatic portal vein* which conveys blood from digestive organs and spleen to the liver.

Portuguese man-of-war. A polymorphic, colonial hydrozoan (*Physalia pelagica*) with a large float from which numerous elongated processes are suspended.

419

position effect. *genetics.* A condition in which the phenotype is altered when the position of the gene in relation to other genes is changed, as by inversion or translocation.

positive. 1. Numerically, anything greater than zero. **2.** *biol.* **a.** Directed toward a stimulus, as a *positive* tropism or taxis. **b.** Designating increased or favorable activity, as a *positive* reaction. **3.** *chem.* **a.** Having less than the normal number of electrons, as a *positive* ion. **b.** Yielding electrons, as a *positive* valence.

possum. The shortened form of opossum.

postabdomen. 1. In primitive arachnids, the posterior portion of the abdomen. **2.** In various crustaceans, the recurved tip of the abdomen, as in cladocerans.

postaxial. 1. Located or situated behind an axis. **2.** Designating the posterior side of a vertebrate limb.

postcardinal. Posterior to the heart.

postcava. The posterior or inferior vena cava.

postecdysis. The final phase of ecdysis during which the endocuticle is secreted and hardening of the exoskeleton occurs.

posterior. 1. *zool.* At or towards the caudal end of the body. **2.** *anat.* At or toward the back or dorsal region of the body. **3.** *bot.* Next to, close to, or toward the main axis; superior.

posterior root. *See* spinal nerve.

postganglionic fibers. Fibers of the autonomic nervous system whose cell bodies are located outside the central nervous system in autonomic ganglia.

postlarva. In crustaceans, the final stage in development in which the individual resembles, in general, the adult but still differs significantly in a number of features.

postmortem. An examination of a body after death; an autopsy.

posture. The attitude or position assumed by a body.

potamophilous. Inhabiting streams or rivers.

potamoplankton. The plankton of running waters; rheoplankton.

potash. POTASSIUM CARBONATE.

potassium. A metallic element of the alkali group, symbol K, at. no. 19.

potassium bicarbonate. A slightly alkaline salt, $KHCO_3$.

potassium carbonate. A strongly alkaline salt, K_2CO_3; potash.

potassium hydroxide. A white solid, KOH, which, in water, forms a strongly alkaline, highly caustic solution.

potato. The edible tuber of the plant *Solanum tuberosum* or the plant which produces it; the white or Irish potato. *See* sweet potato.

potato beetle. The Colorado potato beetle (*Leptinotarsa decemlineata*), a yellow beetle with black stripes, a serious pest to potato plants. Also called potato bug.

potato blight. One of a number of diseases of the white potato, especially the late blight, caused by a fungus (*Phytophthora infestans*), a downy mildew.

potato bug. *See* potato beetle.

potency. **1.** *physiol.* The ability of a male to perform the sexual act. **2.** *embryol.* The extent to which an egg, blastomere, or a part of the embryo is capable of developing under a certain set of conditions. **3.** *phar.* The power or effectiveness of a drug.

potential. *See* action potential.

potentiation. Increasing the effectiveness of a drug by combining it with another.

potoo. A nonmigratory forest bird (*Nyctibius griseus*) of Central and South America.

potter wasp. *See* mason wasp.

potto. A small, lorislike, African primate (*Perodicticus potto*) of the family Lorisidae.

pouch. **1.** A sac or receptacle, as **a.** the marsupial *pouch* of the Metatheria; **b.** the cheek *pouch* of a rodent; **c.** the buccal food *pouch* or gular *pouch* of a bird; **d.** the brood *pouch* of a bivalve. **2.** *anat.* A pocketlike cavity, as the rectouterine *pouch*. **3.** *embryol.* An outpocketing, as a pharyngeal *pouch*.

poulard. A castrated pullet or hen.

poult. The young of domesticated fowl, especially a young chicken or turkey; also the young of game birds.

poultry. Domesticated birds raised for the production of eggs or meat, as chickens, turkeys, ducks, geese.

pout. One of several freshwater catfishes, especially *Ameiurus nebulosus*, the horned pout.

pouter. A domestic pigeon which has the ability to inflate its crop; also called blowing pigeon.

powdery mildew. One of a number of ascomycete fungi which live on the surface of leaves, flowers, and fruits giving infected regions a powdery appearance. All are obligate parasites; some are the cause of serious diseases of fruits and grains.

P-P factor. The pellagra-preventative factor. *See* niacin.

praeputium. The prepuce.

prairie. A large, treeless area of level or rolling land covered with grass.

prairie chicken. The pinnated grouse (*Tympanuchus cupido*), an upland game bird of the Mississippi valley, the males possessing inflatable, orange-colored air sacs.

prairie dog. A stout-bodied, burrowing rodent of the family Sciuridae, especially those of the genus *Cynomys*, common throughout the western plains of the United States where they exist in large colonies.

prairie wolf. *See* coyote.

prawn. One of a number of edible, shrimplike, decapod crustaceans, especially those of the genera *Penaeus* and *Palaemonetes*.

preadaptation. The possession of a trait or traits which are not necessarily and often not advantageous to an organism in its present environment but would be advantageous in a different environment.

preaxial. 1. Located or situated in front of an axis. **2.** Pertaining to or located on the anterior side of a vertebrate limb.

precava. The anterior or superior vena cava.

precipitin. An antibody which develops in response to a specific antigen, a precipitinogen, and which brings about the precipitation of the antigen.

precocial. Designating birds whose young, upon hatching, are covered with down, are active and able to move about and feed themselves. *Compare* altricial.

precocious. 1. *zool.* More fully developed than usual; prematurely developed. **2.** *bot.* Ripening or fruiting before the usual time.

precursor. That which gives rise to something else; something essential for the development of something more complex.

precystic. Preceding the formation of a cyst.

predacious or **predaceous.** Predatory; preying upon other animals.

predation. The act of capturing and killing other animals for food.

predator. A predatory organism; a beast of prey.

predatory. Predacious; living by killing and eating other animals.

preen. To dress or trim the feathers, beak, or fur; used principally with reference to birds.

preformation. *embryol.* A formerly held theory that an individual exists preformed within the ovum or spermatozoan and that development results from the unfolding and growth of previously developed parts. *Compare* epigenesis.

preganglionic fibers. Fibers of the autonomic nervous system whose cell bodies are located in the brain or spinal cord.

preganglionic neuron. The first of two neurons in the efferent pathway of impulses from the brain or spinal cord to a visceral effector organ (gland, smooth muscle, cardiac muscle). *See* postganglionic fibers.

pregnant. The condition of being with child or bearing young: GRAVID.

prehallux. A rudimentary digit on the tibial side of the tarsus in frogs and other salientians; calcar.

prehensile. Adapted for grasping or clasping, especially by wrapping around, as the *prehensile* tail of a monkey.

prehensor. An appendage bearing a poison claw on the first trunk segment of a centipede; a maxilliped.

prementum. In insects, the distal portion of the labium usually bearing the ligula and labial palps.

premolar. In mammals, a tooth located between the canine and the first molar tooth; a bicuspid.

premunition. A state or condition in certain parasitic infections in which the host is immune to reinfection as long as the parasites remain within the body.

prenatal. Existing before birth.

prepubis. One of a pair of bones in monotremes and marsupials which extend forward from the pubis serving to support the ventral body wall and marsupium; a marsupial bone.

prepuce. The foreskin, a fold of skin which covers the glans penis or glans clitoridis.

prepupa. A pharate pupa; a pupa which is retained within the old larval skin.

pressor. Inducing an increase in blood pressure.

pressoreceptor. A receptor which responds to an increase in blood pressure, as those located in the carotid sinus or arch of the aorta: BARORECEPTOR.

presumptive. *embryol.* A term applied to tissues or cells before differentiation indicating that, in the course of normal development, the embryonic structure would follow a certain pre-determined course, as designating certain blastomeres as *presumptive* ectoderm.

prevernal. Of the early spring, with reference to flowering.

¹prey. An animal that is seized by another and eaten.

²prey. To engage in predaceous activity, seizing and killing other animals.

prezygapophysis. The anterior or superior zygapophysis of a vertebra; the anterior articular process.

priapulid. Any of a group of marine worms of the phylum Priapulida, sometimes considered as a class of the Aschelminthes. Formerly was an order of the class Gephyrea, phylum Annelida.

pricket. A male deer in its second year.

prickle. A sharp-pointed spine or process.

prickly pear. A cactus of the genus *Opuntia* or the fruit it produces.

prim. The common privet.

primary. 1. First, with reference to time, order, importance, or development. **2.** *ornith.* Pertaining to or designating the outermost of the large wing feathers (remiges), especially those attached to the bones of the wrist and hand.

primate. Any mammal of the order Primates which includes the lemurs, tarsiers, monkeys, apes, and man.

primipara. A woman who is giving birth or has given birth to a child for the first time.

primite. The anterior of two sporadins.

primitive. 1. Occurring early in evolutionary development; simple, unspecialized; primordial. **2.** Occurring early in development; simple; elemental. **3.** Uncivilized; undeveloped.

primitive streak. A longitudinal thickening on the dorsal surface of the embryo in birds and mammals where involution of the chordamesoderm occurs, considered homologous to the blastopore of amphibians.

primocane. The first year's cane in a bramble.

primordial. Original, first-formed; primitive.

primordium. *embryol.* The first indication of the development of an organ or structure: ANLAGE.

principle. A constituent in a mixture or compound which gives to the mixture or compound its characteristic properties.

prismatic. Shaped like a prism; angulate with flat sides.

prismatic layer. The middle layer in the shell of a mollusk.

prison flower. A flower that traps insects.

privet. A shrub of the genus *Ligustrum* widely used for ornamental purposes, especially hedges. Also called prim.

proboscidian. Any ungulate mammal of the order Proboscidia which includes the elephants and their relatives, the extinct mastodons and mammoths, characterized by possession of a long, muscular proboscis or trunk.

proboscis. 1. The trunk of an elephant. 2. The elongated snout of various animals as the mole, tapir, shrew. 3. Any of a number of elongated and usually tubular structures extending from or capable of being extended from the head or body of various invertebrates.

proboscis worm. A nemertean.

procambium. The cells which give rise to primary xylem and phloem.

procarp. A specialized branch in red algae from which a carpogonium develops.

procaryotic. Pertaining to cells in which the nucleus is not separated from the cytoplasm by a nuclear membrane and hereditary material is not organized into chromosomes. *Compare* eucaryotic. *See* moneran.

procercoid. The first larval stage of certain tapeworms, as the fish tapeworm, which develops in copepods following ingestion of a ciliated coracidium. *See* plerocercoid.

process. 1. A series of related activities, as the *process* of digestion. 2. A series of phenomena exhibiting continuous change, as the *process* of growth. 3. A prominence, outgrowth, or projecting part.

prochordate. A protochordate.

procoelous. Concave at the anterior end, with reference to the centra of vertebrae.

proctodeum. 1. *embryol.* An ectodermal pit in the anal region which gives rise to the anal canal of vertebrates or the hind intestine of an insect. 2. The cloaca or the terminal portion of the intestine in vertebrates.

proctodone. In insects, a hormone produced by epithelial cells of the hindgut which activates the neurosecretory cells of the brain.

proctostome. Term proposed for the single opening of a gastro-vascular cavity which serves both as a mouth and anus, as in coelenterates.

procumbent. *bot.* 1. Lying flat on the ground, prostrate. 2. Trailing, but not taking root, said of stems.

procuticle. The innermost, chitin-containing layer of the cuticle of arthropods, lying beneath the epicuticle.

producer. An organism which can manufacture organic com-pounds from simple inorganic substances, as a green plant. *Compare* consumer.

proembryo. In seed plants, the embryo before differentiation into suspensor and embryo proper.

proenzyme. The zymogen or inactive form of an enzyme.

proestrus. The period preceding estrus.

profile. The outline of a structure as seen from the side, espe-cially the face.

profundal. 1. Of or pertaining to the deepest portion of a body of water. 2. Deep, with reference to blood vessels and nerves.

progametangium. The fertile tip of a conjugating hypha before the formation of a gametangium, as in molds of the genus *Mucor*.

progenetic. Becoming sexually mature before reaching the final host, with reference to the metacercariae of certain flukes.

progeny. Offspring; descendants.

progestational. Preceding gestation or pregnancy.

progesterone. A steroid sex hormone produced by the corpus luteum of the ovary and, during pregnancy, by the placenta. With estrogen, it induces changes in the uterine endometrium which prepares it for implantation of the blastocyst and maintains these changes during early stages of pregnancy. Also, with estrogen, it stimulates the development of the mammary gland during pregnancy. Commonly called the pregnancy hormone.

progestogen. General term for any substance which induces progestational changes in the uterus.

proglottid. One of the segments comprising the body (strobila) of a tapeworm.

prognathous. 1. Having an upper jaw which projects beyond the facial profile. 2. *entomol.* Having a head with its long axis horizontal and mouth parts located anteriorly. *Compare* hypognathous.

prognosis. The prediction or forecast of the course of a disease, its duration, effects, and prospects of recovery.

progoneate. Having the genital opening located anteriorly, as in millipedes.

progravid. Before pregnancy; progestational.

prohaptor. An attachment organ located at the anterior end of a monogenetic fluke.

projection. *physiol.* The mental process by which sensations are referred to the point of stimulation, or in case of exteroceptors, to the source of the stimulating agent outside the body, as in visual sensations.

prolactin. A lactogenic hormone produced by the anterior lobe of the pituitary gland which stimulates development of the crop gland in pigeons and induces lactation in mammals. Also called lactogen, mammotrophin, galactin, lactogen, and luteotrophin.

prolegs. Short, fleshy appendages on the abdominal segments of the larvae of lepidopterans, each bearing at its tip hooks or crochets.

proliferation. Multiplication; an increase, usually rapid, as in **a.** the number of cells in a developing embryo; **b.** the number of parts, as proglottids in a tapeworm; **c.** the number of individuals, as in a colony.

proliferous. 1. *bot.* Reproducing by buds, bulbils, gemmae, or other vegetative structures. 2. *zool.* Proliferating, as production of numerous branchlets by corals.

prolific. Fruitful; productive of numerous offspring.

proloculum. The first chamber in a multilocular foraminiferan.

promeristem. The primordial or embryonic meristem which contains the initials or foundation cells of new organs or tissues.

pronation. Turning the hand or forearm so that the palm faces downward or backward. *Compare* supination.

prone. Prostrate; lying face downward. *Compare* supine.

pronephros. The head kidney, the most anterior of a series of embryonic excretory structures occurring in the embryos of all amniotes, and becoming the functional kidney in cyclostomes and some teleost fishes.

prong. A slender, pointed structure, especially one of the tips of an antler.

pronghorn. A swift-running ruminant (*Antilocapra americana*) formerly abundant throughout the western United States; the pronghorn antelope.

pronotum. The dorsal portion of the prothorax in an insect.

pronucleus. Either of the two haploid nuclei (male or female) of a fertilized ovum.

pronymph. A first stage nymph which has just emerged from its egg case.

propagate. 1. To cause to increase or multiply; to breed. **2.** To disseminate or spread about.

propagation. The act of propagating.

propagule. A propagulum.

propagulum. A propagable shoot, especially a short, flat branch of certain algae which becomes detached and gives rise to a new plant; a propagule.

propalinal. Pertaining to the forward and backward movement of the jaws in mastication.

properdin. A euglobulin in mammalian blood which serves as a nonspecific immune substance.

prophage. A nonpathogenic form of phage within its bacterial host.

prophase. The first phase of mitosis during which the chromatin becomes organized into chromosomes, centrosomes move to the opposite poles, spindle fibers develop and become attached to the chromosomes, and the nuclear membrane disappears. A similar stage occurs in cell divisions in meiosis.

prophylaxis. 1. The prevention of disease. **2.** The use of drugs or other agents to prevent infection.

prophyllum. A bracteole.

propodeum. In hymenopterous insects, the first abdominal segment which is united to the thorax and from which the petiole extends.

propodium. The anterior portion of the foot in certain burrowing gastropods.

propolis. A dark, waxy material gathered by bees from the buds of trees, especially poplars, which is used for stopping up cracks in the hive.

proprioceptive sense. The sense of position or movement; the kinesthetic or muscle sense.

proprioceptor. A sensory receptor located in a muscle, tendon, joint, or the ear which is stimulated by movement, change in position, muscle stretch or tension.

prop root. 1. A stiff, aerial root which functions as a brace in support of a tall, columnar stem, as in maize. **2.** One of the numerous, aerial roots which hold a plant erect, as those of the mangrove.

prosencephalon. The forebrain.

prosenchyma. *bot.* Plant tissue composed of elongated parenchyma cells with tapering ends.

prosodus. A narrow canal between the incurrent channel and a flagellated chamber, as in certain leuconoid sponges.

prosoma. The cephalothorax of an arthropod.

prosopyle. In a sponge, the opening between an incurrent and radial canal.

prostaglandin. One of a number of biologically active lipids present in seminal, menstrual, and amniotic fluids and various mammalian tissues. They stimulate smooth muscle contraction, inhibit lipolysis, platelet aggregation, and gastric secretion.

prostal. A spicule which projects from the surface of a sponge.

prostate gland. An accessory reproductive organ in the male of various animals, as trematodes, annelids, mollusks, and mammals. In mammals, its secretion alkalinizes the seminal fluid and activates the sperm.

prostheca. A fringed plate on the inner surface of the mandible of certain insects.

prosthetic group. In an enzyme system, the nonprotein component or coenzyme.

prostomium. The portion of the head which projects forward and overhangs the mouth, as in various annelids.

prostrate. 1. *zool.* Lying stretched out at full length, either prone or supine. 2. *bot.* Flat on the ground; procumbent.

protandry. 1. *zool.* A condition in monoecious animals in which sperm are produced before eggs, or sperm and eggs are produced alternately, as in the oyster. 2. *bot.* A condition in which anthers mature and shed pollen before the stigma of the same flower is ready to receive the pollen.

protaspis larva. An early larval stage of a trilobite.

[1]protean. Variable; assuming various forms and shapes.

[2]protean. A derived protein, resulting from the action of water, dilute acids, or enzymes on proteins, as fibrin from fibrinogen.

protease. An enzyme which acts on proteins. *See* proteinase.

protein. One of a group of nitrogenous substances which, upon hydrolysis, yield amino acids; present in all animal and plant cells and characteristic of living matter. They comprise three types: simple (albumins, globulins), derived (proteoses, peptones, peptides), and conjugated (nucleoproteins, glycoproteins).

proteinase. A proteolytic enzyme; an enzyme which acts on proteins, especially native proteins, as pepsin, trypsin.

protelean. Designating parasitism by the embryonic or larval stages of an organism, as by the larvae of ichneumon flies.

proteolysis. The breakdown of proteins into simpler substances, usually the result of the action of proteolytic enzymes.

proteose. One of a group of derived proteins resulting from the hydrolysis of proteins. They are intermediate between proteins and peptones.

proteroglyph. Any snake of the group Proteroglypha, venomous snakes with fixed fangs located anteriorly in the upper jaw. Includes the families Hydrophiidae and Elapidae.

Proterozoic era. The second geologic era characterized by the coming into existence of simple multicellular plants and animals.

prothallial. Of or pertaining to a prothallus.

prothallus. A green, heart-shaped, gamete-producing plant of ferns and their allies; prothallium.

prothoracic gland. A gland in insects which secretes ecdysone, the molting hormone.

prothorax. The most anterior of the three divisions of the thorax of an insect.

prothrombin. A substance present in blood plasma which is essential for clotting.

protist. Any organism of the Protista, a major taxonomic group comprising all algae except the blue-greens, fungi, slime molds, and protozoa, characterized by possession of a true nucleus and chromosomes and usually unicellular reproductive structures.

protocaryotic. PROCARYOTIC.

protocercal tail. A primitive type of tail, as that in cyclostomes, in which notochord extends to the tip and dorsal and ventral portions are approximately equal.

protochlorophyll. The precursor of chlorophyll, present in dark-grown seedlings.

protochordate. Any chordate of the group Acrania; a tunicate or lancelet.

protocneme. One of the six pairs of primary septa or mesenteries of an anthozoan coelenterate.

protocoel. The coelomic cavity of a protostome.

protoconch. 1. The first shell formed by a larval gastropod. 2. A calcareous chamber in the first-formed portion of the shell of an ammonite.

protocone. 1. The single cusp of a primitive reptilian tooth. 2. The inner of three cusps of an upper molar tooth.

protoconid. The outer of three cusps of a lower molar tooth.

protoderm. A layer of dividing cells in the apical meristem which gives rise to the epidermis of mature structures.

protogyny. In a monoecious organism, a condition in which the female element matures before the male element. *Compare* protandry.

protomerite. The anterior portion of a gregarine.

proton. A positively charged, subatomic particle which forms the nucleus of the lightest hydrogen isotope (H^1), its mass being 1847 times that of an electron. Protons, with neutrons, form the nucleus of an atom, the number of protons equaling the atomic number of the atom.

protonema pl. **protonemata. 1.** A branched, filamentous structure in a moss which develops from a spore and by budding gives rise to leafy, gametophyte moss plants. **2.** A similar structure in algae which develops from a zygote.

protonephridium. A primitive type of excretory tubule which opens externally by a nephridiopore but terminates internally in a flame bulb or solenocyte, as in flatworms, rotifers, and some nemerteans.

protonymph. The first of three developmental stages in mites, ticks, and pseudoscorpions. *Compare* deutonymph tritonymph.

protonymphon. The larval stage of a pycnogonid, with three pairs of legs.

protophloem. The first phloem elements to develop and reach maturity.

protophyte. Any member of the division Protophyta, comprising bacteria and viruses.

protoplasm. The viscid substance which comprises a living cell; the substance with which living processes, as metabolism, growth, irritability, and reproduction are associated; the physical basis of life.

protoplast. The organized, living protoplasm of a cell.

protopod. A protopodite.

protopodite. The basal portion of a crustacean appendage consisting typically of two parts, a coxopodite and basipodite.

protosome. The most anterior of three body regions in many deuterostomes and certain lophophorates. *Compare* mesosome, metasome.

protostele. A primitive type of stele consisting of a central strand of primary xylem surrounded by a cylinder of phloem and lacking a central column of pitch.

protostome. Any member of the Protostomia, a large group of coelomate animals in which the mouth, and sometimes the anus, develops from the region of the embryonic blastopore, as in annelids, mollusks, and arthropods. *Compare* deuterostome.

prototherian. A mammal of the subclass Prototheria comprising a single order, the Monotremata, which includes the egg-laying mammals.

prototroch. 1. A band or several bands of ciliated cells which encircle a trochophore larva just above its equator. 2. A similar band in the pilidium larva of nemerteans.

protoxylem. The earliest primary xylem which differentiates during the period of organ elongation.

protozoan. Any member of the phylum Protozoa comprising all unicellular or acellular animals, some forming colonies but lacking differentiation of tissues. It includes rhizopods, flagellates, sporozoans, and ciliates.

protozoea larva. A stage in the development of certain crustaceans which follows the nauplius stage, as in shrimps and prawns.

protozoology. The study of protozoans.

protozoon pl. **protozoa.** A protozoan.

protraction. The act of drawing out, protruding, or extending. *Compare* retract.

protractor. A muscle which causes a part to be protruded, as the *protractor* muscle of a bivalve.

proturan. Any insect of the order Protura which includes the telsontails, primitive insects with simple metamorphosis.

proventriculus. 1. In birds, the glandular or true stomach located between the crop and the gizzard. 2. In annelids, the crop, which lies anterior to the gizzard. 3. In insects, the gizzard, located posterior to the crop.

provitamin. 1. A substance which, when activated or chemically changed, becomes an active vitamin. 2. The precursor of a vitamin, as carotene which is converted in the liver to vitamin A.

proximal. Nearest to the body or point of attachment. *Compare* distal.

¹prune. A plum with high sugar content which can be dried without fermentation occurring.

²prune. To cut off surplus branches of trees or shrubs.

psalterium. The omassum or manyplies, the third chamber in the stomach of a ruminant.

psammon. The microfauna or microflora present in the spaces between sand grains of beaches.

psammophilous. Sand-loving; living or growing in sandy places.

psammophyte. A plant which grows in sand or on sandy soils.

psammous. Of or pertaining to sand; of the nature of sand; sabulous.

pseudoallelism. A condition in which groups of alleles show recombination with each other, that is, two or more genes (duplicate genes) act on the same trait and occupy closely adjacent loci.

pseudobranch. A structure which is not a true gill but functions as one, as that on the side of the foot in pulmonate gastropods.

pseudochitin. An organic substance, tectin, present in the shells of certain protozoans.

pseudocoel. The body space of an invertebrate, differing from a true coelom in that it lacks a peritoneal lining.

pseudocoelomate. An invertebrate of the group Pseudocoelomata, comprising the phyla Entoprocta, Aschelminthes, and Acanthocephala characterized by possession of a pseudocoel.

pseudocolony. A group of individuals united by tubes or stalks of dead material but not organically connected, as certain protozoans.

pseudodominance. The unexpected appearance of a recessive trait due to the loss of a dominant allele.

pseudogamy. Gynogenesis; a form of parthenogenesis in which development of the ovum is initiated by the male gamete which does not enter the egg, or if it does enter, its nucleus does not fuse with the female nucleus.

pseudohermaphrodite. An individual whose external genitalia resemble those of one sex but whose gonads are those of the opposite sex.

pseudometamerism. False segmentation.

pseudoperianth. A tubular structure enclosing the archegonium of certain liverworts.

pseudophyllidean. A tapeworm of the order Pseudophyllidea which includes a number which infest mostly fish-eating mammals and some birds, as the fish tapeworm in man.

pseudoplasmodium. In slime molds, the structure formed by the union of myxoamoebae.

pseudopod. 1. A pseudopodium. 2. A blunt locomotor process, as that on the larva of a dipterous insect. 3. An ambulatory bud of a redia.

pseudopodium. 1. *zool.* A blunt and usually temporary extension of the cytoplasm of a cell, as that seen in rhizopod protozoans, amebocytes of sponges, or macrophages of mammals. **2.** *bot.* A slender, setalike stalk which forms the axis of the gametophyte in certain bryophytes and mosses.

pseudopupa. A coarctate larva, a quiescent, pupalike stage preceding the pupa stage in certain coleopterans.

pseudoscorpion. A false scorpion, a small arachnid of the order Pseudoscorpionida (Chelonethida) with large pedipalps resembling those of true scorpions.

pseudotrachea. A fine, trachealike structure, as one of the numerous food channels on the tip of the proboscis of various dipterans or the fine invaginations of the respiratory cavity of an isopod.

pseudovelum. A false velum, an internally projecting flange on the margin of the umbrella of sertain scyphozoan jellyfishes.

pseudozoea. An erichthus.

psilopaedic. Gymnopaedic; naked at hatching, with reference to birds.

psilophyte. A primitive plant of the division Psilophyta, comprising the simplest vascular plants, as *Psilotum* and *Tmesipteris*.

psittaceous. Parrotlike.

psittacine. Of or pertaining to parrots.

psittacosis. Ornithosis; parrot fever, an acute, infectious febrile disease in man caused by a polytropic virus transmitted by parrots, parakeets, and lovebirds.

psocid. Any insect of the order Psocoptera (Corrodentia) which includes the bark lice and book lice.

psoroptid. Of, pertaining to, or caused by mites of the genus *Psoroptes.*

psorosis. A virus-caused disease of citrus trees characterized by formation of concavities and blind pockets on trunks and larger limbs.

psychogenic. Originating in the mind.

psychosomatic. Of or pertaining to the mind and body, especially with respect to the influence of the mind or emotions upon the functioning of the organs and organ systems in health and disease.

psychrophil. An organism which grows best at a low temperature, as certain bacteria, fungi, and algae whose optimum temperature range is from 5° to 10°C.

ptarmigan. A species of grouse of the genus *Lagopus*, of mountainous regions of the northern hemisphere.

PTC. Phenylthiocarbamide, *q.v.*

pteridophyte. Any plant of the Pteridophyta, formerly a division of vascular plants which included the ferns, horsetails, and club mosses.

pterocarpous. With winged fruit.

pterocaulous. With a winged stem.

pterodactyl. A flying reptile of the Mesozoic era.

pteropod. An opisthobranch gastropod of the order Thecosomata, with winglike muscular flaps (parapodia) along the sides of the foot; a sea butterfly.

pteropsid. Any tracheophyte of the subphylum Pteropsida, which includes the ferns.

pterosaur. An extinct flying reptile of the order Pterosauria which includes the pterodactyls.

pterospermous. With winged seeds.

pterygiophore. A cartilaginous or bony structure in the fin of a fish.

pterygium. 1. The limb of a vertebrate. **2.** A lateral lobe at the tip of the snout in certain beetles.

[1]pterygoid. Winglike, as the *pterygoid* process of the sphenoid bone.

[2]pterygoid. A dermal bone in lower vertebrates forming a part of the upper jaw and roof of the mouth.

pterygopodial gland. A mucous gland located at the base of the clasping organ in skates and rays.

pterygote. Any insect of the subclass Pterygota which includes all winged insects, although wings are sometimes reduced or absent.

pteryla. A feather tract, an area on the skin of a bird from which contour feathers arise. *Compare* apterium.

pterylosis. The plumage of a bird, with special reference to the distribution of feathers.

ptilinum. A bladderlike structure at the anterior end of the pupa of many dipterous insects which functions in the rupture of the pupa case.

ptilopaedic. Covered with down at hatching, with reference to birds.

ptilosis. The plumage of birds.

ptomaine. An amino compound formed in the decomposition of protein, usually resulting from the action of putrefying bacteria.

ptosis. The abnormal drooping or lowering of an organ or structure, as the eyelid.

ptyalin. An amylase present in saliva which hydrolyzes starch to maltose and dextrins.

PU. Pregnancy urine, a source of hormones (estrogens and chorionic gonadotrophins).

puberty. The period in life when sexual maturity is attained, that is, when functional sperm and ova begin to be produced and, in the female, menstrual cycles occur.

pubes. 1. The hair covering the pubic region in man. **2.** The pubic region.

pubescence. 1. The state of being pubescent. **2.** The fine hairy covering of **a.** *bot.* the leaves and/or stems of plants; **b.** *zool.* the bodies of certain animals, as insects.

pubescent. 1. Covered with fine hair or down. **2.** Having reached the age of puberty.

pubic symphysis. The ventral union of the two pubic bones.

pubis pl. **pubes.** The os pubis or pubic bone, the most anterior or ventral of the three bones forming the os coxae (innominate or hipbone).

pudendum. The vulva or external genitalia of a female.

puerperal. Of or pertaining to childbirth or labor.

puff adder. 1. The hognose snake. **2.** A short, thick-bodied, short-tailed African viper (*Bitis arietans*).

puffball. A basidiomycete fungus of the order Lycoperdales, with a large, globose fruiting body, as those of the genus *Lycoperdon*.

puffbird. One of a number of South American birds of the family Bucconidae, with a large head and short tail, as *Notharcus*.

puffer. One of a number of marine fishes of the order Plectognathi which can inflate their body so that it assumes a spherical form, as the globefish (*Spheroides maculatus*). Also called blowfish.

437

puffin. A seabird of the family Alcidae, related to the auks and murres, as *Fratercula* and *Lunda*.

pullet. A young hen, especially one less than a year old.

pullulate. To propagate, especially to breed or multiply rapidly.

pullus. A young bird before acquiring its first complete plumage.

pulmocutaneous. Of or pertaining to the lungs and skin.

pulmonary. Of or pertaining to the lungs.

¹**pulmonate.** Possessing lungs or lunglike organs.

²**pulmonate.** Any mollusk of the Pulmonata, a subclass or order of the class Gastropoda which includes freshwater and terrestrial snails and slugs in which the mantle cavity functions as a respiratory chamber.

pulp. 1. *zool.* A soft, spongy tissue, as that which **a.** fills the pulp cavity of a tooth; **b.** forms the central substance of an intervertebral disk; **c.** forms the substance of the spleen. **2.** *bot.* **a.** The soft, succulent portion of a fruit. **b.** The pith of certain stems. **c.** A mixture of plant fibers from which paper is made.

pulsate. To beat, throb, or contract rhythmically.

pulsating vacuole. A contractile vacuole, as in protozoans.

¹**pulse.** The edible seeds of various legumes, as beans, peas, or the plants producing these seeds.

²**pulse. 1.** The regular, periodic increase in tension in the wall of an artery resulting from the beat of the heart. **2.** A repeated increase or decrease in a population, as a plankton *pulse*. **3.** A repeated increase in oxygen concentration, as occurs in certain bog lakes.

pulveraceous. Covered with dust or powder.

pulverulent. 1. Of a powdery nature. **2.** Appearing as though covered by dust or powder.

pulvilliform. Resembling a pulvillus; lobelike, padlike, cushionlike.

pulvillus pl. **pulvilli.** A pad or lobe underlying each tarsal claw, as in most dipterans.

pulvinate. 1. Shaped like or resembling a cushion. **2.** Having a pulvinus.

pulvinus. An enlarged or swollen region at the base of a leaf or leaflet which functions in leaf movements.

puma. *See* cougar.

pumpkin. A vine (*Cucurbita pepo*) or the large, orange-colored, gourdlike fruit it produces.

pumpkinseed. A freshwater sunfish (*Lepomis gibbosus*).

punctate. Marked by or bearing minute dots, depressions, pits, or punctures.

puncticulate. Minutely punctate.

punctiform. Having the form of a point or dot, with reference to colonies of bacteria.

pungent. 1. Ending in a sharp point; piercing. **2.** Acrid, with reference to odors.

punkie. A no-see-um. *See* midge, b.

pup. A young dog or seal; the young of most carnivores.

pupa. In insects with complete metamorphosis, a dormant, inactive stage between the larva and the adult or imago. It is usually enclosed within a protective structure as an earthen cell, cocoon, or puparium.

puparium. A hardened larval skin which encases a pupa, as in dipterans.

pupate. To transform into a pupa.

pupfish. A small, freshwater fish of the genus *Cyprinodon*, of the southern United States and Mexico.

pupil. The opening in the iris of the eye through which light enters.

pupiparous. Producing larvae which are nearly ready to pupate, as dipterans of the Pupipara, a group comprising the louse flies, bee lice, and bat ticks (flies).

pure. Free from impurities or contamination.

pure culture. *microbiol.* A culture which contains organisms of only one species.

pure line. *genetics.* **1.** All the progeny from a single homozygous individual reproducing asexually, as in the protozoa, or sexually as by self-fertilization in plants. **2.** A completely homozygous strain or race, as one resulting from intensive inbreeding.

purine. A substance, $C_5H_4N_4$, not occurring naturally but parent to many substances of biological significance, as uric acid, caffeine, theobromine, and adenine. *See* purine bases.

purine bases. Adenine and guanine, which in combination with a carbohydrate, pyrimidine bases, and phosphoric acid form nucleic acids. *See* deoxyribonucleic acid, ribonucleic acid.

Purkinje cells. Specialized neurons of the cerebellar cortex with a flaskshaped body and dendrites showing extensive arborization.

Purkinje fibers. Atypical muscle fibers found in the impulse-conducting system of the mammalian heart.

purulent. Consisting of, containing, or forming pus.

pus. A semifluid, yellowish substance consisting principally of serum, leukocytes, bacteria, and tissue debris usually the result of inflammatory processes.

pussy willow. A shrub (*Salix discolor*) of the family Salicaceae, with long, cylindrical, silky aments.

pustular. 1. *med.* Resembling or of the nature of pustules. **2.** *bot.* Having small elevations resembling blisters.

pustulate. PUSTULAR.

pustule. A small elevation of the skin containing pus; a pimple.

pusule. In dinoflagellates, a large vacuole opening by a canal into a flagellar pore and functioning as a digestive organelle.

putamen. 1. *anat.* The lateral portion of the lenticular nucleus of the cerebral hemisphere. **2.** *bot.* **a.** The shell of a nut. **b.** The hard, stony endocarp of a stone fruit, as a peach.

putrefy. To become putrid; to rot.

putrescent. Becoming putrid.

putrid. Rotten; decomposed; having a foul odor resulting from decomposition and decay.

pycnidium. 1. In lichens, a small cavity in which spores (spermatia) are produced. **2.** A flask-shaped fruiting body in certain ascomycetes and Fungi Imperfecti in which conidia (pycnidiospores) are produced.

pycniospore. 1. A spore produced by a pycnium, as in the wheat rust. **2.** A spermatium, a gamete of either plus or minus strain.

pycnium. A flask-shaped fruiting body in certain rusts which produces pycniospores; a spermagonium.

pycnogonid. An arthropod of the class Pycnogonida (Pantopoda) which includes the sea spiders.

pycnosis. The shrinkage and condensation of nuclear material into a dense, structureless mass, as occurs in degenerating or dying cells; pyknosis.

pygal. Of or pertaining to the posterior portion of the back.

pygal plate. The most posterior plate in the carapace of a turtle.

pygidium. The caudal, posterior portion of the body in various invertebrates, as the tagma of a trilobite, the terminal tergite of an insect, and the terminal segment of an annelid.

[1]**pygmy** also **pigmy. 1.** One of a race of dwarf people inhabiting central Africa. **2.** Any small person or dwarf.

[2]**pygmy.** Small or diminutive.

pygostyle. A bone at the caudal end of the vertebral column in birds which supports the tail feathers; consists of 6-10 fused vertebrae.

pyknosis. *See* pycnosis.

pylangium. The caudal portion of the conus arteriosus in an amphibian heart.

pyloric. Of or pertaining to the pylorus.

pyloric sphincter. A ring of circular muscles which encircles the alimentary canal at the pylorus and regulates the passage of food from stomach to the intestine. Also called pyloric valve.

pylorus. In vertebrates, **a.** The opening from the stomach to the duodenum; **b.** the mucous membrane and muscle tissue (pyloric sphincter) which surround the opening of stomach to duodenum.

pyogenic. Of or pertaining to the formation of pus.

pyramid. A cone-shaped structure, as **a.** the *pyramids* of the medulla oblongata containing efferent fibers from the cerebral cortex; **b.** the *pyramids* of the kidney containing collecting ducts; **c.** one of the five calcareous plates comprising an Aristotle's lantern.

pyramidal. Having the form of a pyramid.

pyramid shell. A small, cone-shaped gastropod of the family Pyramidellidae.

pyrene. A nutlet, especially the stone of a drupelet, as that of the fruit of the huckleberry.

pyrenoid. A highly refractive protein body present in the chloroplasts of lower organisms, as algae. It serves as a center for the deposition of starch.

pyrethrum. 1. One of a number of chrysanthemums cultivated as a source of insecticides. **2.** The dried flower heads of these plants, used in the manufacture of insecticides.

pyridoxine. Vitamin B_6, present in yeast, whole grain cereals, and liver; a deficiency results in dermatitis, heart disorders, and various degenerative changes.

pyriform. Pear-shaped.

pyrimidine base. A nitrogenous base present in nucleotides, as cystosine, urocil, thymine. *See* nucleoside.

pyruvic acid. An organic acid, $CH_3COCOOH$, which occupies a key position in carbohydrate metabolism, a number of reactions being involved in its formation and its breakdown. *See* Kreb's cycle.

python. One of a number on nonvenomous constrictors of the Old World of the genus *Python*, family Pythonidae; includes the largest of living snakes, some reaching a length of 30 feet.

pyxidate. Having a lid or covering, said of seed capsules.

pyxis. A capsule with circumscissile dehiscence, the upper portion acting as a lid separating during dehiscence, as in plantain. Also called pyxidium.

Q

Q fever. A tick-transmitted, rickettsial disease caused by *Coxiella burnetii*, affecting wild and domestic animals and man.

quadrat. An area of vegetation, usually one square meter chosen at random and compared with a similar area in another region.

quadrate. Square or four-sided.

quadrate bone. In most vertebrates, the ossified posterior end of the pterygoquadrate cartilage of the skull with which the lower jaw articulates. In mammals, it gives rise to the incus.

quadriceps femoris. In man, the large muscle comprising the rectus femoris, vastus lateralis, vastus medialis, and vastus intermedius which forms the anterior portion of the thigh. It extends the leg and flexes the femur.

quadruped. A four-footed animal.

quadruplet. One of a set or group of four of a kind, especially one of four individuals which develop from a single pregnancy.

quagga. A South African wild ass (*Equus quagga*) related to the zebra; now considered extinct.

quahaug. *See* quahog.

quahog. A hard-shelled clam (*Venus mercenaria*) of the East coast.

quail. 1. An Old World, migratory, gallinaceous game bird (*Coturnix coturnix*). **2.** A gallinaceous game bird of North America, especially *Colinus virginianus*, the bobwhite, and *Lophortyx californica*, the California valley quail.

qualitative inheritance. Inheritance involving characters which can be sharply distinguished from each other.

quantitative inheritance. Inheritance involving characters such as size, weight, height, degree of pigmentation, etc., in which there are varying degrees of phenotypic expression.

Quaternary period. A geological period which includes both the Pleistocene and Recent periods comprising the second portion of the Cenozoic era, characterized by the rise of man and modern mammals.

queen. The fully developed, egg-laying female of a social insect.

quetzal. A large, Central American trogon (*Pharomacrus mocinno*) with brilliantly colored plumage and long tail coverts.

quill. 1. The proximal, hollow portion of the shaft of a feather; calamus. **2.** A feather, especially one of the large feathers of the wing or tail. **3.** A stiff spine of a hedgehog or porcupine.

quilled. Designating a ligulate floret which has become tubular.

quillwort. A lycopod of the order Isoetales.

quinary. In fives; quintuple.

quinate. Arranged in groups or sets of five.

quince. A rosaceous tree (*Cydonia oblonga*) or the large, round, acid fruit it produces.

quinine. A bitter alkaloid obtained from the bark of the cinchona tree (*Cinchona*), used medicinally as an antimalarial agent.

quinnat salmon. The chinook, king, or tyee salmon (*Oncorhynchus tschawytscha*), an important commercial species of the Pacific coast.

quinoa. A South American herb (*Chenopodium quinoa*) whose seeds are widely used for food.

quintan. Occurring every fifth day, as a fever.

R

rabbit. A small, leaping mammal of the order Lagomorpha, with long ears and a short tail. *See* snowshoe rabbit, jackrabbit, swamp rabbit, cottontail.

rabbit fever. TULAREMIA.

rabies. An acute, infectious disease of animals, especially carnivores, caused by a neurotropic virus, transmissible to man. Also called hydrophobia.

raccoon. A small, carnivorous mammal (*Procyon lotor*) of the family Procyonidae, commonly called coon; an important fur-bearing and game mammal.

race. A subdivision of a species.

raceme. A simple, elongated, indeterminate inflorescence in which stalked flowers are borne on a central axis, the lowermost having longer pedicels and maturing before those nearest the apex, as the mustard flower.

racemose. 1. *bot.* Pertaining to or of the nature of a raceme. **2.** *anat.* Resembling a bunch of grapes, as a *racemose* gland.

rachilla. *bot.* A small, secondary axis, as in the spikelet of a grass.

rachis. 1. *anat.* The spinal column. **2.** *zool.* The shaft of a feather **3.** *bot.* The main stalk or axis of a structure, as an inflorescence.

radial. 1. Extending outward from a central axis. **2.** *anat.* Pertaining to the radius, a bone of the forearm. **3.** *zool.* Pertaining to a ray, as in echinoderms.

radial canal. 1. A canal extending outward from a central cavity, as in sponges or medusae. **2.** In sea stars, a canal of the water-vascular system which extends peripherally in each ray.

radiale. A cartilage or bone of the carpus which articulates with the radius; in man, the navicular bone.

radial symmetry. The type of symmetry in which a number of similar parts radiate from or are uniformly distributed about a central axis.

radiant. 1. Shining, brilliant, glowing. **2.** Emitting rays or electromagnetic waves. **3.** *bot.* Diverging from a central axis.

radiant energy. Energy emitted and propagated in the form of electromagnetic waves, as long and short radio waves, infrared rays, visible light rays, ultraviolet rays, roentgen rays, and gamma rays.

radiate. 1. To emit rays, as heat or light rays; to shine. **2.** To diverge from a central point or region. **3.** Bearing ray flowers, as in composites.

radiation. The act or process of radiating. *See* adaptive radiation.

¹radical. *bot.* Pertaining to or proceeding from the root or basal portion of a plant.

²radical. *chem.* A group of atoms which acts as a unit in chemical reactions, as the monovalent hydroxyl radical, -OH.

radicant. Rooting, with reference to roots developing from stems or leaves.

radicle. 1. *anat.* A small root, as that of a nerve. **2.** *bot.* The portion of a plant embryo which develops from the primary root.

radioactive. Capable of emitting radiant energy, especially the spontaneous emission of particles (alpha, beta) and gamma rays from the nucleus of an atom.

radioactive carbon. Carbon 14 (C^{14}), present in the atmosphere in small amounts and incorporated into carbon compounds in photosynthesis. On the basis of its half life of 5568 years, the age of a substance containing it can be estimated, hence it is referred to as a "carbon clock."

radioautograph. A photograph taken by radiation from a radioactive substance, used in determining the presence and location of radioactive substances in tissues. Also called autoradiograph.

radiobiology. That division of biology which deals with the effects of radiation on living matter.

radiogenetics. That branch of biology which deals with the effects of radiation upon heredity.

radiolarian. A marine protozoan of the order Radiolaria (Radiolarida) with a body consisting of a central capsule and possessing a skeleton of silica or strontium sulfate; many possess axopodia.

radiolarian ooze. A soft deposit covering vast areas of the ocean floor, consisting principally of the skeletons of radiolarians.

radiole. A pinnate process on the crown of certain polychaetes, as fanworms.

radioreceptor. A receptor which is stimulated by radiant energy, as the eye.

radioulna. The fused radius and ulna, as in the frog.

radish. The fleshy root of the radish plant (*Raphanus sativus*) or the plant which produces it, of the family Cruciferae.

radium. A highly reactive, radioactive element, symbol Ra, at. no. 88.

radius. 1. The more anterior of the two bones in the forelimb of a tetrapod; in man, the bone lying lateral to the ulna. **2.** One of the radial members of a radially symmetrical structure or organism. **3.** The third longitudinal vein in the wing of an insect.

radix pl. **radices.** A root, as **a.** the root of a plant; **b.** the root of a hair.

radula. A ribbon- or platelike structure bearing rows of chitinous teeth present in the floor of the pharynx of most mollusks except bivalves. It functions as a rasping and chewing organ.

radula sac. A ventral diverticulum of the buccal cavity in gastropods in which rows of radular teeth are formed, replacing those which are lost.

radula tooth. A long, hollow tooth of the radula of certain gastropods which functions as a poisonous, harpoonlike structure, as in cone shells.

ragi. A tall grass (*Eleusine corocana*), an important food crop of southeast Asia; raggee, raggi, raggy. Also called korakan, finger millet, African millet.

ragweed. One of several composites of the genus *Ambrosia*, its pollen being an important antigen and a common cause of hay fever.

rail. A plump, chickenlike, marsh bird of the family Rallidae, as *Rallus elegans*, the king rail.

rain tree. A huge tree (*Samanea saman*) of tropical America which produces edible pods used for stock food.

raisin. A special type of dried grape with a high sugar content.

ram. A male sheep or goat.

ramal. Of or pertaining to a ramus or a branch.

rambler. A plant which clambers, especially a climbing rose.

ramentum. One of the numerous, thin, hairlike scales borne on the leaves and young shoots of ferns.

ramet. A single member of a clone.

rami. Plural of RAMUS.

ramiform. Like a branch; branching.

ramify. To branch; to spread out.

ramose. Branched; having many branches.

ramulose. Having many branches, especially small branches.

ramus pl. **rami.** 1. A branch of an artery, vein, or nerve. 2. A portion of a bone which projects from the body, as the *ramus* of the mandible.

range. 1. In statistics, the difference between the lowest and the highest in a set of values. 2. A series of things in a row, as a mountain *range*. 3. The area inhabited by a group of organisms.

rape. 1. A herbaceous plant (*Brassica napus*) grown principally as a forage crop. 2. The pomace of grapes after expression of the juice.

raphe. 1. *anat.* A seamlike fusion or union of the two halves of a structure, usually marked externally by a ridge or groove, as the median *raphe* of the scrotum. 2. *bot.* **a.** A longitudinal groove or cleft in the valve of a diatom. **b.** A longitudinal ridge on a seed which develops in an anatropous ovule, marking fusion of the funiculus and integument, as in the castor bean.

raphides sing. **raphid.** Minute crystals, usually of calcium oxalate, formed within plant cells, considered to be metabolic waste products.

raptorial. 1. Living on prey, with special reference to certain birds. 2. Pertaining to birds of prey; adapted for seizing prey. 3. Having feet adapted for seizing and tearing prey.

rasorial. Scratching the ground in search of food, as by gallinaceous birds.

raspberry. A berry (drupelet) produced by a rosaceous shrub (*Rubus*) or the shrub which produces it.

rassenkreis. German for *race-circle*, a species consisting of a series of subspecies which may interbreed, but with species at the two ends of the series differing to such an extent that they cannot interbreed.

447

rastellus. A row of short spines on the basal segment of the chelicera of a burrowing spider.

rat. 1. One of a number of long-tailed rodents of the genus *Rattus* and related genera of the family Muridae, as *R. norvegicus*, the brown or Norway rat. **2.** One of a number of similar rodents, as the cotton rat, muskrat, wood rat.

ratel. The honey badger (*Mellivora capensis*) of Africa and southern Asia; called grave-digger in India.

rat flea. A parasitic flea (*Xenopsylla cheopis*) which transmits the causative organisms of plague and murine typhus.

Rathke's pouch. A dorsal outpocketing of the stomodeum which gives rise to the anterior and intermediate lobes of the hypophysis.

ratite. 1. Having a flat, keelless sternum. *Compare* carinate. **2.** One of several flightless birds with poorly developed wings and an unkeeled sternum, as the ostrich, rhea, cassowary, emu, and the extinct moas.

ratoon. A secondary shoot which sprouts from the root of a perennial plant, as in the second year's growth of sugar cane.

rat-tailed maggot. The aquatic larva of flies of the genus *Eristalis*, family Syrphidae, which possess a long, extensible breathing tube.

rattan. A climbing palm of the genus *Calamus* or material obtained from its stem, used extensively in wickerwork.

rattle. A sound-producing structure on the end of the tail of a rattlesnake, composed of several loosely interlocking segments,

rattler. A rattlesnake.

rattlesnake. One of a number of sluggish, thick-bodied, venomous snakes of the family Crotalidae, characterized by possession of a rattle. They comprise two genera: *Sistrurus* (massasaugas and ground rattlers) and *Crotalus* (the timber, diamondback, and prairie rattlesnakes).

raven. A crowlike bird (*Corvus corax*) of the Northern Hemisphere.

¹ray. 1. A beam of electromagnetic radiations, as a light *ray*, roentgen *ray*. **2.** *bot.* **a.** A ray flower. **b.** A branch of an umbel, corymb, or similar inflorescence. **c.** A radiating, linear group of cells, as a pith *ray*. **3.** *zool.* **a.** A primary division of a radiate animal. **b.** A slender cartilaginous or bony supporting struc-

448

ture, as a fin *ray*. **c.** An extension of the medulla of the kidney into the cortex, forming a medullary *ray*.

²**ray.** A flat-bodied elasmobranch fish with large pectoral fins and slender, whiplike caudal region, as *Raja*, the common *ray* or skate. *See* electric ray, stingray.

ray-finned fish. A bony fish of the subclass Actinopterygii with fins supported by fin rays.

ray flower. 1. An outer or marginal flower or floret in the head of a composite. *Compare* disk flower. **2.** An entire head which lacks disk flowers.

reaction. 1. *chem.* **a.** The interaction between two or more substances. **b.** Any chemical change. **2.** *bacteriol.* The state of the medium with reference to pH (hydrogen ion concentration). **3.** *biol.* A response to a stimulus.

realm. One of the six major faunal areas of the globe, namely Nearctic, Neotropical, Palearctic, Ethiopian, Oriental, and Australian.

recapitulation. *biol.* The appearance in the development of an individual of transitory or temporary structures which resemble or correspond to functional structures possessed by ancestors.

recapitulation theory. The biogenetic law, a principle that an individual in its development recapitulates or goes through stages which the species has gone through in its evolutionary development; sometimes expressed as "ontogeny recapitulates phylogeny."

Recent epoch. An epoch in the Quaternary period following the Pleistocene extending from the end of the last ice age; the Age of Man.

receptacle. 1. *zool.* A structure which receives and retains something, as a seminal *receptacle*. **2.** *bot.* **a.** In rockweeds, the swollen tips of branches bearing sex organs. **b.** In liverworts, a disc-shaped structure which bears the archegonium or antheridium. **c.** In seed plants, the tip of a peduncle or pedicel on or around which floral parts develop; in a composite, the structure to which flowers of the head are attached.

receptor. 1. *zool.* and *anat.* A structure which responds to a stimulus, as **a.** a sensory end organ or ending of a peripheral nerve; **b.** a neuroepithelial cell, as an olfactory cell; **c.** a part of a sense organ which contains receptor cells, as the organ of

Corti of the ear; **d.** a special sense organ, as the eye or ear.
2. *bacteriol.* A part of a cell which combines with the antigen to
form an antibody.

recess. A depression, fossa, cleft, or cavity.

¹recessive. Hidden, not obvious.

²recessive. 1. A recessive character. **2.** An individual exhibiting
a recessive trait or traits.

recessive character. In genetics, a trait which, in the presence of
a dominant gene, fails to express itself; one which appears only
in a homozygous individual or one lacking the dominant gene.

recessive gene. One which can produce its effects only in a
homozygous individual or in the absence of a dominant gene.

reclinate. Bent downward.

reclining. *bot.* Bending down; turning away from the perpendicular.

recombination. In genetics, the appearance in offspring of new
combinations of traits not exhibited by the parents.

rectal. Of or pertaining to the rectum.

rectal gills. Tracheal gills located on the internal surface of the
rectum, as in the nymphs of dragonflies.

rectal gland. A gland in elasmobranch fishes which opens into
the large intestine; it functions in the excretion of salt.

rectrices. Plural of RECTRIX.

rectrix pl. **rectrices.** An elongated tail feather of a bird.

rectum. The portion of the large intestine extending **a.** in insects,
from colon to the anus; **b.** in amphibians, from the small
intestine to the cloaca; **c.** in mammals, from the colon to the
anal canal.

red algae. Algae of the division Rhodophyta (Rhodophyceae)
with accessory phycobilins in addition to caretenoids and
chlorophyll and food reserves in the form of floridian starch.

redbird. 1. The cardinal (*Richmondena cardinalis*), of the family
Fringillidae. **2.** Any of a number of birds predominately red in
color, as the scarlet tanager.

red body. A vascular body in the inner surface of the swim
bladder of certain fishes, as the eel.

redbud. The Judas tree (*Cercis canadensis*), a legume.

red bug. A harvest mite or chigger. *See* chigger.

red corpuscle. An erythrocyte.

redd. A spawning area or nest prepared by certain fishes for the deposition of eggs.

red deer. 1. The common deer of Europe (*Cervus elaphus*). **2.** The white-tailed deer in its summer coat.

redhead. A diving duck (*Aythya americana*) with a round, red-brown head.

redhorse. A sucker of the genus *Moxostoma*, a coarse food fish of North America.

Redi, Francesco. An Italian naturalist, 1626–1697, noted for his experiments disproving the theory of the spontaneous generation of life.

redia. The second larval stage in the development of a digenetic trematode found in snails or bivalves. Redia arise from sporocysts and in turn give rise to daughter rediae or cercariae.

redox reaction. A chemical reaction involving both reduction and oxidation.

Red Poll. One of a breed of hornless, dual-purpose cattle of Great Britain.

redpoll. A finch of the genus *Acanthis*, of northern regions.

red snow. Snow in arctic or mountainous regions with a pinkish coloration due to multiplication of algae or protozoans, as various hematochrome-bearing Phytomastigina.

redstart. A small wood warbler (*Setophaga ruticilla*) with flight resembling a butterfly.

red tide. The recurrence of enormous numbers of dinoflagellates, especially *Gonyaulax* and *Gymnodinium*, in waters off the coasts of Florida and California, resulting in reddish hue of waters by day and luminescence by night.

reducer. *See* decomposer.

reductase. An enzyme which catalyzes a reduction reaction.

reduction. 1. *biol.* The reduction in number of chromosomes from diploid to haploid number, as occurs in meiosis. **2.** *chem.* A chemical reaction in which one or more electrons are gained by the substance reduced; the addition of hydrogen atoms or the loss of oxygen atoms. *Compare* oxidation.

red water. 1. Water containing large numbers of organisms containing a red pigment. *See* red tide. **2.** Urine containing blood; hematuria.

red water fever. TEXAS FEVER.

redwood. *See* sequoia.

reed. One of a number of tall, bamboolike grasses or their slender, jointed stems.

reedbird. *See* bobolink.

reedbuck. An African antelope of the genus *Redunca*.

reef. A ridge of rocks, coral, or sand close to the surface of water. *See* coral reef.

reeve. A female sandpiper (*Philomachus pugnax*). *See* ruff.

reflex. A reflex action.

reflex action. An involuntary, stereotyped, and usually purposeful response or reaction resulting from nervous impulses initiated by a stimulus, as the knee jerk.

reflex arc. The nervous pathway from point of stimulation to responding organ, comprising a receptor, afferent neuron, reflex center in brain or spinal cord, efferent neuron, and an effector organ.

reflexed. Bent or turned downward or backward; turned in the opposite direction.

reforest. To reestablish forests on land from which timber has been removed, as by fire or cutting.

refract. To bend; to subject to refraction as light rays.

refraction. The bending or deflection from a straight course of rays of light, heat, or sound, as occurs when such pass from a medium of one type into that of another, as *refraction* of light rays by a lens.

refractive. Having the power to refract.

refractory. 1. Obstinate; intractable; tending to resist the effects of; incapable of responding to stimuli. 2. *med.* Failing to respond to treatment.

refractory period. The period following stimulation during which a muscle or nerve fiber is incapable of responding.

refugium pl. **refugia.** An area, usually more or less isolated, which, with respect to fauna and flora, has remained relatively unchanged in contrast to surrounding areas which have been markedly affected by environmental changes; a refuge.

regeneration. 1. In higher organisms, the replacement of lost tissues or structures as occurs in the repair of wounds or the replacement of parts, as a claw, feather, tail, or appendage. 2. In invertebrates and lower plants, the restoration of a complete functional individual from a part other than a reproductive cell or structure.

region. A division of a zoological realm characterized by a distinctive type of fauna or flora.

regma. A form of schizocarp in which segments open elastically.

regress. To go backward; to return to a former state or condition.

regular. 1. According to custom, habit, or rule. 2. Uniform in size, shape, or structure. 3. *bot.* Having members of a floral whorl, especially the corolla, of similar shape and equally spaced about the center of the flower.

regulative development. That which occurs in eggs with indeterminate cleavage in which loss of one or more blastomeres does not adversely affect the development of the embryo, organ-specific regions being established fairly late, as in mammalian ova.

regurgitation. 1. The casting out of undigested or partially digested food from the stomach. 2. The backward flow of blood through a defective heart valve.

reindeer. One of the domesticated strains of *Rangifer*, especially *R. tarandus* of the Northern Hemisphere.

reindeer moss. A lichen (*Cladonia rangiferina*) of northern regions.

rejuvenation. The restoration of youth, strength, and vigor; making young again.

relaxin. An ovarian hormone which, in certain pregnant mammals, relaxes joints of the pelvic girdle and dilates the uterine cervix.

release factor. A substance produced by neurosecretory cells of the hypothalamus which regulates the production and release of hormones by the anterior pituitary.

relic. A survivor or vestige; a relict.

relict. That left over as a survivor, as **a.** a survivor or remnant of a once successful and flourishing group, as *Sphenodon, Psilotum;* **b.** a group existing in a local area widely separated from others of the same or closely related species.

remex pl. **remiges.** A flight feather of a bird, including primaries, those attached to the manus, and secondaries, those attached to the ulna.

remicle. The reduced outermost primary.

remiges. Plural of REMEX.

remora. A shark sucker or suckerfish with a modified dorsal fin by which it attaches itself to sharks and other fishes. Common genera are *Remora* and *Echeneis* of the order Discocephali.

renal. Of or pertaining to the kidney.

renal corpuscle. A filtering structure in the kidney of amniotes, consisting of a glomerulus surrounded by Bowman's capsule; Malpighian corpuscle.

renal portal vein. A vein which carries blood from the posterior portion of the body to the kidney, as in fishes, amphibians, reptiles, and birds.

renal tubule. A nephron.

renette. A large gland cell in nematodes opening to the outside, functioning as an excretory structure.

reniform. Shaped like a kidney.

renin. An enzyme produced in the kidney which, when released into the bloodstream, acts on a blood protein with the resultant production of angiotensin, a pressor substance which induces vasoconstriction.

rennet. 1. A substance containing rennin, obtained from the lining of a calf's stomach. **2.** The abomasum, the fourth chamber in the stomach of a ruminant.

rennin. An enzyme present in the gastric juice of a calf which coagulates milk; absent in the gastric juice of adult humans.

reovirus. One of a group of orphan viruses found in the respiratory and digestive tracts of man.

repand. Having a wrinkled, wavy, or undulated margin, as a leaf or bacterial colony.

repellent. An agent used to repel noxious organisms as insects, ticks, and mites.

repent. Creeping, prostrate, with roots at the internodes.

replete. *See* honey ant.

repletum. A fruit whose valves remain attached together by threads following dehiscence, as in orchids.

¹replicate. To duplicate or reproduce.

²replicate. Folded over backwards upon itself.

replication. The act of replicating; the doubling of genetic material (DNA) at each cell division.

replum. The framework of the placenta which remains after the valves fall away in dehiscence, as in plants of the mustard family.

reproduce. 1. To produce again. **2.** To bring into existence another of the same kind or an individual that will give rise to one of the same kind.

reproductive organ. 1. An organ which functions in the process of reproduction. **2.** *zool.* The primary *reproductive organs* are the gonads (testes and ovaries); secondary or accessory reproductive organs include glands, ducts, storage reservoirs, and copulatory structures which function in reproductive activities or development of the young. **3.** *bot.* In sexual reproduction, the principal organs are gametangia (antheridia and archegonia); in asexual reproduction, sporangia.

reproductive system. All the organs concerned with the production of gametes, the bringing of the gametes together (fertilization), and the development of the young.

reptile. Any member of the class Reptilia, comprising poikilothermous vertebrates with four pentadactyl limbs and body usually covered with scales or scutes. Living members include snakes, lizards, turtles, crocodiles, and the tuatara; fossil reptiles include the dinosaurs, ichthyosaurs, and pterodactyls abundant during the Mesozoic era.

repugnatorial gland. A gland in an animal which produces a substance which has repellent properties, as the stink gland of hemipterans.

repulsion. In genetics, the tendency of linked genes to separate in the formation of gametes; the opposite of coupling.

resect. To cut away.

resection. Surgical removal of a portion of an organ.

reservoir. A cavity where something is stored or kept in reserve; a cisterna.

reservoir host. In parasitology, a species which serves as a host of an organism which is pathogenic in or normally inhabits individuals of other species, as wild animals in Africa serving as a *reservoir host* for human trypanosomes.

residual. That pertaining to a residue or that left behind.

residue. That which remains after a part has been taken away; the remainder.

resin. A solid or semisolid exudate of a plant, either natural or induced, usually soluble in alcohol or ether but insoluble in water. Resins are usually oxidized terpenes of volatile oils.

resin duct. A duct in a stem or leaf of a conifer which is involved in the secretion and conduction of resins.

resinous. Resembling or having the properties of a resin.

resistance. Opposition to a force or action; the capacity to resist the action or effects of a certain activity, as *resistance* to disease.

resorb. To take in again.

resorption. 1. The act of resorbing. 2. The removal of tissues by absorption, as *resorption* of the tail of a tadpole during metamorphosis.

respiration. 1. The interchange of gases between an organism and its environment. 2. The act of breathing or drawing in and expelling air from a cavity, as the lungs; inspiration and expiration; inhaling and exhaling; such comprising *external respiration*. 3. The liberation of energy within, and its utilization by, a cell, whether occurring aerobically or anaerobically, such comprising *internal respiration*.

respiratory. Of or pertaining to respiration.

respiratory organ. An organ or structure which functions in gaseous exchange, as the integument, gills, trachea, tracheal gills, diffusion lungs (book lungs of arachnids, mantle cavity of gastropods), gas bladders and lungs of fishes, alveolar lungs of mammals, and various specialized structures.

respiratory pigment. 1. A pigment which has the ability to transport oxygen and carbon dioxide, as hemoglobin, hemocyanin, chlorocruorin, hemerythrin. 2. A blood pigment.

respiratory quotient. The ratio of CO_2 discharged to the volume of O_2 taken in, obtained by dividing the volume of CO_2 exhaled by the volume of O_2 inhaled.

respiratory tree. 1. In holothuroids, a treelike structure attached to the cloaca into which water can be pumped. 2. In mammals, the trachea, bronchi, and bronchioles taken as a whole.

respiratory tube. 1. A tube leading to a respiratory organ, as in various arthropods. 2. In lampreys, the tube through which water passes from the pharynx to the gill sacs.

response. An activity initiated by or resulting from a stimulus, as in animals, movement, secretion, conduction; in plants, various tropisms, movement, etc.

¹rest. *physiol.* To cease activity; repose, sleep, quiescence.

²rest. *anat.* A mass of epithelial cells which persists after cessation of development; called fetal or embryonic *rest.*

resting cell. 1. A cell not in active division; one that is in the interphase between two mitotic divisions. 2. *bacteriol.* a. A cell freed from the medium upon which it was grown. b. An inactive cell; one not multiplying.

resting egg. In rotifers and other invertebrates, a thick-walled, haploid egg which is resistant to unfavorable conditions and usually requires a long period for development.

resupinate. Inverted; turned in a direction opposite from normal; upside down.

resurrection plant. A club moss (*Selaginella lepidophylla*).

ret. *See* retting.

rete pl. retia. An interlacing network or plexus of vessels or nerve fibers.

rete mirabile pl. retia mirabilia. A small, dense network of blood vessels, as that present in the red body of a teleost fish.

retia. Plural of RETE.

reticular. In the form of a network.

reticular connective tissue. A type of connective tissue consisting of reticular fibers enclosing primitive reticular cells and macrophages, as that comprising lymphatic tissue and organs.

reticular formation. A structure in the brainstem consisting of a network of nerve cells and their fibers which monitors incoming sensory impulses and controls various muscular and glandular responses.

reticulate. 1. Resembling or having the form of a network. 2. *bot* Net-veined, with reference to veins in a leaf.

reticuloendothelial system. Collectively, the fixed phagocytic cells of the body identified by their ability to take up certain colloidal dyes, as trypan blue. It includes macrophages, histiocytes, Kupffer's cells, dust cells, and reticular cells of lymphatic tissue.

reticulopodium. A branching and anastomosing rhizopodium, as in foraminiferans.

reticulum. 1. A netlike structure. 2. A delicate interlacing network of fibers. **3.** The second division of the stomach of a ruminant; honeycomb.

retina. 1. The innermost, nervous layer of the vertebrate eye upon which the image is focused; contains rods and cones, the sensory receptors for light. **2.** A similar layer in the eye of a cephalopod. **3.** A structure containing sensory visual cells in the ocelli of various arthropods.

retinaculum. 1. *zool.* A retaining band or structure, as the curved setae on the forewing of lepidopterans, the carpal ligament in man, or the hamula or tenaculum of a springtail. **2.** *bot.* **a.** The structure in orchids or milkweeds to which pollinia are attached. **b.** A curved, hooklike funicle, as in certain fruits.

retinene. A decomposition product of rhodopsin formed in the retina when rhodopsin in the rods is exposed to light.

retinula. A group of elongated sensory cells in the ommatidia or ocelli of insects.

retort cells. Elongated, flask-shaped, hyaline cells, each with an apical pore, in cortex of sphagnum mosses.

retract. To withdraw; to draw back in. *Compare* protraction.

retreat. A hiding place, den, or other structure into which an animal can withdraw for protection.

retreat makers. Caddis flies.

retrofection. Reinfection, with reference to the reentry of eggs and larvae of pinworms.

retrograde. Characterized by retrogression. *See* retrogress.

retrogress. To go backward; to proceed from a higher to a lower state of development or from a specialized condition to a generalized.

retroperitoneal. Behind the peritoneum.

retropharyngeal. Behind the pharynx.

retrorse. Directed or turned backward or downward. *Compare* antrorse.

retroversion. State of being turned backward or downward.

retting. A process employed in the preparation of flax, involving soaking followed by rotting and decomposition during which fibers are freed.

retuse. Having a rounded or obtuse apex with a shallow notch, with reference to leaves.

reversible. Capable of going in the opposite direction.

reversion. 1. The appearance of a hereditary trait which resembles that of a grandparent or a more remote ancestor. **2.** A backward change in which an organism assumes a condition or takes to an environment occupied by its ancestors, as a land mammal reverting to an aquatic life.

revolute. 1. *bot.* Rolled backward from the apex or with margin turned toward the lower side, with reference to vernation. **2.** *zool.* Turned backward or downward.

rhabd. A monaxon spicule of a sponge.

rhabdite. A rod-shaped rhabdoid present in the epidermis of turbellarians; also present in some nemerteans.

rhabditiform larva. A type of nematode larva in which the esophagus has a thick anterior portion connected by a narrow isthmus to a posterior bulb, as in the hookworm.

rhabdocoel. A turbellarian flatworm with a simple, unbranched digestive cavity.

¹rhabdoid. Rodlike.

²rhabdoid. A rod-shaped structure present in the epidermis and sometimes the mesenchyme of a turbellarian flatworm.

rhabdome. A rod-shaped refractive body in the midaxis of an ommatidium or in the retina of an ocellus, each surrounded by retinacular cells.

rhabdomere. A component of a rhabdome.

rhachis. *See* rachis.

rhagon. A type of sponge with a broad base and tapering to a tip which bears an osculum; each possesses a leuconoid type of canal system.

rhammite. A long, slender, and sometimes sinuous rhabdoid.

rhamphotheca. The covering of the entire bill of a bird.

rhea. A large ratite bird of the genera *Rhea* and *Pterocnemia* of South America resembling an ostrich but smaller, with three toes, and a fully feathered neck; hunted for sport.

rhebok. A South African antelope (*Pelea capreolus*).

rheobase. The minimal or liminal electric stimulus which will induce a response.

rheoreceptor. A sense organ which is stimulated by currents or movement of water. *See* lateral line system.

rheotaxis. The reaction of an organism in response to a moving current of water.

459

rhesus monkey. *See* macaque.

Rh factor. One of several antigens naturally present in or on erythrocytes, responsible for certain adverse reactions which occur following the injection of anti-Rh serum; originally found in the rhesus monkey but present also in man.

rhinal. Of or pertaining to the nose.

rhinencephalon. The portion of the vertebrate brain concerned with olfactory sensations. It includes the olfactory bulb and tract, hippocampal formation, and associated structures.

rhinoceros. A large, heavy-bodied, thick-skinned perissodactyl mammal of the genera *Rhinoceros*, inhabiting Asia, and *Diceros*, inhabiting Africa.

rhinoceros beetle. A large scarab beetle (*Dynastes*) with a large, prominent horn on the head or pronotum.

rhinophore. One of the second pair of tentacles of a nudibranch mollusk.

rhinotheca. The sheath or covering of the upper mandible (maxilla) of a bird.

rhipidium. A fan-shaped cyme.

rhizanthous. Flowering or apparently flowering from the root.

rhizine. A bundle of hyphae by which a lichen is attached to its substrate.

rhizocaline. A substance thought to be present in stems which, interacting with auxin, induces the development of roots.

rhizocarpous. Having a perennial root but an annual stem.

rhizocephalan. A degenerate, parasitic crustacean of the order Rhizocephala, as *Sacculina* which lives in and on crabs.

rhizogenic. Root-producing, as certain cells.

¹rhizoid. Rootlike, with reference to bacterial colonies.

²rhizoid. A slender, rootlike filament which attaches the mycelium of some fungi or the gametophyte of mosses, liverworts, and ferns to the substrate.

rhizom. The nutritive network which develops from the kentrogen larva of *Sacculina* within a crab.

rhizome. 1. *bot.* A rootstock, a rootlike stem which grows on or under the ground producing stems and roots. 2. *zool.* A stolon.

rhizomorph. A fine, rootlike strand of hyphae which attaches the mycelium of various fungi to its substrate.

rhizophore. A branch of a stem in a club moss which grows downward into the ground and develops roots at its tip.

rhizoplast. In flagellates, a threadlike structure which connects the basal granule to the nucleus.

rhizopod. 1. Any protozoan of the class Sarcodina (*Rhizopoda*). **2.** A rhizopodium.

rhizopodium. A branching, filamentous, and anastomosing pseudopodium, as in various foraminiferans.

rhizosphere. The region immediately surrounding the roots of a plant.

rhodophyte. Any member of the division Rhodophyta which comprises the red algae.

rhodopsin. Visual purple, a pigment present in rod cells, essential for vision in dim light. Upon exposure to light it is bleached, dissociating into retinene and opsin.

rhombencephalon. The hindbrain.

rhombogen. A stage in the development of certain mesozoans which follows the nematogen phase, as in dicyemids.

rhomboid. Having the shape of a rhomboid, a four-sided figure with opposite sides equal and parallel but two of the angles oblique.

rhopalium. A club-shaped structure bearing sense organs on the margin of the bell in scyphozoan jellyfishes.

rhubarb. A common garden plant of the genus *Rheum*, grown for its succulent, acid leafstalks.

rhynchocephalian. A primitive reptile of the order Rhynchocephalia, comprising a single living species, *Sphenodon punctatum*, the tuatara of New Zealand.

rhynchocoel. A closed, fluid-filled cavity in nemerteans which contains the proboscis.

rhynchus. The oral sucker of a digenetic trematode.

rhythm. A sequence of events or activities repeated at regular intervals.

rhytidome. The outer bark.

rib. 1. *anat.* One of a series of curved, elongated, cartilaginous or bony structures in a vertebrate, the proximal ends of which are attached to the vertebral column; distal ends may be free or attached to the sternum. **2.** *zool.* See costa. **3.** *bot.* The principal or primary vein of a leaf; the midrib.

ribbed. Bearing ribs or riblike structures.

ribbonfish. Any of a number of deep-sea fishes with a long, narrow, and compressed body, as the dealfish, (*Trachypterus*) of the order Allotriognathi.

ribbon worm. A nemertean.

riboflavin. A vitamin of the B complex essential for cellular respiration; widely distributed in plant and animal tissues and synthesized by bacteria within the body. Also called vitamin B_2, vitamin G, lactoflavin.

ribonuclease. An enzyme which hydrolyzes RNA, splitting the phosphate bonds linking the adjacent nucleotides.

ribonucleic acid. RNA, one of a number of nucleic acids which, upon hydrolysis, yield ribose; of importance in the synthesis of proteins. Three types are: messenger (mRNA), of nuclear origin; transfer (tRNA), which conveys amino acids; ribosomal (rRNA), located in the ribosomes of the cytoplasm.

ribose. A pentose sugar, $C_5H_{10}O_5$, present in riboflavin, ribonucleic acid, and various nucleotides.

ribose nucleic acid. Ribonucleic acid (RNA), *q. v.*

ribosome. One of the numerous ultramicroscopic bodies (microsomes) present in the cytoplasm of cells and usually associated with the endoplasmic reticulum. They are rich in RNA and function in protein synthesis.

rice. A cereal grass (*Oryza sativa*) extensively cultivated as a food crop; collectively, the grain or seeds of this plant.

ricebird. Any of a number of birds which frequent rice fields, especially the bobolink.

rice rat. A moderate sized rodent (*Oryzomys palustris*) of marshy areas frequenting rice and grain fields of southeastern United States.

ricin. A toxic albumin present in the seeds of castor bean plant.

rickets. A deficiency disease of children characterized by defective bone formation due to a deficiency of vitamin D which impairs absorption of calcium and phosphorus.

rickettsia. A microorganism of the group Rickettsiales, intermediate between viruses and bacteria. All are obligate, intracellular parasites and are common in various arthropods; some are pathogenic and the causative agents of various diseases as typhus, Rocky Mountain spotted fever, scrub typhus, Q fever, and rickettsialpox.

rictus. **1.** *zool.* The gape of the mouth, with reference to birds. **2.** *bot.* The mouth of a bilabiate corolla.

right whale. A whalebone whale of the family Balaenidae differing from other whales in absence of dorsal fin and grooves on throat and chest, as *Balaena mysticetus*, the Greenland right whale.

rigor mortis. The condition following death in which muscles become stiff and rigid.

rimose. Rimous; having numerous clefts or fissures, as the bark of trees.

rind. A thin, tough, external covering or layer, as that on the surface of various fruits and plant stems.

rinderpest. An infectious viral disease of herbivores characterized by inflammation of the digestive tract.

ring. A circular structure. *See* annual ring, fairy ring.

ring canal. A circular canal, as that **a.** in the water-vascular system of an echinoderm; **b.** in the tentacle of an ectoproct.

ringed. Resembling, marked by, or possessing a ringlike structure or marking.

ringent. Gaping, with reference either to **a.** *zool.* the valves of bivalve; **b.** *bot.* an open, bilabiate corolla.

Ringer's solution. An isotonic solution of three chlorides. For human tissues, it is composed of 8.6 g. of sodium chloride, 0.30 g. of potassium chloride, and 0.33 g. of calcium chloride in each 1000 ml. of water.

ring gland. In dipterous insects, the ring of Weissman, composed of the corpus cardiacum, corpus allatum, and prothoracic gland fused to form a neurosecretory structure surrounding the aorta. Also called retrocerebral complex.

ringhals. A South African cobra (*Haemachates haemachatus*), a spitting snake, capable of ejecting venom a considerable distance.

ring-tailed cat. A raccoonlike mammal (*Bassariscus astutus*) of the family Procyonidae of southwestern United States and Mexico; also called cacomistle.

ringworm. An infectious disease of the skin and its appendages (hair, nails), caused by various fungi, as *Microsporum*, *Tricho-*

phyton and *Epidermophyton.* Also called dermatomycosis, trichomycosis, tinea.

riparian. Pertaining to the bank or shore of a river, lake, or stream.

riparious. Growing close to a river or stream.

rivulose. Having small sinuous channels or markings.

RNA. Ribonucleic acid, *q. v.*

roach. 1. One of several species of freshwater fishes of western United States, especially those of the genus *Hesperoleucus,* family Cyprinidae. **2.** A freshwater cyprinid fish (*Rutilus rutilus*) of Europe. **3.** A cockroach.

roadrunner. A cuckoo (*Geococcyx californianus*) of the family Cuculidae, which runs on the ground, common throughout western United States. Also called chaparral cock or hen.

robber fly. A predaceous fly (*Asilus snowi*) of the family Asilidae.

robin. 1. An American thrush (*Turdus migratorius*), a familiar songbird of the family Turdidae; commonly called robin red-breast. **2.** A European thrush (*Erithacus rubecola*).

rock barnacle. A sessile barnacle of the genus *Balanus.*

rock bass. A pan fish (*Ambloplites ruprestris*) of the sunfish family Centrarchidae, common throughout eastern United States.

rock crab. A hard-shelled crab of the genus *Cancer.*

rockfish. One of a number of fishes that inhabit rocky habitats, especially marine fishes of the family Scorpaenidae.

rockweed. One of a number of brown algae which grow attached to rocks along the shore line, as *Fucus, Ascophyllum.*

Rocky Mountain goat. A bovine ruminant (*Oreamnos americanus*) of western North America.

Rocky Mountain sheep. *See* bighorn.

rod. The elongated, dendritic portion of a rod cell present in the retina of the eye; a receptor for light stimulated by rays of low intensity. *See* rhodopsin; *compare* cone.

rodent. Any mammal of the order Rodentia (Glires), which includes rats, mice, squirrels, chipmunks, woodchucks, gophers, porcupines, and beavers, characterized by two chisel-shaped incisors in each jaw and absence of canine teeth. Comprise over 6500 species, the majority of living mammals.

rod organ. A structure in holozoic euglenoids which functions in the ingestion of food, as in *Peranema*.

roe. 1. The eggs of a fish, especially when still contained within the ovarian membrane. **2.** The eggs or ovaries of certain invertebrates, as the coral of a lobster. **3.** A roe deer.

roebuck. The male of the roe deer.

roe deer. A small deer (*Capreolus capreolus*) of northern Europe and Asia.

roentgen. An international unit of quantity of roentgen or X-rays; symbol r.

roentgen rays. X-rays; radiant energy of short wave length produced by high speed electrons striking a metal target in a vacuum.

¹rogue. A chance variation or deviation from a standard variety or type, especially one of an inferior nature.

²rogue. To remove inferior, worthless, or undesirable plants from a growing crop, especially from a bed of seedlings.

roller. 1. One of a variety of pigeons which rolls or tumbles in flight. **2.** A European bird of the family Coraciidae which turns in flight like a tumbler pigeon. **3.** An insect or an arachnid which rolls leaves. *See* leaf roller.

rolling. The turning in of the margins of leaves to prevent water loss. *See* motor cells.

rook. A crowlike bird (*Corvus frugilegus*) common throughout Europe.

rookery. The breeding grounds or place for **a.** a colony of rooks; **b.** any gregarious bird, as the penguin; **c.** aquatic mammals, as the fur seal.

¹roost. To sit, rest, or sleep on a pole or a limb of a tree, said of birds at night.

²roost. 1. A place where birds rest at night. **2.** A group of birds roosting together.

rooster. The male or cock of domestic fowl.

root. 1. *anat.* The basal portion of an organ or structure which is embedded in underlying tissues or by which a structure is attached, as the root of a tooth, hair, tongue, or spinal nerve; a radix. **2.** *bot.* **a.** The part of the main axis of a vascular plant which, with its branches, grows downward into the soil. It originates from the hypocotyl, lacks nodes and internodes, and functions in anchorage, absorption and conduction of water and minerals, and in some cases for food storage. **b.** A similar structure which grows from a stem or leaf which serves for support or attachment, as the prop *roots* in maize or the clinging *roots* of ivy.

root cap. A mass of loosely arranged cells which covers and protects the tip of a growing root.

root climber. A plant which develops adventitious roots from the stem for support and attachment, as English ivy.

root hair. One of the numerous microscopic, hairlike processes which grow outward from the epidermal cells in the root hair zone of young, actively growing roots, greatly increasing the surface for absorption of water and minerals.

rootlet. A small root.

root nodule. A knotlike structure or tubercle which develops on the roots of various legumes as a result of the growth of symbiotic nitrogen-fixing bacteria.

root pressure. The pressure resulting from osmotic pressure on or in a root system which causes sap to ascend in a stem.

rootstock. A rhizome.

root system. All the roots of a plant taken collectively, comprising one of three types: fibrous, tap, and fascicled.

root-tipping. The development of roots at the tips of stems, as in plants with arching or trailing stems which root upon-contact with the ground.

rorqual. A whalebone whale of the genus *Balaenoptera*, as *B. physalus*, the common rorqual, also called finner whale, razorback finback.

rose. A plant of the genus *Rosa* of the family Rosaceae, an erect or climbing shrub with prickly stem and showy flowers.

rose family. The Rosaceae, a family of angiosperms comprising over 2500 species, many of commercial importance, as the rose, peach, apple, cherry, strawberry, hawthorn, and blackberry.

rosette. 1. Any of a number of structures in which the parts have a circular, radiate arrangement, as **a.** *bot.* a cluster of leaves arising from a short stem; **b.** *zool.* the fringed, posterior holdfast organ of certain cestodarians, or in crinoids, an ossicle on the aboral surface. **2.** One of several plant diseases resulting from virus infection or mineral deficiency.

466

rosorial. Of or pertaining to gnawing animals.

ross. The roughened external surface of bark.

rostellate. Having a small beak or rostrum.

rostellum. 1. *zool.* A small projection or rostrum, as **a.** the projecting anterior end of certain flagellates. **b.** The beak of certain sucking insects. **c.** A rounded process on the tip of the scolex of certain tapeworms; may be retractile and surrounded by recurved hooks. **2.** *bot.* A sterile, beaklike lobe on the stigma of an orchid.

rostral. Of or pertaining to a beak or rostrum.

rostrate. Having a beak or rostrum.

rostrum. A beak or beaklike projection, as **a.** a middorsal projection on the head of a rotifer; **b.** the median process extending anteriorly from the cephalothorax of a malacostracan; **c.** the plates opposite the carina in a sessile cirriped; **d.** a prominence at the posterior end of the shell of a cuttlefish; **e.** the cone-shaped, basal portion of the proboscis in dipterans; **f.** in hemipterans, the beak projecting posteriorly from the head; **g.** in weevils, the elongated vertex bearing at its tip the mouth parts; **h.** in mites, the capitulum or gnathosoma; **i.** in vertebrates, an anterior extension of the cranium in the shark, paddlefish, sawfish, and sturgeon; **j.** an anterior extension of certain bones, as the basisphenoid.

rosula. A rosette.

rosulate. In the form of or resembling a rosette.

¹rot. To undergo decay or decomposition.

²rot. One of a number of plant diseases caused by bacteria or fungi, as brown *rot* of fruit.

rotate. Wheel-shaped; flat and circular in outline, with reference to the corolla of certain flowers.

rotator. A muscle which rotates a part or turns a structure on its long axis.

rotifer. Any of a number of minute, microscopic invertebrates of the Rotifera (Rotatoria), a class of the Aschelminthes or a phylum which includes the wheel animalcules with a body bearing a ciliated disc (corona or wheel organ), and a trunk containing a mastax and bearing a posterior foot.

rotting. The process of decomposition or decay brought about by extracellular enzymes of various bacteria and fungi which dissolve pectin, reducing plant tissues to a soft, moist, pulpy mass. *See* rot.

rotula. 1. One of the five radial plates in Aristotle's lantern. **2.** The patella or kneecap.

rotund. Spherical or rounded in outline.

rough. Having a surface marked by irregularities, ridges, or projections; not smooth.

roughage. 1. Food with a high content of undigestible material. **2.** The undigested food of the large intestine, consisting principally of cellulose.

rough colony. *bacteriol.* A type of colony marked by a flat, uneven surface containing abnormal cells which cling together after fission and differ physiologically from cells of contrasting smooth colonies.

rouleau pl. **rouleaux.** A row of red blood cells which resembles a roll or stack of coins.

roundworm. A nematode.

rove beetle. A slender, elongate beetle of the family Staphylinidae found principally around decaying matter, dung, and carrion.

royal jelly. A substance produced by the pharyngeal glands of worker bees which, when fed to the female larvae, results in their development into queens.

RQ. Respiratory quotient, *q. v.*

rubber. An elastic substance prepared from latex, the milky juice of various plants of the tropics, especially *Hevea brasiliensis.*

rubescent. Becoming red.

rubythroat. The common hummingbird (*Archilochus colubris*).

ruddy. Red or reddish.

ruddy duck. A surface-feeding duck (*Oxyura jamaicensis rubida*).

rudiment. 1. An organ or structure which is just beginning to develop or one in which development has been arrested; an anlage or primordium. **2.** A vestige or the remains of a structure which was functional in an earlier stage of evolutionary development.

ruff. 1. A ring or collar of elongated or modified feathers about the neck of certain birds. **2.** A small European perch (*Acerina* *cernua*). **3.** The male of the sandpiper (*Philomachus pugnax*). *See* reeve.

ruffed grouse. An upland game bird (*Bonasa umbellus*), males of which are noted for their drumming.

rufous. Reddish or brownish red.

ruga pl. **rugae.** A fold, wrinkle, or ridge.

rugose. 1. Wrinkled or ridged. **2.** Pertaining to corals of the order Rugosa, an extinct group.

rumen. The paunch, the first of four chambers of the stomach of a ruminant.

ruminant. 1. An animal which chews its cud. **2.** An even-toed hoofed mammal of the suborder Ruminantia with a stomach of four chambers (rumen, reticulum, omasum, and abomasum) that swallows its food unchewed, then regurgitates it, chews it thoroughly, and reswallows it. Common ruminants are the camel, pronghorn, giraffe, deer, and cattle.

¹ruminate. *zool.* To chew the cud.

²ruminate. *bot.* Appearing as though chewed.

rump. 1. The region of the buttocks in a quadruped. **2.** *ornith.* The region between the back and upper tail coverts; the region of the sacrum.

run. The movement of a school of fishes up a river to spawn, as a salmon *run*.

runcinate. Deeply incised with lobes or segments directed backwards, as the leaf of a dandelion.

runner. 1. *bot.* A slender, prostrate branch (stolon) which takes root near its tip and forms buds which develop into new plants, as in the strawberry. **2.** One of a number of carangid fishes, as the leather jack and jurel.

runt. A small individual; one that is smaller than the average of its kind.

rupicolous. Living in or frequenting rocks or rocky areas.

rupture. 1. A breaking apart or separation. **2.** *med.* A hernia.

rupturing. *bot.* Bursting irregularly.

rush. Any of a number of grasslike plants of the rush family Junaceae, as *Juncus*, the bog rush and *Luzula*, the wood rush.

russet. A brown, roughened area on the skin of various fruits resulting from insect injury, disease, spray damage, or some other causative factor.

rust. 1. Any of a number of parasitic fungi of the order Uredinales which cause brown or rust-colored spots to develop on the stems and leaves of infected plants, as the *stem rust* of wheat. **2.** A disease caused by a rust fungus.

rut. A recurrent period of intense sexual activity in the males of various mammals as cattle, sheep, deer and related forms; heat. *Compare* estrus.

rutabaga. A turniplike vegetable (*Brassica napobrassica*) grown for its root and the lower portion of the stem; used extensively for stock feed.

rye. 1. A hardy annual grass (*Secale cereale*), a minor cereal, grown extensively in Europe for its seeds, used for food. Also grown for hay, winter pasturage, and for cover to prevent erosion. **2.** The grain (seeds) of the rye plant.

S

saber-toothed tiger. An extinct, catlike mammal characterized by long, curved, thin, upper canine teeth, once common throughout Europe and North America.

sable. A carnivorous, fur-bearing mammal (*Martes zibellina*) related to the martens, inhabiting northern Europe and Asia, or its fur or pelt.

sable antelope. A large, African antelope (*Hippotragus niger*) with glossy black body and recurved horns.

sabulous. Sandy, gritty.

sac. A pouch or baglike cavity.

saccate. Shaped like a sac or pouch.

saccharase. An enzyme which hydrolyzes disaccharides to monosaccharides.

saccharide. A simple sugar or a compound composed of simple sugars; a carbohydrate.

saccharose. Cane sugar: SUCROSE.

sacculated. Consisting of or possessing sacs or saclike compartments.

saccule. **1.** A little sac. **2.** A cavity, as that of the green gland in crustaceans or the coxal gland in scorpions. **3.** A cavity of the membranous labyrinth located in the vestibule of the internal ear. In its wall are the maculae acoustica, sensory receptors responding to changes in the position of the head. Also called sacculus.

sacculus. A saccule.

saccus vasculosus. A thin-walled vascular sac forming the distal end of the infundibulum in fishes and amphibians.

sacellus. A seed consisting of a one-seeded pericarp surrounded by a hardened calyx.

sac fungi. Fungi of the order Ascomycetes.

sacral. Of or pertaining to the sacrum or the region of the sacrum.

sacrum. A triangular bone of the vertebral column consisting of three to five ankylosed vertebrae located between the lumbar vertebrae and caudal vertebrae; in man, between the lumbar vertebrae and coccyx. It articulates with the hip bones.

sage. A mint (*Salvia officinalis*), a shrublike plant highly esteemed as a spice and a source of sage oil used in perfumery.

sagebrush. One of a number of species of *Artemisia*, especially *A. tridentata*, a shrub which covers vast areas of western United States.

sage grouse. A grouse (*Centrocercus urophasianus*) formerly common in sagebrush areas of the United States. Females are called sage hens, males sage cocks.

sagittal. **1.** Pertaining to or resembling an arrow in shape. **2.** Of or pertaining to a sagittal plane or section.

sagittal plane. In a bilaterally symmetrical animal, a vertical plane through the body dividing it into right and left portions.

sagittal section. One cut through a sagittal plane.

sagittate. *bot.* Shaped like an arrowhead, with reference to leaves which are elongate and triangular with two basal lobes directed backward.

sago. A starch obtained principally from stems of the sago palm, used extensively for food and as a demulcent.

sago palm. Any palm which yields sago, especially *Metroxylon sagu*, cultivated extensively in Malaya and Indonesia.

saiga. An antelope (*Saiga tatarica*) of central Asia with an over-developed and mobile nose.

sailfish. A large, marine fish of the genus *Istiophorus* with a dorsal fin of enormous size; a relative of the swordfish.

saki. A large, New World monkey of the genera *Pithecia* and *Chiropotes*, with a bushy, nonprehensile tail and long, curly hair.

salamander. A tailed amphibian of the subclass Caudata (*Urodela*), small lizardlike forms but lacking scales; larvae aquatic with gills, adults terrestrial, as the tiger salamander (*Ambystoma tigrinum*).

[1]salientian. Of or pertaining to the Salientia (*Anura*), a taxon (subclass or order) of the class Amphibia which includes the tailless forms, as frogs and toads.

[2]salientian. Any amphibian of the Salientia.

saline. Salty; consisting of or containing salt, especially sodium chloride.

saline solution. A solution of sodium chloride. A *physiological saline solution* is one that is isotonic, as a 0.7% solution for amphibians or a 0.9% solution for mammals.

saliva. The fluid secreted by salivary glands and discharged into the mouth cavity. In man, it contains ptyalin, an amylase.

salivary gland. A gland which secretes saliva, present in various invertebrates, as insects, and in terrestrial vertebrates. In man, the principal salivary glands are the parotid, submandibular (submaxillary), and sublingual.

salmon. 1. A large, soft-finned, anadromous fish (*Salmo salar*) of the East Coast. **2.** One of a number of fishes of the genus *Onchorhynchus* of the North Pacific which ascend freshwater streams, especially the Columbia River, to spawn, as *O. tschawytscha*, the chinook, quinnat, or tyee salmon, and *O. kisutch*, the coho salmon, now established in the Great Lakes.

salp. A free-swimming, marine tunicate of the class Thaliacea, especially those of the genus *Salpa*.

salpinx pl. **salpinges.** A trumpet-shaped tube, as the Eustachian or Fallopian tube.

salt. **1.** Sodium chloride, NaCl. **2.** *chem.* **a.** A substance other than water resulting from the reaction between an acid and a base. **b.** A compound of a metal or a positive radical and a non-metal or negative radical. **c.** A substance which, in solution, provides a cation other than the hydrogen ion and an anion other than the hydroxyl ion. **d.** An electrovalent or an ionic compound which is crystalline.

saltation. **1.** A jump or a leap. **2.** An abrupt change. **3.** A mutation.

saltatory. **1.** Pertaining to leaping or dancing. **2.** In evolution, occurring by sudden, abrupt changes.

salt gland. A gland located above the orbit in certain marine reptiles and birds which discharges its secretion containing large quantities of salt into the nasal cavity.

salverform. Having a slender tube which expands abruptly into a flat limb, as the flower of the primrose or phlox.

samara. A simple, dry, indehiscent fruit, usually one-seeded and bearing wings, as that of the maple or elm; also called key.

sambar. An Indian deer (*Rusa unicolor*).

sandalwood. Wood obtained from an Oriental tree (*Santalum album*) or the tree that produces it, the source of sandalwood oil used in perfumery and medicine; also the source of a dye.

sand dab. A flatfish of the genera *Limanda* or *Citharichthys*.

sand dollar. A flat, disc-shaped echinoderm of the class Echinoidea, as *Echinarachnius*.

sanderling. A small sandpiper (*Crocethia alba*) frequenting sandy flats and beaches.

sand flea. **1.** The chigoe or jigger (*Tunga penetrans*). **2.** A beach flea.

sand fly. A small, bloodsucking fly of the genus *Phlebotomus*, a transmitter of leishmaniasis and other diseases to man.

sandhill crane. A large crane (*Grus canadensis*) of North America.

sand hopper. A beach flea.

sandpiper. One of a number of shorebirds of the family Scolopacidae, as the spotted sandpiper (*Actitis macularius*).

sandworm. Any of a number of polychaete annelids which burrow in the sand, as *Neanthes*, *Nereis*.

sanguine. Red, blood-colored; resembling blood.

sanguivore. An animal which feeds upon blood, as an insect, bat.

San Jose scale. A scale insect (*Aspidiotus perniciosus*), a serious pest of fruit and ornamental trees and shrubs.

S-A node. *See* sinoatrial node.

sap. Plant juice or the fluid substance of a plant, especially the fluid which circulates through the vascular tissues. *See* cell sap.

saponification. **1.** The act or process of making soap, that is, the hydrolysis of fats by alkalies resulting in the formation of glycerol and soap (an alkali salt of a fatty acid). **2.** The hydrolysis of an ester by any means.

saponify. To undergo saponification; to convert into soap.

saprobe. A saprotroph.

saprobic. SAPROTROPHIC.

saprogenic. Causing decay or putrefaction; saprogenous.

sapropelic. Pertaining to bottom deposits in lakes which are rich in decomposing organic matter.

saprophagous. Feeding on decaying matter.

saprophyte. A plant which obtains its nourishment from dead organic matter, as most fungi; saprotroph.

saprotroph. A saprotrophic organism.

saprotrophic. Designating a type of nutrition in which an organism obtains its nourishment from dead or decaying matter.

saprovore. A saprozoan.

saprozoan. An animal which feeds upon dead or decaying matter.

sapsucker. A woodpecker of the genus *Sphyrapicus*.

sapwood. The outer and softer wood (xylem) in stems which functions in conduction, food storage, and in providing mechanical support. *Compare* heartwood.

sarcina. The characteristic arrangement of bacterial cells in an eight-celled, cube-shaped packet.

sarcobiont. An organism which feeds upon flesh.

sarcocarp. The mesocarp.

sarcodinian. A protozoan of the class Sarcodina which includes the ameba and related forms, characterized by possession of pseudopodia.

sarcolemma. The thin membrane closely investing a striated muscle fiber, consisting of the plasma membrane and associated reticular fibers.

sarcoma. A malignant tumor composed principally of non-epithelial cells.

sarcomere. A contractile unit of a striated muscle fiber, comprising the region between two Z bands or discs.

sarcophagid. A flesh fly.

sarcophagous. Feeding upon flesh.

sarcoplasm. The fluid portion of a striated muscle fiber as distinguished from the myofibrils.

sarcoptid. A mite of the family Sarcoptidae which includes the itch mites, the causative agents of scabies and mange in domestic animals and man.

sarcosome. A mitochondrion of striated muscle fibers.

sarcostyle. A nematophore.

sarcous. Of or pertaining to flesh or muscle.

sardine. One of a number of small fishes of the genus *Sardinella*, used extensively for food.

sargasso. The gulfweed, a seaweed of the genus *Sargassum*, a brown alga, forming the major portion of the vegetation of the Sargasso Sea.

sarment. A runner or stolon.

sarmentose. Possessing runners or stolons.

sartorius. A long, slender muscle on the inner surface of the thigh, a flexor of the leg; also a flexor and external rotator of the thigh.

satellite. 1. A small structure attached to a larger structure. 2. A satellite cell.

satellite cell. A neuroglial cell which lies close to the cell body of a neuron, as oligodendroglial cells of the brain and spinal cord or capsular cells of a spinal ganglion.

saturniid. A moth of the family Saturniidae which includes the giant silkworm moths.

satyr. A small to medium sized butterfly of the family Satyridae, as the wood nymph (*Minois alope*).

sauger. A pike perch (*Stizostedion canadense*) resembling the walleye but smaller.

saurian. Of or pertaining to the Sauria (Lacertilia), a suborder of the Squamata comprising the lizards.

saurischian. An extinct dinosaur of the order Saurischia which includes a number of carnivorous, bipedal forms some of which attained large size, as *Tyrannosaurus*.

sauropsidan. A reptile or bird.

savanna. A type of grassland in dry tropical and subtropical regions containing scattered trees and shrubs, as in southeastern United States, especially Florida.

sawbuck. A turtle of the genus *Graptemys*, as *G. oculifera*, the ringed sawbuck of the southern states.

sawfish. A large, elasmobranch fish (*Pristis pectinatus*) with an elongated rostrum bearing sharp, toothlike processes on each side.

sawfly. One of a large number of hymenopterous insects with a well-developed ovipositor used in depositing eggs in the host plant, as those of the families Tenthredinidae (typical sawflies) and Cephidae (stem sawflies).

sawyer. A wood-boring beetle of the family Cerambycidae, especially those of the genus *Monochamus*.

saxicolous. Living in, on, or among rocks; saxicoline.

scab. 1. *bot.* **a.** A roughened, crustlike area on a plant resulting from a bacterial or fungus infection. **b.** The disease of which scab is a symptom. **2.** *zool.* **a.** The dried exudate which covers the surface of a lesion. **b.** One of a number of infections of domestic animals caused by scab mites.

scabellum. The expanded distal portion of a halter.

scabies. An infectious skin disease caused by the itch mite (*Sarcoptes scabiei*); also called the itch or seven year itch.

scab mite. One of a number of mites which cause scab or mange, especially those of the genera *Psoroptes* or *Chorioptes* which infest domestic animals.

scabrous. Having a rough surface; rough to the touch; scaly.

scala media. The cochlear duct, the middle canal of the cochlea; filled with endolymph and containing the organ of Corti.

scalariform. Having transverse bars or markings; ladderlike.

scala tympani. A canal of the cochlea extending from its tip to the round window which is filled with perilymph.

scala vestibuli. A canal of the cochlea extending from the vestibule to the tip of the cochlea where it is continuous with the scala tympani.

scald. 1. A disease of fruits in which superficial browning occurs, thought to be due to immaturity at harvesting. **2.** A fungus disease of forage grasses characterized by scaldlike blotches.

scale. 1. *zool.* **a.** A thin, flattened, platelike structure forming a part of the surface covering of various vertebrates, as fishes and reptiles. **b.** A flat, modified hair characteristic of lepidopterous insects; also present on other insects. **c.** The hardened, waxy covering of a scale insect. **d.** Scale insects considered collectively. **2.** *bot.* **a.** A thin, membranous, chafflike structure, as a woody bract or a microsporophyll of a conifer. **b.** A flattened trichome. **c.** A degenerate or modified leaf, as a bud *scale.* **d.** A glume. **e.** A thin outgrowth on the lower surface of the gametophyte of various bryophytes.

scale insect. One of a number of small, homopterous insects of the family Coccidae, comprising three groups; mealy bugs, soft scales, and armored scales. Females feed on plant juices and secrete a wax which hardens, forming a protective covering. Some are serious plant pests; others the source of commercial products, as cochineal, shellac, and China wax.

scale leaf. A modified leaf which is scalelike, as a bud scale.

scale moss. A leafy liverwort of the Jungermanniales which bears delicate, scalelike leaves.

scale worm. A polychaete of the family Aphroditidae with platelike scales (elytra) on the dorsal surface of the body.

scallop. 1. A marine, bivalve mollusk of the genus *Pecten* which swims by opening and closing the valves of its radially ribbed shell. **2.** The adductor muscle of these mollusks used as food. **3.** One of the valves of this mollusk.

scalp. The hairy integument covering the cranium.

scandent. Climbing, with reference to plants.

scansorial. 1. Having the habit of climbing. **2.** Adapted for climbing, as the feet of reptiles and birds.

scape. 1. *bot.* The leafless peduncle of a flower arising at or near the ground, as in the tulip. **2.** *zool.* **a.** The basal segment of an insect antenna. **b.** The peduncle of a halter. **c.** The shaft of a feather.

scaphium. *See* Weberian ossicles.

scaphognathite. A flattened process on the second maxilla of a decapod crustacean which functions in the creation of a respiratory current over the gills; a bailer.

¹**scaphoid.** Boat-shaped.

²**scaphoid.** The navicular bone of the carpus and tarsus.

scaphopod. Any mollusk of the class Scaphopoda which includes the tooth shells or tusk shells.

scapiflorous. Having flowers borne on a scape.

scapiform. Resembling a scape; scapose.

scapose. SCAPIFORM.

scapula. **1.** A bone of the pectoral girdle with which the forelimb articulates; in man, the shoulder blade. **2.** *entomol.* A parapsis.

scapus. The collar (parapet) of a sea anemone.

scar. **1.** A mark, as that which remains on the skin after the healing of an injury; a cicatrix. **2.** *bot.* **a.** A mark on a stem marking the former attachment of a leaf, bud, flower, or fruit. **b.** The hilum of a seed. **3.** *zool.* A mark on the inner surface of a bivalve shell marking point of attachment of a muscle.

scarab. A dung beetle, especially the Egyptian sacred scarab beetle (*Scarabaeus sacer*).

scarabaeid. Any lamelliform beetle of the family Scarabaeidae which contains the tumblebugs, scarabs, and leaf chafers.

scarify. **1.** To make small cuts or scratches in the skin. **2.** To slit the outer seed coat in order to speed germination.

scarious. Thin, dry, and membranous.

scat. Excrement, dung, feces, ordure.

scatology. The study of excrement, especially for determination of nature of food.

scatophagous. Feeding upon dung; coprophagous.

scatoscopy. The examination or inspection of feces, especially for diagnostic purposes.

scaup. A diving duck of the genus *Aythya*, commonly called bluebill.

scavenger. **1.** An animal which devours dead animals or feeds on dead organic material. **2.** A phagocytic cell, as a macrophage or leukocyte.

¹**scent.** To perceive through stimulation of olfactory receptors; to smell.

²**scent.** **1.** The odor left by an animal by which it may be traced or tracked down. **2.** A distinctive smell, odor, or fragrance.

scent gland. A modified sweat or sebaceous gland which produces an odorous secretion which serves as an attractant, a repellent, or for defensive purposes, as that in many mammals. *See* musk gland.

schemochrome. A structural color, as that produced by uneven surfaces; an iridescent color.

schistosome. A blood fluke of the genus *Schistosoma* which lives as a parasite in the blood vessels of man and domestic animals.

schistosome dermatitis. *See* swimmer's itch.

schistosomiasis. Infestation by blood flukes of the genus *Schistosoma* resulting from entry of cercariae directly through the skin.

schizocarp. A dry fruit consisting usually of two carpels which, when mature, split apart forming one-seeded indehiscent halves, each called a mericarp, as in the carrot family.

schizocoel. A coelom which arises by a splitting of the lateral plates of the mesoderm, as in arthropods and mollusks.

schizocoelomate. An animal which possesses a schizocoel.

schizogenous. Arising by fission or splitting.

schizogony. A form of asexual reproduction in which multiple division occurs, resulting in numerous individuals (schizonts), as in the malarial parasite (*Plasmodium*) and various coccidians.

schizomycete. Any member of the class Schizomycetes which includes the bacteria.

schizont. 1. A cell which reproduces by schizogony. **2.** In the malarial organism (*Plasmodium*), an amebalike trophozoite which divides into many (6-36) daughter merozoites.

schizopetalous. With split or divided petals.

schizophyte. Any member of the division Schizophyta which contains the classes Schizomycetes (bacteria) and Schizophyceae (blue-green algae); the fission plants.

schizopod larva. A mysis larva; one having exopods on the thoracic appendages.

schizozoite. A cell resulting from schizogony, as occurs in certain gregarines

school. A large number of aquatic organisms swimming together, as a *school* of fishes. *See* gam, pod.

Schwann's cells. Cells which make up the neurilemma (Schwann's sheath) of an axon.

sciatic. Of, pertaining to, or near the ischium, as *sciatic* nerve.

scientific method. A method utilized in acquiring knowledge and establishing laws and principles pertaining to natural phenomena based on observation and experimentation.

scientific name. *See* binomial nomenclature.

scion. A bud or shoot which is removed from a plant and prepared for grafting onto another plant.

sciophilous. Shade-loving; preferring shade.

scirrhous. Hard, knotty.

scirrhus. A hard, indurated organ or structure.

scissile. Easily split or separated into parts.

scissortail. A flycatcher (*Muscivora forficata*) with a long, scissorlike tail.

sciurid. Any rodent of the family Sciuridae which includes the squirrels, chipmunks, and marmots.

sciuroid. 1. Squirrellike. **2.** *bot.* Resembling in structure or appearance a squirrel's tail; curved and bushy.

sclera. The dense, fibrous outer coat of the posterior portion of the eye, continuous anteriorly with the cornea; the "white" of the eye.

scleral ring. A series of bony plates lying in the anterior half of the sclera of the eye of certain fishes, reptiles, and birds.

sclereid. An elongated sclerenchyma cell with heavily lignified walls found in the cortex of stems and the hard shells of fruits and seed coats; sometimes present in the pith of stems and the leaves of certain plants; a stone cell, rod cell, or bone cell of a plant.

sclerenchyma. The strengthening or supporting tissue of higher plants consisting of cells with thick, lignified walls; includes sclereids and fibers.

sclerite. 1. A hardened, platelike portion of the exoskeleton of an arthropod bounded by sutures and/or membranous areas. **2.** A spicule of a sponge.

scleroblast. A mesenchymal cell which secretes or forms spicules, as in sponges.

scleroid. Hard; indurated.

sclerophyllous. Possessing leaves with an excessive amount of sclerenchyma, as in certain desert plants.

scleroprotein. One of a large class of proteins found in connective tissue, as collagen, elastin, and keratin; an albuminoid.

sclerosis. 1. An abnormal hardening or thickening of body tissues. **2.** *bot.* The hardening and thickening of a cell wall by lignification.

sclerotic. 1. *anat.* Of or pertaining to the sclera. **2.** *bot.* With hardened cell walls, as *sclerotic* cells (sclereids).

sclerotium. 1. In fungi, a hardened, compact mass of mycelium which gives rise to fruiting bodies. *See* ergot. **2.** In slime molds, a hard, brittle, or waxy mass formed upon drying, as in *Lycogala.*

sclerotome. The portion of a mesodermal somite which gives rise to the vertebrae and ribs.

sclerous. Hard, dense, indurated.

scobiform. Resembling sawdust.

scobina. The rasplike rachilla of the spikelet of certain grasses.

scobinate. Rough; possessing a rasplike surface.

scolex pl. **scolices. 1.** The anterior end of a tapeworm comprising a holdfast organ usually provided with anchoring structures, as bothria, bothridia, suckers, or hooks by which a tapeworm attaches itself to its host. **2.** A similar structure in a larval tapeworm, as a cysticercus, coenurus, or hydatid.

scolophore. A specialized sensillum (chordotonal organ) in an insect.

scolops. A sensory rod of a chordotonal organ.

scolus. A spiny process on the larva of certain moths.

scopa. *See* pollen brush.

scopula. 1. A dense tuft of hairs on the tip of the tarsus in a spider. **2.** An adhesive disc in a peritrich. **3.** The structure which secretes the stalk in certain ciliates.

scorbutic. Of or pertaining to scurvy.

scorpioid. 1. *zool.* Of, pertaining to, or resembling a scorpion. **2.** *bot.* With reference to a circinately coiled inflorescence in which flowers are two-ranked and borne alternately on the right and left.

scorpion. An arachnid of the order Scorpionida with an elongated, segmented body terminating in a curved, poison sting; pedipalps are elongated and end in chelicera. Sting is dangerous but rarely fatal to man.

scorpion fish. A mail-cheeked fish (*Scorpaena*) of the order Scleroparei.

scorpion fly. A medium-sized, slender-bodied insect of the order

Mecoptera, males possessing an abdomen resembling that of a scorpion, as *Panorpa*.

scoter. A diving duck of the genera *Melanitta* and *Oidemia* of northern regions.

scotopia. Vision in dim light in which shape and form can be distinguished but not color; involves a minimal amount of light energy and rods as receptors.

scotopic. Of or pertaining to scotopia.

scotopsin. A visual purple, as rhodopsin, present in rod cells.

scouring rush. A horsetail.

scrapie. An infectious, virus-caused disease of sheep.

screamer. A large, aquatic, South American bird of the family Anhimidae.

screech owl. A small owl (*Otus asio*) with conspicuous, erectile ear tufts.

screwworm. The larva of certain flies which deposit their eggs in wounds with resulting myiasis, as those of the genus *Callitroga*, a serious pest to livestock.

scrobe. A groove of the rostrum of beetles for reception of the scape of the antenna.

scrobiculate. Having numerous small hollows or shallow depressions; pitted.

scrotiform. Pouch-shaped.

scrotum. 1. The external pouch which contains the testes, present in most mammals. **2.** The peritoneal investment of the testis in an insect.

scrub. Vegetation consisting chiefly of shrubs and stunted trees, or a tract covered with such.

scud. An amphipod crustacean, as *Gammarus*.

sculpin. A large-headed, scaleless fish of the family Cottidae including the genera *Myxocephalus*, a deep-water marine form; *Scorpaena*, the scorpionfish of California; and *Cottus*, a fresh-water form.

scup. A porgy (*Stenotomus*), a marine food fish.

scurf. Thin, dry, epidermal scales, especially that of the scalp; dandruff.

scurvy. A deficiency disease resulting from lack of vitamin C, characterized by bleeding of the skin and mucous membranes, swollen, hemorrhagic gums, and defective teeth and bone formation.

scut. A short, erect tail, as that of a rabbit.

scutate. **1.** *bot.* Shield-shaped. **2.** *zool.* Possessing scutes.

scute. A thin, flat, bony or horny plate or scale, as those of the exoskeleton of fishes, snakes, and turtles.

scutellate. **1.** Shaped like a small dish or platter. **2.** Covered with small plates or scales, with reference to the tarsus of a bird. **3.** Having a scutellum, as in certain insects.

scutelliform. Shaped like a large, flat plate.

scutellum. **1.** A shield-shaped structure. **2.** *bot.* **a.** The cotyledon in the seed of various monocots. **b.** A caplike structure on the tip of the endosperm in the seeds of cycads. **3.** *zool.* **a.** One of the two sclerites of the alinotum of the thorax of an insect. **b.** A scale on the tarsus or toes of a bird.

scutiform. Shield-shaped; peltate.

scutum. **1.** *bot.* The broadened apex of the style in milkweeds. **2.** *zool.* A scute or hard, platelike structure, as **a.** the second sclerite in the notum of an insect; **b.** one of the plates of the carapace of a barnacle; **c.** the dorsal shield of a hard tick.

scyphistoma. A stage in the development of a jellyfish consisting of a sessile, polyplike form which, by strobilization, gives rise to several ephyrula. *See* ephyra.

scyphozoan. A coelenterate of the class Scyphozoa which includes the true jellyfishes, free-swimming medusae lacking a velum, as *Aurelia.*

scyphus. **1.** A cup-shaped corolla. **2.** The expanded cup-shaped portion of a podetium in certain lichens.

sea anemone. An anthozoan coelenterate, typically a sessile, solitary polyp often beautifully colored, as *Metridium.*

sea bass. One of a number of fishes of the family Serranidae, especially *Centropristes striatus,* a food and game fish of the Atlantic.

sea bear. A fur seal.

sea beef. Chitons prepared for human consumption.

sea biscuit. A cake urchin, an echinoid echinoderm.

sea bottle. A large, bladderlike, primary cell of a green alga (*Valonia*) with an enormous vacuole.

sea butterfly. A pteropod.

sea cow. A manatee or dugong of the order Sirenia.

sea cucumber. An echinoderm of the class Holothuroidea with an elongated body covered by a leathery integument containing minute endoskeletal ossicles or spicules, as *Thyone, Cucumaria.*

sea dragon. A peculiar fish (*Phyllopteryx*) of the order Solenichthyes, its body bearing numerous outgrowths resembling seaweeds.

sea elephant. An elephant seal (*Mirounga*) of the family Phocidae, males possessing an inflatable snout or proboscis.

sea fan. A fan-shaped gorgonian, especially one of the genus *Gorgonia.*

sea feather. An anthozoan of the order Pennatulacea with a form resembling a feather, as *Pennatula;* a sea pen.

sea fig. An herb (*Mesembryanthemum*) with conspicuous, water-storing leaves.

sea fir. A hydrozoan (*Abietinaria abietina*).

sea gooseberry. A ctenophore (*Pleurobrachia*) of the order Cydippida.

sea hare. An opisthobranch gastropod with large, earlike tentacles, as *Aplysia, Tethys.*

sea horse. A peculiar fish (*Hippocampus*) with an elongated head bent at a right angle to the body. They possess a prehensile tail, eggs are carried in a brood pouch by the male, and they swim in a vertical position.

seal. One of several marine carnivores of the suborder *Pinnipedia* comprising two families, the Otariidae or eared seals (sea lion and northern fur seal), and Phocidae, the earless or hair seals (harbor seal and sea elephant).

sea lamprey. *See* lamprey.

sea lettuce. A seaweed of the genus *Ulva*, a green alga.

sea lily. A stalked crinoid.

sea lion. A large, eared seal of the family Otariidae, as *Zalophus californianus*, the California sea lion, or *Eumetopias jubata*, Steller's sea lion.

sea louse. A marine copepod (*Lepeophtheirus*) of the order Caligoida, parasitic on marine fishes.

seam. A junction or line of fusion; a suture, a raphe.

sea mat. An encrusting layer of colonial ectoprocts.

sea mew. A sea gull (*Larus canus*).

sea moss. Any of a number of red algae noted for their color and graceful form, as *Dasya*, *Ceramium*, and others, mostly tropical forms.

sea mouse. A flat polychaete (*Aphrodite*) with hairlike covering of notopodial setae.

seam squirrel. The cootie or human body louse.

sea mussel. An edible marine bivalve of the genus *Mytilus*.

sea nettle. One of several scyphozoan jellyfishes noted for their dangerous stings, as *Dactylometra* or *Chrysaora* of the Atlantic coast.

sea onion. A liliaceous plant (*Urginea maritima*) of the Mediterranean coasts, the source of a number of glucosides, some of medical importance; also the source of red squill, a rodenticide.

sea otter. A fur-bearing carnivore (*Enhydra lutris*) of north Pacific coasts, a swimming, marine animal with webbed feet.

sea palm. A kelp (*Postelsia palmaeformis*) of north Pacific coasts.

sea pansy. A disc-shaped alcyonarian coelenterate (*Renilla*).

sea parrot. A puffin.

sea peach. A tunicate (*Halocynthia pyriformis*).

sea pen. A sea feather.

sea pork. A tunicate (*Amaroucium stellatum*).

sea purse. The horny egg case of various elasmobranchs.

sea raven. A sculpin (*Hemitripterus americanus*), a large, grotesque fish of the North Atlantic.

sea robin. A bottom-living, marine fish of the genus *Prionotus*.

sea serpent. A sea snake.

sea slug. A nudibranch.

sea snail. 1. A snailfish. 2. A marine gastropod.

sea snake. One of a number of swimming, venomous reptiles of the family Hydrophiidae, as *Hydrophis* and *Hydrus*, common in tropical waters.

seasonal. Of, pertaining to, correlated with, or related to a season or seasons.

sea spider. A long-legged marine arthropod of the class Pycnogonida.

sea squirt. An ascidian or tunicate.

sea star. A starfish.

485

sea turtle. 1. A large marine turtle, as *Chelonia mydas*, the green turtle. **2.** The loggerhead or hawksbill turtle.

seat worm. *See* pinworm.

sea urchin. A globe-shaped echinoderm of the class Echinoidea with a test of calcareous plates bearing long, movable spines, as *Arbacia*.

sea walnut. A ctenophore.

sea wasp. A coelenterate of the order Cubomedusae, so called because of the virulence of its sting.

seawater. Saline water occupying the ocean basins, of variable salt content but averaging 3.5%. Percentages of the principal salts are: $NaCl$–2.35%, $MgCl_2$–0.5%, Na_2SO_4–0.4%, $CaCl_2$–0.11%, KCl–0.07%, $NaHCO_3$–0.02%; other salts are present in traces.

seaweed. Any of the large plants which grow in the sea, especially various marine algae as the rockweeds, kelps, sea lettuce, dulse, and similar plants.

sea whip. Any of the horny corals of the order Gorgonacea, especially those with long, flexible, axial rods bearing many polyps.

sebaceous. Of, pertaining to, or secreting sebum; resembling tallow or fat.

sebaceous gland. A cutaneous gland which secretes sebum, usually a saccular, holocrine gland, its duct opening into a hair follicle.

sebum. The fatty secretion of a sebaceous gland which oils the hair and lubricates the surface of the skin.

secodont. Pertaining to or designating teeth with sharp cutting edges.

[1]secondary. 1. After the first (primary) in importance. **2.** Occurring after the first, as *secondary* development. **[2]secondary.** A secondary feather; a flight feather attached to the ulna.

secretagogue. 1. Any substance which induces secretion. **2.** A substance present in certain foods, as meat extracts, which increases the rate of secretion of digestive glands.

secretary bird. A long-legged, African bird of prey (*Sagittarius serpentarius*) which feeds principally upon reptiles.

secrete. To produce a secretion.

secretin. A hormone produced by the duodenal mucosa which increases the rate of secretion of pancreatic juice; the first hormone to be discovered (1902).

secretion. 1. The act of secreting; the process by which a gland forms or elaborates, from materials supplied to it by the blood or surrounding fluids, a specific substance or substances which are discharged through a duct or into the blood or surrounding medium, these substances then being utilized by the body or excreted. **2.** The secretory product of a gland. **3.** *bot.* The elaboration by plant cells of substances such as resins, mucilages, gums, oils, or nectar.

secretory. Of or pertaining to secretion.

¹section. To cut through or divide.

²section. A cut or slice, as that of a tissue or organ.

sectorial. With sharp cutting edges, as teeth; carnassial.

secund. Unilateral, directed to one side only; occurring on one side, as the flowers in certain racemes or spikes.

sedentary. 1. Permanently attached to the substratum; not free-moving. **2.** Remaining in the same place; not migratory.

sedge. One of a number of grass- or rushlike herbs of the family Cyperaceae, as *Cyperus, Carex.*

sediment. That which settles to the bottom, as in a flask or lake.

seed. 1. *bot.* In flowering plants, a mature or ripened ovule consisting of an embryo enclosed within a seed coat or coats. In addition, stored food in the form of endosperm may be present **2.** *zool.* **a.** Semen or milt. **b.** Young oysters used for transplanting.

seed coat. The outer covering of a seed, consisting of an outer tegmen and an inner testa; the part which develops from the integuments.

seed fern. A pteridosperm, a fossil plant of the order Cycadofilicales, class Gymnospermae, the oldest and most primitive of seed plants.

seed leaf. A cotyledon.

seedless. Lacking seeds. *See* parthenocarpy.

seedling. 1. Any plant, especially a young plant which develops from a seed. **2.** A plant grown in a nursery for transplanting. **3.** A tree under three feet in height.

seed oyster. A young oyster, especially one used for transplanting.

seed plant. **1.** Any plant which produces seeds, comprising the ginkgo, cycads, conifers, monocots, and dicots. **2.** A spermatophyte; a gymnosperm or angiosperm. **3.** Any flowering plant.

seed shrimp. Any crustacean of the class Ostracoda.

seed stalk. The funiculus or stalk by which a seed is attached to the wall of a mature ovary.

seed tick. The six-legged larva of a tick.

segment. **1.** A part of an organism or a structure which is marked off or separated from adjacent parts, as a *segment* of a leg, worm, or stem. **2.** A metamere. **3.** A somite.

segmentation. **1.** The state of being divided into segments; metamerism. **2.** *embryol.* CLEAVAGE. **3.** Something divided into parts, as the division of a nucleus in schizogony.

segmentation cavity. The blastocoel.

segmenter. In *Plasmodium*, a mature schizont in which the chromatin division is complete.

segregation. **1.** Something in isolation. **2.** *genetics.* The separation of members of allelomorphic pairs of genes during meiosis, resulting in each gamete receiving only one of each pair.

seiche. A local periodic rise and fall in the water level of a lake or landlocked sea.

¹selachian. Of or pertaining to the Selachii, an order of the class Chondrichthyes, which includes the sharks, skates, and rays.

²selachian. Any member of the Selachii.

selaginella. Any member of the genus *Selaginella*, a lycopod, which comprises the little club mosses.

selection. In evolution, any process by which certain individuals or groups are favored and survive and propagate while others fail to do so and become extinct.

selenium. A nonmetallic element, symbol Se, present in soils in parts of western United States. It accumulates in certain plants, as *Astragulus*, a vetch, which, when eaten by domestic animals, causes alkali poisoning or blind staggers.

selenodont. Pertaining to a type of dentition characteristic of ruminants, in which teeth possess vertical, crescent-shaped folds of enamel enclosing the softer dentine.

self-differentiation. The capacity of a part of the embryo to develop independently of other parts. *See* determination.

self-fertilization. The union of male and female gametes produced by the same individual.

self-pollination. Pollination effected by pollen produced by an anther of the same flower or by pollen produced by another flower of the same plant.

self-sterility. 1. A condition in hermaphroditic organisms in which ova are incapable of being fertilized by the individual's own sperm. **2.** The condition in plants in which self-pollination fails to bring about development of ovules.

sella turcica. A depression on the upper surface of the sphenoid bone which lodges the pituitary gland. Also called hypophyseal fossa.

sematic. Functioning as a warning or danger signal, as certain colors, odors, or sound devices.

semen. The viscid, whitish fluid produced by the male which contains the spermatozoa.

semicanal. A canal open on one side; a sulcus.

semicell. One of two halves of a desmid separated by a constriction, the sinus, and connected by a narrow isthmus which contains the nucleus.

semicircular canal. One of the three curved, bony canals of the osseous labyrinth of the internal ear, each filled with perilymph and housing the corresponding semicircular duct.

semicircular duct. One of the three, curved semicircular ducts lying within the corresponding semicircular canal and forming a part of the membranous labyrinth. Each is filled with endolymph and has an expanded end, the ampulla, within which are located the cristae ampullares containing hair cells sensitive to movement.

semilethal. Partly lethal, with reference to genes which **a.** produce death in at least 50% of a population; **b.** cause death but not until some of the afflicted individuals reach maturity and produce offspring.

¹semilunar. Resembling a half-moon in shape.

²semilunar. Old term for lunate bone.

semilunar cartilage. A meniscus, one of two crescentric discs of fibrocartilage of the knee joint.

semilunar valve. A heart valve consisting of three pocket-shaped cusps (valvules) which prevent backflow into the heart from the pulmonary artery or aorta.

semilunate. Crescent-shaped; SEMILUNAR.

seminal. Of or pertaining to seed or semen.

seminal duct. A sperm-conveying duct, especially the ductus deferens.

seminal fluid. The semen.

seminal leaf. A cotyledon.

seminal receptacle. A saclike structure in various invertebrates which stores sperm received from the male; a spermatheca.

seminal root. An embryonic root present in a seed.

seminal vesicle. 1. A dilated portion of or a saclike structure connected with the sperm duct which stores sperm preliminary to their discharge from the body, as in various invertebrates. **2.** In mammals, a glandular diverticulum of the ductus deferens which produces a viscid secretion rich in glucose which forms a part of the semen.

semination. The discharge of sperm from the testes. *Compare* inseminate.

seminiferous. 1. *bot.* Producing or bearing seed. **2.** *zool.* Producing or conveying sperm or semen.

seminiferous ampulla. A spherical, sperm-producing structure in the testes of lower vertebrates, as fishes, amphibians.

seminiferous tubule. A sperm-producing tubule in the testis of various vertebrates.

semipalmate. With toes partially webbed, as in certain shore-birds.

semipermeable. Partially permeable, with reference to the cell or plasma membrane, through which certain molecules or ions pass readily but others do not.

semisterile. With reduced reproductive capacity.

senescence. The state of being old or aged.

senile. Of, pertaining to, or characteristic of old age.

senna. A leguminous plant (*Cassia*) or a cathartic drug obtained from it.

sensation. A feeling; a state of consciousness or an awareness resulting from stimulation of a sensory receptor or arising subjectively from an unknown cause.

sense. The faculty by which an organism is aware of an environmental change registered through a sensory receptor, as the *sense* of sight or muscle *sense*.

sense club. *See* lithostyle.

sense organ. 1. A receptor. **2.** An organ which contains sensory cells which, when stimulated, give rise to a sensation, as the eye, ear.

sensibility. The ability to experience sensations; sense perception.

sensillum pl. **sensilla.** A sense organ or sensory receptor in insects and other arthropods.

sensitive plant. A tropical legume (*Mimosa pudica*) whose leaflets fold when touched.

sensitivity. The quality of being sensitive or responsive.

sensitization. The process by which an individual is rendered sensitive to an allergen, usually a protein. *See* allergy.

sensorium. The region of the brain concerned with the development of sensations, comprising principally the cerebral cortex and thalamus.

sensory. Of, pertaining to, or concerning sensations or the senses.

sensory area. A region in the cerebral cortex in which impulses arising in sensory receptors come into consciousness, as the somesthetic, visual, and auditory receptive areas.

sensory ganglion. A ganglion on the dorsal root of a spinal nerve or on a cranial nerve which contains the cell bodies of sensory neurons.

sensory neuron. An afferent neuron which conveys impulses from a receptor to the brain or spinal cord.

sepal. A modified leaf forming one of the components of the calyx of a flower.

sepaloid. Resembling a sepal.

sepia. 1. A cuttlefish of the genus *Sepia*. **2.** A rich brown, melanin pigment obtained from the ink of various cuttlefishes.

sepicolous. Inhabiting or living in hedges.

sepsis. A condition resulting from the presence and spread of pathogenic bacteria, or their products, or both in the blood and tissues.

septa. Plural of SEPTUM.

septal. Of or pertaining to a septum.

septate. Divided into parts or divisions by one or more septa.

septenate. Divided into seven parts.

septic. Of or pertaining to sepsis.

septicemia. The presence of and spread of pathogenic organisms throughout the body by way of the blood stream; blood poisoning; bacteremia.

septicidal. *bot.* Dehiscing along or through the partitions which separate the locules of a capsule or fruit.

septifragal. Designating a type of dehiscence in which the valves break away from the partitions, which remain attached to the axis.

septum pl. **septa.** A wall or membrane which divides a cavity or a structure into two or more parts; a dissepiment.

sequela. 1. That which follows or is a consequence of. **2.** *med.* An abnormal condition resulting from a disease.

sequoia. A coniferous tree of the genus *Sequoia*, as the giant sequoia (*S. gigantea*) and the redwood (*S. sempervirens*) of California, the oldest and largest of living things; estimated age, 3500 years; height, 275 feet.

seral. Of or pertaining to a sere.

sere. *ecol.* A complete series of successional changes in a community from the initial stage through transitional stages to the climax. *See* hydrosere, lithosere, xerosere.

serial. Arranged in a row or sequence; following one another in succession.

serial homology. That exhibited by a series of homologous structures arranged serially, as seen in the paired appendages of a crayfish and related crustaceans, in which all have the same basic structure but are modified according to the functions they perform. *See* homology.

seriate. Arranged in a series.

sericate. SILKY.

sericeous. Silky; covered with fine silky hairs closely appressed to the surface.

sericulture. The raising of silkworms for the production of silk.

serine. An amino acid, $C_3H_7NO_3$.

serosa. 1. A serous membrane. **2.** A membrane which envelops the embryo of an insect. **3.** The chorion of reptiles and birds.

serotinal. Developing later in the season than is customary; late in the day or season; serotinous.

serotonin. 5-hydroxytryptamine (5-HT), a vasoconstrictor agent present in the tissues of vertebrates, some invertebrates, and a few plants; considered to be a neurohumoral substance similar to epinephrine.

serous. 1. Thin and watery, like serum. 2. Pertaining to, producing, or containing serum.

serous cavity. A cavity lined with a serous membrane.

serous cells. Gland cells which produce a thin, watery secretion, as those of the salivary glands.

serous fluid. A thin, watery fluid, as that present in a serous cavity.

serous membrane. A membrane which lines a serous cavity and covers the enclosed visceral organs, as the pleura, pericardium, and peritoneum; a serosa.

serow. A heavy, ungainly, goatlike, bovine mammal (*Capricornis*) of eastern and southern Asia.

serpent. A snake, especially a large one.

serpentine. Sinuous, like a snake; winding, twisting.

serpent star. An echinoderm of the class Ophiuroidea.

serrate. 1. *zool.* Notched; having teeth like a saw. 2. Designating a type of leaf in which the points of the marginal teeth are directed toward the apex.

serrula. A comblike structure, as that on the chelicera of a pseudoscorpion or on the pedipalp of a spider.

serrulate. Finely serrate; bearing minute teeth.

Sertoli cell. A sustentacular cell in the epithelial lining of a seminiferous tubule of a mammalian testis.

serum. 1. The clear fluid which exudes from a clot in the coagulation of blood, consisting of plasma minus fibrogen. 2. A biological product obtained from the blood of animals or humans that is used for prophylactic, diagnostic, or therapeutic purposes. 3. A clear fluid which moistens the surfaces of serous membranes and fills spaces within serous cavities.

serval. A wild cat of Africa (*Leptailurus serval* or *Felis serval*) a spotted cat with long ears lacking tufts.

sesame. An herb (*Sesamum indicum*), the source of sesame oil and sesame seeds used for flavoring.

sesamoid bone. One which develops in a tendon, as the patella.

sessile. 1. *bot.* Attached without a stalk, petiole, or peduncle, as a *sessile* leaf. **2.** *zool.* **a.** Attached; nonmotile; sedentary. **b.** lacking a stalk or peduncle, as a *sessile* tumor.

sessoblast. An internal bud (statoblast) in an ectoproct which remains attached to the parent.

seston. All bodies which swim or float in water. Includes bioseston and abioseston.

seta pl. **setae. 1.** *zool.* A stiff bristle or bristlelike structure, as the chitinous bristles of an annelid, the hairs of a caterpillar, the tactile hairs of a crustacean, or the fringe of the wing of thrips. **2.** *bot.* **a.** A bristle. **b.** The stalk which supports the capsule in a bryophyte.

setaceous. 1. Bristlelike. **2.** Bearing bristles. **3.** Composed of bristles.

setiform. Shaped like a bristle.

setose. Bearing many bristles. SETACEOUS.

setula. A small, short hair or bristle.

setule. A minute bristle, as those on the margin of the filtering setae of a crustacean.

setulose. Having many small bristles or setulae.

sex. 1. One of two types of organisms, male and female, as distinguished by the nature of the reproductive cells produced, males producing sperm or spermatozoa and females producing eggs or ova. **2.** The total of the characteristics, both anatomical and physiological, which distinguish a male from a female.

sex cell. A male or female gamete; a spermatozoan or sperm cell; an ovum or egg; a macro-, micro-, or isogamete.

sex chromatin. A small chromatin body present at the periphery of the nucleus in certain somatic cells, as neurons and neutrophils; present in the cells of females, absent in males.

sex chromosome. One of a pair of chromosomes (XX in the female, XY or XO in the male) which play a primary role in the determination of sex. In birds, some insects, and fishes, the males possess two X-chromosomes, the females having a single X with or without the Y.

sex determination. The establishment at fertilization of conditions within a zygote which result in an individual developing into a male or female, usually the result of chromosomal constitution. This is referred to as genetic sex.

sex differentiation. The process by which a sexually indifferent embryo develops into either a male or female.

sex gland. A gonad.

sex hormone. 1. *zool.* A hormone produced by the gonad or other organs which is responsible for the development of secondary sex characteristics and accessory reproductive organs and structures. *See* androgen, estrogen, progesterone. **2.** *bot.* A diffusible substance produced by some algae and molds which influences the development of the sexes.

sex-limited character. One which is expressed in only one sex, although its genes are present in the autosomes of both sexes, as milk production in cattle.

sex-linked inheritance. The transmission of traits whose genes lie in or on the sex chromosomes, resulting in the character appearing more frequently in one sex than the other, as hemophilia in man.

sex organ. A reproductive organ.

sex ratio. The ratio of males to females; for man, it is approximately 105 males to 100 females at birth.

sex reversal. The transformation of an individual of one sex into an individual of the opposite sex.

sex transformation. SEX REVERSAL.

sexual. Of, pertaining to, or associated with sex.

sexual generation. In alternation of generations, the generation which produces gametes; in plants, the gametophyte.

sexual intercourse. Coition: COPULATION. *Compare* amplexus.

sexual reproduction. Reproduction which involves sex cells or gametes.

sexual selection. A theory proposed by Charles Darwin to account for extreme differences between the sexes based on the assumption that individuals possessing the most pronounced secondary sex characters usually win or are selected by individuals of the opposite sex for mating.

shad. A herring (*Alosa sapidissima*) which spawns in fresh water; an important food fish.

shade plants. Green plants which are able to live under conditions of low light intensity, hence tolerate a considerable amount of shade.

shaft. The main axis or trunk of an elongated structure, as **a.** the diaphysis of a long bone; **b.** the rachis of a feather; **c.** the exposed portion of a hair.

shag. A cormorant.

shagreen. 1. Sharkskin, bearing sharp-pointed placoid scales. **2.** In Russia, untanned leather.

shank. 1. In man, the portion of the lower limb between knee and ankle. **2.** In some vertebrates the part between the fetlock and the joint above.

shark. Any of a number of voracious, predatory, elasmobranch fishes, as the dogfish shark (*Squalus* or *Mustelus*), great white shark (*Carcharodon*), hammerhead shark (*Sphyrna*), sand shark (*Carcharius*), and whale shark (*Rhineodon*).

shark sucker. A remora.

sharptail. Common name for the pintail or sharp-tailed grouse.

shearwater. A gull-like bird of the genus *Puffinus*, related to the fulmars and petrels.

sheath. 1. An investing covering or protective envelope. **2.** *zool.* **a.** A tubular fold or sac into which a protrusible organ, as a penis, proboscis, or cirrus, is retracted. **b.** The protective covering of a blood vessel, nerve, muscle, or tendon. **c.** The delicate covering of a nerve fiber or a microscopic structure, as a mitochondrion. **3.** *bot.* **a.** The gelatinous covering of certain algal cells or filaments. **b.** The base of a leaf which envelops a stem, as in grasses.

sheep. A horned ruminant of the genus *Ovis*, family Bovidae, which includes many domesticated varieties raised for their flesh (mutton), wool, and hides. There are also a number of varieties of wild sheep. *See* bighorn.

sheep louse. A biting louse (*Bovicola ovis*).

sheepshead. 1. A marine food fish (*Archosargus probatocephalus*) of the Atlantic and Gulf coasts. **2.** The freshwater drum (*Aplodinotus grunniens*) of southern United States and Mexico.

sheep tick. The sheep ked (*Melophagus ovinus*), a small wingless, dipterous insect of the family Hippoboscidae, parasitic on sheep.

sheldrake. One of several European surface-feeding ducks of the genera *Casarca* and *Tadorna*.

shell. 1. *zool* **a.** The hard, rigid exoskeleton or covering of various animals, as that of mollusks, crustaceans, and reptiles. **b.** The hard, calcareous covering of an egg. **2.** *bot*. The hard, outer covering of a fruit or seed, especially a nut; a husk or pod.

shellac. A lac resin, a substance prepared from stick-lac, a resinous substance secreted by the lac insect.

shellfish. Any aquatic invertebrate possessing a shell, especially any edible mollusk or crustacean, as oysters, clams, lobsters, shrimps.

shell gland. 1. The region of the oviduct in sharks or of the uterus in domestic fowl which secretes the shell of the egg. **2.** Term formerly applied to Mehlis' gland of trematodes and cestodes. **3.** The nidamental gland of a cephalopod.

shell membrane. A parchmentlike membrane underlying the calcareous shell in the eggs of various birds.

shield. A protective structure, as a plate, scute, or carapace.

shin. 1. The sharp, anterior margin of the tibia. **2.** In cattle, the lower portion of the forelimb.

shinbone. The tibia.

shiner. Any of a number of freshwater minnows of the family Cyprinidae, especially those of the genus *Notropis*.

shipworm. One of a number of wood-boring bivalves, especially *Teredo navalis*, which causes extensive damage to wooden vessels and piles of wharves. Also called pileworm.

shock. A state of collapse often following hemorrhage or severe physical or emotional trauma characterized by a marked fall in blood pressure; weak, rapid pulse; pale, cold skin; cyanosis; and thirst, these symptoms resulting from an inadequate volume of circulating blood.

shoot. 1. In a young seed plant, the main portion showing above the ground comprising the stem and leaves, taken collectively. **2.** A young branch with its leaves.

shoot system. In a seed plant, the stem with its branches and leaves.

shorebird. One of a nunber of long-legged birds which frequent sea beaches and shallow inland waters, especially those of the suborder Charadrii, which includes oyster-catchers, sandpipers, snipes, avocets, stilts, phalaropes, plovers, turnstones, killdeer, woodcocks, yellowlegs, and curlews; commonly called waders.

short-day plant. A plant which will flower only if the daily period of illumination is less than a particular critical length (less than 12 hours), as most plants which bloom normally in the early spring or late fall.

shoulder. The region where the forelimb or arm joins the body.

shoulder blade. The scapula.

shoulder girdle. The pectoral girdle.

shoulder joint. The articulation between the scapula and humerus, a ball-and-socket joint.

shoveller. A surface-feeding duck (*Spatula* (*Anas*) *clypeata*) with a long, broad bill.

shrew. 1. A small, mouselike mammal of the order Insectivora, most belonging to the genus *Sorex* which includes the pygmy shrew (*S. minutus*), the smallest mammal known, weighing less than one ounce. **2.** A primitive primate, the tree shrew.

shrike. A predaceous, oscine bird of the family Laniidae which feeds on insects and smaller birds, sometimes impaling its prey on thorns; also called butcher-bird.

shrimp. 1. One of a number of marine, decapod crustaceans of the suborder Natantia, especially those of the genus *Crago*, used extensively for food. **2.** Any of a number of crustaceans which resemble true shrimps, as the brine shrimp, fairy shrimp, and many others.

shrub. A bush; a low-growing, perennial plant, usually with several main stems arising near the ground.

siamang. A long-armed, tailless primate (*Symphalangus* (*Hylobates*) *syndactylus*) of Sumatra and the Malay Peninsula, the largest of gibbons.

Siamese twins. Twins which are united; conjoined or parabiotic twins.

siblings. Children of the same parents, either male or female.

sibs. SIBLINGS.

sicklebill. Any of a number of birds with a strongly curved bill, as the curlew.

sidewinder. A western rattlesnake (*Crotalus cerastes*).

sieve area. A perforated region of a sieve plate.

sieve cell. 1. A phloem cell of a gymnosperm and simpler vascular plants. **2.** A sieve tube element.

sieve element. A sieve cell of a sieve tube element.

sieve plate. A perforated structure separating the ends of sieve cells or sieve tube elements in angiosperms.

sieve tube. A conducting structure in phloem consisting of a series of sieve tube elements joined together end to end and separated by sieve plates.

sieve tube element. One of a series of elongated cells which form the units of a sieve tube, the cytoplasm of one element being continuous with that of adjoining elements through pores in the sieve plate.

sight. Vision; the sense of seeing.

sigmoid. Shaped like the letter S; curved in two directions.

sign. A physical manifestation of a disease. *Compare* symptom.

signal mark. A conspicuous marking which serves as a warning signal to other members of a herd, as the white hair on the rump of a pronghorn.

sika. A Japanese deer (*Cervus nippon*).

silica. Silicon dioxide, SiO_2, the principal constituent of sand. It occurs naturally as quartz, flint, opal, and in other forms comprising over 27% of the earth's crust. It is an important component of the skeletons of diatoms, radiolarians, and sponges.

siliceous. Of, pertaining to, or containing silica.

silicle. A short silique.

silicon. A nonmetallic element, symbol Si, at. no. 14. *See* silica.

silique. A many-seeded fruit in which the two valves which split from the bottom are separated by a replum or false partition, characteristic of the mustard family, Cruciferae.

silk. 1. A fine, lustrous fiber produced by the silk glands of various insects, especially lepidopterans, and used in the construction of cocoons. **2.** The fiber produced by silkworms and used for weaving into fabrics. **3.** A fine fiber produced by spiders and used in making webs, egg sacs, insect traps, retreats, shelters, and aerial floats. **4.** *bot.* The elongated style of a single pistillate flower on the ear of maize.

silkworm. The larva of certain moths, especially *Bombyx mori*, raised for the production of silk. *See* sericulture.

499

silkworm disease. Pebrine, caused by a microsporidian (*Nosema bombycis*) discovered by Pasteur in 1870.

silky. 1. Resembling silk in appearance or texture. **2.** *bot.* SERICEOUS.

Silurian period. A period of the Paleozoic era between the Ordovician and Devonian periods, characterized by the rise of land plants and fishes.

silverfish. 1. A primitive, wingless insect (*Lepisma saccharina*) of the order Thysanura, a common household pest; also called bristletail. **2.** One of a number of fishes, as the tarpon, silverside.

silverside. A freshwater fish of the family Atherinidae, especially those of the genera *Labidesthes* and *Menidia*, the latter inhabiting brackish water.

simuliid. Any dipterous insect of the family Simuliidae which includes the black flies and buffalo gnats.

sing-sing. *See* defassa.

sinistral. Of or pertaining to the left side; inclined or turning to the left; left-handed. *Compare* dextral.

sinistrorse. Turning or twining to the left. *Compare* dextrorse.

sinoatrial. Pertaining to the sinus venosus and the atrium.

sinoatrial node. The S-A node, a mass of specialized cardiac muscle in the wall of the right atrium which is the source of impulses which initiate each heartbeat; the pacemaker of the heart.

sinuate. 1. Wavy, tortuous. **2.** *bot.* Having a wavy margin with deep indentations, with reference to leaves.

sinuous. 1. Serpentine; bending in and out; winding. **2.** *bot.* SINUATE.

sinus. 1. *zool.* A cavity, hollow, recess, channel, or space. **2.** *bot.* A deep cleft or recess, as that separating **a.** two lobes or divisions of a leaf; **b.** two semicells of a desmid.

sinus gland. A neuroendocrine gland in certain crustaceans.

sinusoid. A minute blood channel lacking an endothelial wall which connects arterioles and venules in organs, as the liver, adrenal gland, bone marrow, and hypophysis.

sinus venosus. A chamber in the heart of lower vertebrates which receives venous blood and passes it to the atrium; present as a temporary structure in the embryos of higher vertebrates.

siphon. 1. A tube or tubelike structure for drawing in or expelling fluids, as that in bivalve mollusks, cephalopods, lepidopterans, and tunicates. **2.** In elasmobranchs, a blind, muscular sac which functions in the ejection of sperm. **3.** A feeding polyp (gastrozooid) in siphonophores. **4.** An elongated tube or a constricted portion of the intestine in echinoids.

siphonaceous. Tubular, with reference to filaments of certain algae.

siphonapteran. Any insect of the order Siphonaptera which includes the fleas, small, wingless, leaping insects which live as ectoparasites on birds and mammals.

siphonoglyph. A ciliated groove along one or both sides of the gullet in a sea anemone.

siphonophore. A coelenterate of the order Siphonophora which includes swimming and floating, colonial, polymorphic hydrozoans, as the Portugese man-of-war.

siphonostele. Vascular tissue in the form of a hollow cylinder with pith occupying the central portion.

siphonous. Designating a type of algae in which the thallus consists of a filamentous structure composed of elongated, multinucleate cells.

siphonozooid. A modified polyp which functions in the creation of water currents, as in polymorphic siphonophores.

siphuncle. A slender, tubular extension of the visceral hump in a chambered nautilus which extends through openings in the septa to the innermost chamber of the shell.

sipunculid. Any invertebrate of the phylum Sipunculida which includes the peanut worms, a small group of marine worms related to the annelids, as *Sipunculus, Phascolosoma*.

siren. An eel-shaped, tailed amphibian of the family Sirenidae, as *Siren lacertina*, the mud eel of southern United States.

sirenian. Any mammal of the order Sirenia which includes the manatees and dugongs, large, marine, herbivorous vertebrates.

sirenin. A substance (ectohormone) produced by female gametes of the water mold (*Allomyces*) which attracts male gametes.

sirup. *See* syrup.

sisal. A fiber obtained from various species of *Agave*, especially henequen or Yucatan sisal, used principally for binder twine.

siskin. A small songbird of the finch family, as *Spinus pinus*, the pine siskin.

skate. A ray. *See* ²ray.

skeletal. Of or pertaining to the skeleton.

skeletal muscle. A muscle or muscle tissue which is attached to or moves a part of the skeleton; in vertebrates, striated or striped muscle.

skeletal system. In vertebrates, the structures composed of bone or cartilage which form a framework giving general form to the body and serving for support, protection, and attachment of muscles. It also is a source of blood cells.

skeleton. The rigid or semirigid supporting framework or protective structures of an organism or a part. *See* endoskeleton, exoskeleton.

skimmer. 1. A dragonfly of the family Libellulidae. 2. A marine bird (*Rynchops*), as the black skimmer of the east coast.

skin. 1. The integument or surface covering of a vertebrate, consisting of two layers, the epidermis and dermis (corium). 2. The surface covering of any organism. 3. *bot.* An investing membrane, as the rind or peel of a fruit.

skink. Any of a number of small lizards of the family Scincidae, as *Eumeces*, *Lygosoma*, and *Neoseps* of the United States.

skipjack. 1. One of a number of fishes which swim at the surface of the water leaping in and out, as the bonito and related forms. 2. A click beetle.

skipper. 1. A small, stout-bodied butterfly of the family Hesperiidae characterized by short, rapid, and erratic flight. 2. A marine fish of the family Scombresocidae, with long, toothless jaws.

skipping blades. Narrow, elongated appendages (paddles) forming two clusters located at the posterior end of certain rotifers.

skua. A large, predatory marine bird of the genera *Catharacta* and *Stercorarius*, related to the jaegers.

skull. In vertebrates, the cartilaginous and/or bony framework of the head which encloses and protects the brain and sense organs and forms the supporting framework of the nasal and buccal cavities; consists of the cranium and facial bones.

skunk. A fur-bearing mammal of the family Mustelidae, noted for the possession of perianal glands which secrete an offensive, nauseating, odorous fluid used for defense, as *Mephitis mephitis*, the striped skunk of the United States.

skunk bear. The wolverine.

skunk cabbage. A perennial plant (*Symplocarpus foetidus*) of eastern United States, of the arum family, with a strong odor like that of a skunk; also a related plant (*Lysichitum americanum*) of the western states.

skylark. The Old World lark (*Alauda arvensis*).

slavery. The subjugation of individuals of one species by those of another, as that practiced by ants of the genus *Polyergus* which raid the nests of other ants.

sleep. A natural, periodic state of diminished activity with lowered metabolic rate, dulling of sensations, and finally loss of consciousness, during which fatigue is relieved and the body rested.

sleeper. A marine, percomorph fish (*Eleotris*) of the family Eleotridae.

sleeping sickness. 1. Trypanosomiasis, a chronic disease resulting from infection by protozoans of the genus *Trypanosoma*, as African sleeping sickness, caused by *T. gambiense* and *T. rhodesiense* and transmitted by the tsetse fly, and South American trypanosomiasis (Chagas' disease), caused by *T. cruzi* and transmitted by reduviid bugs. **2.** Encephalitis lethargica, an epidemic disease of viral origin thought to be transmitted by arthropods.

sleep movement. A nyctinasty, as occurs in certain plants in which the leaflets fold together at night or on cloudy days, as in clovers.

slider. A freshwater turtle of the genus *Pseudemys*, especially *P. scripta*, the pond slider.

slime. A viscid, mucous secretion, especially that of the cutaneous glands of various animals, as snails, slugs, various annelids, hagfishes, catfishes, and amphibians.

slime bacteria. Myxobacters, bacteria of the order Myxobacteriales, which produce a slimy growth, common in the soil, rotting material, and organic wastes.

slimeball. A mass of cercariae formed in the respiratory cavity of a snail which, when discarded, is transported by ants to their nest, the metacercariae developing in ants, which, when eaten by sheep, develop into a liver fluke (*Dicrocoelium*).

slime bodies. Discrete bodies of a viscous, proteinaceous substance commonly present in the sieve tube cells and companion cells of dicots.

slime eel. The hagfish (*Myxine limnosa*).

slime fungi. The Myxomycetes or slime molds.

slime gland. 1. Any gland which produces slime. 2. In *Peripatus*, one of a pair of glands opening on the oral papillae from which slime can be forcibly ejected.

slime mold. Any fungus of the division Myxomycota comprising two groups, the true slime molds of the class Myxomycetes characterized by a naked, motile, multinuclear assimilative stage called a plasmodium, as *Physarium*, and the cellular slime molds of the classes Acrasiomycetes and Labyrinthulomycetes.

slime tube. A tube formed around the anterior segments and clitellum of an earthworm which, on being discharged, forms a cocoon within which fertilized eggs develop.

slipper animalcule. A ciliate protozoan of the genus *Paramecium*.

slipper shell. Any gastropod of the genus *Crepidula*.

slit sense organ. A lyriform organ of an arachnid.

sloth. A slow-moving, arboreal edentate of Central and South America with long claws by which it hangs downward from branches, as the ai or three-toed sloth of the genus *Bradypus*, or the unau or two-toed sloth of the genus *Choloepus*.

sloth bear. A bear (*Melursus ursinus*) of India and Ceylon.

slough. 1. The cast-off skin of an animal, as that shed by a snake. 2. *med*. A mass of necrotic tissue separated from living tissue.

slowworm. A limbless lizard (*Anguis fragilis*) of the family Anguidae, the blindworm of the Old World.

slug. 1. One of a number of terrestrial, pulmonate gastropods resembling a land snail in structure but with shell much reduced or absent, as *Arion, Limax, Testacella*. 2. The larva of a sawfly which resembles a tiny slug. 3. A pseudoplasmodium.

slug caterpillar. The larva of lepidopteran insects of the family Limacodidae which are sluglike in appearance and movement.

smear. A preparation made for microscopic examination by taking a small amount of the substance, as blood, seminal fluid, or bacterial culture, and spreading it thinly on a slide. *See* squash.

smegma. The cheeselike material which collects between the glans penis or glans clitoridis and the prepuce, secreted by sebaceous glands.

smell. **1.** To perceive through the olfactory sense or organs of smell. **2.** To detect a scent, as to *smell* a flower.

smelt. One of a number of small fishes of the family Osmeridae, especially those of the genus *Osmerus* used extensively for food.

smoke tree. One of several trees of the genus *Cotinus*, its flowers forming finely plumose panicles suggestive of smoke.

smolt. The young of salmon when about two years old as they descend streams leading to salt water. *Compare* grilse.

smooth. **1.** Having an even surface or margin; devoid of irregularities. **2.** *bot.* Lacking projections or hairs; glabrous.

smooth colony. *bacteriol.* A type of colony characterized by a smooth surface, the individuals of which exhibit a sensitivity to bacteriophage, resistance to phagocytosis, and are good immunizing forms.

smooth muscle. A type of involuntary muscle composed of elongated, nonstriated, spindle-shaped cells found in the walls of tubes and hollow organs, as the alimentary canal, blood vessels, urinary and gall bladder.

smut. A disease caused by a smut fungus or the fungus which causes the disease.

smut fungus. One of a number of basidiomycete fungi which are parasitic on higher plants, especially grasses. Some produce black, sootlike masses of spores (loose smut); others form smut balls containing covered or kernel smut. Important genera are *Ustilago* and *Tilletia*.

snail. A mollusk of the class Gastropoda, especially those with a well-developed spiral shell.

snailfish. A sea snail, a fish of the family Liparididae, with a sucker formed from pelvic fins, as *Liparis*.

snake. **1.** An elongated, creeping reptile of the order Squamata, suborder Serpentes (Ophidia). **2.** One of a number of snakelike animals, as the glass snake, congo snake, horsehair snake.

snakebird. A water turkey.

snakeroot. One of a number of plants which reputedly possess curative properties for snakebites, as the Virginia snakeroot (*Aristolochia serpentaria*).

snapdragon. A garden plant of the genus *Antirrhinum*.

snapper. 1. An animal that snaps, as a snapping turtle or snapping beetle. 2. An important food and game fish of the family Lutjanidae, as *Lutjanus griseus*, the gray snapper. However some species are highly poisonous, as the Cuban snapper.

snapping beetle. A click beetle.

snapping shrimp. The pistol shrimp (*Alpheus*), in which the rapid and forceful closure of the movable finger of the cheliped produces a snapping sound.

snapping turtle. A large, predaceous, freshwater turtle of the family Chelydridae, comprising two species, *Chelydra serpentina*, the common snapping turtle, and *Macrochelys temmincki*, the alligator snapping turtle.

snipe. One of a number of shorebirds of the family Scolopacidae, as Wilson's snipe (*Capella* (*Gallinago*) *gallinago*) of the United States and the European jacksnipe (*Limnocryptes minimus*).

snook. A food and game fish of the family Centropomidae often ascending rivers of southern United States, as *Centropomus undecimalis* of Florida. Also called robalo.

snout. 1. A long, projecting nose, as that of swine. 2. An anterior extension of the head, as seen in various animals, as weevils; a rostrum.

snout beetle. A beetle of the Rhynchophora characterized by possession of an elongated beak or snout, as the curculios and weevils.

snowbird. A finch of the genus *Junco*.

snow bunting. A finch of the genus *Plectrophenax*.

snowflake. A snow bunting.

snow flea. A springtail (*Achorutes nivicolus*) often appearing on the surface of the snow.

snow goose. A species of goose of the genus *Chen*, as the greater snow goose (*Chen hyperborea*).

snow leopard. The ounce (*Uncia uncia*) of central Asia.

snowshoe rabbit. The varying hare (*Lepus americanus*) with summer coat reddish brown, winter coat white with black ear tips; common in northern United States and Canada.

506

snowy. White; resembling snow in color.

snowy owl. A large owl (*Nyctea nyctea*) of northern regions.

sobole. A shoot, especially one coming from the base of the stem near the ground.

soboliferous. Producing shoots near the ground and forming a clump, said of certain shrubs and trees, as the lilac.

social. Living together in organized communities.

social insects. Termites, ants, bees, and wasps which develop highly organized colonies consisting of groups of individuals differing in structure and function which live and work together as a unit. Castes (queens, drones, workers, and soldiers) may exist.

social parasitism. Nest parasitism, a mode of life of certain birds, as cuckoos, cowbirds, and some ducks in which nest-building is dispensed with and eggs are laid in the nest of other birds, called hosts or fosterers which care for the young when hatched.

society. 1. A closely knit group of individuals within a species which interbreed and associate together, exhibiting mutual dependence upon other members of the group, as a school of fishes, flock of birds, or herd of deer. **2.** A highly organized group of individuals, as a colony of ants, bees, wasps, or termites. *See* social insects.

soda. Sodium carbonate.

soda ash. Commercial sodium carbonate (anhydrous).

soda lime. A mixture of caustic soda (sodium hydroxide) and calcium hydroxide (quicklime), usually in granular form, used to absorb carbon dioxide.

sodium. A white, metallic element of the alkali group, symbol Na, at. no. 11.

sodium bicarbonate. A white, crystalline powder, $NaHCO_3$, a mild alkali; called baking or cooking soda.

sodium carbonate. Soda, Na_2CO_3. Its hydrated form is called sal soda or washing soda.

sodium chloride. Common salt, $NaCl$, one of the principle electrolytes in body fluids.

sodium hydroxide. Caustic soda, $NaOH$, a strong alkali.

soft palate. The velum palati, a thin, musculomembranous partition which forms the posterior portion of the roof of the mouth. It closes the posterior openings to the nasal cavity during swallowing movements.

soft-shell or **soft-shelled.** Having a soft shell or lacking a hard one, a state often following molting.

soft-shelled clam. *Mya arenaria*, a burrowing clam with thin, whitish shell.

soft-shelled crab. A crab which has just molted, especially the blue crab.

soft-shelled turtle. A freshwater turtle (*Trionyx muticus*) with a flat, oval, leathery carapace, commonly called flapjack.

softwood. 1. Any wood that is soft in consistency and easily worked. **2.** *forestry.* The wood of a coniferous tree. *Compare* hardwood.

soil. The loose material on the surface of the earth in which plants grow, consisting of disintegrated rock and organic material (humus). It contains water, dissolved substances, as mineral salts, and living organisms (protozoans, algae, bacteria, fungi, and various metazoans, as nematodes, annelids).

soil profile. The various layers of soil, as seen in a vertical section obtained by making a boring, or as observed in a deep cut through the soil.

sol. The liquid or fluid phase of a colloidal system in which water forms the continuous phase. *Compare* gel.

solanaceous. Of or pertaining to the nightshade family, Solanaceae, a large family of herbs, shrubs, and trees, including many of economic importance, as the genus *Solanum* which includes the white potato. It includes a number of plants which are the source of drugs, as belladonna, hyoscyamus, and stramonium.

solar. Of, pertaining to, or from the sun, as *solar* energy.

solarization. Inhibition of photosynthesis resulting from exposure to light of high intensity.

solar plexus. The coeliac plexus, composed of sympathetic ganglia and nerve fibers which innervate visceral organs of the abdominal cavity; located behind the stomach near the base of coeliac artery.

solation. Conversion into a fluid or sol state, with reference to colloidal solutions.

soldier. A member of a caste of termites having a greatly enlarged head and mandibles; a sterile male which functions in defense of the colony.

¹**sole.** **1.** The plantar or under surface of the foot. **2.** The trivium of a holothuroid. *See* trivium. **3.** *bot.* The basal end of a carpel; that farthest from the apex.

²**sole.** One of a number of flatfishes, especially *Solea solea*, of European waters, a valuable food fish. Most American species belong to the genera *Symphurus* and *Achirus*.

solehorn. The subunguis.

solenia. Gastrodermal tubes which connect the polyps of colonial corals.

solenocyte. An excretory cell in polychaetes which possesses a flagellum as the current-producing structure; also present in lancelets.

solenodon. A small, ratlike, insectivorous mammal (*Solenodon cubanus*) of Cuba and Haiti.

solifugid. A solpugid.

solitaire. **1.** A large, flightless bird (*Pezophaps solitaria*) related to the dodo, formerly inhabiting the island of Rodriguez, extinct since 1750. **2.** A thrush (*Myadestes townsendi*), Townsend's solitaire.

solitary. **1.** Existing singly or alone. **2.** *bot.* Not occurring with other individuals of the same kind.

solpugid. An arachnid of the order Solpugida (Solifugae) which includes the sun spiders, tropical arachnids with enormous chelicerae.

soluble. Capable of being dissolved or going into solution.

solution. A homogeneous system resulting from the mixing of a solute, a solid, liquid, or gas with a liquid, the solvent. In a true solution, the molecules of the solute are uniformly dispersed among those of the solvent.

solvent. A liquid in which a substance, the solute, is dissolved.

soma. **1.** All cells and tissues of a body except the reproductive cells. **2.** The axial portion of an organism (head, neck, trunk, tail), excluding the appendages. **3.** The anterior portion of an appendiculate fluke, as a hemiurid. **4.** *psychol.* The body, as distinguished from the mind.

somatic. **1.** Of or pertaining to body cells in contrast to reproductive cells. **2.** Of, or pertaining to the body wall and appendages as distinguished from visceral structures, as *somatic* nerves.

somatocyst. A structure at the tip of a gastrovascular canal in a siphonophore, usually containing oil droplets.

somatogamy. The union of plus and minus mycelia, as in fungi.

somatoplasm. The soma.

somatopleure. *embryol.* The ectoderm and somatic mesoderm combined from which the lateral portions of the body wall are formed. *Compare* splanchnopleure.

somatotrophic hormone. The growth hormone, STH, secreted by the anterior lobe of the pituitary. Also called somatotrophin.

somesthetic area. A region in the parietal lobe of the cerebral cortex which receives sensory impulses from cutaneous and proprioceptive receptors and brings them into consciousness.

somite. 1. A body segment or metamere. 2. *embryol.* One of a series of paired, blocklike, mesodermal segments which lie alongside the notochord and neural tube of a developing vertebrate embryo.

sonar. An apparatus for detecting the position of objects through recording sound waves reflected by them, as that utilized by bats in avoiding obstructions or detecting presence of food objects by the emission of ultrasonic waves.

songbird. Any of a large number of birds of the order Passeriformes, suborder Oscines, which possess a syrinx enabling them to produce beautiful sounds.

soor. The Indian or crested wild boar (*Sus cristata*).

soporific. Causing or inducing sleep.

sora. A small, plump rail (*Porzana carolina*) of North America.

soredium. An asexual reproductive structure in lichens consisting of a group of algal cells enclosed within some hyphae.

sorghum. 1. A cultivated grass (*Sorghum vulgare*) grown for stock feed and forage and used in the manufacture of syrup, brushes, and paper. 2. A syrup obtained from sorgo or sweet sorghum.

sori. Plural of SORUS.

soricine. 1. Shrewlike. 2. Of or pertaining to the family Soricidae, which includes the true shrews.

sorophore. A stalk supporting a sorus in certain slime molds.

sorosis. A fleshy, multiple fruit, as a mulberry.

sorrel. One of a number of plants with a sour juice, as the sheep sorrel (*Rumex acetosella*) or the wood sorrel (*Oxalis*).

sorus pl. **sori.** 1. A cluster of sporangia in ferns. 2. A mass or cluster of spores, as that produced by certain rusts and smuts.

¹**sound.** 1. A sensation resulting from the stimulation of auditory receptors. 2. A form of vibrational energy transmitted through a medium, as air, water, or bone, which is capable of stimulating auditory receptors.

²**sound.** A long, narrow instrument used for introduction into a cavity or passageway for exploration, dilatation, and other purposes.

sound production. 1. In insects, sound produced by stridulation, vibration of wings or tymbals, tapping of the head, or friction of body parts. 2. In fishes, **a.** by rubbing teeth, bones, or spines against each other; **b.** by expulsion of air from the air bladder or through the anus. 3. In land vertebrates, **a.** by vibrating structures within a larynx or syrinx; **b.** by drumming with the wings.

sour. 1. Having a tart taste, as that of vinegar or unripe fruits. 2. Acid in reaction, as *sour* soil or *sour* milk.

sow. The adult female of swine.

sow bug. A wood louse, a land crustacean of the order Isopoda, as *Oniscus*.

soybean. A legume (*Glycine max*) indigenous to Asia but now extensively grown in the United States, its seed the source of soybean oil, milk, flour, and meal. Also grown for soil improvement.

space biology. The study of living things above the atmosphere or above an altitude of 50,000 ft., at which height environmental hazards of consequence to humans are encountered; exobiology, extraterrestrial biology.

spadefish. 1. A deep-bodied, spiny-finned food fish (*Chaetodipterus faber*) of the Atlantic coast. 2. Common name for the paddlefish.

spadefoot toad. A burrowing toad of the family Pelobatidae, as *Pelobates* of the Old World; *Scaphiopus* of the United States.

spadiceous. 1. Light brown or chestnut colored. 2. Resembling or bearing flowers on a spadix.

spadix. 1. *bot.* An indeterminate inflorescence consisting of a spike or head with a fleshy axis usually surrounded by a spathe, as in the calla lily. **2.** *zool.* The central core of a sporosac or the manubrium of a sessile medusa in hydroids.

Spanish moss. An epiphyte (*Tillandsia usneoides*) of the family Bromeliaceae, which hangs in long tufts from trees, especially in swamps. Also called black moss, long moss, or old-man's beard.

Spanish needles. Elongated achenes with barbed awns produced by *Bidens bipinnata* or closely related plants (beggar-ticks); the plants which produce them.

sparganosis. Infestation in man by the plerocercoid larva of tapeworms of the genus *Spirometra*, a tapeworm of dogs, cats, and other carnivores.

sparganum. The plerocercoid larva of tapeworms of the genus *Spirometra* normally found in frogs, snakes, and amphibious mammals. *See* sparganosis.

sparrow. 1. A weaver finch of the family Ploceidae, especially *Passer domesticus*, the house or English sparrow. **2.** One of a large number of seed-eating birds of the family Fringillidae belonging to many different genera.

sparrow hawk. 1. A small hawk (*Falco sparverius*) of the family Falconidae. **2.** A European hawk (*Accipiter nisus*) and related species.

spasm. A sudden, involuntary muscle contraction.

spat. The young of oysters and other bivalves.

spathe. A large bract or a pair of bracts which form an enclosing sheath of an inflorescence, as in a spadix.

spatulate. 1. Shaped like a spatula, **2.** Broad and flat at the tip and narrowing at the base, as certain leaves. **3.** Spoon-shaped.

¹spawn. To produce and deposit eggs, with reference to aquatic animals.

²spawn. 1. The eggs of various aquatic animals, as fishes, oysters, especially when produced in large numbers. **2.** *bot.* Dried manure and other organic matter containing the mycelium of mushrooms, used in starting new beds.

spay. To remove the ovaries of a female animal; to castrate a a female.

spear. A young, pointed shoot or sprout, as in asparagus.

spearmint. An herb (*Mentha spicata*) of the family Labiatae which produces an aromatic oil used extensively for flavoring.

special. 1. Of or pertaining to a species. **2.** Unique, distinctive, unusual.

special creation. Creationism; concerning the origin of species, the doctrine that existing species of animals and plants did not evolve from common ancestors but were created separately and distinct from other species.

specialization. 1. The act or process of becoming specialized. **2.** A specialized adaptation or adjustment to environment.

specialized. 1. Adapted for the performance of a specific function. **2.** Adapted or adjusted to a specific type of habitat or environment.

speciation. The process by which a new species comes into existence; the origin of new species.

species. A taxon forming a basic taxonomic group comprising a division of a genus consisting of a group of individuals of common ancestry which closely resemble each other structurally and physiologically and, in nature, interbreed, producing fertile offspring.

specific. 1. Of, pertaining to, or characteristic of a species, as s*pecific* differences. **2.** Restricted to, definite, precise.

specific gravity. Relative density; the weight of a given volume of a substance compared to that of an equal volume of water at 4°C. (39.2°F.). The specific gravity of water equals 1. Abbreviated sp. gr.

spectrum. 1. A series of images formed when light rays are dispersed and their constituents arranged in order of their wavelength. The *visible spectrum*, seen when white light is passed through a prism, consists of colors from red to violet with wave lengths of 8000 A° (red) to 4000 A° (violet). **2.** In a general sense, a series of units or entities possessing a common property arranged in a graded fashion, as an *antibiotic spectrum*, which is the range of effectiveness of an antibiotic against various types of bacteria.

speculum. 1. A brightly colored area on the wings of certain birds; formerly called mirror. **2.** A clear area on the wings of certain lepidopterous insects.

sperm. 1. A mature male gamete; a spermatozoan. **2.** The seminal fluid or semen.

513

spermaceti. A waxy substance obtained from the head of a sperm whale used in the preparation of cerates and ointments.

spermagonium. A pycnium.

spermary. An organ which produces male gametes; a testis or antheridium.

spermatangium. A male gamete-producing structure in certain algae.

spermatheca. A pouch or sac in the female for reception and storage of sperm; a seminal receptacle, as in insects.

spermatia. Nonmotile male gametes produced in a spermagonium, as in rusts.

spermatic. Of or pertaining to sperm or a sperm-producing structure.

spermatic cord. A cord in mammals which extends from the deep inguinal ring to the testis, containing blood and lymph vessels, nerves, and the ductus deferens.

spermatid. A haploid cell formed from the division of a secondary spermatocyte which transforms into a functional spermatozoan.

spermatocyte. One of two generations of cells (primary and secondary spermatocytes) which occur in the development of spermatozoa.

spermatogenesis. The process by which functional, motile, haploid spermatozoa are formed from diploid spermatogonia. Cells involved include spermatogonia, primary spermatocytes, secondary spermatocytes, spermatids, and spermatozoa, in the process of which meiosis occurs. *See* meiosis.

spermatogonium. A primordial germ cell which gives rise to a primary spermatocyte.

spermatoid. Resembling a spermatozoon or sperm cell.

spermatophore. A compact mass or packet of spermatozoa which is transferred to the female, as in various invertebrates (crustaceans, snails, squids, leeches, insects, mites) and some vertebrates (salamanders).

spermatophyte. A seed-producing plant; a gymnosperm or angiosperm.

spermatozeugma. A mass of sperm attached to a central cell, as in certain aphallate gastropods. Also spermiozengma.

spermatozoid. A motile, male gamete or sperm produced by an antheridium; an antherozoid.

spermatozoon pl. **spermatozoa.** The mature, male gamete or reproductive cell of animals, a haploid cell usually possessing a flagellum. In certain invertebrates (nematodes, arthropods), a flagellum is lacking. *See* sperm cell.

sperm cell. A male gamete or reproductive cell, as a microgamete, sperm, spermatozoon, spermatozoid, or the generative cell of a germinating pollen grain.

spermiation. The release of mature sperm from the Sertoli cells of the testis.

spermicidal. Destroying sperm.

spermiducal glands. Prostate glands, especially those of oligochaete annelids.

spermiogenesis. The transformation of a spermatid into a motile spermatozoon.

sperm nucleus. 1. *zool.* The nucleus of a spermatozoon which, in the zygote, becomes the male pronucleus. **2.** *bot.* One of two male nuclei within a growing pollen tube.

spermophile. The thirteen-striped ground squirrel (*Citellus tridecemlineatus*).

sperm plug. A mass of coagulated semen in the vagina of a rat; also called copulation plug.

sperm sac. A receptacle in fishes which receives sperm from the seminal vesicle and in turn discharges them into the urinogenital sinus.

sperm whale. A cachalot, a large whale (*Physeter catodon*), the source of spermaceti and ambergris; also *Kogia breviceps*, the pygmy sperm whale, a related form.

sphacelated. Decayed, withered.

sphacelial. Of or pertaining to decay or withering.

sphaeridium. A rounded or oval, pedunculated sense organ in echinoids.

sphagnicolous. Inhabiting sphagnum moss, as rotifers.

sphagnum moss. A peat or bog moss of the genus *Sphagnum*, noted for its ability when dry to absorb large quantities of water.

¹sphenoid. Wedge-shaped.

515

²**sphenoid.** The sphenoid bone, located in the base of the cranium.

sphenopsid. Any plant of the Sphenopsida (Sphenophyta) which includes the horsetails or scouring rushes, abundant during the Paleozoic era but with only one genus (*Equisetum*) now extant.

spheridium. *See* sphaeridium.

sphincter. A ringlike band of smooth muscle tissue which surrounds a tube or natural opening, which, upon contraction, constricts the passageway.

sphinx moth. A hawkmoth of the family Sphingidae.

spicate. Having the form of a spur or spike.

spiciform. Shaped like a spike.

spicose. Bearing spikes.

spiculate. Possessing or covered with spicules.

spicule. **1.** A sharp, pointed, siliceous or calcareous body, as **a.** those forming the endoskeleton of sponges, corals, and certain protozoans; **b.** those formed in the development of bone. **2.** One of a pair of curved copulatory structures in a male nematode; a spiculum.

spiculum. A copulatory spicule, as in nematodes.

spider. An arachnid of the class Araneae.

spider crab. A decapod crustacean of the family Majidae with long slender legs, as *Macrocheira*, the giant crab of Japan, with a leg span of eight feet.

spider mite. A red mite of the family Tetranychidae.

spider monkey. A New World monkey of the genus *Ateles*, with a slender body, long legs, and an elongated, prehensile tail.

spider wasp. A wasp of the family Pompilidae which kills or paralyzes spiders which it uses as food for its young.

spiderwort. A dicot plant of the genus *Tradescantia*.

spike. **1.** *bot.* A type of inflorescence in which sessile or nearly sessile flowers are borne on an elongated central axis. **2.** *physiol.* A sharp, pointed wave, as seen in a recording of action potentials. **3.** *zool.* **a.** A young mackerel. **b.** The single, unbranched antler of a young male deer.

spikelet. A small or secondary spike, as those formed in the flowers of grasses and sedges.

spike potential. The action potential of a nerve, as seen in an oscillograph.

spin. To form a thread or threads, as by a spider in spinning a web or a caterpillar in the formation of a cocoon.

spinach. A pot herb (*Spinacea oleracea*) cultivated extensively.

spinal. Of or pertaining to the spine or vertebral column.

spinal accessory nerve. The accessory nerve, the eleventh cranial nerve (XI).

spinal animal. An experimental animal in which the spinal cord has been transected above the lumbosacral enlargement.

spinal canal. The canal within the vertebral column which contains the spinal cord; the vertebral canal.

spinal column. The vertebral column or backbone.

spinal cord. An elongated structure composed of nervous tissue extending posteriorly from the brain and contained within the spinal canal. It functions as a reflex center and a conducting pathway.

spinal nerve. Any of the paired nerves which leave the spinal cord through the intervertebral foramina, each connected by a posterior or dorsal root containing afferent fibers and an anterior or ventral root containing efferent fibers.

spindle. **1.** A spindle-shaped or fusiform structure. **2.** In mitosis a bundle of delicate, curved fibrils which extend between the two poles or asters.

spindle cell. A thrombocyte in the blood of vertebrates other than mammals.

spindle fibers. *See* spindle.

spindle-shaped. Fusiform; elongated and tapered at each end.

spine. **1.** *bot.* A stiff, sharp-pointed outgrowth or process, as those on the stems of cacti. **2.** *zool.* A sharp, pointed process, as those of sea urchins, insects, or fishes. **3.** *anat.* **a.** A sharp, pointed process, especially one on a bone. **b.** The spinal column.

spinescent. Terminating in a spine; bearing a spine.

spiniferous. Bearing spines; spiny.

spinneret. An organ or structure which produces a thread or threads of silk formed from the secretion of the silk glands, as **a.** that on the ventral surface of the abdomen of spiders; **b.** that on the tip of the labium of the larvae of lepidopterous insects; **c.** that at the posterior end of certain nematodes.

spinose. Spinelike; bearing or possessing spines.

spinous. Spiny; possessing spines.

spinule. A small or minute spine.

spinulose. Bearing small spines or spinules.

spiny. Covered with or bearing many spines.

spiny anteater. *See* anteater.

spiny dogfish. An elasmobranch (*Squalus acanthias*).

spiny-headed worm. Any member of the phylum Acanthocephala.

spiny lobster. The sea crayfish (*Panulirus argus*) of the Pacific coast; *Palinurus* of European waters.

spiracle. 1. One of a pair of openings on the head of an elasmobranch through which water enters. 2. The external opening of the branchial chamber in tadpoles. 3. In air-breathing arthropods, an external opening through which air enters the tracheal tubes; a stigma. 4. The blowhole of a cetacean.

spiral. 1. Winding; encircling; coiling about a center or central axis. 2. Helical, resembling the thread of a screw.

spiral cleavage. Cleavage in which the cells of the upper quartet in the eight-celled stage lie above and between the cells of the lower quartet, as in annelids, mollusks, polyclads, and nemerteans.

spiral intestine. An intestine which possesses a spiral valve.

spiral valve. 1. A spiral, longitudinal fold or valve in the conus arteriosus or truncus arteriosus of lung fishes and amphibians which directs deoxygenated blood into the pulmonary arteries. 2. A spiral, longitudinal fold in the intestine of elasmobranch fishes.

spire. 1. The pointed tip of a structure, as the horn of a ruminant. 2. The pointed portion of a spiral gastropod shell; the upper whorls from the apex to the last body whorl. 3. *bot.* A slender, tapering leaf or stem, as in grasses.

spirea. A flowering shrub (*Spiraea*) of the rose family.

spiricle. A minute, coiled, threadlike, hydroscopic structure in the coat of certain seeds, as *Ruellia*.

spirillum. A spiral-shaped bacterium, especially those of the genera *Vibrio* and *Spirillum*, of the family Spirillaceae.

spirochete. Any bacterium of the order Spirochaetales, flexible, spiral-shaped bacteria, as those of the genera *Treponema* and *Borrelia*, causative agents of syphilis and relapsing fever. Also spirochaete.

spirometer. An apparatus for measuring the volume of air inhaled and exhaled.

spirotrich. Any protozoan of the subclass Spirotricha characterized by well-developed buccal ciliature, as *Stentor*, *Spirostomum*.

¹spit. To discharge from the mouth, saliva or other material.

²spit. The material ejected from the mouth; spittle.

spitting snake. A snake, as the cobra, which is capable of ejecting venom usually toward the victim's eyes for a considerable distance (up to six feet).

spittle. 1. SALIVA. **2.** The frothy mass produced by and enclosing the nymphs of spittlebugs.

spittlebug. A froghopper, a small, leaping, homopterous insect of the family Cercopidae, the nymphs of which feed upon various herbs and grasses, forming frothy masses of spittle.

splake. The hybrid offspring of a cross between the brook trout and lake trout.

splanchnic. Of or pertaining to the viscera.

splanchnocoel. The coelomic space between the visceral and parietal layers of mesoderm which gives rise, in mammals, to the serous cavities.

splanchnocranium. The portion of the skull derived from the visceral (branchial) skeleton.

splanchnopleure. *embryol.* The combined layers of splanchnic mesoderm and endoderm.

splay. To expand or spread out.

splayfoot. An abnormally expanded and flat foot.

spleen. A large organ of the lymphatic system located near the stomach, which functions as a blood reservoir, blood and lymph filter, a hemopoietic organ, and as a source of antibodies. It is the chief organ involved in the destruction of worn-out erythrocytes.

spleenwort. One of a number of ferns of the genus *Asplenium*.

splenial bone. A dermal bone present in the lower jaw of certain amphibians, sometimes bearing teeth, as in the salamander (*Siren*).

splenic. Of or pertaining to the spleen.

splenic corpuscle. A nodule of lymphatic tissue in the white pulp of the spleen. Also called Malpighian corpuscle.

519

spodogram. The image of a slice of tissue following micro-incineration.

sponge. 1. Any animal of the phylum Porifera. **2.** The porous, horny skeleton of various poriferans, especially *Spongia* and *Hippiospongia* which provide the bath sponge of commerce.

spongin. A proteinaceous, fiberlike material which forms the skeleton of horny sponges.

spongocoel. The central cavity of a simple sponge opening to the outside through the osculum. Formerly called gastral cavity, cloacal cavity, paragaster.

spongy. Of the nature of or resembling a sponge in appearance or texture.

spongy bone. Cancellous bone, containing irregular spaces filled with bone marrow.

spongy parenchyma. In a leaf, a layer of cells lying below the palisade layer, loose-textured and containing many air spaces. It is a part of the mesophyll.

spontaneous. Originating from within; not the result of outside forces or influences.

spontaneous generation. The theory that living things originate from nonliving, inorganic matter; abiogenesis. *Compare* biogenesis.

spoonbill. 1. A wading bird of the family Threskiornithidae with a long, spatulate bill, as *Ajaia ajaja*, the roseate spoonbill. **2.** The paddlefish.

sporadic. Occurring here and there or occasionally, without continuity with others of the same kind.

sporadin. The trophozoite stage of a gregarine, consisting of a primite and satellite.

sporangiole. A one or few-spored sporangium, as in certain phycomycetes.

sporangiophore. A stalklike structure in fungi which bears spores or sporangia.

sporangiospore. A spore produced by a sporangium.

sporangium. A specialized structure in which spores are produced; a spore case. *See* microsporangium, megasporangium, zoosporangium.

spore. 1. An asexual reproductive structure consisting of one or a few cells which is capable of giving rise to a new organism without gametic union. **2.** In protozoans, a cell resulting from repeated or multiple fission of a sporont, often protected by a resistant membrane. *See* sporocyst. **3.** *bacteriol.* A dormant body resistant to drying, sunlight, and heat into which a bacterium may be transformed. *See* endospore.

spore mother cell. The last generation of cells in a sporophyte which, by division (meiosis or mitosis), gives rise to two or more haploid spores.

sporidium. A basidiospore which develops from a promycelium, as in smuts.

sporiferous. Spore-producing.

sporine state. A sessile state in certain protists in which cell division occurs during vegetative existence.

sporling. A young plant which develops from a spore.

sporoblast. One of four cells which develop from an oocyst, as in coccidians.

sporocarp. A spore-producing structure in water ferns.

sporocyst. 1. In protozoans, **a.** a resistant spore containing sporozoites, as in most sporozoans; **b.** the protective wall within which sporozotes are formed. **2.** A stage in the development of a digenetic fluke consisting of a saclike structure arising from a miracidium and giving rise to rediae (in some cases daughter sporocysts or cercariae); occurs in snails or bivalves.

sporocyte. 1. *bot.* A spore mother cell. **2.** *zool.* A binucleate cell produced by certain myxosporidians.

sporogenesis. The production of spores.

sporogenous. Producing spores, said of tissues or organs.

sporogonium. The sporophyte of liverworts and mosses.

sporogony. The formation of spores; sporogenesis.

sporont. A stage in the development of certain protozoans, sometimes the zygote, which gives rise to spores, as in myxosporidians.

sporophore. 1. The fruiting body of fleshy fungi; the structure which produces and bears spores. **2.** A spore-producing structure in mycetozoans.

sporophyll. A leaf or leaflike structure which bears sporangia. *See* megasporophyll, microsporophyll.

sporophyte. In plants which exhibit alternation of generation, the diploid plant which produces haploid meiospores. *Compare* gametophyte.

sporoplasm. The protoplasm of a spore.

sporosac. A reduced type of gonophore in hydroids in which sex cells ripen directly on the sides of the blastostyle.

sporozoan. Any protozoan of the class Sporozoa, which comprises parasitic forms which produce or bear spores. Locomotor organs are lacking and reproduction is principally by schizogony. Many are pathogenic, causing diseases as coccidiosis, Texas fever, and malaria.

sporozoite. The active invasive form of various sporozoans, usually developing from a sporoblast.

sport. A sudden deviation from type; a mutation.

sporulation. The act or process of producing spores; the discharge or liberation of spores.

spot. 1. A small area differing in color or texture from the surrounding area. **2.** *bot.* A definite diseased area, or the disease of which it is a symptom, as *leaf spot.*

spotted. Marked by spots; mottled.

spotted fever. One of a group of typhuslike diseases caused by rickettsias transmitted by ticks, as Rocky Mountain spotted fever.

¹spout. To eject liquid forcibly in a jet.

²spout. The material discharged through the fused nostrils or blowhole of a whale.

spouting. The discharge by a whale of the contents of the lungs through the blowhole. *See* spout.

spp. Abbreviation for species (plural).

sprain. The twisting of a joint resulting in the stretching, tearing, or rupture of the ligaments.

sprat. A small European herring (*Clupea sprattus*).

spray. 1. A group of small branches with their foliage, especially when horizontal, as in the hemlock. **2.** An aqueous solution distributed in fine droplets or particles, as an insect *spray.*

spray zone. In marine ecology, the region above high tide line which is repeatedly moistened by the spray of ordinary surf.

spreader. 1. An agent added to an insect spray in order to secure uniform coverage of the foliage. **2.** *bacteriol.* An organism

522

which spreads very rapidly over the culture medium, as the Hauch or H form of *Proteus*.

spreading factor. HYALURONIDASE.

sprig. A small shoot or twig.

springbok. A South African gazelle (*Antidorcas marsupialis*); also springbuck.

springhaas. The spring hare (*Pedetes capensis*) of South Africa; also called cape jumping hare.

springtail. Any insect of the order Collembola with a springing organ by which it propels itself through the air.

springwood. The inner layer of the annual ring in a dicot.

¹**sprout.** The shoot of a plant, especially the first from a seed or the first growth of a root or tuber.

²**sprout.** To bring forth a new shoot; to germinate.

spruce. 1. One of a number of coniferous trees of the genus *Picea*, family Pinacea. **2.** Term loosely applied to other conifers, as Douglas *spruce* (Douglas fir) and hemlock *spruce* (hemlock).

spruce budworm. The larva of a moth (*Choristoneura fumiferana*), a serious pest of spruce, fir, balsam, and other conifers.

spruce grouse. The Canada spruce grouse (*Canachites canadensis*), also called spruce partridge.

spur. 1. *zool.* A short, spinelike projection, as **a.** that on the leg of the cock in domestic fowl; **b.** that on the legs of various insects, as the pollen *spur* in the honeybee. **2.** *bot.* A hollow, tubular extension of a flower usually functioning as a nectar-producing or storing structure, as in the columbine.

spurious. False, with reference to a structure which resembles another but differs morphologically, as a *spurious fruit*, *spurious claw* in spiders, or *spurious wing* or alula in birds.

squab. The nestling or young of a pigeon.

squama. A flat plate or scalelike structure; a squame.

squamate. Scaly.

squamation. The arrangement of scales, as in a fish.

squame. 1. A squama. **2.** A flat, pointed process which forms the exopodite of the antenna in decapod crustaceans.

squamellate. Having small, secondary scales.

squamosal. A membrane bone of the skull in lower vertebrates located behind the orbit, as in the frog.

squamous. 1. Covered with, composed of, or bearing scales. **2.** Flat, scalelike.

squamule. A small lobe of a squamulose lichen.

squamulose. 1. Having minute scales. **2.** Designating a type of thallus in lichens composed of many small, flat lobes.

squarrose. 1. Rough, with diverging scales or processes. **2.** *bot.* Having parts spreading and recurved at their ends, as bracts surrounding an inflorescence.

¹squash. The fruit of one of several species of the vine *Cucurbita* or the plant which produces it.

²squash. In microtechnique, a preparation made by placing material upon a slide, covering with a coverslip, and applying pressure.

squash bug. A hemipterous insect (*Anasa tristis*) of the family Coreidae, a serious pest of squash, pumpkin, cucumber, and melon vines.

squawfish. One of several cyprinid fishes of the genus *Ptychocheilus* of the western United States.

squealer. 1. Any of several birds, as the Eastern harlequin duck, the golden plover, or the yellow-bellied sapsucker. **2.** A young pigeon, quail, or grouse.

squid. One of a number of dibranchiate cephalopods with an elongated body bearing a modified foot of ten arms and possessing an internal shell called a pen. The common squid (*Loligo*) averages 12 in. in length. The giant squid (*Architeuthis princeps*) is the largest invertebrate, reaching a length of 50 ft.

squill. 1. The sea onion (*Uriginea maritima*), a liliaceous, bulbous plant of the Mediterranean coasts, the source of red squill, a rodenticide. **2.** *Scilla*, an Old World genus of bulbous plants of the lily family.

squirrel. Any of a large number of small to medium-sized rodents of the family Sciuridae, most arboreal, diurnal animals with a large, bushy tail. Common representatives are the red, fox, and gray squirrels of the genus *Sciurus*. The North American red squirrel or chickaree belongs to the genus *Tamiasciurus*, flying squirrels to the genera *Glaucomys* and *Pteromys*, ground squirrels or gophers to the genus *Citellus*.

stab culture. *bacteriol.* A culture made by plunging an inoculation needle deep into a mass of solid medium, such as gelatin.

stadium pl. **stadia.** The interval between two successive molts, used especially with reference to insects but also applied to other invertebrates.

stag. 1. The adult male of members of the deer family; a hart. **2.** The male of various mammals castrated after maturity.

stain. In microscopy, a dye or dyestuff applied to cells or tissues in order to increase the visibility of the various constituents or for the identification of various chemical constituents.

¹stalk. 1. *bot.* **a.** The main supporting axis or stem of a plant. **b.** An elongated supporting structure, as a peduncle, pedicel, stipe, filament, or seta. **2.** *zool.* A similar structure in an animal, as **a.** the *stalk* of a hydroid, barnacle, brachiopod, or crinoid; **b.** a connecting structure, as the optic, hypophyseal, or body *stalk*.

²stalk. To approach stealthily under cover, as when a predator *stalks* its prey.

stalk cell. One of two cells which develop from the generative cell within the pollen tube of a gymnosperm.

stallion. An uncastrated male horse, especially one kept for breeding purposes.

stamen. The pollen-producing structure of a flower, consisting typically of an anther borne on the tip of a filament; an androsporophyll or microsporophyll.

staminate. 1. Possessing only stamens and lacking pistils, with reference to flowers. **2.** Possessing only microsporophylls, with reference to strobili or cones. **3.** Bearing only staminate flowers or cones, with reference to plants.

stamineal. Pertaining to or consisting of stamens.

staminode. A sterile or nonproductive stamen.

staminodium. A staminode.

standard. A banner or vexillum. *See* banner.

stapes. The third of a series of ear ossicles located in the middle ear, commonly called the stirrup.

staphylococcus. Any bacterium of the genus *Staphylococcus* which form masses resembling a bunch of grapes. Many are pathogens, being the causative agents of boils, carbuncles, abscesses, and septicemia.

starch. A polysaccharide $(C_5H_{10}O_5)_n$, composed of many glucose units and present in leucoplasts and chloroplasts of plant cells and in the cotyledons and endosperm of seeds. It occurs usually in the form of grains which have a characteristic form and appearance in different plants. It is an important constituent in animal food and has many commercial uses. *See* glycogen, floridian starch.

starch sheath. In young stems of angiosperms, the innermost layer of the cortex containing many large starch grains, considered as homologue of the endodermis.

starfish. A sea star, an echinoderm of the class Asteroidea, with a characteristic star-shaped body, as *Asterias forbesi* of the East Coast.

stargazer. A bottom-dwelling, marine fish (*Astroscopus guttatus*) of the family Uranoscopidae, with eyes on the dorsal surface of the head and directed upward.

starling. An Old World passerine bird of the family Sturnidae, especially *Sturnus vulgaris*, introduced into the United States in 1890 and now widely distributed.

starter. A pure culture inoculum added to milk or cream to bring about desired fermentations, as in the production of butter, cheese, and fermented milks.

starvation. The state or condition of being starved.

starve. **1.** To become weak or die of hunger. **2.** To become weak or disabled by want of any kind.

stasad. A plant which lives in stagnant water.

stasis. The slowing down or cessation of flow of a substance through a tube, as blood, lymph, bile, or intestinal contents.

static. **1.** At rest or equilibrium. *Compare* kinetic. **2.** Undergoing little or no change.

station. *bot.* A particular locality for a given plant.

statoacoustic. Pertaining to the senses of equilibrium and hearing.

statoacoustic nerve. *See* vestibulocochlear nerve.

statoblast. A sessoblast or flotoblast, an internal bud formed in ectoprocts.

statocyst. An organ of equilibrium consisting typically of a fluid-filled sac containing statoliths, movement of which stimulates sensory receptors, present in various invertebrates.

statolith. A calcareous body present within a statocyst, or, in crustaceans, a grain of sand placed within a statocyst. *See* statocyst, otolith.

statoreceptor. A receptor for the sense of position or equilibrium, as a statocyst, macula, or crista.

statospore. A type of resting spore produced by certain algae, as in the Chrysophyta.

status quo hormone. SQH, a hormone secreted by the corpora allata in certain insect nymphs and larvae which prevents premature metamorphosis.

steapsin. A lipase present in pancreatic juice.

stearic acid. A fatty acid, $CH_3(CH_2)_{16}COOH$, found principally in animal fats.

stearin. A glycerol ester of stearic acid present in many animal fats.

steatopygia. Excessive enlargement of the buttocks due to accumulation of fat, as in Hottentot women of Africa.

steenbok. A small African antelope (*Raphicerus campestris*); also steinbok.

steep. To soak in a liquid for the purpose of cleansing, softening, or extracting a principle.

steer. A young, castrated, male ox, especially one castrated before maturity and raised for beef.

stegocephalian. A heavily armored, fossil amphibian of the subclass Stegocephalia, considered to be the ancestor of modern amphibians and reptiles.

stegosaur. A heavily armored, herbivorous dinosaur of the suborder Stegosauria.

steinbok. *See* steenbok.

stele. The central cylinder consisting of vascular tissue, pith, and pith rays within the cortex of the root or stem of a vascular plant.

stellate. Star-shaped; with parts radiating from a center.

stelliform. STELLATE.

stem. **1.** The main axis of a vascular plant; the ascending portion which bears buds in contrast to the descending portion which bears roots. **2.** A part which supports a structure, as that of a leaf or fruit. *See* petiole peduncle, pedicel. **3.** In a phylogenetic tree, the main stalk or one of its principal branches.

stem cell. An undifferentiated cell from which specialized cells arise.

stemma. 1. An ocellus, especially the lateral ocellus of an arthropod. **2.** A facet of an ommatidium.

stem mother. In parthenogenetic animals, a female which develops from a zygote and then gives rise to a number of parthenogenetic generations, as in aphids.

stem reptiles. The cotylosaurs, thought to be the ancestors of reptiles, birds, and mammals.

stenobathic. Having a restricted depth range, with reference to marine animals.

stenobenthic. Restricted to a narrow depth range on the bottom of the ocean.

stenophagous. Subsisting on a single kind or a limited variety of foods.

stenopodium. A type of biramous appendage in which the exopodite and endopodite are attached to a common stem, the protopodite, as a swimmeret of a crayfish.

stenosis. A narrowing or constriction of a tube, duct, or orifice.

stenothermous. Intolerant of wide variations in temperature.

stenotopic. Intolerant of environmental changes, hence restricted to a narrow geographic range. *Compare* eurytopic.

steppe. A vast area of treeless grassland in southeastern Europe and western Asia.

stercoraceous. Of, pertaining to, or consisting of dung or feces.

stercoral pocket. A diverticulum on the dorsal surface of the rectum in spiders.

stercoricolous. Living in or frequenting dung.

stereoblastula. A blastula which lacks a blastocoel.

stereocilia. Nonmotile cilia, as those on epithelial cells lining the epididymis.

stereogastrula. A gastrula with a solid interior, as in sponges.

stereognosis. The ability to perceive objects through senses other than vision.

stereome. The collenchyma and sclerenchyma combined.

stereoscopic. Of or pertaining to three-dimensional vision, with perception of depth.

stereoscopic microscope. One with two sets of lenses.

stereotaxis. A taxis in which an animal responds to contact with a solid object; thigmotaxis.

stereotropism. A tropism in which a plant responds to contact with a solid object; thigmotropism.

sterigma pl. **sterigmata.** One of four slender stalks at the tip of a basidium, each of which bears a basidiospore, as in mushrooms.

sterile. **1.** Nonfertile; incapable of producing functional gametes. **2.** Aseptic; free from microscopic organisms, especially pathogens. **3.** Lacking reproductive structures, as a *sterile* frond.

sterility. The state or condition of being sterile.

sterilization. The act or process of sterilizing.

sterilize. To render or make sterile, as **a.** to deprive of the power of reproduction, as by removal of testes or ovaries in animals or stamens and pistils in plants; **b.** to render free of microorganisms.

sterlet. A small sturgeon of western Asia.

sternebra. A segment of the sternum before complete ossification.

sternite. The sternum of an arthropod or one of its sclerites.

sternum. **1.** The breastbone of a tetrapod, a ventral structure of bone and cartilage with which the pectoral girdle or ribs or both articulate. **2.** The sclerotized ventral portion of a segment in an arthropod; also called sternite.

steroid. Any of a group of compounds which include the sterols, bile acids, sex and adrenocortical hormones, D vitamins, certain heart poisons (glycosides), toad poisons, and sapogenins. Chemically, all contain the cyclopentanoperhydrophenanthrene ring.

sterol. One of a class of lipids which contains the steroid nucleus. They are nonsaponifiable alcohols, many occurring naturally, as zoosterols (of animal origin), mycosterols (from fungi, especially yeast), and phytosterols (of plant origin).

stichocyte. One of a number of cells comprising a stichosome.

stichosome. A long column of stichocytes which encloses the slender esophagus of certain nematodes, as *Trichurus*.

stick insect. *See* walking stick.

stickleback. A small, scaleless fish of the family Gasterosteidae with dorsally projecting, erectile spines, inhabiting fresh and brackish waters, as *Apeltes, Eucalia, Pungitius.*

stickseed. A weedy herb (*Lappula*) whose nutlets are armed with barbed prickles.

sticktight. The bur marigold (*Bidens cerbus*) and related species whose achenes bear barbed awns which cling tightly to fur, wool, and clothing.

sticktight flea. *Echidnophaga gallinacea,* a parasite of poultry, dogs, and cats.

stigma pl. **stigmata. 1.** *anat.* **a.** A small mark, spot, scar, or blemish; a red spot on the skin. **b.** A clear area on an ovary marking position of a follicle. **2.** *zool.* **a.** A colored, light-sensitive eyespot in certain protozoans. **b.** A cuticular thickening or opaque spot on the wings of certain insects. **c.** A gill slit of a tunicate. **d.** A spiracle of an arthropod. **3.** *bot.* The portion of the pistil which receives the pollen; the expanded apex of the style.

stigmatic. Of or pertaining to stigma.

stigmatoid tissue. Tissue in a pistil which facilitates the growth of the pollen tube from stigma to ovule; also called transmitting tissue or pollen-transmitting tracts.

stilt. A long-legged marsh bird with an upcurved bill, as *Himantopus mexicanus,* the black-necked stilt.

stilt bristles. Elongated bristles in nematodes for attachment or locomotion.

stimulate. To initiate activity; to excite; to induce functional activity.

stimulus. 1. Any agent which acts as an excitant or irritant. **2.** An environmental change which induces or brings about a response in a cell or organism.

sting. 1. An acute, sharp, burning sensation, as that resulting from the piercing of the skin by a venom-producing organ, nematocysts, or nettles. **2.** A sharp, pointed weapon of offense or defense, as **a.** the *sting* of a bee; **b.** the sharp, curved barb of a scorpion.

stingaree. A stingray.

stinger. 1. An organism capable of stinging. **2.** A stinging organ. *See* sting, 2.

stinging cells. Nematocysts or nettle cells of coelenterates.

stinging hair. A stiff, glandular hair which secretes an irritating fluid, as those of **a.** plants of the nettle family, Uricaceae; **b.** various caterpillars of the order Lepidoptera.

stingray. One of several large rays of the family Dasyatidae which possess a slender, whiplike tail armed with serrated spines with poison glands at their bases; a stingaree.

stinkbug. One of a number of hemipterous insects of the family Pentatomidae which emit a distinctive, disagreeable odor.

stinkhorn. One of a number of foul-smelling, gasteromycete fungi, especially those of the order Phallales; a carrion fungus.

stinkpot. A musk turtle (*Sternotherus odoratus*) of the southern states.

stint. A small sandpiper of Britain of the genus *Calidris*.

stipe. A short stalk or supporting structure, as **a.** the stalk of a mushroom; **b.** the stalk of a rockweed; **c.** the petiole of a fern; **d.** the caudex of a tree fern; **e.** the stalk supporting a pistil in seed plants.

stipel. An appendage of a leaflet corresponding to a stipule; a stipellum.

stipellum. A stipel.

stipes pl. **stipites.** *zool.* A stalk or supporting structure, as **a.** the second segment of an insect maxilla; **b.** the peduncle of a compound eye.

stipitate. Having a stipe.

stipula. A newly sprouted feather.

stipular. Of, pertaining to, possessing, or originating from stipules.

stipulate. Bearing stipules.

stipule. One of a pair of leaflike appendages at the base of a petiole of a leaf.

stipulose. Possessing stipules.

stirrup. The stapes, an ear ossicle.

stoat. The common European weasel (*Mustela erminea*), especially with reddish-brown summer coat. *See* ermine.

stock. A strain of closely related individuals of common ancestry.

stolon. 1. *bot.* **a.** A slender, horizontal branch or shoot at the base of a plant which gives rise to new shoots. **b.** In certain fungi, a horizontal hypha which gives rise to new plants where it comes into contact with the substratum. **2.** *zool.* **a.** A root-like outgrowth at the base of a sessile coelenterate which gives rise to new individuals by budding; a rhizome. **b.** In tunicates, a tubular outgrowth which gives rise to a new individual.

stoloniferous. Bearing or producing stolons.

stolonization. 1. The formation of stolons. **2.** In annelids, the formation of new individuals at the posterior end, sometimes in chains, as in syllids.

stoma pl. **stomata. 1.** *bot.* A minute opening in a leaf or stem through which gases pass. **2.** *zool.* **a.** A minute opening, as that in the mesentery of an anthozoan or in the serous membrane of a mammal. **b.** The pharynx or buccal capsule of a nematode.

stomach. 1. *anat.* A saclike cavity located between the esophagus and small intestine, as in man and most vertebrates. **2.** *zool.* A saclike dilatation of the alimentary canal in various invertebrates.

stomatic. Of, pertaining to, or possessing a stoma; stomatal.

stomatopod. A mantis shrimp of the order Stomatopoda, as *Squilla.*

stomatous. Possessing a stoma or stomata.

stomium. *bot.* **1.** A region in the wall of a fern sporangium marked by lip cells where rupture occurs at maturity. **2.** The opening between two pollen locules at dehiscence.

stomodeum. 1. *zool.* **a.** The gullet of an anthozoan. **b.** The foregut of an arthropod. **2.** *embryol.* An external depression lined with ectoderm located ventral to the head. Upon rupture of the pharyngeal membrane, it becomes continuous with the foregut.

stone. 1. *anat.* A calculus. **2.** *bot.* The hard, inner portion of a stone fruit or drupe.

stone canal. In various echinoderms, the slightly S-shaped canal leading from the madreporite to the ring canal.

stonecat. A small catfish of the genus *Noturus.*

stone cell. A brachysclereid, a cell present in stony or fibrous tissue, as that in fruits, seeds, and nuts.

stonecrop. A flowering plant (*Sedum*) of the order Rosales.

stone fly. Any insect of the order Plecoptera.

stone fruit. A pyrene or drupe; any fruit with a hard, stony endocarp, as peach, plum, or cherry.

stoneroller. A small, cyprinid minnow (*Campostoma anomalum*) with a long intestine wound many times about its air bladder.

stonewort. A green alga of the genus *Chara*, with incrustations of calcium carbonate.

stony coral. A madrepore.

stool. **1.** The basal portion of a plant, as a tree stump, from which shoots arise. **2.** A pole to which a bird acting as a decoy is fastened. **3.** The evacuation of the bowels or the material evacuated; the feces.

storax. **1.** A tree or shrub of the genus *Styrax*. **2.** An aromatic balsam (Levant storax) obtained from *Liquidambar orientalis*, a tree of southwestern Asia Minor, used in medicine and perfumery; American storax or styrax is obtained from the sweet gum (*L. styraciflua*).

stork. A large, Old World, wading bird of the family Ciconiidae, as *Ciconia ciconia*.

¹strain. A subdivision of a race or species which possesses a distinctive characteristic or trait, as a resistant *strain* of bacteria.

²strain. **1.** To filter or remove by filtration. **2.** To apply excessive tension or stretch to a part; to overuse.

strangle. To compress the windpipe until asphyxiation occurs; to suffocate, as by water entering the windpipe.

strangleweed. DODDER.

strangulated. **1.** *bot.* Constricted at irregular intervals. **2.** *med.* Having circulation or passage of substances through a tube stopped by compression.

stratification. The state or condition of being in layers.

stratified. Arranged in strata or layers.

stratum pl. **strata.** A layer.

strawberry. The juicy, red fruit of the genus *Fragaria* of the rose family, or the plant which produces it.

streaking. *bacteriol.* The preparation of a streak plate.

streak plate. *bacteriol.* A culture of organisms prepared by passing a transfer loop back and forth on the surface of a solid medium, as agar, a technique for culturing isolated colonies of a mixed culture.

strepsipteran. A stylopid, an insect of the order Strepsiptera comprising insects which parasitize other insects, as *Stylops*, a parasite of bees.

533

streptococcus pl. **streptococci.** Any bacterium of the genus *Streptococcus*, gram-positive bacteria of the coccus type arranged in beadlike chains.

streptomycin. An antibiotic produced by *Streptomyces griseus*, an actinomycete.

stria pl. **striae.** A narrow streak, line, band, or mark, especially when arranged in parallel fashion.

striate. 1. STRIATED. 2. Parallel, with reference to veins in a leaf.

striated. 1. Marked by cross bands or stripes. 2. Marked by fine parallel lines, grooves, or ridges.

striated border. A layer of microvilli of the free border of epithelial cells, especially absorptive cells of the intestine.

striated muscle. Muscle composed of fibers bearing transverse striations, as skeletal and cardiac muscle.

striation. One of a series of parallel bands or lines; a stria.

strict. *bot.* With straight and upright habit with little or no branching; erect.

stricture. A constriction or narrowing of a tube, canal, or hollow organ.

stridulate. To produce sounds by stridulation.

stridulation. The production of a shrill, high-pitched sound, as by various insects, especially orthopterans.

stridulatory organ. A sound-producing structure in an insect consisting of two body parts which are rubbed against each other, as the inner surface of the femur against the forewing in locusts.

striga pl. **strigae.** A sharp, appressed hairlike scale or bristle.

strigeid. A fluke of the families Strigeidae, Diplostomatidae, and others which possesses a peculiar holdfast organ; parasitic in aquatic birds and fish- and frog-eating mammals.

strigiform. Of or belonging to the order Strigiformes which includes the owls, comprising two families, the Tytonidae (barn owls) and Strigidae (all other owls).

strigose. Covered with bristles or strigae.

strike. Infestation of the skin and wool of sheep by maggots.

strip. 1. To express the contents of a canal or duct by milking movements. 2. To remove leaves from stalks, as from tobacco plants.

strobila pl. **strobilae.** A linear chain of similar structures, as

a. the stage in development of a scyphozoan jellyfish consisting of disclike ephyrula; **b.** the body of a tapeworm consisting of a chain of proglottids.

strobilaceous. Of, pertaining to, or resembling a cone.

strobilation. STROBILIZATION.

strobile. A strobilus.

strobilization. Asexual reproduction by transverse fission resulting in the formation of a strobila.

strobilocercus. A cysticercus (bladderworm) with a chain of undeveloped segments.

strobilus pl. **strobili.** *bot.* A strobile, as **a.** a cone or a conelike aggregation of sporophylls, as in the horsetails and club mosses; **b.** a cone, as in gymnosperms; **c.** a spikelike inflorescence consisting of imbricated scales or bracts, as in the hop.

stroma pl. **stromata. 1.** *anat.* **a.** The connective tissue framework of an organ in contrast to the parenchyma. **b.** The framework of certain cells, as an erythrocyte. **2.** *bot.* **a.** The material of a chloroplast which lies between the grana. **b.** In certain fungi, a mass of cells, sometimes mixed with host cells, which may give rise to spores.

strombus. A spirally coiled seedpod of a legume.

strongyle. 1. A straight sponge spicule rounded at both ends; a monaxon. **2.** A nematode of the family Strongylidae.

strongyliform larva. A larva of a nematode which possesses a long cylindrical esophagus with a terminal bulb not sharply demarcated. *Compare* rhabditiform larva.

strophiole. A caruncle.

structure. 1. The mode of construction, form, makeup, or arrangement of parts of an organ or organism. **2.** Any part of a living thing, especially one which has a specific function; an organ or organelle.

struggle for existence. The conflict between an organism and its environment; the struggle to stay alive or survive. *See* natural selection.

struma. 1. *bot.* A swollen mass at the base of the capsule in certain mosses. **2.** *med.* GOITER.

stud. 1. A studhorse or stallion. **2.** Any male animal kept for breeding.

studfish. A killifish of the genus *Fundulus.*

stunt. A plant disease characterized by retarded development.

sturgeon. A primitive chondrostean fish of the family Acipenseridae, including several species of *Acipensor*, valued for its flesh and roe, the latter the source of caviar.

stylar. Of or pertaining to a style.

stylar column. The united styles of a compound pistil.

stylate. Possessing or resembling a style or stylet.

style. 1. *bot.* The slender, elongated portion of a pistil which bears at its tip the stigma. 2. *zool.* **a.** A slender, bristlelike process; a stylus. **b.** A slender process at the tip of the antenna of an insect. **c.** A conical mass of mucus in the style sac of a mollusk. **d.** A transparent, flexible rod (crystalline style) secreted by the style sac in certain pelecypods and gastropods. It projects into the stomach, where it is rotated against the gastric shield, in which process enzymes are liberated. **e.** An upright, pointed process in the gastropore of millepores of the order Stylasterina.

style sac. A sac in the anterior end of the stomach of certain mollusks which secretes the crystalline style.

stylet. A sharp, bristlelike structure or appendage, as **a.** those on the ventral surface of the abdomen of bristletails; **b.** the sharp, piercing mouthparts of a sucking insect, as a hemipteran; **c.** the pointed structure within the proboscis of a nemertean, the oral sucker of a cercaria, or the the buccal capsule of a nematode.

styloid. Long and pointed.

stylopid. A strepsipteran, especially one of the genus *Stylops*.

stylopodium. A disclike enlargement at the base of the style, as in flowers of the parsley family (Umbelliferae).

stylostome. A minute, tubular cavity in the skin filled with semidigested tissue debris upon which redbugs (mites) feed.

stylus. In symphylans, an eversible, pointed process in the body wall at the base of each leg.

styrax. *See* storax.

suaveolent. Fragrant, sweet-smelling.

subalpine. Immediately below the timberline.

subapical. Below or beneath an apex.

subchelate. Lacking an opposable jaw, with reference to chelicera.

subclimax. The stage preceding the climax stage for a particular climate.

subcosta. A longitudinal vein in an insect wing lying between the radius and the costa.

suberin. A fatty or waxy substance present in cork tissue which functions as a waterproofing agent.

suberization. The deposition of suberin in the walls of plant cells resulting in the development of cork tissue.

suberose. Corklike in texture.

sub-erose. Appearing as if slightly gnawed. *See* erose.

subgerminal cavity. The blastocoel or segmentation cavity in a meroblastic ovum undergoing cleavage.

subherbaceous. Herbaceous at first, but later becoming woody as the season advances.

subiculum. A dense mass of fungal mycelium on the surface of the host plant.

subimago. In mayflies, the first of two winged instars following emergence of the nymph from the water.

subinferior. Somewhat inferior; half inferior, with reference to the ovary of a flower.

subjective. Pertaining to or arising within an individual, as *subjective* symptoms.

sublethal. Less than lethal; not fatal.

subliminal. Below the threshold, as a *subliminal* stimulus.

sublingual. Beneath the tongue.

submental. Beneath the chin.

submentum. The basal portion of the labium in insects.

submerged. Covered with water; growing under water.

submicroscopic. Below the limit of resolution of an optical microscope; too small to be seen by such an instrument.

submucosa. A layer of connective tissue which lies beneath a mucous membrane.

subneural gland. In tunicates, a glandular structure located ventral to the nerve ganglion, its duct opening into the pharynx; the neural or adneural gland.

subpetiolar. Under a petiole and often enclosed or enveloped by it, with reference to buds.

subsere. A secondary succession or a plant succession on a denuded area.

537

subshrub. A perennial with a stem which is woody at its base.

subsidiary. Secondary, auxiliary, tributary, accessory.

subspecies. A subdivision of a species consisting of a group of individuals, usually a geographic race, which differs slightly from other groups (subspecies) of the same species but between which interbreeding is possible.

substrate. 1. The substance acted upon by an enzyme. **2.** The solid material upon which an organism lives or to which it is attached. **3.** A substratum.

substratum. A substrate.

subtend. *bot.* **1.** To stand below and close to. **2.** To be adjacent to. **3.** To enclose within its axil, as a bract which *subtends* a flower.

subterminal. Located near the end but not at the extreme end, with reference to bacterial spores.

subterranean. Underground; beneath the surface of the earth.

subterrestrial. SUBTERRANEAN.

subtilin. An antibiotic obtained from *Bacillus subtilis*, effective against gram-positive bacteria.

subtropical. Nearly tropical, with reference to regions near the tropical zone or regions almost tropical in their climate.

subulate. Awl-shaped; narrow and tapering from base to apex.

subumbrella. The concave, oral surface of the bell (umbrella) of a medusa or jellyfish.

subungual. Below or beneath a claw, nail, or hoof.

subunguis. The solehorn or ventral plate of a claw (unguis).

succession. *ecol.* The sequence of communities which replace one another in a given area. A *primary succession* is that occurring in an area previously unoccupied, as a rock surface; a *secondary succession* is that occurring in an area from which a community has been removed or destroyed, as a cut-over forest. *See* climax community.

succubous. A type of leaf insertion in which the upper margin of a leaf is covered by the lower margin of the leaf directly above it, as in the leafy liverworts.

[1]succulent. Juicy; possessing tissues which store water; soft and thick in texture.

[2]succulent. A succulent plant, as a cactus.

succus entericus. Intestinal juice, secreted principally by glands of the small intestine.

suck. To take nourishment or draw milk from the breast; to draw into the mouth or digestive tract by producing a vacuum.

sucker. 1. *zool.* **a.** A structure, usually cup-shaped, for food-getting or for attachment or both, as that of a tapeworm, fluke, leech, or squid. *See* acetabulum. **b.** One of a number of carplike, freshwater fishes of the family Catostomidae, as *Catostomus*, the common sucker. **2.** *bot.* A shoot which develops from the roots or lower part of the stem.

suckfish. A remora.

suckle. To nurse at the breast or the udder.

suckling. A nursling; a child or animal before it is weaned.

sucrase. Saccharase, an enzyme which hydrolyzes sucrose to glucose and fructose. Also called invertase.

sucrose. A disaccharide, $C_{12}H_{22}O_{11}$, obtained from the sugar-cane or sugar beet, commonly called cane or beet sugar, widely used as a sweetening agent and preservative.

suctorial. Adapted for sucking.

suctorian. A protozoan of the class Suctoria, order Suctorida, ciliate protozoans but lacking cilia in the adult state; sessile protozoans with tentacles for attachment or piercing prey, as *Podophyra*, a stalked form.

sudd. A floating mass of dense vegetation, sometimes blocking navigable streams.

sudoriferous gland. A sweat gland; a sudoriparous gland.

suet. Fat in the abdominal cavity of domestic animals, especially that about the kidneys.

suffrutescent. Of the nature of a shrub; low and woody.

suffruticose. SUFFRUTESCENT.

sugar. Any carbohydrate with a sweet taste, especially sucrose. *See* mono-, di-, and polysaccharide.

sugar beet. A variety of the common beet (*Beta vulgaris*) whose white roots are the source of beet sugar.

sugarcane. A tall, grass plant (*Saccharum officinarum*), stems of which are the source of cane sugar.

sulcate. With grooves or furrows.

sulcus pl. **sulci.** A groove or furrow.

sulfa drug. A drug of the sulfonamide type possessing marked bacteriostatic properties. *See* sulfonamide.

sulfonamide. One of a number of drugs derived from sulfanilamide and used in the treatment of infectious diseases. Important ones are sulfadiazine, sulfamerazine, sulfamethazine, sulfacetamide, sulfisoxazole, sulfisomidine, and sulfamethizole.

sulfur. A solid, yellow, nonmetallic element, symbol S, at. no. 16; a common constituent of many proteins.

sulfur bacteria. One of several species of bacteria which store or metabolize elemental sulfur or its inorganic compounds, especially *Beggiatoa, Thiobacillus*.

sumac. A shrub (*Rhus*) of the family Anacardiaceae, including *R. vernis*, poison sumac, a cause of contact dermatitis.

summation. *physiol.* **1.** The increased response resulting from two or more rapidly repeated stimuli **2.** A response resulting from two subliminal stimuli.

summer kill. The partial or complete destruction of the fish population of a pond or lake during the summer season due to reduced oxygen content of the warm waters.

summerwood. The outer layer of an annual ring consisting principally of small, thin-walled elements. *See* annual ring, springwood.

sun animalcule. A heliozoan, as *Actinophrys sol*, with radiating axopodia.

sundew. An insectivorous plant (*Drosera*), a bog plant whose circular leaves bear long, glandular hairs ("tentacles") to which insects adhere.

sunfish. 1. A large, ocean sunfish (*Mola mola*) sometimes reaching a weight of 2000 lbs. **2.** Any of a number of small, freshwater fishes of the family Centrarchidae, as *Lepomis*, and other genera.

sunflower. A composite (*Helianthus*) with large, conspicuous, many-flowered heads.

sun plant. A plant in which photosynthesis occurs under conditions of high light intensity, hence is intolerant of shade, as maize.

sun star. An asteroid echinoderm with numerous radiating arms, as *Solaster*, with 7 to 14 arms.

superciliary. Of or pertaining to the eyebrows.

superfecundation. The fertilization of two or more ova liberated at the same time following successive acts of coitus but not necessarily involving the same male.

superfemale. *genetics.* In *Drosophila*, a weak, feeble female of low vitality with three X-chromosomes but the normal number of autosomes.

superfetation. The development of a second fetus after one has already started development in the uterus.

superficial. Of, pertaining to, or near the surface. *Compare* deep.

superficial cleavage. Cleavage in which the dividing cells form a surface layer about a central, uncleaved yolk, as in an insect egg.

superior. Higher, uppermost, above, or more elevated in position.

supermale. *genetics.* In *Drosophila*, a weak, feeble male of low vitality possessing one X-chromosome and an extra set of autosomes.

supernumerary. More than the usual or normal number, as *supernumerary* digits.

superposition image. One formed in compound eyes in dim light when pigment is retracted in which corresponding parts of images formed by different ommatidia are superimposed. *See* mosaic image.

super-regeneration. The development of extra organs or parts as a result of regeneration.

supersonic. Having a velocity exceeding the speed of sound (760 mph.). *Compare* ultrasonic.

supination. The act of assuming or being brought to a supine position.

supine. 1. Lying on the back with the face directed upward. **2.** With reference to the hand, with the palm forward or upward. *Compare* prone.

supplemental. Extra, additional.

suppressor. That which restrains, inhibits, counteracts, or prevents an action, as a *suppressor* gene.

suppuration. The formation of pus.

suprabranchial. Above the gills; epibranchial.

suprabranchial cavity. In bivalves, a channel lying dorsal to each gill which receives water from water tubes and conveys it to the excurrent siphon.

suprarenal. Above the kidney.

suprarenal gland. The adrenal gland.

suprascapula. A bone in the pectoral girdle of certain amphibians which extends dorsally and medially from the scapula, as in the frog.

supravital staining. Staining by the addition of dyes to a medium containing cells which have been removed from an organism, as the study of blood cells to which neutral red or Janus green has been added.

sural. Of or pertaining to the calf of the leg.

surculose. Producing suckers.

surcurrent. Having winged expansions extending up the stem from the base of a leaf.

surface tension. The tension exerted by the surface of a liquid resulting from cohesive forces between the molecules.

surfactant. An agent which lowers surface tension, as soap, bile, and certain detergents.

suricate. A South African burrowing mammal (*Suricata suricatta*) of the family Viverridae; the grey meerkat.

Surinam toad. An aquatic toad of South America, males of which carry the eggs attached to their back.

surra. Trypanosomiasis of horses, camels, and other animals caused by *Trypanosoma evansi*, transmitted principally by biting flies.

survival of the fittest. *See* natural selection.

susceptible. 1. Capable of being influenced or affected by an agent or environmental condition. **2.** Liable to infection if exposed to the causative agent of an infectious disease.

suspended animation. A state or condition resembling death in which vital activities, as respiration, are temporarily stopped, as in near drowning.

suspension. *chem.* A heterogeneous system consisting of finely divided particles of a solid temporarily suspended in a liquid, as soil particles in water.

suspensor. *bot.* A connecting or supporting structure, as **a.** a hyphal branch which supports a conjugating isogamete, as in molds; **b.** a group of cells which places the embryo closer to a food supply, as in club mosses and gymnosperms; **c.** a chain of cells which attaches the embryo to the wall of the embryo sac, as in angiosperms.

suspensory ligament. A group of delicate fibers encircling the lens of the eye by which it is attached to the processes of the ciliary body.

sustentacular cells. Cells which serve for suspension or support, as the non-nervous cells of the olfactory epithelium and taste buds or Sertoli cells of the seminiferous tubules.

suture. 1. *anat.* **a.** A seamlike line of fusion between two cranial bones. **b.** The joint itself, a synarthrosis. **2.** *zool.* **a.** A groove or membranous area between two sclerites of an arthropod exoskeleton. **b.** The junction line between two elytra of a beetle. **c.** A line on a shell of an ammonite marking junction of septum and outer shell. **3.** *bot.* **a.** A line or seam marking the union of two contiguous parts. **b.** The line of dehiscence of a fruit or capsule.

swale. 1. A shaded place. **2.** A low, moist area with rank vegetation.

¹**swallow.** A passerine bird of the family Hirundinidae, as the barn swallow (*Hirundo rustica*).

²**swallow.** To pass food from the mouth cavity to the crop or stomach.

swallow's nest. In the Far East, the edible nest of a swiftlet composed of gumlike saliva. *See* swiftlet.

swallowtail. A large butterfly of the genus *Papilio*, with a backwardly projecting process on each wing. Its larva is the celery or parsley worm.

swamp. An area of wet, spongy ground saturated with water.

swan. A large, heavy-bodied, long-necked bird of the family Anatidae, especially those of the genus *Cygnus*, as *C. columbianus*, the whistling swan.

¹**swarm. 1.** A large number of insects, as bees, ants, or termites in the process of swarming. **2.** A large group of free-swimming or floating cells or minute organisms.

²**swarm.** To collect together and leave the hive or nest to found a new colony, as the honeybee.

swarm cell. 1. A swarm spore. **2.** One of the numerous motile conidia formed within and released from trichomes of iron and manganese bacteria.

swarming. The sudden appearance of large numbers of individuals, as **a.** the release of swarm spores by fungi and protozoans; **b.** the appearance of free-swimming sexual individuals of the palola worm; **c.** the rapid spreading of bacteria over a moist surface, as *Proteus;* **d.** the departure of the queen bee with a large number of workers from the nest; **e.** the appearance of large numbers of winged termites or ants.

swarm spore. A motile zoospore of certain mycetozoans, especially when produced in large numbers.

¹sweat. 1. To perspire or discharge sweat from a sweat gland, or the material discharged. **2.** To exude moisture in small drops, as seen in green plants or in the ripening of cheese.

²sweat. Moisture which has condensed in small droplets.

sweat bee. One of several species of bees which are attracted to people who are perspiring, as *Halictus*, of the family Halictidae.

sweat gland. A sudoriferous gland, one of the numerous, coiled, tubular glands of the skin which produce sweat.

sweet. 1. Having an agreeable taste, as that of sugar. **2.** The opposite of sour or bitter. **3.** Not acid, as a *sweet* soil.

sweetbread. The thymus and pancreas of an animal, especially a calf, when used for food.

sweet gum. A tree (*Liquidambar styraciflua*) of the witch hazel family, or its reddish-brown wood used for furniture; also called red gum. Also the source of storax, a medicinal product.

sweet potato. A twining, trailing, perennial vine (*Ipomoea batatas*) with fleshy, adventitious roots rich in starch and sugar. *See* yam.

swellfish. A puffer.

swift. A rapidly moving animal, as **a.** the chimney swift (*Chaetura pelagica*), a swallowlike bird of the family Apodidae; **b.** a lizard of the genus *Sceloporus;* **c.** a ghost moth of the family Hepialidae.

swiftlet. A swift of East Asia and Oceania of the genus *Collocalia* which produces large quantities of saliva which is used in nest building. *See* swallow's nest.

swimmeret. A pleopod, an unspecialized, biramous appendage of various crustaceans, as those on the abdomen of a lobster or crayfish.

swimmer's itch. Cercarial dermatitis caused by invasion of the skin by foreign schistosome cercariae.

swimming bell. A swimming medusa of a siphonophore; a nectocalyx.

swine. Any member of the family Suidae, especially those of the genus *Sus* which comprises all true pigs. Term is applied collectively to all pigs and hogs, especially those of domestic varieties. *See* boar, sow, gilt, barrow.

swordfish. A large, ocean food fish (*Xiphias gladias*) of the family Xiphiidae, with an extremely long, swordlike upper jaw.

swordtail. A tropical freshwater fish (*Xiphophorus helleri*) with a greatly elongated caudal fin; popular as an aquarium fish.

syconium. A composite, fleshy fruit borne on an enlarged receptacle, as a fig.

syconoid. A type of canal system in sponges in which the incurrent canals parallel the radial canals, with openings called prosopyles connecting the two, as in *Scypha*.

syllid. An epitokous, polychaete worm (*Syllis ramosus*) of the family Syllidae.

sylvan. Of or pertaining to trees or woods.

sylvatic plague. Plague among rodents.

symbiont. Either of two organisms living together in symbiosis.

symbiosis. A mode of life in which two organisms of different species live in intimate association with each other. Depending on the nature of the association, the relationship is designated mutualism, commensalism, parasitism, or phoresis. *See* consortium, symphilism.

symbiotic. Of or pertaining to symbiosis.

symmetrical. 1. Possessing or exhibiting symmetry. **2.** *bot.* Regular as to number of parts as in an actinomorphic flower which has the same number of parts in each circle.

symmetrogenic. Designating a type of binary fission in which an individual divides longitudinally giving rise to mirror image daughter cells, as in *Euglena*.

symmetry. A likeness or correspondence in size, shape, or structure of parts of an organism or of a part, as a flower. Types include bilateral, radial, and universal. *See* asymmetrical, biradial symmetry.

sympathetic nervous system. A division of the autonomic nervous system which includes preganglionic neurons whose cell bodies lie in the thoracic and lumbar regions of the cord. Fibers pass out through thoracic and lumbar nerves and synapse with postganglionic neurons whose cell bodies are located in vertebral and prevertebral ganglia. Postganglionic fibers innervate involuntary effectors (glands, smooth and cardiac muscle). Norepinephrine is the neurohumoral agent liberated at synapses and most effector endings. Also called thoracolumbar division.

sympatric. Occupying the same or overlapping areas, said of species. *Compare* allopatric.

sympetalous. Having petals united, especially at their bases; gamopetalous.

sympetaly. Condition in which the petals of a flower are united; gamopetaly.

symphile. The subject organism in symphilism. *See* guest.

symphilism. A form of symbiosis in which one organism takes care of and protects another, as when ants care for and transport aphids and in return the aphids provide honeydew.

symphylan. An arthropod of the class Symphyla, small, worm-like animals with 12 pairs of legs, as *Scolopendrella*.

symphysis. 1. *anat.* A synarthrosis type of joint in which two bones are connected by fibrocartilage, as the *pubic symphysis*. 2. *bot.* The coalescence or fusion of parts, as in sympetaly.

sympodial. A method of branching in which growth is principally from lateral buds several nodes behind the apex, as in the catalpa. *Compare* monopodial.

symptom. Any perceptible change, usually subjective, indicative of a disease or pathological condition. *Compare* sign.

synacme. SYNANTHESIS.

synandrium. An androecium in which anthers are coherent.

synandry. A condition in which anthers are united.

synangium. 1. *bot.* A mass of fused sporangia, as in some ferns. 2. *zool.* The distal, muscular portion of the conus arteriosus of a frog. *See* pylangium.

synantherous. A condition in which anthers are fused or joined together.

synanthesis. A condition in which stamens and pistil mature simultaneously.

synanthous. Condition in which flowers and leaves appear at the same time.

synapse. The point of contact between the axon of one neuron and the cell body or dendrites of another; the point at which an impulse in one neuron initiates an impulse in the next of a series of neurons.

synapsid. Having a single temporal fenestra, with reference to the skull of mammallike reptiles of the order Synapsida and true mammals.

synapsis. The coming together in pairs of homologous chromosomes during meiosis.

synapticula. Cross bars between adjacent gill bars, as in lancelets.

synarthrosis. A type of joint lacking a joint cavity in which bones are bound firmly to each other, permitting little or no movement. *See* suture, symphysis, syndesmosis.

syncarpous. Having united carpels, with reference to an ovary of a flower with two or more carpels.

synconium. *See* syconium.

syncope. Fainting; temporary loss of consciousness resulting from reduced blood flow to the brain.

syncytium. A multinucleated mass of protoplasm as that comprising the vegetative body of a slime mold or a striated muscle fiber. *Compare* coenocyte.

syndactyly. The condition in which two or more digits are fused.

syndesis. SYNAPSIS.

syndesmosis. A type of joint, a synarthrosis, in which bones are bound together by a fibrous membrane.

syndrome. A group of symptoms and signs which, when taken together, characterize a pathological condition.

synechthran. *See* guest.

synecology. The division of ecology which deals with groups of organisms associated together as a unit; community or population ecology.

syneresis. The contraction of a gel accompanied by exudation of a fluid, as in the coagulation of blood or milk.

synergid. One of two or three cells or nuclei which, with the egg, comprise the egg apparatus of the female gametophyte or embryo sac in an angiosperm.

547

synergism. The working together or harmonious action of two or more agents, as muscles, microorganisms, drugs, by which a result is obtained or an effect produced which could not be accomplished by the action of one alone.

synergistic. Of, pertaining to, or involving synergism.

syngamy. The union or fusion of gametes in sexual reproduction: FERTILIZATION.

syngen. A specific mating type, as occurs in certain ciliates, conjugation being restricted to members of the opposite type.

syngenesious. Having anthers united in the form of a cylinder about the style, as in the composites.

synizesis. A contracture or closure, as the constriction of the pupil of the eye or condensation of chromatin at the side of a nucleus during synapsis.

synkaryon. A fusion nucleus as that in protozoans following conjugation or in a zygote following fertilization.

synoecy. The complicated interrelationships between organisms living together in complex communities.

synoekete. *See* guest.

synonym. *taxonomy.* One of two or more scientific terms which are applied to the same taxon.

synonomy. A complete list of scientific names applied to a taxon, preferably with data as to author, date of publication, reference, and type locality.

synovia. Synovial fluid, a clear, viscous fluid normally present in a synovial cavity of a joint, a tendon sheath, or a bursa.

synsacrum. In birds, a bone consisting of the fused posterior thoracic, lumbar, sacral, and some caudal vertebrae.

synsepaly. The condition in which the calyx is composed of fused or united sepals; gamosepaly.

synthesis. The building up or production of a compound by bringing together or combining simpler elements or constituents, as the *synthesis* of proteins by plants.

synthetic. Prepared artificially; not natural.

syntype. One of several specimens from which a species is described when the holotype is not designated by the author.

syphilis. A chronic, infectious venereal disease caused by a spirochete (*Treponema pallidum*).

548

syrinx. In birds, the organ of sound production located at the lower end of the trachea.

syrup also **sirup. 1.** A thick, sweet liquid. **2.** A concentrated fruit juice. **3.** A sweetened medicinal product. *See* molasses.

system. 1. A methodical arrangement of objects or units. **2** *biol.* **a.** A group of organs or structures which function together as a unit, as the *root system* of a plant or the *digestive system* of an animal. **b.** The orderly arrangement of taxonomic groups, as the *binomial system* of nomenclature.

systematic. 1. Well-organized: systematized. **2.** Of or pertaining to systematics or classification.

systemic. Of, pertaining to, or affecting systems, as a *systemic* disease.

systemic circulation. The circulation of blood through arteries to the body tissues and back through veins to the heart. *Compare* pulmonary circulation.

systemics. Taxonomy; the study of classification.

systole. 1. *physiol.* The contraction phase of the heartbeat. *Compare* diastole. **2.** In protozoa, the contraction of the contractile vacuole and discharge of its contents.

syzygy. 1. The union of gamonts in pairs in certain protozoans, as *Eimeria*. **2.** The end-to-end union of two sporodins (trophozoites), as in gregarines.

T

tabanid. A bloodsucking, dipterous insect of the family Tabanidae which includes the genera *Tabanus* (horseflies) and *Chyrsops* (deerflies).

tabasheer. A siliceous concretion in the joints of bamboo.

table. A thin, flat plate of bone, as that on the surfaces of a cranial bone.

tabula. A horizontal, calcareous plate traversing certain cavities in a millepore or stony coral.

tachina fly. A dipterous insect of the family Tachinidae whose larvae parasitize the larvae of other insects, as *Winthemia*.

tachyauxesis. Rapid growth, especially with reference to a part which grows more rapidly than the organism as a whole.

tachyblastic. Developing immediately, with reference to eggs. *Compare* opsiblastic.

tachycardia. Excessive rapidity of heartbeat.

tachygenesis. **1.** Rapid or accelerated growth or development. **2.** The deletion or omission of embryonic or larval stages, as occurs in various invertebrates.

tachypnea. Excessive rapidity of respirations.

tachytelic. Evolving rapidly with resultant rapid speciation.

tactic. Of or pertaining to a taxis.

tactile. Of or pertaining to the sense of touch.

tactile organ. A sensory receptor which responds to touch or contact, as Meissner's corpuscle in man.

tadpole. The larva of a frog or toad, an aquatic form with gills; a pollywog.

taenia. **1.** A band or bandlike structure. **2.** A tapeworm of the genus *Taenia*.

taenidium. A spiral band or thickening in the wall of an insect trachea.

taeniola. An endodermal ridge in the planula larva of a scyphozoan.

tagged. Bearing an identifying element or compound which enables a specific substance (atom or compound) to be traced, as CO_2 containing heavy oxygen used in experiments involving photosynthesis.

tagma pl. **tagmata.** A division of the body of an arthropod, as the head, thorax, abdomen, or cephalothorax.

tahr. A beardless, wild goat, as *Hemitragus jemlaicus*, of southern Asia.

taiga. A northern, coniferous forest, especially one in Eurasia.

tail. **1.** *zool.* The elongated, posterior portion of the body of an animal, as **a.** in vertebrates, the caudal extension of the vertebral column with its muscles and associated structures; **b.** in birds, a group of feathers, attached fanlike, extending caudally over the pygostyle; **c.** in invertebrates, any long, flexible, terminal appendage extending caudally. **2.** *anat.* A slender, terminal portion of an organ or structure, as the *tail* of the pancreas.

tailed. 1. *zool.* Having a tail. **2.** *bot.* Having a long, taillike appendage, as the caudal appendage of an anther.

tail fan. In various crustaceans, a swimming organ consisting of the telson and two uropods.

tail fin. The caudal fin, as in fishes.

tailorbird. One of several Asian or African passerine birds of the genera *Sutoria* and *Orthotomus* which sew the edges of leaves together to form a pocket in which a nest is built.

takahe. A large, flightless bird (*Notornis mantelli*) of New Zealand, nearly extinct.

takin. A goat antelope (*Budorcas*) of the highlands of central Asia.

talon. A sharp claw, especially that of a bird of prey.

talonid. The posterior portion of the crown of a lower molar tooth, bearing two cusps, the hypoconid and entoconid.

talus. A tarsal bone which articulates with the tibia and fibula; astragulus.

tamandua. An arboreal anteater (*Tamandua tetradactyla*) of Central and South America.

tamarack. *See* larch.

tamarau. A small, black buffalo (*Bubalus mindorensis*) of the Phillipines; timarau.

tamarin. A small, South American monkey (*Mystax*) related to the marmosets.

tamarind. A tropical tree (*Tamarindus indica*) or the fruit it produces which is widely used in the Far East.

tambour. *physiol.* A drumlike device used in recording changes in pressure.

tanager. A brightly colored passerine bird of the family Thraupidae, as the scarlet tanager (*Piranga olivacea*).

tandem. One after another, as damselflies when mating.

tangerine. A small, round orange, with reddish-orange color, easily removable peel, and segments which separate easily.

tangoreceptor. A touch receptor; a receptor which is stimulated by physical contact, as Meissner's corpuscle; a tactile organ.

tannic acid. A tannin obtained from nut galls, used medicinally as an astringent.

tannin. A bitter substance present in the bark, fruits, and other parts of plants, with styptic and astringent properties and used in the tanning of leather.

tapaculo. The turco, a small, terrestrial bird (*Pteroptochos*) inhabiting forests of Central and South America.

tapetum. **1.** *bot.* The inner nutrient layer of **a.** the sporangium of ferns; **b.** the pollen sac of angiosperms; **c.** the inner integument of an embryo sac. **2.** *zool.* **a.** A postretinal membrane in the indirect eye of an arthropod. **b.** A light reflecting layer in the choroid of the eye, as the *tapetum lucidum* of elasmobranchs, *tapetum fibrosum* of ungulates, and *tapetum cellulosum* of carnivores and lower primates. **c.** A layer of nerve fibers in the roof and lateral wall of a lateral ventricle of the brain.

tapeworm. A parasitic flatworm of the subclass Cestoda with a body consisting of a scolex and a chain of proglottids. Adults live in the intestine of vertebrates; larvae live in intermediate hosts. *See* cysticercus, hydatid.

tapioca. A starchy preparation obtained from the root of the cassava.

tapir. A large, almost tailless ungulate of the family Tapiridae inhabiting Central and South America and Southeast Asia, as *Tapirus terrestris*, the Brazilian tapir.

taproot. The primary root of a plant which grows directly downward, giving off lateral branches.

tarantula. **1.** Any of a number of large, hairy spiders of the family Theraphosidae, including the genera *Dugesiella* and *Eurypelma* of the southwestern United States. **2.** The wolf spider (*Lycosa tarantula*) of Europe.

tardigrade. A water bear, an invertebrate of the class Tardigrada, which includes a number of minute, aquatic or semi-aquatic animals of uncertain affinities variously considered to be a phylum, an order of Arachnida, or a class of the Arthropoda. Common genera are *Echiniscus* and *Macrobiotus*.

target organ. *endocrin.* An organ or structure upon which a hormone has a specific effect.

tarpon. A large, marine, game fish (*Tarpon atlanticus*) of the family Megalopidae, of the Atlantic coast.

tarsal. Of or pertaining to a tarsus.

tarsal glands. The Meibomian glands of the eyelid, sebaceous glands whose ducts open within the inner border of each lid.

tarsal plate. A plate of dense connective tissue forming the supporting framework of an eyelid.

tarsier. A small, nocturnal, arboreal primate (*Tarsius spectrum*) of the Phillipines and neighboring regions.

tarsometatarsus. A compound bone in the foot of a bird which bears the toes, formed by a fusion of the distal tarsal bones with the metatarsals.

tarsus. 1. In vertebrates, the portion of the foot between the leg and metatarsals; the seven bones comprising the ankle: calcaneus, talus, cuboid, navicular, and three cuneiforms. **2.** The terminal portion of the leg of an arthropod. **3.** The tarsal plate of an eyelid.

tartareous. Having a loose, rough, and crumbly surface, with reference to lichens.

tartaride. A schizomid, an arachnid of the order Schizomida.

Tasmanian devil. A small, carnivorous marsupial (*Sarcophilus harrisii*) of Tasmania, with an exaggerated reputation for ferocity.

Tasmanian wolf. *See* thylacine.

tassel. The male or staminate inflorescence of maize and related plants.

taste. The sensation resulting from stimulation of taste buds, by which such qualities as sweetness, saltiness, sourness, and bitterness are recognized.

taste bud. A gustatory receptor; a sensory receptor containing gustatory cells located principally on the tongue.

tatouay. The broad-banded armadillo (*Cabassous unicinctus*) of South America.

tattler. A sandpiper (*Heteroscelus incanum*), the wandering tattler.

tatuasa. The giant armadillo (*Priodontes gigas*) of Brazil.

tautog. The blackfish (*Tautoga onitis*), an edible fish of the East Coast.

taxis. 1. *biol.* The movement of a motile organism in response to a stimulus, which can be away from (negative) or toward (positive) the stimulus. *Compare* tropism. **2.** *med.* The restoration to normal position of an organ or part, as in the reduction of a hernia.

taxon. A taxonomic group of any rank or size.

taxonomic. Of or pertaining to taxonomy.

taxonomy. The science of classification; the arrangement of animals and plants into groups based on their natural relationships; also called systematics.

tea. The dried leaves of the shrub *Camellia* (*Thea*) *sinensis*, extensively cultivated in Asia and used in the preparation of a caffeine- and tannin-containing beverage.

teacher bird. The ovenbird.

teak. The wood of the tree *Tectona grandis*, noted for its beauty and durability, an important commercial wood of East Asia.

teal. One of several species of small, short-necked, surface-feeding ducks of the genus *Anas*.

tear. A drop of the secretion of the lacrimal gland.

tease. To separate into small parts or shreds; to tear to pieces.

teat. An elongated protuberance through which the young of mammals suck or draw milk from the udder or mammary gland. *See* nipple.

tectin. Pseudochitin, a substance forming the skeleton of radiolarians.

tectorial. In the nature of or serving as a roof or covering.

tectorial membrane. A thin, jellylike membrane which overlays and is in contact with the hair cells of the organ of Corti of the inner ear.

tectrix pl. **tectrices.** A wing covert or tail covert of a bird.

tectum. A roof or covering.

teeth. Plural of TOOTH.

tegmen pl. **tegmina.** **1.** A coat or covering; an integument. **2.** The leatherlike covering of the calyx of a crinoid. **3.** The thickened, hardened, front wing of an orthopteran.

tegmentum. **1.** A covering. **2.** *zool.* **a.** The surface layer of the shell plate of a chiton. **b.** In vertebrates, the dorsal portion of the cerebral peduncles of the midbrain. **3.** *bot.* A layer of scales which encloses a bud.

tegmina. Plural of TEGMEN.

tegula. A small, scalelike sclerite which lies over the base of the front wing in certain insects.

tegument. An integument.

tela. A weblike tissue or structure.

tela choroidea. A membranous structure lined with ependymal epithelium which forms the roof of the third and fourth ventricles of the brain.

teleceptor. A distance receptor, as the eye, ear, or nose; a teloreceptor.

telegony. An erroneous belief that mating with a male influences the future offspring of that female even though subsequent matings are with other males.

telencephalon. The anterior portion of the embryonic forebrain from which the rhinencephalon, corpora striata, and cerebral cortex develop.

teleoptile. A mature or adult type of feather. *Compare* neoptile.

teleost. Any fish of the class Osteichthyes (Teleostomi), characterized by possession of a bony skeleton.

teleutospore. A teliospore.

teliosorus. A telium.

teliospore. A heavy-walled, two-celled spore of the resting type produced late in the growing season by certain rusts and smuts. It gives rise to basidiospores at the beginning of the next season.

telium pl. **telia.** The fruiting body of rusts and smuts in which teliospores are formed; a teliosorus.

teloblast cell. In protostomes, one of the two cells derived from the fourth blastomere from which the mesoderm develops.

telodont. With extremely large mandibles, as the stag beetle.

telolecithal. Designating ova in which a large amount of yolk is concentrated at the vegetal pole, as in the eggs of reptiles and birds.

telome. A single terminal segment of a branching axis, with reference to the evolutionary development of plants. *See* mesome.

telophase. The final stage in mitosis or meiosis during which two new nuclei are formed, and cytokinesis occurs.

telopodite. A sickle-shaped copulatory structure on the gonopod of a diplopod in the base of which sperm are stored.

teloreceptor. *See* teleceptor.

telotaxis. A type of taxis in which an animal responds to only one of several stimuli.

telotroch. **1.** In a trochophore, a ring of cilia which encircles the embryo just above the anus. **2.** A free-swimming individual produced by a stalked ciliate.

telson. **1.** In crustaceans, the median, terminal portion of the abdomen, as in the crayfish. **2.** In arachnids, **a.** the terminal spine of the king crab; **b.** the curved, dorsal spine or sting of a scorpion. **3.** The terminal abdominal segment in a primitive insect or an insect embryo.

telsontail. An insect of the order Protura.

template. A pattern or mold which serves as a guide or base for work to be performed.

temple. The region of the head above the zygomatic arch.

temporal. Of or pertaining to the temple or temporal bone.

temporal bone. A bone forming a portion of the side and base of the cranium, composed of four portions: tympanic, squamous, mastoid, and petrous.

temulentous. With jerky, irregular movements, as those of an intoxicated person.

tenaculum. **1.** *zool.* A clasping or holding structure, as **a.** a glandular, adhesive structure on sea anemones; **b.** the structure which clasps the furcula in springtails; **c.** in fishes, a ligamentous structure which holds the eyeball in place. **2.** *bot.* **a.** A holdfast in certain algae. **b.** A structure surrounding the ostiole of an ascus in certain fungi.

tendinous. **1.** Pertaining to or of the nature of a tendon. **2.** Composed of tendons.

tendon. An inelastic band or cord of dense connective tissue by which a muscle is connected to a bone, cartilage, or other structure.

tendril. In climbing plants, a slender, elongated structure which, serves as an organ of attachment; may be a modified stem, leaf, or stipule.

tension. **1.** The act of stretching or state of being stretched; as muscle *tension.* **2.** The partial pressure of a gas.

tent. A cocoonlike shelter or a tentlike nest, as that constructed by various insects, as the tent caterpillar.

tentacle. **1.** *zool.* Any of the various slender, elongated, flexible, unsegmented processes usually located at the oral or anterior end of an animal, functioning as sensory, food-getting, defensive, or attachment structures. **2.** *bot.* A glandular, adhesive structure, as that on the leaf of the sundew.

tentacular bulb. In coelenterates, an enlargement at the base of a tentacle which functions in digestion and nematocyst formation, as in various medusae. It may bear ocelli and other sensory structures. Also called ocellar bulb.

tentaculocyst. In scyphozoans, a minute tentacle which functions as a sense organ.

tentaculum. A tentaclelike organelle located between the feathery membranelles in certain loricate protozoans.

tent caterpillar. One of several moths whose larvae construct a tentlike nest which is used as a shelter, as *Malacosoma americanum*, the Eastern tent caterpillar.

tentillum. A branch of a tentacle.

tentorium. 1. *zool.* The endoskeleton of the head of an insect. **2.** *anat.* A fold of the dura mater located between the cerebral hemispheres and the cerebellum. Also called tentorium cerebelli.

teosinte. A tall grass (*Euchlaena mexicana*) of Mexico and Central America, related to maize.

tepal. A division of the perianth when petals and sepals are indistinguishable from each other.

teratology. The branch of embryology which deals with malformations; the study of abnormal development.

tercel. The male of certain hawks, especially the peregrine falcon; also tercelet.

terebra. The serrated, boring or piercing ovipositor or sting of an insect, as that of various hymenopterans.

teredo. *See* shipworm.

terete. Circular in cross section.

tergite. A sclerite of the tergum.

tergum pl. **terga. 1.** The dorsal surface of a body somite or segment, as in arthropods. **2.** One of a pair of plates covering the anterior lateral surface of a barnacle.

terminal. 1. Pertaining to, placed at, or forming an end or extremity. **2.** *bot.* Located at or growing from the apex of a branch or stem, as a bud or flower. **3.** *zool.* **a.** Being the last of a series. **b.** Located at the extreme end of a body or structure.

terminology. A system of special names or terms used in a specific field; nomenclature. *See* binomial nomenclature.

termitarium. A complex structure which houses a termite colony.

termite. Any of a number of wood-eating insects of the order Isoptera; a white ant.

termone. A sex-determining chemotactic substance in protozoa.

tern. A gull-like bird of the family Laridae comprising the genus *Sterna* and related genera.

ternary. Trimerous; composed of or proceeding by threes.

ternate. In threes; trimerous.

terrapin. One of a number of freshwater turtles, especially the diamondback terrapin (*Malaclemys terrapin*), highly esteemed for its flesh; also the painted terrapin (*Chrysemys picta*) and Spanish terrapin (*Clemmys leprosa*). *See* turtle, tortoise.

terrarium. An enclosure for keeping terrestrial organisms.

terrestrial. **1.** Adapted to and living on land; not aquatic. **2.** *bot.* Growing on the ground.

terricolous. Living or dwelling in or on the ground.

territory. *ecol.* An area to which animals (individuals, pairs, or family groups) confine their activities and guard closely against intrusion from other members of the same group, as manifested by some arthropods, fishes, birds, and mammals.

tertian. Occurring every other day, as *tertian* malaria, with reference to chills.

tertiary. Third in order, rank, or formation.

Tertiary period. A major division of the Cenozoic Era comprising the Pliocene, Miocene, Oligocene, Eocene, and Paleocene epochs characterized by the rise of modern mammals and flora.

tesselate. In the form of a checkerboard; marked by squares.

¹test. *zool.* **1.** A shell or covering of various animals, as that of testaceans or echinoids. **2.** The tunic of a tunicate. **3.** A girdlelike prototroch of a trochophore, as in various mollusks.

²test. A trial, examination, or procedure used to identify or to determine the presence or absence of, or the amount of a substance; to detect alterations in function; or to determine a specific condition or situation, as a *test* for blood sugar, liver function *test*, or pregnancy *test*.

testa. The outer covering or coat of a seed.

testacean. A rhizopod protozoan of the order Testacea, characterized by possession of a shell or test, as *Arcella*.

testaceous. 1. Of or pertaining to a shell or test. **2.** Brick red or brownish yellow in color.

test cross. *genetics.* The mating of a hybrid with a recessive parental type, used for determining linkage and crossover values; a backcross.

testicle. A testis.

testiculate. 1. Shaped like a testis. **2.** *bot.* Ovoid and solid, as the tubers of certain orchids.

testis pl. testes. The male gonad which at maturity produces male gametes or spermatozoa. In mammals, it is the source of androgens. *See* antheridium.

testosterone. An androgenic hormone produced by the mammalian testis.

¹testudinate. A turtle; a chelonian.

²testudinate. Arched like the carapace of a turtle.

tetanus. 1. *physiol.* A steady, contracted state of a muscle. **2.** *med.* An infectious disease caused by *Clostridium tetani;* lockjaw.

tetany. A condition characterized by painful, tonic spasms of muscles due to lowered blood calcium and elevated pH.

tetra. Any of a number of brightly-colored, tropical fishes of the family Characinidae, a common inhabitant of aquaria.

tetrabranchiate. A mollusk with two pairs of gills, as the pearly nautilus, of the subclass Tetrabranchia (Nautiloidea).

tetracotyle. Having two regular suckers and two pseudosuckers, with reference to cercariae.

tetrad. 1. A group of four. **2.** *zool.* A group of four chromatids (two pairs of bivalent chromosomes) which is formed during meiosis. **3.** *bot.* A group of four cells, as the four microspores formed in the production of pollen.

tetradynamous. Having six stamens, four long and two short.

tetrafossate. With four suckers on the scolex, with reference to tapeworms.

tetragonal. Having four angles.

tetramerous. Composed of four units or parts; in multiples of four.

tetrandrous. Possessing four stamens.

tetrapetalous. Possessing four petals.

tetraploid. An individual with four sets of chromosomes, hence with a chromosome number of 4n.

tetraploidy. Possessing four times the normal haploid number of chromosomes.

tetrapod. 1. An individual with four limbs. 2. Any vertebrate of the Tetrapoda, a group which includes all vertebrates having paired appendages other then fishes. It comprises the Amphibia, Reptilia, Aves, and Mammalia.

tetrapterous. Having four wings or winglike appendages.

tetrarch. A type of xylem which has four rays extending outward from the center of the vascular cylinder.

tetrasepalous. Having four sepals.

tetrasomic. In aneuploidy, having four of one pair of chromosomes. Chromosome number is 2n + 2.

tetraspore. 1. One of four spores of a tetrad. 2. A nonmotile, asexual spore formed in groups of four, as in the red algae.

tetrastachyous. Having four spikes.

tetrastichous. Four-ranked; arranged in four vertical rows, said of flowers.

tetraxon. A sponge spicule having four rays radiating from a common point but not in the same plane; tetractine, quadriradiate.

Texas fever. An infectious disease of cattle caused by a protozoan (*Babesia bigemina*) transmitted by a tick. Also called redwater fever, hemoglobinuric fever, tick fever.

thalamus. 1. *anat.* A division of the diencephalon consisting of a number of nuclei which function as relay centers for sensory impulses passing to the cerebral cortex. It is also a center for crude, uncritical sensations and is involved in affective states of an individual. 2. *bot.* The receptacle of a flower; a torus.

thalessa. A large ichneumon fly (*Megarhyssa macrurus*) with an extremely long ovipositor by which eggs are laid in the larva of the pigeon tremex.

thaliacean. A tunicate of the class Thaliacea which includes the free-living, pelagic, chain tunicates, as *Salpa*, *Pyrosoma*, *Doliolum*.

thalloid. Resembling a thallus.

thallophyte. 1. Any plant of the Thallophyta, formerly a taxon which included the algae and fungi. **2.** A simple plant; one lacking true roots, stems, or leaves.

thallose. With a simple, undifferentiated plant body.

thallus. A simple plant body not differentiated into root, stem, or leaves, as that characteristic of the algae and fungi.

thanatoid. Resembling death.

thanatosis. Feigning death.

theca. 1. *zool.* A sheath or covering, as **a.** the skeletal cup of a coral; **b.** an armor of plates in pelmatozoans (extinct echinoderms); **c.** a structure enclosing a hydranth or gonophore in colonial coelenterates; **d.** a sheath enclosing an ovarian follicle. **2.** *bot.* A capsule, spore case, or similar enclosing structure.

thecate. Possessing a theca, with reference to hydrozoans.

thecodont. Designating a type of dentition in which the teeth are set in sockets in the jaw bones.

theine. CAFFEINE.

thelephorous. Bearing nipplelike prominences or projections.

thelytoky. Parthenogenesis in which only females are produced; thelyotoky.

thenar. 1. The palm of the hand. **2.** The fleshy prominence at the base of the thumb.

theory. 1. The abstract principles of a science as distinguished from basic or applied science. **2.** A reasonable explanation or assumption advanced to explain a natural phenomenon but lacking confirming proof. *See* hypothesis.

theraspid. One of an extinct group of mammallike reptiles of the Triassic period thought to be ancestral to modern mammals.

¹therian. Of or pertaining to the subclass Theria which includes the true mammals.

²therian. Any member of the Theria, comprising mammals which bring forth their young alive and possess mammary glands with nipples.

thermal. Of or pertaining to heat or warmth.

thermal stratification. *limnology.* A condition in deep lakes

561

following the spring overturn in which layers of water differing in temperature exist. *See* epilimnion, thermocline, hypolimnion.

thermocline. The middle layer of water in thermal stratification.

thermoduric. Capable of enduring high temperatures, with reference to bacteria.

thermolabile. Changed or destroyed by heat, said of enzymes.

thermometer. An instrument or device for measuring temperature. *See* centigrade thermometer; Fahrenheit thermometer.

thermonasty. A plant movement in response to a change in temperature.

thermoperiodicity. The effects of changes in temperature between day and night periods upon plant activity, especially growth.

thermophilic. Heat-loving, with reference to organisms which thrive at high temperatures (50°–70° C.), as some algae and bacteria.

thermophyllous. Producing leaves in summer.

thermophyte. A heat-tolerant plant.

thermoreceptor. A sensory receptor which is stimulated by changes in temperature, as an end bulb of Krause or brush of Ruffini.

thermostable. Not readily affected by heat; resistant to heat up to 100° C.

thermotropism. The bending of a part of a plant in response to heat.

therophyte. An annual, a plant which completes its development in one season.

thiamine. Vitamin B_1, present in the seed coats of cereal grains and in yeast. It plays an important role in the metabolism of carbohydrates, a deficiency resulting in nervous and circulatory disorders.

thigh. The portion of the lower extremity or hindlimb between the pelvic girdle and knee.

thigmotaxis. The response of an animal to tactile stimuli.

thigmotropism. The response of a plant to mechanical contact, as the curving of tendrils about objects touched or the turning away by roots from objects in the path of growth.

thoracic. Of or pertaining to the thorax.

thoracic duct. In mammals, the principal lymph duct opening

562

into the left subclavian vein near its junction with the internal jugular vein. It receives lymph from all the body except the upper right quadrant.

thorax. **1.** In mammals, the chest or portion of the body between the neck and abdomen. **2.** In insects, the portion of the body between the head and abdomen, bearing the legs and wings. **3.** In various animals, the region of the body immediately posterior to the head, often fused with the head to form the cephalothorax.

thorn. A sharp-pointed process or projection.

thorn apple. **1.** The red haw or the fruit of the hawthorn (*Crataegus*). **2.** The jimsonweed or stramonium (*Datura*).

thrasher. A thrushlike bird of the family Mimidae, as the brown thrasher (*Toxostoma rufum*).

thread. **1.** The coiled tube of a nematocyst. **2.** A spirally coiled structure within a thread cell of a hagfish. **3.** Any of a number of natural filaments, especially those produced by spinnerets.

thread cell. A unicellular gland of a hagfish which produces spirally coiled threads which can be discharged for a considerable distance.

threshold. The lowest limit at which a certain phenomenon will occur, as the *threshold* of consciousness or renal *threshold*.

thrips. A minute insect of the order Thysanoptera.

throat. **1.** *bot.* The opening into a gamopetalous corolla; the point of junction between the tube and limb. **2.** *anat.* **a.** The anterior portion of the neck containing the lower portion of the pharynx, larynx, and portions of the trachea and esophagus. **b.** In common usage, the fauces and pharynx.

thrombin. A substance in shed blood formed from prothrombin which converts fibrinogen to fibrin, thus inducing clotting.

thrombocyte. **1.** In mammals, a blood platelet. **2.** In vertebrates below mammals, a spindle cell of the blood.

thromboplastin. A substance released from disintegrating blood platelets and injured tissue which, in the presence of calcium ions, converts prothrombin to thrombin.

thrombosis. The formation of a thrombus.

thrombus. A blood clot formed within a blood vessel or the heart.

throwback. A reversion to an ancestral type.

[1]**thrush.** One of a number of small to medium-sized passerine birds of the family Turdidae, as *Hylocichla mustelina*, the wood thrush.

[2]**thrush.** *med.* A throat infection caused by a fungus (*Candida albicans*).

thumb. The pollex, the short first digit on radial side of the hand; the corresponding digit in other animals.

thunderworm. *See* worm lizard.

thylacine. A carnivorous marsupial (*Thylacinus cynocephalus*) of Tasmania, also called the Tasmanian wolf.

thyme. A low shrub (*Thymus*) of the mint family, with aromatic leaves used in seasoning.

thymine. A pyrimidine base, $C_5H_6N_2O_2$, present in DNA.

thymus. An unpaired organ composed of lymphatic tissue located in the upper thorax and neck that produces lymphocytes and is the source of cells which leave the organ and lodge in peripheral lymphatic organs. It is possibly the source of hormones essential for the production of immunologic substances and is large in young animals but slowly atrophies after sexual maturity.

thyroid. Of, pertaining to, produced by, or located near the thyroid gland.

thyroid gland. A large, bilobed, endocrine gland lying in the neck alongside the larynx and trachea that produces a hormone, thyroxine, and other iodine-containing compounds which regulate metabolic rate, electrolyte balance, and other functions. In amphibians, it controls metamorphosis. Thyroid deficiency in early life results in cretinism.

thyrotrophic. Thyroid-stimulating; also thyrotropic.

thyrotrophin. A thyroid-stimulating hormone (TSH) produced by the anterior lobe of the pituitary. Also thyrotropin.

thyroxine. An iodine-containing compound which, with 3, 5, 3'-triiodothyronine, is the active principle or hormone secreted by the thyroid gland.

thyrse. A paniclelike inflorescence in which the main axis is indeterminate and lateral axes determinate, as in the lilac.

thysanopteran. Any insect of the order Thysanoptera, which includes the thrips, fringe-winged insects with conelike mouth parts, as *Thrips tabaci*, the onion thrips.

thysanuran. A primitive, wingless insect of the order Thysanura which includes the bristletails, as the silverfish (*Lepisma*) and firebrat (*Thermobia*).

tibia. 1. The innermost and larger of the two bones of the leg, lying medial to the fibula; the shinbone. **2.** In arthropods, the fourth segment of the leg located between femur and tarsus.

tibiofibula. The fused tibia and fibula as in the frog.

tibiotarsus. The tibia and some fused tarsals comprising the principal leg bone of a bird, located between femur and tarsometatarsus.

tick. 1. A bloodsucking arachnid of the order Acarina comprising two families, the Argasidae (soft ticks) and Ixodidae (hard ticks). They parasitize warm-blooded animals, are annoying pests, and serve as reservoirs and transmitting agents for the causative agents of a number of diseases affecting man and domestic animals. **2.** A louse fly.

tickbird. The oxpecker, an African bird (*Buphagus*) which feeds on ixodid ticks, especially those infesting the rhinoceros.

tick fever. Any of a number of diseases resulting from the bite of a tick, as Rocky Mountain spotted fever, Colorado tick fever, relapsing fever, Texas fever, and tularemia.

tidal. 1. Of, pertaining to, or associated with tides, as *tidal* rhythms. **2.** Periodically rising or falling, or flowing in or out, as *tidal* air.

tide. 1. The periodic rise and fall of the sea level along coasts occurring twice each lunar day (24 hrs., 51 min.). **2.** *physiol.* A periodic increase in the pH of urine (alkaline *tide*), as occurs following ingestion of food.

Tiedemann's bodies. Nine vesicles located on the inner margin of the ring canal of a starfish; they produce amebocytes.

tiger. A large, carnivorous mammal (*Panthera* (*Felis*) *tigris*) of the cat family, widely distributed throughout southern Asia.

tiger beetle. One of a number of brightly colored, predaceous beetles of the family Cicindelidae, especially those of the genus *Cincidela*.

tiger cat. 1. A marsupial (*Dasyurus maculatus*) of southern Australia. **2.** The ocelot. **3.** The margay.

tiger salamander. A spotted salamander (*Ambystoma tigrinum*),

565

a neotenic species common throughout the United States and Mexico. *See* axolotl.

tigroid bodies. *See* Nissl bodies.

tigrolysis. Chromatolysis or dissolution of Nissl bodies in a nerve cell.

tilefish. A deepwater food fish (*Lopholatilus chamaeleonticeps*) of the Atlantic coast.

tiller. A sprout or branch which grows from the base of a monocot, especially those of the grass family.

timberline. The upper limit of tree growth on mountainsides or in high latitudes.

timothy. A meadow grass (*Phleum pratense*).

tinamou. A neotropical, nonmigratory, forest bird of the family Tinamidae, of Mexico south to Argentina, as *Phychotus rufescens*.

tinctorial. Of or pertaining to colors, staining, or dyeing.

tine. A slender, pointed process, as the prong of an antler.

tip. The extreme end of a structure; the apex or summit.

tiphad. A pond plant.

tiphophilous. Pond-loving.

tissue. An aggregation of cells of more or less similar structure and function together with their intercellular material. Principal animal tissues are epithelial, connective, muscular, and nervous; plant tissues include meristematic and permanent tissue (parenchyma, collenchyma, sclerenchyma, cork, xylem, phloem).

tissue culture. The cultivation of cells or small masses of tissue outside of the body.

[1]**tit.** TITMOUSE.

[2]**tit.** TEAT; NIPPLE.

titi. A small South American monkey of the genus *Callicebus*.

titmouse. A small, active, woodland bird of the family Paridae, as *Parus bicolor*, the tufted titmouse; also called tomtit or peterbird.

toad. One of a number of tailless, leaping amphibians of the order Salientia, most of which are dry-skinned terrestrial forms except during the breeding season. True toads belong to the family Bufonidae, as *Bufo americanus*.

toadfish. A small, bottom-living, venomous, marine fish of the family Batrachoididae, as *Opsanus tau* of the Atlantic coast; also called oysterfish.

toadstool. A poisonous, inedible mushroom.

tobacco. One of several species of solanaceous plants of the genus *Nicotiana*, especially *N. tabacum*, whose leaves are widely used for smoking and chewing. *See* nicotine.

tobacco hornworm. *See* hornworm.

tobacco mosaic disease. A disease of tobacco plants caused by a virus. It was the first disease for which a filterable agent was found to be the causative agent (Iwanowski, 1892) and its virus was the first to be purified in crystalline form (Stanley, 1935).

tocopherol. One of a group of substances possessing vitamin E properties, a deficiency resulting in impairment of reproductive functions.

toddy cat. *See* musang.

toe. **1.** A digit of the foot. *See* hallux. **2.** A process on the foot of a rotifer.

tolerance. The act of tolerating or enduring, as *tolerance* of shade or of a drug. In ecology, the degree of *tolerance* is expressed by the prefixes *steno-*, meaning narrow, and *eury-*, meaning wide.

tolerate. To bear, withstand, or endure without injury or harm.

tomato. A solanaceous garden plant (*Lycopersicon esculentum*) or the fruit it produces.

tomato worm. *See* hornworm.

tomentose. Covered with a dense mat of woollike hairs; densely pubescent or woolly.

tomentulose. Finely or delicately tomentose.

tomentum. A fine, densely-matted pubescence or woollike hairiness.

tomium. The cutting edge of the bill of a bird.

tomite. A free-swimming, nonfeeding stage in the life cycle of certain holotrich protozoans that follows the protomite stage.

tomont. A stage in the life cycle of a holotrich protozoan in which multiple division occurs, resulting in the formation of many small, ciliated individuals. Usually occurs in encysted forms.

tomtit. *See* titmouse.

tone. **1.** A distinct musical or vocal sound or the quality of the sound. **2.** The normal state of the body or one of its parts. **3.** In muscles, a state of slight tension or resistance to stretch; also called tonus.

tongue. **1.** In vertebrates, a muscular structure in the mouth, usually movable and protrusable, which functions in food-getting, manipulation, tasting, and swallowing. It may also serve other functions, as touch, olfaction, and sound production. **2.** In invertebrates, a comparable organ, as **a.** the hypopharynx of a grasshopper; **b.** the coiled proboscis of a moth or butterfly; **c.** the labium of a honeybee. **3.** *bot.* A ligule or ligula.

tongue bar. In lancelets, a secondary gill bar which divides each primary gill cleft.

tongue worm. **1.** A pentastomid. **2.** An acorn worm or enteropneustid.

tonoplast. A specialized layer of protoplasm surrounding a water vacuole of a cell; also called vacuolar membrane.

tonsil. A structure composed of masses of lymphatic tissue located in the walls of the pharynx and in the root of the tongue. *See* adenoid.

tonus. *See* tone.

tooth pl. **teeth.** **1.** In vertebrates, a hard cornified or calcified structure in the mouth used primarily for cutting and mastication of food, but also functioning as offensive and defensive weapons, for grasping and holding prey, and for injecting venom. A *true tooth*, as in a mammal, consists of a root, neck, and crown, the crown consisting of a surface layer of enamel overlying a layer of dentine, the whole enclosing a pulp cavity. *See* dentition, dental formula. **2.** In invertebrates, a structure in or about the mouth which resembles a tooth, as the teeth of Aristotle's lantern of echinoids or the gastric mill of a crustacean. **3.** One of a pair of structures on the hinge of a lamellibranch shell which function in locking the valves. **4.** *bot.* **a.** A small, pointed, marginal process of a leaf. **b.** A toothlike process, as in the toothed fungi. **c.** A pointed process of a peristome of a moss capsule.

toothed. Bearing teeth; dentate; having small pointed projections; having indentations forming a notched edge.

toothed fungi. Fungi of the family Hydnaceae in which the hymenium covers the surfaces of downward-projecting spines or teeth.

tooth shell. A burrowing marine gastropod of the class Scaphopoda, as *Dentalium*, with a long, pointed shell open at both ends.

tope. A small shark (*Galeorhinus galeus*) of European waters.

topi. An East African antelope (*Damaliscus corrigum*).

topiary. Designating trees or shrubs which are trimmed or trained to grow in unusual or ornamental shapes.

topminnow. One of a number of small, surface-feeding fishes, as *Fundulus* and *Gambusia*, used in mosquito control.

topographical. Pertaining to the configuration of a surface.

topotype. *taxonomy.* A specimen of a species collected at the type locality.

top shell. A primitive, top-shaped marine gastropod of the order Archeogastropoda, related to the limpets, as *Astraea, Trochus.*

tormogen cell. A hypodermal cell in insects which produces the articular membrane of the setae.

tornaria larva. The pelagic, ciliated larva of various enteropneustans which bears a marked resemblance to the auricularia larva of echinoderms.

tornote. A sponge spicule, lance-headed at each end.

torose. 1. Bulging, knobbed. **2.** *bot.* Cylindrical, with constrictions.

torpedo ray. An electric ray of the genus *Torpedo.*

torpid. Inactive; dormant; lacking in energy, as a hibernating animal.

torpor. A state of sluggishness or inactivity, as in estivation: SUSPENDED ANIMATION.

torques. A collarlike structure of hair, feathers, or modified integument which encircles the neck, usually of a distinctive color or structure.

torsalo. The larva of *Dermatobia hominis*, a skin bot.

torsion. A turning or twisting, as occurs in snails when the visceral mass is twisted on the head-foot.

torso. The trunk; the body lacking head and limbs.

tortoise. A turtle, especially a land turtle, as *Testudo*, the giant land tortoise of the Galapagos Islands or *Gopherus* of the southern states.

torula. One of the false or pseudoyeasts which cause undesirable fermentations.

torulus. A socket which lodges the basal end of the antenna of an insect.

torus. 1. *anat.* An elevation; a small rounded protuberance. 2. *bot.* **a.** The receptacle of a flower; thalamus. **b.** The thickened, central portion of a pit membrane in a pit-pair.

totipalmate. With four toes united by a web, as in birds.

toucan. A brilliantly-colored, fruit-eating bird of tropical America possessing an enormous but light and thin-walled beak, as *Rhamphastos cuvieri*, of the family Rhamphastidae.

touch. The tactile sense; the sense of feeling.

touraco. A brightly-colored, African bird of the family Musophagidae.

towhee. A finch of the family Fringillidae, as *Pipilo erythrophthalmus*, the chewink.

toxic. Of or pertaining to a poison; poisonous.

toxin. A poisonous substance produced by an animal or plant cell, which, within an animal, induces the formation of antibodies. *See* endotoxin, exotoxin, phytotoxin, zootoxin.

toxoid. A toxin which has been modified by physical or chemical agents so that its toxic properties are lost but its antigenic properties remain intact, as the *toxoids* of tetanus and diphtheria.

trabecula. 1. *anat.* A bar, band, septum, or partial septum which forms the framework of an organ or structure, as a lymph node or bone. 2. *bot.* A plate of cells traversing a cavity.

trace element. An element which is essential for normal activities of an organism, but required only in minute amounts, as **a.** for plants, boron, copper, zinc, and manganese; **b.** for animals, aluminum, boron, copper, zinc, silicon, cobalt, iodine, fluorine, manganese, selenium, and possibly others.

tracer. A radioactive isotope or other substance whose course can be traced within the body or in biochemical reactions.

trachea. 1. *zool.* **a.** In vertebrates, a cartilaginous tube leading from larynx to the bronchi through which air passes to the lungs. **b.** In air-breathing arthropods, one of the minute air tubes which conveys air to the tissues. **c.** In land snails, a tubular extension of the mantle cavity. **2.** VESSEL 2.

tracheal. Of, pertaining to, or resembling a trachea.

tracheal gill. A gill containing trachea and tracheoles, as in the naiads, larvae, and pupae of aquatic insects, as dragonflies and mayflies.

tracheal system. A respiratory system in which air is conveyed through minute tubes (trachea, tracheoles) to the tissues, as in various terrestrial arthropods, as insects, spiders.

tracheary. Of or pertaining to xylem elements which resemble animal tracheae, as tracheids and vessel members.

tracheid. In vascular plants, an elongated, dead cell with a pronounced cavity or lumen forming an element of xylem. Its lignified walls are usually thick and pitted; it functions in water conduction and support.

tracheole. One of the numerous, minute branches of a trachea in an arthropod.

tracheophyte. Any plant of the division Tracheophyta which includes the vascular plants comprising the ferns and fernlike plants and the seed plants.

trachycarpous. Rough-skinned, with reference to fruits.

trachyspermous. Possessing a rough coat, with reference to seeds.

tract. **1.** A pathway or course, as the alimentary *tract*. **2.** A bundle or group of nerve fibers in the spinal cord or brain.

tractellum. The anterior flagellum of a flagellate, used primarily for pulling.

traction. The act of drawing or pulling.

tragacanth. A gum with marked water-absorbent properties obtained from various Asiatic plants of the genus *Astragulus*.

tragus. **1.** *anat.* **a.** A process anterior to and projecting over the external acoustic meatus. **b.** A hair of the external acoustic meatus. **2.** *zool.* In insectivorous bats, a flap that can be drawn over the ear; an earlet.

trailing. Prostrate but not taking root.

trait. A distinguishing character or feature, as a hereditary *trait*.

trama. In a basidiocarp, the supporting structure between two hymenial layers.

transaminase. An enzyme which catalyzes a transamination reaction.

transamination. The transfer of an amino group (NH_2) from one compound to another.

transcription. The production of messenger RNA from DNA bearing encoded genetic information, catalyzed by an enzyme, transcriptase.

transduction. *genetics.* The process by which, in bacteria, genetic material (DNA) is transmitted from one cell to another through the agency of a bacteriophage.

¹**transect.** To cut across.

²**transect.** A line, belt, or strip of vegetation used for study purposes.

transection. A cut or section made across the long axis of a structure.

transformation. *genetics.* A process by which bacteria grown on a nutrient medium containing a transforming principle (DNA) extracted from killed bacteria are transformed or changed to a different type, such involving the transfer of genetic material.

transformer. *ecol.* An organism, as a nitrate or nitrite bacterium, which converts simpler substances into compounds which can be used by plants in the synthesis of more complex compounds, as proteins.

transfusion. The introduction of whole blood, plasma, or saline solution into a blood vessel of the body.

translation. The process by which amino acids are combined into a polypeptide chain as directed or specified by mRNA.

translator. A structure at the tip of fused anthers on certain flowers to which masses of pollen grains become attached, as in the milkweed.

translocation. 1. *physiol.* a. The movement of substances from one location to another. b. The movement of ions, atoms, or molecules through cell membranes. 2. *genetics.* A type of chromosome aberration in which a portion of a chromosome is transferred to a nonhomologous chromosome, or its reverse.

translucent. Semitransparent; transmitting light rays only partially.

transmigration. The movement from one place to another, as ions across a cell membrane.

transmission. The passage from one place to another.

transmitting tissue. *See* stigmatoid tissue.

transmutation. The change of a species or type to another: EVOLUTION

transparent. Permitting the passage of light rays; capable of being seen through.

transpiration. The emission of water vapor from the surface of an organism, as that lost through leaves or through the skin.

¹**transplant.** To remove and place in another location where growth may take place, as seedlings or tissue.

²**transplant.** A piece of tissue or a part of an organism which is removed and transferred to another site, either in the same individual (autoplastic or homoplastic *transplant*) or in another individual (heteroplastic *transplant*).

transport. The movement of substances from one place to another. *See* active transport, passive transport.

transposition. A change in position, especially the reversal or transfer to the opposite side.

transudate. The product resulting from transudation.

transudation. The passage of a substance through pores of a membrane or interstices of a tissue, as blood plasma through a capillary wall.

transverse. Crosswise; at right angles to the long axis.

trap. *bot.* A structure which imprisons an insect or small organism, as the modified leaf of an insectivorous plant.

trap flower. A prison flower.

trap hairs. Hairs which prevent an insect from leaving a flower or leaf.

trap-prison. *See* prison flower.

trauma. **1.** An injury or wound, especially one induced by a physical agent, as a cut, bruise, or burn. **2.** A severe psychic or emotional shock.

traumatic. Of, pertaining to, or resulting from trauma.

traumatism. **1.** An abnormal condition resulting from trauma. **2.** *bot.* An abnormal growth resulting from injury.

traumatropism. The abnormal sensitiveness of certain plant structures to injury.

tread. To copulate or to cover, with reference to male birds.

tree. **1.** A woody, perennial plant with a single main axis or trunk which bears branches. **2.** Something resembling a tree in shape or form.

573

tree fern. One of a number of giant ferns with well-developed, treelike trunks, as those of the families Cyatheaceae and Marattiaceae.

tree frog. *See* tree toad.

treehopper. One of a number of small, leaping insects of the order Homoptera with a well-developed pronotum giving them a humpbacked appearance, as the buffalo treehopper (*Ceresa bubalus*).

tree of heaven. A rapid-growing Asiatic tree (*Ailanthus altissima*) with ill-smelling staminate flowers.

tree shrew. A small, arboreal, rodentlike mammal of the family Tupaiidae, of Southeast Asia, considered to be the most primitive of primates, as *Tupaia*, *Ptilocercus*.

tree toad. A small, leaping amphibian of the family Hylidae with expanded adhesive discs on their toes, as *Hyla versicolor*; a tree frog.

trematode. A parasitic flatworm of the class Trematoda; a fluke.

trepan. *See* trephine.

trepang. Dried sea cucumbers used as food. Also called beche-de-mer.

¹trephine. An instrument for making an opening into the skull.

²trephine. To perforate the skull with a trephine.

treppe. *physiol.* A gradual increase in the extent of muscle contraction up to a certain limit following rapidly repeated stimuli. Also called staircase phenomenon.

triad. A closely associated or related group or set of three objects, things, or individuals.

triadelphous. With stamens arranged in three bundles or sets.

triandrous. Possessing three stamens.

triangulate. Having three angles.

triarch. Designating a protoxylem with three rays.

Triassic period. The first and earliest period of the Mesozoic era, characterized by the rise of giant reptiles.

tribe. 1. A division of a family consisting of related genera. 2. *anthropol.* A social group consisting of a number of related families or clans.

tribocytic organ. A glandular, adhesive organ in holostome flukes located immediately posterior to the acetabulum, as in strigeids.

tricamerous. Having three loculi, said of fruits.

tricarboxylic acid cycle. *See* Kreb's cycle.

tricarinate. Having three keels or angles.

tricarpellate. Having three carpels.

triceps. A muscle with three heads, as the *triceps* brachii.

trichina worm. A parasitic nematode (*Trichinella spiralis*) which infests man, pigs, rats, and other vertebrates. Adults live a short while in the intestine; embryos encyst in skeletal muscle.

trichite. A tubular trichocyst containing a coiled thread.

trichoblast. A simple hairlike branch in certain red algae.

trichobothrium. A sensory hair or a cluster on the tarsus of a spider.

trichocyst. 1. One of the numerous oval or rod-shaped structures in the ectoplasm of various ciliates which function as defensive or anchoring structures. **2.** A hairlike structure in certain cryptomonads and chloromonads capable of being discharged; an ejectosome.

trichogen. A cell in the hypodermis of an insect which gives rise to a hair or seta.

trichogyne. A hairlike receptive cell at the tip of a female gametangium in certain algae and fungi.

trichome. 1. *bot.* An epidermal hair or hairlike process. **2.** *microbiol.* A uniseriate, multicellular organism in which individual cells can be noted without staining.

trichopteran. Any insect of the order Trichoptera which includes the caddis flies.

trichothallic. Designating growth of a filamentous thallus, as in certain brown algae.

trichotomous. In three parts or divisions; three-forked.

triclad. A free-living, turbellarian flatworm of the order Tricladida with a three-part gastrovascular cavity, as *Planaria, Dugesia*.

tricolpate. With three grooves, with reference to pollen grains.

tricostate. With three ribs or costae.

tricuspid. With three cusps or leaflets.

tricuspid valve. A heart valve located between the right atrium and right ventricle.

tricussate. On groups of three; ternate, with reference to whorls of leaves.

tridactyl. With three digits; tridigitate.

trident. TRIDENTATE.

tridentate. With three teeth, toothlike processes, or points.

tridigitate. With three digits or fingerlike processes; tridactyl.

tridynamous. With six stamens (three long, three short).

trifid. Divided into three parts; consisting of three lobes or divisions.

trifoliate. With three leaflets, said of leaves.

trigeminal nerve. The fifth cranial nerve (V) consisting of three main divisions, the ophthalmic, maxillary, and mandibular. On its root is the semilunar or Gasserian ganglion.

trigonal. Triangular.

trigone. 1. *anat.* A triangular area in the wall of the bladder between the orifices of the two ureters and the urethra. 2. *bot.* In liverworts, a triangular thickening of the cell wall at each corner of a cell.

trigonid. A well-developed triangular area on a lower molar tooth.

trigonous. Triangular; with three corners or three angles.

trihybrid. A hybrid heterozygous for three pairs of genes.

trijugate. With three pairs of leaflets.

trilobate. With three lobes.

trilobite. A primitive, marine arthropod of the subphylum Trilobitomorpha, an extinct group widely distributed during the Paleozoic era.

trilobite larva. The larva of the king crab.

trilocular. Having three cavities or compartments.

trimonoecious. Having staminate, pistillate, and perfect flowers on one plant.

trimorphic. Existing or occurring in three types or forms.

trinervate. Having three nerves, veins, or ribs.

trinomial nomenclature. An extension of the binomial system of nomenclature by which a subspecies or variety is designated, as *Pyrus malus albiflorus*, the white-flowered apple.

trioecious. Producing staminate, pistillate, and perfect flowers on different plants within a species.

tripartite. Divided into three parts or lobes.

tripetalous. With three petals.

tripinnate. Thrice pinnate.

triple fusion. In angiosperms, the union of **a.** one of the sperm nuclei of a pollen tube with an ovum which gives rise to a diploid zygote and, **b.** the other sperm nucleus with the two endosperm nuclei to form the triploid endosperm of a seed. Also called double fertilization.

triploblastic. Composed of three primary germ layers: ectoderm, mesoderm, and endoderm.

triploid. Having three times the haploid number of chromosomes; a triploid individual.

triploidy. The condition in which an individual has three of each kind of chromosome; chromosome number = 3n.

tripterous. Three-winged, said of fruits and seeds.

tripton. ABIOSESTON.

triquetral. The cuneiform bone of the carpus, located between lunate and pisiform.

triquetrous. Having three salient angles, said of triangular stems with convex sides.

triradiate. Having three rays or radiating parts.

trisomic. Having one extra chromosome, the chromosome number being 2n + 1.

tristichous. Arranged in three vertical rows.

triternate. Thrice ternate; three times three.

tritonymph. The last of three stages in the development of certain mites.

triturate. To crush or pulverize; to grind to a powder.

triungulin. The free-living, first larval stage of strepsipterans and certain coleopterans which undergo hypermetamorphosis.

trivium. **1.** In asteroids, the three arms opposite the madreporite. **2.** In holothuroids, the three ambulacral areas forming the ventral surface.

trochal. Resembling a wheel.

trochal disc. The flattened, ciliated, anterior end of a rotifer.

trochanter. **1.** One of two large processes on the proximal end of the femur. **2.** The second segment in the leg of an arthropod, located between the coxa and femur.

trochlea. A pulleylike structure, as **a.** a process at the distal end of the humerus; **b.** a ligamentous ring in the orbit through which passes the tendon of the superior oblique muscle.

trochlear nerve. The 4th cranial nerve (IV) supplying the superior oblique muscle of the eye.

577

trochophore. The ciliated, free-swimming, pear-shaped larva of a number of invertebrate groups, especially annelids and mollusks; a trochosphere.

trochosphere. A trochophore larva.

trochus. The inner, ciliated ring of a trochal disc of a rotifer.

troglobiont. A cave-dweller; an organism which spends a portion of or its entire life in caves.

troglobiotic. Pertaining to cave life.

trogon. A tropical bird of the family Trogonidae, noted for its brilliant plumage. *See* quetzal.

trombiculid. A mite of the family Trombiculidae which includes the red bugs or chiggers.

trophallaxis 1. The mutual exchange of food between adults and larvae, as in certain hymenopterans. **2.** The transfer of secretions which serve as food and regulatory substances between adults and young, as in termite colonies.

trophi. 1. The mouthparts of an insect including their appendages. **2.** The sclerotized jaws of a rotifer located on the inner wall of the pharynx or mastax.

trophic. 1. Pertaining to growth or nutrition. **2.** Pertaining to that which influences growth and development or stimulates functional activity, as a *trophic* hormone.

trophic level. A stage in the food chain of an ecosystem in which organisms obtain their food in the same number of steps or in the same general manner.

trophoblast. The surface layer of cells of a mammalian blastocyst from which the chorion develops.

trophocyte. 1. A nutritive cell or one that stores food, as those of the fat body of an insect. **2.** A nurse cell.

trophoderm. 1. The portion of the trophoblast which comes in contact with the uterine endometrium. **2.** In mammals, the serosa or extraembryonic somatopleure.

trophont. A vegetative stage in the development of certain parasitic ciliates.

trophosome. Collectively, the nonsexual structures of a hydrozoan coelenterate. *See* gonosome.

trophozoite. The vegetative phase in the life cycle of a protozoan during which asexual reproduction occurs.

trophozooid. A gastrozooid.

tropic. Of or pertaining to a tropism.

tropical. Of or pertaining to the tropics.

tropic bird. A large seabird of the family Phaethontidae, as *Phaethon lepturus*, the white-tailed tropic bird with extremely long middle tail feathers.

tropism. An involuntary response to a stimulus exhibited by plants and sedentary animals in which a bending, turning, or growth occurs, as a phototropism, geotropism, or hydrotropism. The response may be positive (towards) or negative (away from) the stimulus. *Compare* taxis.

tropophilous. Thriving under alternating environmental conditions, as dry-moist habitats or hot-cold seasons.

tropophyte. A tropophilous plant, as a deciduous tree which is mesophytic during the summer and xerophytic during the winter.

troupial. A tropical bird of the family Icteridae, especially *Icterus icterus*, an oriole of Central and South America.

trout. **1.** One of a number of food and game fishes of the salmon family, Salmonidae, especially those of the genera *Salmo* and *Salvelinus*. **2.** One of a number of other fishes which resemble the true trouts.

truffle. An edible, ascomycete fungus of the genus *Tuber*, which produces tuberlike, subterranean ascocarps, located with the aid of trained dogs and pigs.

trumpeter. Any of several animals which makes a trumpetlike sound, as **a.** a large, long-legged South American bird of the family Psophiidae; **b.** the trumpeter swan; **c.** several marine fishes, as the trumpeter perch; **d.** a breed of domestic pigeons.

trumpeter swan. A North American wild swan (*Cygnus buccinator*), now nearly extinct.

trumpet-shaped. Tubular, with one end narrow, the other widely expanded.

truncate. **1.** Having a square end, or appearing as though the apex or tip of a tapering structure had been broken or cut off. **2.** Lacking a normal tip or apex.

truncus arteriosus. **1.** In lungfishes and amphibians, the anterior end of the conus arteriosus. **2.** In vertebrate embryos, the tube leading from the bulbus arteriosus to the aortic sac and aortic arches. **3.** The ventral aorta.

trunk. **1.** *bot.* The main stem of a tree. **2.** *anat* and *zool.* **a.** The body without head or limbs; the torso. **b.** The thorax and abdomen of an arthropod. **c.** The main portion of a structure, as a nerve *trunk*. **d.** A proboscis, as that of an elephant.

trunkfish. A plectognath fish of the family Ostraciidae, as *Ostracion*, with angular body and heavy protective armor; boxfish.

trypanosome. Any flagellate protozoan of the genus *Trypanosoma*, including blood parasites infesting man and other vertebrates, the causative agent of a number of diseases, as African sleeping sickness, Chaga's disease, surra, nagana, and dourine.

trypanosomiasis. Any disease or condition resulting from infection with trypanosomes.

trypsin. A proteolytic enzyme formed from trypsinogen secreted by the pancreas.

trypsinogen. The precursor or zymogen of trypsin secreted by the pancreas.

tryptophan. An essential amino acid, $C_{11}H_{12}N_2O_2$.

tsetse fly. One of a number of species of bloodsucking, African flies of the genus *Glossina* which act as vectors of trypanosomes. *See* trypanosome.

TSH. Thyroid-stimulating hormone, a thyrotrophic hormone secreted by the anterior lobe of the pituitary.

tuatara. A primitive, rhynchocephalian reptile (*Sphenodon punctatum*) of New Zealand.

tube. **1.** Any hollow, cylindrical structure. **2.** *bot.* The basal portion of a calyx or corolla in a synsepalous or sympetalous flower. **3.** *anat.* The uterine or auditory *tube*. **4.** *zool.* A cylindrical burrow or house constructed by various invertebrates, as tube worms.

tube cell. *See* tube nucleus.

tube foot. A tubular extension of the water-vascular system of an echinoderm, which functions in locomotion and grasping; a podium.

tube nucleus. A nucleus at the tip of a pollen tube which functions in the growth and development of the tube.

tuber. **1.** A short, thick, underground stem usually bearing minute buds or eyes, as the white potato. **2.** A swollen end of a stem or a knoblike structure near the edge of a thallus, in various hepatics.

tubercle. A small, rounded prominence or excrescence, as **a.** *anat.* that on the crown of a molar tooth or on a bone; **b.** *bot.* a nodule on the root of a legume; **c.** *med.* a specific lesion produced by a tubercle bacillus. *See* tuberculosis.

tuberculate. Bearing tubercles or knoblike excrescences.

tuberculosis. An infectious disease caused by the tubercle bacillus (*Mycobacterium tuberculosis*), characterized by the formation of tubercles in the lungs or other tissue.

tuberculum. A tubercle.

tuberiferous. Producing or bearing tubers.

tuberoid. Resembling a tuber, as the thick, fleshy roots of certain orchids.

tuberosity. A prominence on a bone; a large tubercle.

tuberous. Pertaining to, resembling, bearing, or producing tubers.

tube worm. One of a number of polychaete annelids which live in tubes which they rarely leave, as those of the orders Sabellidae and Serpulidae (fan worms and feather duster worms).

tubicolous. Tube-dwelling; living in a tube.

tubifacient. Secreting or constructing a tube.

tubular. 1. Having the shape of a tube. **2.** Composed of or possessing tubes.

tubule. A small tube or a minute, tube-shaped structure.

tubuloacinar. Possessing both secretory tubules and acini, with reference to compound glands.

tubuloalveolar. Possessing both secretory tubules and alveoli, with reference to compound glands.

tubulose. Tubular; made up of tubular structures.

tuco-tuco. A small, burrowing, South American rodent (*Ctenomys*). Also tuca-tuca.

tuft. A small cluster of flexible structures, as hairs, feathers, or blades of grass, closely associated at their bases but with free ends spread apart.

tufted. Bearing or possessing a tuft; growing in tufts or clusters.

tui. The parson bird (*Prosthemadera novaseelandiae*), a honey eater of New Zealand, noted for its powers of mimicry.

tularemia. An infectious disease caused by a bacillus (*Pasteurella*

581

tularensis) acquired by handling infected animals or through the bite of an arthropod. Also called rabbit fever.

tulip. A European bulbous herb (*Tulipa*) of the lily family, cultivated for its showy, single flowers.

tumblebug. A scarab beetle, as *Canthon* or *Deltochilum*, which feeds on dung. Working in pairs, tumblebugs roll pieces of dung into a ball, bury it, and deposit their eggs in or on it. Also called dung beetles.

tumbler. A variety of domestic pigeon which tumbles or somersaults in flight.

tumbu fly. A dipterous insect (*Cordylobia anthropophaga*) whose larvae infest the skin of man, causing boillike lesions.

tumefacient. Swollen; tending to produce swelling.

tumid. Swollen, inflated, distended.

tumor. A swelling, especially that resulting from the growth of new tissue; a neoplasm.

tuna. A large food and game fish of the family Thunnidae, order Percomorphi, as *Thunnus*, *Neothunnus*, *Euthynnus*. Also called tuna fish, tunny. *See* albacore.

tundra. The level or slightly undulating treeless plains of northern regions.

tung oil. A fixed oil obtained from the nuts of the tung oil tree (*Aleurites fordii*) of central and western China; an important drying oil.

tunic. A coat or covering, as **a.** *bot.* the investing covering of a bulb or the integument of a seed; **b.** *zool.* the covering (test) of a tunicate.

tunica. 1. An enveloping coat or covering; a tunic. **2.** One of the three layers comprising the wall of a blood vessel.

¹**tunicate.** Any member of the subphylum Urochordata (*Tunicata*); a sea squirt.

²**tunicate** or **tunicated. 1.** Possessing a tunic. **2.** Possessing concentric layers, as the bulb of an onion.

tunicine. A polysaccharide resembling cellulose, present in the tunic of tunicates and in tubes housing sessile hemichordates.

tunny. A tuna.

tupelo. 1. The black gum tree (*Nyssa sylvatica*). **2.** The sour gum or water tupelo (*N. aquatica*).

tur. A Caucasian wild goat (*Capra caucasica*) of the central region.

turbellarian. A free-living flatworm of the class Turbellaria. *See* planaria.

turbid. Cloudy, not clear.

turbidimetry. Determination of the number of bacteria by estimation of the degree of turbidity of a suspension.

¹**turbinate.** 1. *bot.* Top-shaped; inversely conical. 2. *zool.* Spiral-shaped with whorls decreasing in size from base to apex, as certain shells.

²**turbinate.** *anat.* A turbinate bone. *See* concha.

turbin shell. A turbinate or top-shaped, herbivorous gastropod of the family Turbinidae inhabiting warm waters.

turbot. A European flounder (*Scopthalmus maximus*), a highly esteemed food fish.

turco. *See* tapaculo.

turgid. 1. Abnormally swollen, distended, or inflated. 2. Tightly drawn or stretched by pressure from within.

turgor. The state of being turgid; turgescence.

turgor movement. Movement resulting from changes in turgor pressure, as the opening and closing of stomata.

turgor pressure. Hydrostatic pressure within a cell resulting from endosmosis or imbibition of water by the protoplasm.

turion. A young shoot or sucker, as in asparagus.

turkey. 1. A large American, gallinaceous bird of the family Meleagrididae, especially *Meleagris gallopavo*, now rare in the wild state. Many domestic strains have been developed. 2. The ocellated turkey (*Agiocharis ocellata*) of Mexico and Central America.

turnip. A biennial herb (*Brassica rapa*) with edible root and leaves. *See* rutabaga.

turnstone. A stout-bodied, marine shorebird of the family Charadriidae, especially those of the genus *Arenaria*.

turpentine. An oleoresin obtained from various conifers, especially the longleaf pine (*Pinus palustris*), the source of turpentine oil, rosin, and other products.

turreted. Having the shape of a turret; long-spired, as certain gastropod shells.

turtle. Any reptile of the order Chelonia including both terrestrial and aquatic forms with a body enclosed in a shell composed of a dorsal carapace and a ventral plastron. *See* terrapin, tortoise.

turtledove. 1. The Eastern mourning dove (*Zenaidura macroura*). **2.** An Old World wild dove (*Streptopelia*) and related species.

tusk. 1. A greatly elongated and enlarged tooth, as that seen in the elephant, walrus, wild boar, and other mammals. **2.** Any large, projecting tooth.

tusk shell. A tooth shell.

tussock. A tuft of hair, grass, twigs, or similar structures.

tussock moth. A moth of the family Lymantriidae whose larvae possess tufts of hairs projecting from the body, as the gypsy and browntail moths.

twig. A small branch, as that of a tree, nerve, or blood vessel.

twin. 1. One of two individuals born at the same birth of the same parent. **2.** One of a pair of similar structures.

twine. To coil or twist about a structure.

twiner. A plant whose stem twines about an object for support.

twining. Winding about or twisting together; growing in a spiral fashion about an object, characteristic of certain climbing plants.

twinning. 1. The production of twins. **2.** The production of twin objects, as kernels of a nut.

tylarus. A pad on the undersurface of the toe of a bird, especially hawks.

tylosis pl. **tyloses.** A bladderlike outgrowth of a parenchyma cell which grows into and plugs a water-conducting element of wood.

tymbal. A vibrating structure in the sound-producing organ of a cicada; a timbal.

tympanal organ. The auditory organ of an insect.

tympanic. Of or pertaining to the tympanum or tympanic cavity.

tympanic cavity. The cavity of the middle ear; the tympanum or eardrum.

tympanic membrane. 1. The drum membrane, which separates the external acoustic meatus from the tympanic cavity. **2.** In common usage, the eardrum.

tympanum. 1. The middle ear; the tympanic cavity or eardrum.

584

2. In birds, a resonating chamber in the bronchotracheal type of syrinx. **3.** In insects, the auditory organ or eardrum, as that on the first abdominal segment in a locust.

type. An individual or a group of individuals which serves as a basis for the name of a taxonomic category (taxon), as a specimen which bears the name of a species.

type specimen. A single specimen which was before the original describer of a new species and designated by the original author as the type; the only specimen existing at the time of the original description; also called holotype. *See* allotype, neotype, paratype, syntype.

typhlosole. **1.** A longitudinal infolding of the wall of the intestine in annelids or in isopods. **2.** The spiral valve of a cyclostome or elasmobranch.

typhoid fever. An acute infectious disease caused by a bacillus (*Salmonella typhosa* or *Eberthella typhi*), characterized by marked involvement of lymphatic tissues.

typhus. An infectious disease caused by rickettsias transmitted by arthropods, as *epidemic typhus*, transmitted by the human body louse; *endemic* (*murine* or *shop*) *typhus* transmitted by the rat flea; *scrub typhus*, transmitted by the chigger.

tyrannosaur. A bipedal, carnivorous dinosaur, as *Tyrannosaurus rex*, of the Cretaceous period.

Tyrian purple. A highly prized purple dye obtained from the adrectal glands of certain gastropods (*Murex, Purpura, Mitra*) and used by the ancient Greeks and Romans.

tyrogenous. Produced by or originating in cheese.

tyrosinase. An enzyme which catalyzes the formation of melanin from tyrosine.

tyrosine. An amino acid, $C_9H_{11}NO_3$, widely distributed in nature; also prepared synthetically.

tyrothricin. An antibiotic polypeptide mixture isolated from soil bacteria of the *Tyrothrix* type effective against gram-positive bacteria, some fungi, and protozoans.

U

uakari. A bald-headed monkey of the genus *Cacajao* of the Amazon forests.

ubiquitous. Widespread; occurring everywhere.

udder. A large, pendulous mammary gland, especially one with two or more teats, as in the cow.

ulcer. A break in the continuity of an epithelium of the skin or mucous membrane accompanied by degeneration and necrosis of underlying tissues.

uliginose. 1. Swampy; oozy. **2.** Inhabiting or growing in swamps.

ulna. 1. The medial bone of the forearm in man. **2.** The corresponding bone in the forelimb of other tetrapods.

ulnare. In birds, the posterior of the two carpal bones in the wing; in mammals, the triquetral or cuneiform bone, a carpal bone.

ulotrichous. Having woolly, curly hair.

ultracentrifuge. A high-speed centrifuge with speeds up to 100,000 rpm and exerting forces up to several hundred thousand times the force of gravity.

ultrafiltration. Filtering utilizing filters by which particles of colloidal size can be separated from those of molecular size in true solutions.

ultramicroscope. 1. A slit ultramicroscope by which colloidal particles not visible by an ordinary microscope can be detected. **2.** An electron microscope.

ultramicroscopic. Not capable of being seen by an ordinary microscope.

ultrastructure. Fine structure; the structure of cells and tissues beyond that seen by use of a light microscope; structure elucidated only by use of an electron microscope, or by other methods as X-ray diffraction, polarization, optical analysis, and special physicochemical techniques.

586

ultrasonic. Pertaining to sound waves of frequencies above that of audible sound (20,000 cps), but applied especially to waves of very high frequencies (above 500,000 cps). *Compare* supersonic.

ultraviolet. Beyond the visible spectrum at the violet end.

ultraviolet light. Invisible light consisting of wave lengths from 4000 to 135 A., a normal component of sunlight and essential for the formation of vitamin D in man.

umbel. An indeterminate inflorescence in which a number of pedicels or peduncles of equal length arise from a common point, as in the onion.

umbellate. Pertaining to or producing umbels.

umbelliferous. Producing or bearing umbels.

umbellule. A secondary umbel or umbellet; a small umbel.

umber bird. *See* umbrette.

umbilical cord. 1. *anat.* The cordlike structure which connects the placenta with the fetus. 2. *bot.* A funiculus.

umbilicate. Having a depression in the center resembling an umbilicus.

umbilicus. 1. *zool.* **a.** The navel, a round depressed scar in the midventral line of the abdomen marking the point of attachment of the umbilical cord. **b.** A depression or opening in the shell of certain gastropods about which the whorls coil. **c.** One of two openings in the quill of a feather: the *inferior umbilicus* at the tip of the quill, the *superior umbilicus* at junction of rachis and quill. 2. *bot.* The hilum of a seed.

umbo pl. **umbones.** A boss; a rounded prominence or protuberance, as **a.** that on the valve of a bivalve marking the oldest portion of the shell; **b.** the central portion of the tympanic membrane; **c.** the posterior beak of the ventral valve of a brachiopod.

umbonate. Bearing a conical or rounded boss or umbo.

umbonulate. Bearing a small boss or a nipplelike protuberance.

umbracticolous. Inhabiting shady places.

umbraculiferous. Bearing an umbrellalike structure.

umbraculiform. Shaped like an umbrella.

umbraculum. The wavy, pigmented margin of the iris in the eye of certain mammals, as the camel.

umbrageous. Providing shade.

587

umbrella. 1. The saucer-shaped body of a medusa or jellyfish; the bell. *See* exumbrella, subumbrella. **2.** Ectoderm cells located anterior to the prototroch in the development of a trochosphere.

umbrella bird. A South American bird of the genus *Cephalopterus*, males of which have an umbrellalike crest.

umbrella tree. The American magnolia (*Magnolia tripetala*) with large leaves in clusters resembling an umbrella.

umbrette. The umber bird (*Scopus umbretta*), a storklike wading bird of Arabia and Africa; also called hammerhead or hammerkop.

unarmed. 1. *zool.* Lacking hooks, claws, armor, or other defensive structures. **2.** *bot.* Lacking spines or prickles; pointless, blunt, muticous.

unarmored. Naked; lacking discrete, sculptured, articulated plates, with reference to dinoflagellates.

unau. The two-toed sloth (*Choloepus didactylus*) of tropical America.

uncate. Shaped like a hook; hamate.

unciform. UNCINATE.

uncinate. Bent like a hook; with a hooked tip.

uncinate process. In birds and some reptiles, a thin, flat process projecting backward and upward from a rib which overlies the next succeeding rib, thus strengthening the rib cage.

uncinus pl. **uncini.** A small, hooklike structure, or one bearing hooks.

unctuous. Greasy, oily, fatty.

uncus. A hook or clawlike structure, as **a.** the anterior end of the hippocampal gyrus; **b.** a part of the malleus in the mastax of a rotifer; **c.** a median, hooked process in the male genitalia of lepidopterans.

underfur. The dense, soft fur of animals composed of short, closely packed hairs among which are interspersed longer and coarser hairs which form the surface covering.

undulant. Characterized by undulations or up and down movements.

undulant fever. An infectious disease affecting domestic animals and man caused by bacteria of the genus *Brucella;* brucellosis; malta fever.

undulate. Possessing a wavy surface or margin; repand.

undulating membrane. 1. In ciliates, a thin, vibrating organelle formed of fused cilia located in region of the cytopharynx. **2.** In trypanosomes, a thin, flat, finlike, cytoplasmic process through which the flagellum runs.

ungual. Pertaining to, resembling, or possessing a hoof, claw, or nail.

unguiculate. 1. *zool.* **a.** Pertaining to the Unguiculata, a group of mammals which includes those with claws and nails as distinguished from hoofed mammals (ungulates). **b.** Bearing claws or nails. **2.** *bot.* Contracted at the base into a clawlike structure, said of petals.

unguis. The dorsal, keratinized portion of a claw, nail, or hoof; the portion overlying the subunguis.

ungulate. 1. Pertaining to or belonging to the Ungulata comprising hoofed mammals. **2.** Shaped like a hoof; possessing hoofs.

unguligrade. A type of foot posture in which an animal walks on hoofs, as a horse or cow.

uniaxial. Having a single axis.

unicarpellate. Composed of a single carpel, with reference to fruits.

unicellular. Composed of a single cell.

unicorn. A narwhal.

unicuspid. Having a single point or cusp.

uniflagellate. Having a single flagellum.

unifoliate. Designating a compound leaf which has been reduced to a single leaf, usually the enlarged terminal leaflet, as in the orange.

unigeneric. Monogeneric; composed of a single genus, with reference to a family.

unijugate. Having a single pair of leaflets.

unilateral. One-sided; pertaining to or affecting one side only.

unilobar. Having a single lobe.

unilocular. Having a single chamber, cavity, or loculus.

uninuclear. Possessing only one nucleus.

unionid. A freshwater clam or mussel of the family Unionidae, common throughout the Mississippi valley, as *Unio, Amblema.*

uniovular. From one egg; monozygotic.

uniparous. Having one offspring at a time.

unipetalous. Having a corolla of one petal.

unipolar. 1. Having a single pole. **2.** Having a single process, as a *unipolar* neuron.

unipotent. Giving rise to only one type of cell or tissue.

uniramous. Lacking an exopodite, with reference to crustacean appendages.

uniseptate. With a single septum or partition.

uniseriate. With a single row of cells.

unisexual. 1. With organs of one sex only; dioecious. **2.** Male or female only; staminate or pistillate only.

unistratose. Possessing one layer.

unit character. *genetics.* A trait or character which differs from others by a single gene difference; a trait whose inheritance can be accounted for on the basis of a single pair of genes.

unitunicate. Having a wall consisting of a single layer, with reference to spores.

unitypic. Monotypic; of a single type.

univalent. 1. *biol.* Unpaired at meiosis, as a chromosome which is lacking its synaptic mate. **2.** *chem.* Having a valence of one.

¹univalve. Having one valve.

²univalve. A mollusk with a univalve shell, as a gastropod.

universal. All-inclusive, wide-spread, all-embracing.

universal donor. In blood transfusion, an individual whose blood is of the group O type and can be received by those of all blood groups.

universal recipient. One whose blood is of group AB and Rh positive.

universal symmetry. Symmetry as that possessed by spherical organisms in which any plane passing through the center would divide the body into two similar halves.

universal veil. A thin tissue which completely encloses the basidiocarp of *Amanita muscara*, a poisonous mushroom, in the button stage. *See* volva.

univoltine. Having one brood a year, as various insects and most birds.

unsealed. With margins of the carpels not coherent.

upland game. Game animals inhabiting high grounds above the lowlands along rivers or the ocean, as the bobwhite, pheasant, squirrel, rabbit.

urachis. An embryonic cord or tube which connects the tip of the urinary bladder with the allantoic stalk.

uracil. A pyrimidine base, $C_4H_4N_2O_2$, present in nucleic acids.

urate. A salt of uric acid.

urceolate. Urn-shaped; expanded below and constricted near the mouth.

urchin. 1. A sea urchin. 2. A hedgehog.

urea. A white crystalline substance, $CO(NH_2)_2$, derived from proteins, the principal solid excreted in the urine of man and most mammals. It is formed in the liver from ammonia following the deamination of amino acids.

urease. An enzyme which catalyzes the conversion of urine to ammonia and CO_2.

urediniospore. *See* uredospore.

uredinium. A rust-colored pustule (sorus) formed on a wheat plant infected by the rust fungus. It produces uredospores.

urediosorus. A uredinium.

uredospore. A dikaryotic, one-celled spore formed in the uredinium of certain rusts, as the wheat rust; also urediospore, urediniospore.

ureotelic. Term applied to animals in which urea is the principal nitrogenous product excreted in the urine, as amphibians, mammals. *Compare* ammonotelic, uricotelic.

ureter. The duct which conveys urine from the kidney to the cloaca or bladder.

urethra. The duct in mammals through which urine passes from the bladder to the outside. In males, it traverses the penis and functions also as a sperm duct.

urial. A wild sheep (*Ovis vignei*) of central Asia. Also called shapu.

uric acid. A white, crystalline substance, $C_5H_4N_4O_3$, present in the urine of all carnivorous animals and the principal end product of nitrogen metabolism in uricotelic animals.

uricotelic. Designating animals in which uric acid is the principal nitrogenous product excreted in urine, as insects, gastropods, lizards, snakes, and birds. *Compare* ammonotelic, ureotelic.

urinary. Of or pertaining to urine or to the structu secrete, store, or convey urine.

urinary bladder. A reservoir in which urine is stored

urinary system. The organs concerned with the formation and excretion of urine; in mammals, the kidneys, ureters, bladder, and urethra.

urinate. To discharge urine. *See* micturition.

urine. In vertebrates and some invertebrates, the fluid or semi-fluid product discharged from the urinary or excretory organs; in man, the clear, amber-colored fluid containing water, salts (chlorides, phosphates, sulfates), and various other substances (pigments, hormones, and waste products of metabolism).

uriniferous. Conveying urine, as uriniferous tubule. *See* nephron.

urinogenital. UROGENITAL.

urn. 1. *bot.* The lower portion of the base of a pyxis. 2. *zool.* A ciliated vase- or funnel-shaped, excretory structure present in the body cavity of sipunculids, echiurids, and some holothuroids.

urochordate. Any chordate of the subphylum Urochordata comprising mostly sessile forms with an enveloping covering or tunic; a tunicate, ascidian, or sea squirt.

urochrome. A pigment present in normal urine and responsible for its color.

urodele. A tailed amphibian of the class Caudata (Urodela), as a newt or salamander.

urodeum. In birds and monotremes, the portion of the cloaca which receives the ureters and reproductive ducts.

urogenital. 1. Of or pertaining to the urinary and reproductive organs. 2. Functioning as both an excretory and reproductive structure, as a *urogenital* papilla, pore, or sinus.

urogenital system. The organs which comprise the urinary and reproductive systems; urinogenital system.

urogomphus. A fixed and unjointed cercus in an insect larva.

urohypophysis. A neurosecretory structure located at the posterior end of the spinal cord in fishes; also called urophysis.

urophysis. *See* urohypophysis.

uropod. One of a pair of appendages located on either side of the telson in various crustaceans, as the lobster. *See* tail fan.

uropygial gland. An oil gland located at the base of the uropygium in birds; also called preen gland.

uropygid. A whip scorpion, an arachnid of the order Uropygi.

uropygium. In birds, a pointed prominence at the posterior end of the body containing the caudal vertebrae and pygostyle. It supports the tail feathers.

urosome. The hind body of a copepod, consisting of one or two thoracic segments and the abdomen.

urostyle. **1.** A rodlike bone composed of fused caudal vertebrae present in frogs and toads. **2.** The posterior end of the vertebral column which extends into the dorsal flange of a homocercal tail in fishes.

ursine. Of or pertaining to bears or the bear family, Ursidae.

urticaceous. Of or belonging to the family Urticaceae, a family of herbs, shrubs, and trees which includes the nettles.

urticaria. Inflammation of the skin characterized by redness, development of welts and wheals, and intense itching; hives; nettle rash.

urticating. Causing urticaria, as *urticating* caterpillars whose poison hairs or spines cause a dermatitis.

uta. An iguanid lizard of the genus *Uta* of western North America.

uterine. Of or pertaining to the uterus.

uterine bell. A structure at the posterior end of an acanthocephalan into which eggs are drawn which functions as a sorting device returning immature ova to the body cavity.

uterine horns. The anterior portions of a bipartite or bicornate uterus; also called cornua.

uterine milk. A uterine secretion which bathes developing embryos, as in the opossum.

uterus. **1.** In mammals, the female organ in which the embryo or fetus is nourished and develops; the womb. **2.** In various vertebrates, the expanded caudal end of the oviduct which may function as **a.** a storage place for ova; **b.** an organ for production of the eggshell; **c.** a place for development, as in viviparous forms. **3.** In various invertebrates, a tube or saclike structure in which fertilized eggs are retained before discharge from the body.

utricle. **1.** *anat.* **a.** A saclike structure in the internal ear which contains the maculae utriculi, sensory receptors for position and movement. **b.** A small pocket in the prostatic portion of the male urethra. **2.** *bot.* **a.** An air bladder of a seaweed. **b.** A bladderlike structure in certain fungi, as *Atriplex*. **c.** A small, bladderlike, one-seeded fruit, as in a sedge. **3.** A utriculus.

utricular. Of or pertaining to a utricle or utriculus; inflated; saclike; bladderlike; utriculate.

utriculate. UTRICULAR.

utriculus. A utricle.

uvea. The pigmented, vascular coat of the eye, comprising the choroid, ciliary body, and iris; the tunica vasculosa, the middle layer.

uvula. A pendant, fleshy, conical structure hanging from the free border of the soft palate.

vaccination. 1. Inoculation with a vaccine in order to develop immunity against a specific disease. **2.** Inoculation with the virus of vaccinia in order to develop immunity against smallpox.

vaccine. A suspension of killed or living, attenuated organisms which, when inoculated or taken into the body, acts as an antigen causing the development of antibodies that render the body immune against infection by the specific organism.

vaccinia. Cowpox, a virus disease of cows transmissible to man by contact or vaccination.

vacuolar. Pertaining to, resembling, or composed of vacuoles.

vacuolated. Containing vacuoles or cavities.

vacuole. A space or cavity within a cell, as the cavity of a plant cell containing cell sap or the contractile or food *vacuole* of a protozoan.

vacuome. The vacuoles of a cell together with their cell sap.

vagal. Of or pertaining to the vagus nerve.

vagina. 1. A sheath or covering. **2.** *zool.* and *anat.* A female reproductive duct or canal which receives sperm from the male or an intromittent organ as the penis in copulation, and which functions as a passageway for eggs or a birth canal for passage of young to the outside. **3.** *bot.* The sheathlike portion of a petiole which forms a sheath surrounding the stem, as in certain grasses.

vaginal. Of or pertaining to the vagina.

vaginate. Sheathed; enclosed within a sheath.

vaginiferous. Possessing a vagina or sheath.

vagus nerve. The tenth cranial nerve (X), a mixed nerve arising from the medulla, its fibers widely distributed to the head, neck, thorax, and abdomen. Formerly called pneumogastric nerve.

valence. The combining power of an atom; its capacity to combine with other atoms to form a molecule, expressed in terms of the number of hydrogen atoms or their equivalent with which any atom may combine.

valine. An essential amino acid, $C_5H_{11}NO_2$, present in most proteins.

vallate. Surrounded by a depression outside of which is an elevated rim, as the *vallate* papillae of the tongue.

vallecula. A shallow groove, channel, or depression.

vallecular canal. One of the numerous air canals in the cortex of the stem of *Equisetum*, a horsetail.

valleculate. Having valleculae.

valvate. 1. Having a valve or valves; of the nature of a valve. **2.** Opening by valves or valvelike structures. **3.** Meeting at the edges by structures which do not overlap, as some capsules, anthers, and leaves in bud.

valve. 1. *anat.* A structure which **a.** permits flow in one direction only, preventing flow in the reverse direction, as *valves* in the heart or veins; **b.** directs the flow of a fluid, as a spiral *valve*. **2.** *zool.* **a.** One of the distinct plates or pieces comprising the shell of various invertebrates, as in mollusks, brachiopods, and crustaceans. **b.** A movable jaw of a pedicellaria. **c.** One of the parts of an ovipositor of an insect. **d.** A clasper in a male lepidopteran. **3.** *bot.* **a.** One of the two halves of the cell wall of a diatom. **b.** The partially detached lid of an anther. **c.** One of the segments into which a dehiscent fruit (capsule) splits; one of the halves of a pod.

valvifer. One of two small structures to which the valves of an insect ovipositor are attached.

valvula. A little valve or fold.

valvulae conniventes. Circular folds (plicae conniventes) in the small intestine of a vertebrate.

vampire. One of several bloodsucking bats of the genera *Desmodus* and *Diphylla* of Central and South America. False vampires comprise the genera *Vampyrus* and *Phyllostomus*.

vampire fish. The candiru (*Vandellia cirrhosa*), a small, bloodsucking catfish of the Amazon which invades the urogenital openings of vertebrates. It is the only vertebrate parasite of man.

vanadium. A rare element, symbol V, at. no. 23, absent in most animals but highly concentrated in ascidians, as *Cliona*.

vanadocyte. A green cell in the blood of an ascidian in which vanadium is bound to hydrosulfuric acid.

vane. 1. *zool.* The flat, expanded portion of a feather; web. **2.** *bot.* A vexillum.

vanilla. A climbing orchid (*Vanilla planifolia*) of tropical America, whose fruits, vanilla beans, are the source of vanilla flavoring extract.

variant. An individual or group which differs from others of its own type.

variable. Tending to deviate from type; inconstant.

variation. A divergence, deviation, or change in a trait or character among individuals of a closely related group. *See* mutation.

varicellate. Having small and indistinct ridges or varices, said of shells.

varicose. Swollen, knotted, tortuous, as a *varicose* vein.

variegated. Having streaks or patches of different colors, with reference to leaves.

variety. A group within a species or subspecies which differs in some significant respect from other members of the species.

variola. Smallpox.

variole. A small pit or foveola.

varix pl. **varices. 1.** *zool.* A transverse ridge across the whorls of various gastropod shells marking the former position of the outer lip. **2.** *med.* A tortuous and dilated vein.

varus. Turned abnormally inward, with reference to the lower extremity.

varve. A stratum in the bottom deposits of a lake consisting of two layers, one formed in the spring, the other in the winter.

varying. Exhibiting or undergoing change; differing.

varying hare. *See* snowshoe rabbit.

vas pl. **vasa.** A vessel or canal; a duct.

vas afferens pl. **vasa afferentia.** An afferent vessel, as that leading to a lymph node.

vasa vasorum. Small blood vessels supplying the wall of an artery or vein.

vascular. 1. Of or pertaining to vessels conducting a fluid. 2. Supplied with or containing vessels or ducts, especially blood vessels.

vascular bundle. *bot.* A strand of conducting tissue consisting of primary xylem and phloem and sometimes procambium. When enclosed within a fibrous sheath, it comprises a *fibrovascular bundle*.

vascular cambium. The cambium of a vascular cylinder which gives rise to secondary xylem and phloem.

vascular cylinder. The stele of a root or stem. *See* stele.

vascularization. The formation and development of blood vessels in a tissue or organ.

vascularized. Supplied with blood vessels.

vascular plant. Any plant of the division Tracheophyta characterized by possession of vascular tissue. *Lower vascular plants* include the silopsids, lycopsids, and sphenopsids; *higher vascular plants* include the ferns, conifers, and flowering plants.

vascular ray. A radial extension of the vascular cambium in the stem of a woody plant.

vascular system. 1. *zool.* In vertebrates, the cardiovascular or circulatory system; in echinoderms, the lacunar system. *See* water-vascular system. 2. *bot.* The water- and food-conducting system of a vascular plant. *See* vascular tissue.

vascular tissue. In tracheophytes, the xylem, phloem, and vascular cambium.

vasculum. A metal container used for collecting botanical specimens.

vas deferens pl. **vasa deferentia.** The ductus deferens, the principal duct conveying sperm from the testis to the outside.

597

vasectomy. Removal of a portion of the vas deferens.

vas efferens pl. **vasa efferentia. 1.** One of a group of ducts which carry sperm from the testis to the epididymis or vas deferens. **2.** An efferent vessel leading from a lymph node.

vasoconstriction. Reduction in diameter of a blood vessel.

vasodepressor. That which brings about a lowering of blood pressure by dilatation of blood vessels, as a vasodilator nerve.

vasodilatation. Dilatation or relaxation of blood vessels.

vasomotor. Pertaining to or regulating constriction (vasoconstriction) or dilatation (vasodilatation) of blood vessels, as a *vasomotor* nerve.

vasopressin. A hormone produced by neurons of the hypothalamus but stored and released by the posterior lobe of the hypophysis. It has a pressor effect, elevating blood pressure and an antidiuretic action; also called antidiuretic hormone (ADH). Its trade name is pitressin.

vasotomy. The cutting of or making an incision into the vas deferens.

vector. An organism, usually an arthropod, which transports a pathogen. A *biological vector* is one in which the pathogen undergoes a period of incubation or development.

vector tissue. A special tissue in leeches which serves as a pathway for sperm passing to the ovisacs.

veery. Wilson's thrush (*Hylocichla fuscescens*) of the northeastern United States.

¹**vegetable. 1.** Of, pertaining to, or of the nature of a plant. **2.** Derived from plants, as a *vegetable* oil.

²**vegetable.** A plant, especially a herbaceous plant which is cultivated for food, or the edible portion of such a plant.

vegetal. 1. Of or pertaining to plants. **2.** Pertaining to growth and nutrition; vegetative.

vegetalized. *embryol.* A condition in which the animal pole is weakened and the vegetal pole becomes dominant.

vegetal pole. In an ovum, the pole opposite the animal pole, characterized by accumulation of yolk.

vegetation. 1. All the plants of an area or region. **2.** *med.* A plantlike growth, as a *vegetation* on a heart valve.

vegetative. 1. Resembling a plant, with respect to nutrition and growth. **2.** Pertaining to stem and leaf development in contrast to flower and seed development. **3.** With respect to reproduction in plants, nonsexual in contrast to sexual development. **4.** Inactive, as the *vegetative* nucleus.

vegetative nucleus. 1. The macronucleus or trophic nucleus, as in certain protozoans. **2.** The tube nucleus of a pollen grain. **3.** The nucleus of a cell during interphase.

vegetative propagation. In plants, reproduction by any method other than the use of true seeds, as by **a.** specialized or modified stems (rhizomes, tubers, corms, bulbs, bulblets); **b.** root tubers; **c.** stem, leaf, and root cuttings and layerage; **d.** grafting and budding.

vegetative stage. In protozoans, the trophozoite which reproduces asexually.

veil. 1. *zool.* **a.** A velum. **b.** A caul. **2.** *bot.* A velum.

vein. 1. *zool.* **a.** A blood vessel which transports blood from the tissues to or toward the heart. **b.** A thickened, hollow ridge which serves as a supporting structure in the wing of an insect. **2.** *bot.* A strand of vascular tissue in a leaf, petal, or similar structure.

velamen. 1. *bot.* A thin sheath or covering, especially the surface layer of an aerial root which functions in the absorption of water and minerals, as in an orchid. **2.** *anat.* A covering membrane or velum.

velar. Of or pertaining to a velum.

velarium. A structure resembling a velum present in certain scyphozoan medusae.

velate. Having a veil or velum.

veliger. The free-swimming larva of certain mollusks characterized by possession of a velum.

velum. 1. A thin, membranous structure which resembles a curtain or veil, as the shelflike structure which projects inwardly from the margin of the bell in hydromedusae. **2.** *anat.* The soft palate. **3.** *bot.* **a.** A membrane extending from the stalk to the pileus in certain fungi, as *Amanita*. **b.** A protective membrane covering the sporangium in quillworts.

velutinous. Velvety; covered with a fine, silky pubescence.

velvet. A soft, highly vascular skin which covers the antlers of deer during the growth period and is sloughed off when development is complete.

vena cava pl. **venae cavae.** In air-breathing vertebrates, a large vein which returns blood to the sinus venosus or right atrium of the heart; also called pσstcava.

venation. 1. *zool.* The arrangement of veins, as **a.** in the wing of an insect; **b.** in an organ or a part of the body. **2.** *bot.* The arrangement of veins in a leaf; nervation.

venereal. Pertaining to or related to sexual intercourse.

venesection. Taking blood from the body by cutting into a vein; phlebotomy.

venin. 1. A mixture of venoms from various snakes. **2.** The poisonous principle in a venom. Also venene.

venison. The flesh of deer and related animals.

venom. A toxic substance produced by various animals, as fishes, snakes, lizards, spiders, ticks, scorpions, centipedes, and insects. Its toxic principle may be a neurotoxin, neurocytolysin, hemolysin, hemocoagulin, proteolysin, or cytolysin.

venomous. Poisonous; capable of inflicting a poisonous bite, sting, or injury.

venose. Having numerous and conspicuous veins, said of leaves.

venous. 1. Of or pertaining to a vein or veins. **2.** Designating blood carried by veins.

vent. An outlet, as the anal or cloacal opening in certain vertebrates, as fishes and birds.

venter. 1. *anat.* **a.** The abdomen or belly, or its cavity. **b.** The belly of a muscle. **2.** *bot.* The enlarged, basal portion of an archegonium in which an egg cell develops.

ventilation. *physiol.* The process of renewing or supplying fresh air to the lungs.

ventilator. A worker in a bee colony which, by fanning with its wings, aids in ventilating the nest or hive.

ventral. 1. *zool.* **a.** Of or pertaining to, towards, or near the lower or underneath surface or, in an upright animal as man, the anterior surface; the opposite of dorsal. **b.** Of or pertaining to the abdomen or belly. **2.** *bot.* Of, pertaining to, or designating **a.** the inner or axial surface of a structure, as a petal or carpel; **b.** the upper or axial surface of a typical leaf; **c.** the lower or underneath surface of a dorsal-ventral orientated structure, as a thallus.

ventricle. A cavity of an organ, as **a.** a chamber in the heart which discharges blood into the arteries; **b.** a cavity of the vertebrate brain; **c.** a recess between the true and false vocal folds of the larynx.

ventricose. Inflated, swollen, protruding outward on one side.

ventricous. VENTRICOSE.

ventricular. 1. Of or pertaining to a ventricle. **2.** Bulging or protruding.

ventricular folds. The false vocal folds in the larynx of a mammal.

ventriculus. *zool.* **1.** The midgut or midintestine (stomach) of an insect. **2.** The gizzard or stomach of a bird.

ventrose. Having a belly or a bellylike protrusion; pot-bellied.

venule. A small vein or veinlet.

Venus's-flower-basket. A genus of hexactinellid sponges (*Euplectella*) with a cylindrical skeleton of siliceous fibers resembling spun glass, from waters off the coasts of Southeast Asia.

Venus's-flytrap. An insectivorous plant (*Dionaea muscipula*) with leaves modified for capturing insects.

Venus's girdle. A ctenophore (*Cestum venerum*) with a long, ribbonlike body.

Venus's hair fern. The maidenhair fern (*Adiantum capillus-veneris*).

verdant. Covered with green vegetation.

verdin. A small titmouse (*Auriparus flaviceps*).

verge. The male copulatory organ of certain invertebrates, as a snail.

vermicide. An agent which kills or destroys worms, especially parasitic worms.

vermicular. Wormlike; resembling a worm in form or activities.

vermiculate. Vermicular; wormlike.

vermiculated. 1. Infested with worms. **2.** Having cavities resembling the burrows of worms.

vermiculose. VERMICULAR.

vermiform. Shaped like a worm.

vermiform process. The appendix in man.

vermin. 1. Any of the various small animals that infest the bodies or the habitations of man and domestic animals, especially cockroaches, bedbugs, fleas, lice, mites, mice, or rats. **2.** Any animal which preys upon game.

verminous. Pertaining to, infested with, or caused by vermin.

vermis. A small median lobe of the cerebellum.

vernacular. Local; characteristic of a locality.

vernacular name. The common or local name of an animal or plant.

vernal. 1. Of or pertaining to spring. **2.** *ecol.* Late spring.

vernalization. The induction of early flowering by subjecting slightly germinated seeds to a low temperature (5° C.) just before sowing.

vernation. The arrangement of leaves within a leaf bud.

vernicose. Shiny; appearing as though polished, said of leaves.

verruca. A wart or wartlike elevation.

verrucose. Warty; bearing wartlike elevations.

versatile. Capable of being moved forward or backward or being turned around, as in flowers, a *versatile* anther, or in birds, a *versatile* toe.

versicolor. 1. Possessing various colors; variegated; partly colored. **2.** Changeable in color.

vertebra pl. **vertebrae. 1.** One of the segmentally arranged bony or cartilaginous structures comprising the spinal or vertebral column. **2.** One of the articulating ossicles in the arm of an ophiuroid, as a brittle star.

vertebral. Of or pertaining to a vertebra or the vertebral column; composed of vertebrae.

vertebral canal. The canal within the vertebral column formed by the vertebral foramina of successive vertebrae which lodges the spinal cord, meninges, cerebral spinal fluid, and roots of the spinal nerves.

vertebral column. The spinal column, composed of a series of vertebrae extending from skull to tip of tail, which forms the main supporting axis of the body and encloses and protects the spinal cord.

vertebral ossicle. *See* vertebra 2.

[1]**vertebrate.** Possessing a backbone (spinal or vertebral column).

[2]**vertebrate.** Any chordate of the subphylum Vertebrata (Craniata), comprising all animals with a vertebral column (cyclostomes, fishes, amphibians, reptiles, birds, and mammals).

vertex. 1. The highest point, summit, or apex. **2.** The top of the head.

vertical. 1. Of or pertaining to a vertex. **2.** Upright; perpendicular to the horizontal plane. **3.** Lengthwise or in the direction of the long axis.

verticil. A whorl; a circle of similar parts, as leaves, petals, or inflorescences arising from a point on an axis.

verticillastrate. Arranged in clusters resembling whorls.

verticillate. Whorled; arranged in verticils.

vervet. An African monkey (*Cercopithecus pygerythrus*), a guenon, related to the green monkey.

vesical. Of, pertaining to, or resembling a bladder, especially the urinary bladder.

vesicle. 1. *anat.* A small cavity or sac containing a fluid; a blister. **2.** *embryol.* A closed cavity, as **a.** one of the three primary divisions of the brain; **b.** a lateral outgrowth of the forebrain, the *optic vesicle;* **c.** the otocyst or *auditory vesicle;* **d.** the blastocyst or *blastodermic vesicle.* **3.** *bot.* **a.** A small sac or cavity filled with a fluid. **b.** A bladder. **c.** An air cavity.

vesicular. Of, pertaining to, or composed of vesicles.

vesiculate. 1. Covered with vesicles; containing vesicles. **2.** Bearing wings with reference to pollen grains.

vesper. The evening or eventide.

vespertine. 1. *zool.* Active in the evening: CREPUSCULAR. **2.** *bot.* Blossoming or appearing in the evening.

vespiary. The nest of any of the social wasps (yellow jackets, hornets) or the colony inhabiting it.

vespid. Any hymenopterous insect of the family Vespidae which includes the wasps, solitary and social.

vespine. Of, pertaining to, or resembling a wasp or wasps.

vespoid. Wasplike.

vessel. 1. *zool.* A tube or canal through which fluid passes, as a blood or lymphatic *vessel.* **2.** *bot.* A conducting tube or duct in the xylem of vascular tissue formed of elongated, cylindrical dead cells (vessel members) joined end to end with intervening walls perforated or dissolved away; sometimes called trachea.

vessel member. A vessel element, one of the dead cells which, joined end to end, form a vessel.

vestibule. 1. *anat.* A cavity or space which serves as an entryway or approach, as that of the mouth, nose, ear, or vagina. **2.** *zool.*

The cavity leading to the mouth or buccal cavity in various invertebrates.

vestibulocochlear nerve. The eighth cranial nerve (VIII), a sensory nerve from the internal ear composed of two branches, the cochlear and vestibular. Also called auditory, acoustic, or statoacoustic nerve.

vestige. 1. A small, imperfectly developed, or degenerate structure which represents an organ or structure which was fully developed and functional in an ancestor or in an earlier stage in development. **2.** A rudimentary structure.

vesture. 1. Anything that covers or clothes; a covering. **2.** *bot.* The hairy covering of a leaf or stem.

vetch. A forage legume of the genus *Vicia* or its beanlike seed.

veterinary. A veterinarian or a veterinary surgeon.

veterinary medicine. The science and art of treating diseased and injured animals; the prevention and treatment of disease in animals, especially domestic animals.

vexillate. Having a vexillum.

vexillum. 1. *bot.* A banner or standard. *See* banner. **2.** *zool. See* vane 1.

viable. 1. Capable of growing and developing, said of seeds. **2.** Capable of living, with reference to a fetus removed from the uterus or born prematurely.

viatical. Growing along roadsides or pathways.

vibraculum. A modified zooid of certain operculate ectoprocts which possesses a long bristle or seta which by sweeping movements removes debris from the surface of the colony.

vibrate. To move back and forth; to quiver.

vibrio. A comma- or spiral-shaped bacterium of the genus *Vibrio* which contains the causative organisms of cholera (*V. comma*) and abortion in cattle (*V. fetus*).

vibrissa pl. **vibrissae. 1.** One of the long, stiff hairs present on the snout of various mammals, as the cat; a feeler or whisker. **2.** One of a pair of stout bristles on each side of the face of a dipterous insect.

viceroy. A deep-orange colored butterfly (*Basilarchia* (*Limenitis*) *archippus*) which mimics the monarch butterfly.

villiform. 1. Having the shape and appearance of a villus. **2.** Velvetlike in appearance.

villous. 1. *anat.* **a.** Pertaining to villi. **b.** Covered with or possessing villi. **2.** *bot.* Bearing long and soft, but not matted, hairs; shaggy.

villus pl. **villi.** *anat.* One of the numerous minute, fingerlike processes or papillae, as those on the mucous membrane lining the intestine or on the chorion of an embryo. **2.** *bot.* A soft, straight hair, as those found on various plant structures.

vimen. A twig; a long, flexible branch or shoot.

vine. A plant with a climbing or creeping stem; one whose stem rests upon or is attached to a supporting structure; a rambler or climber.

vinegar. A liquid containing acetic acid, about 6%, resulting from the fermentation of alcoholic liquids, brought about by bacteria of the genus *Acetobacter*.

vinegar eel. The eelworm (*Turbatrix* (*Anguillula*) *aceti*), a nematode commonly found in old vinegar.

vinegarroon. A whip scorpion (*Mastigoproctus giganteus*) of southwestern United States and Mexico.

violet. A common garden plant or flower of the genus *Viola*.

viper. 1. An Old World venomous snake of the family Viperidae, including the genera *Vipera*, *Cerastes*, *Bitis*, and *Causus*. **2.** A venomous snake of the family Crotalidae. *See* pit viper.

viral. Of, pertaining to, or caused by a virus.

vireo. One of a number of small, insectivorous, passeriform birds of the genus *Vireo*, family Vireonidae.

virescence. *bot.* The state or condition of becoming green, as that resulting from the development of chloroplasts in uncolored structures.

virgate. 1. Wand- or rod-shaped; long, straight, and slender. **2.** Bearing many twigs.

virgin. *genetics.* A female which has not been mated to a male.

virgin birth. PARTHENOGENESIS.

virgin forest. The original forest which has never been cut.

virgin soil. Soil which has never been cultivated.

virgula. A transparent, fluid-filled gland which opens on the dorsal surface of the oral sucker in xiphidiocercariae.

virile. Pertaining to the characteristics of the male; having manly qualities.

virilism. The development of male traits and characteristics in a female, as *adrenal virilism* resulting from overproduction of androgens by the adrenal cortex.

virility. The state or quality of being manly; having procreative power.

virion. An infectious particle of a virus consisting of a nucleic acid core surrounded by a protein coat or capsid. The two comprise a nucleocapsid which may be naked or surrounded by an envelope.

virology. The science which deals with viruses and virus diseases.

virulence. The disease-producing capacity of an organism.

virulent. Noxious, malignant, highly infectious.

virus. One of a large group of ultramicroscopic, filterable, disease-producing agents all of which are intracellular parasites. They are large, high-molecular weight nucleoproteins capable of multiplying in living cells. They are the causative agents of many diseases, as smallpox, influenza, hog cholera, rabies, poliomyelitis, and foot and mouth disease in animals, and various mosaic, spot, and breaking diseases in plants.

viscacha. A South American rodent of the genera *Lagidium* and *Lagostomas*, related to the chinchilla.

viscera. Collectively, the internal organs of the body, especially those of the abdominal cavity. *See* viscus.

visceral. 1. Of or pertaining to a viscus or viscera. **2.** Of or pertaining to the branchial region of a vertebrate.

visceral arch. A branchial arch; a gill arch.

visceral cleft. A gill cleft or gill slit.

visceral functions. Those of an involuntary nature and under the control of the autonomic nervous system.

visceral furrow. A branchial groove.

visceral hump. The portion of the body of a gastropod which is contained within the coiled portion of the shell.

visceral mass. The main portion of the body of a bivalve mollusk lying dorsal to the foot.

visceral musculature. 1. The smooth muscles of hollow organs. **2.** The musculature of the heart. **3.** In aquatic vertebrates, the branchial muscles which move the gill arches and the jaws; in terrestrial vertebrates, the muscles derived from these.

visceral pouch. A pharyngeal pouch.

visceral skeleton. The branchial skeleton, comprising, in aquatic vertebrates, the gill arches, jaws, and cartilaginous structures supporting the gills; in terrestrial forms, the bones of the jaws, ear ossicles, hyoid bone, and cartilages of the larynx.

viscid. Thick, sticky, adherent, adhesive, glutinous.

viscosity. A condition or state of a fluid in which there is resistance to flow resulting from internal friction due to cohesion of molecules.

viscous. Thick, sticky; of a viscid or glutinous consistency; possessing viscosity.

viscus pl. **viscera.** Any of the organs contained within one of the body cavities (cranium, thorax, abdomen, pelvis), especially one within the abdominal cavity.

visible. Capable of being seen; obvious; manifest.

vision. 1. The sense of sight; seeing. 2. The sense by which the color, shape, and form of an object can be perceived. 3. The sensation resulting from stimulation of a light receptor, as an eye.

visual. Of or pertaining to the sense of vision.

visual acuity. Sharpness of vision; the ability to distinguish details of an object.

visual pigment. A pigment present in a photoreceptor organ which is essential for the perception of light. *See* rhodopsin, photopsin.

visual purple. RHODOPSIN.

vital. Of, or pertaining to life; essential for life; related to life or living things.

vital capacity. The amount of air that can be expelled from the lungs by the most forceful expiratory effort following the deepest possible inspiration; average amount is about 4 liters.

vitalism. The theory that activities of organisms are due to a vital principle or force and are not simply the result of chemical and physical processes. *Compare* mechanism.

vitamin. One of a group of organic substances which are essential in minute amounts for normal growth, development, and maintenance of normal physiological activities in an organism. They may be present in natural foods, produced by symbiotic organisms, developed from precursors, or be synthesized artificially.

vitamin A. A fat-soluble vitamin necessary for the maintenance of mucous membranes and for the formation of rhodopsin, essential for vision in dim light.

vitamin B complex. A group of vitamins occurring in yeast and liver which includes thiamine (B_1), riboflavin (B_2), niacin, pyridoxine (B_6), pantothenic acid, biotin, folic acid, the cobalamines (B_{12}), inositol, *p*- aminobenzoic acid (PABA), and possibly others.

vitamin B_1. THIAMINE.

vitamin B_2. RIBOFLAVIN.

vitamin B_6. PYRIDOXINE.

vitamin B_{12}. The cobalamines; the extrinsic factor of the antianemia principle; the erythrocyte-maturing factor (EMF).

vitamin C. Ascorbic acid, the antiscorbutic vitamin.

vitamin D. A complex of several antirachitic factors, including D_2 (calciferol) and D_3 (irradiated 7-dehydroxycholesterol), of importance in calcium and phosphorus metabolism.

vitamin E. Tocopherol, the antisterility vitamin.

vitamin K. The antihemorrhagic vitamin, essential for the formation of prothrombin.

vitamin P-P. Niacin, the pellagra-preventative vitamin.

vitellarium pl. **vitellaria.** **1.** A yolk gland, as those found in various invertebrates, as trematodes. **2.** The portion of an insect ovariole which contains developing ova and mature nurse cells.

vitellin. A phosphoprotein, the principal constituent of egg yolk.

vitelline. Of, pertaining to, or producing yolk.

vitelline cells. Cells produced by the vitelline glands of trematodes which contain shell globules utilized in the formation of the eggshell.

vitelline membrane. The membrane surrounding the cytoplasm of an ovum; the cell membrane.

vitellus. The yolk of an egg.

vitrella. In an ommatidium, a group of two to five crystal cells grouped around a crystalline cone which they secrete.

vitreous. Like glass: HYALINE.

vitreous body. The vitreous humor, a transparent, jellylike substance which fills the posterior cavity of the globe of the eye.

vitrodentine. A hard, outer layer of mesodermal origin which covers the dentine of placoid scales and teeth of elasmobranchs.

vitta. An oil tube in the pericarp of fruits of members of the Umbelliferae (parsley family).

vivarium. A place for keeping and raising animals, especially terrestrial animals. *Compare* aquarium, terrarium.

viviparous. 1. *zool.* Giving birth to living young which develop within and are nourished through the uterus. *Compare* ovoviviparous, oviparous. **2.** *bot.* Germination of a seed while still attached to the parent plant, as in the mangrove.

vivisection. Operating upon or cutting into a live animal, especially for investigational purposes and without the use of an anesthetic.

vixen. A female fox.

vocal. Of or pertaining to speech or voice.

vocal cords. Either of two pairs of folds which project from the lateral walls into the cavity of the larynx, as the superior ventricular folds (false vocal cords) and the inferior vocal folds (true vocal cords).

vocal folds. The true vocal cords, a pair of folds on the inner surface of the larynx extending from the thyroid cartilage posteriorly to the arytenoid cartilages, vibration of which results in production of sound.

vocal sac. One of a pair of sacs in the males of various amphibians located beneath the skin and opening into the mouth cavity.

voice. Sounds, especially articulate sounds purposely produced, as in speech or song, resulting from the vibration of the vocal cords (folds). The quality of the sound is determined by resonating chambers (mouth pharynx, nasal cavities) and articulators (lips, teeth, tongue, palate).

void. To empty; to make empty or to evacuate.

[1]**volant.** Flying; capable of moving through the air.

[2]**volant.** A flying individual, as winged flies of the genus *Lipotena*, a parasite of deer.

volar. 1. Of or pertaining to the palm of the hand or the sole of the foot. **2.** Pertaining to or utilized in flight.

volatile. Readily vaporized or converted into a gas; evaporating readily at a low temperature.

vole. One of a number of mouse- or ratlike rodents of the genera

Microtus (meadow mouse, field mouse), *Pitymys* (pine vole), *Arvicola* (water vole), and *Clethrionmys* (bank vole).

volitant. 1. VOLANT. **2.** Capable of moving rapidly.

volition. Acting from choice or from the exercise of the will.

voltine. Pertaining to broods. *See* bivoltine, polyvoltine.

voluble. 1. Turning or rotating readily. **2.** *bot.* Having the habit of twining.

voluntary. Under the control of the will; acting from choice.

¹**volute. 1.** Spiral or scroll-shaped. **2.** *bot.* Rolled up.

²**volute. 1.** *zool.* A turn or whorl of a spiral shell. **2.** A marine gastropod of the family Volutidae, especially one of the genus *Voluta*.

volutin. Inclusion bodies of nucleic acid and related substances present in yeasts and certain bacteria.

volva. A cup-shaped structure at the base of the stipe in certain fleshy fungi, especially agarics. *See* universal veil.

volvate. Possessing a volva.

volvent. An unarmed type of nematocyst which wraps itself about and entangles prey.

volvox. Any flagellate of the genus *Volvox*, a green, spherical, colonial form of the order Volvocales, usually regarded as a green alga; sometimes considered as a protozoan of the order Volvocida.

vomer. A membrane bone of the nasal region, in mammals forming a part of the nasal septum.

vomerine. Of, pertaining to, or borne on the vomer bone, as *vomerine* teeth in amphibians.

vomeronasal organ. Jacobson's organ, one of a pair of diverticula or grooves extending ventrally from the nasal cavity and opening into the mouth cavity in snakes and lizards; rudimentary in birds and mammals.

vomit. To eject forcibly the contents of the stomach.

voracious. Having an insatiable desire for food; ravenous.

vortex pl. **vortices. 1.** A whorl. **2.** A structure having the appearance of being produced by rapid rotation about an axis, as *vortices pilorum* (hair whorls).

vorticella. A stalked, bell-shaped, ciliate protozoan of the genus *Vorticella*.

vorticose. Whirling, having a whorled structure or appearance.

vulpine. Of, pertaining to, or like a fox.

vulture. One of a number of large, soaring, carrion-eating birds of the families Cathartidae (New World vultures) and Accipitridae (Old World vultures). To the former belong the turkey buzzard (*Cathartes aura*), the black buzzard, and the condor.

vulva. 1. The pudendum or external genitalia of a female. **2.** The external genital opening in certain invertebrates, as *Ascaris*.

wader. Any of a number of long-legged birds which frequent shallow waters in search of food, as cranes, herons, sandpipers.

waggle dance. A dance performed by returning worker bees which communicates to other bees information concerning direction and distance of food from the hive. Also called wagging dance.

wagtail. A pipit with a characteristic tail-wagging habit.

waist. 1. The narrowest part of the body between thorax and hips. **2.** A peduncle or petiole in insects or spiders.

walking fern. A fern (*Camptosorus rhizophyllus*) whose turned-down leaf tips, on contact with the substratum, give rise to new plants.

walkingstick. A stick insect, a long, slender orthopteran insect of the family Phasmidae, with a wingless body resembling a stick.

wall. The surface layer or layers of a cell, organ, or organism, especially when of a protective or supportive nature, as a cell *wall*, body *wall*.

wallaby. A small kangaroo, a marsupial of the family Macropodidae, as *Petrogale*, the rock wallaby.

wallaroo. A kangaroo intermediate in size between a kangaroo and wallaby.

walleye. The walleyed perch or pike perch (*Stizostedion vitreum*), a valuable freshwater food and game fish.

wall pressure. The outside pressure on the contents of a plant cell.

walnut. A hardwood tree of the genus *Juglans*, or its fruit, a nut, or its wood.

walrus. A large, marine mammal of the genus *Odobaenus* characterized by a whiskered, blunt muzzle and elongated ivory tusks that lives in herds in polar waters.

waltzer. One of a special strain of mice or rats which walks in a bizarre fashion due to an inherited defect in the inner ear.

wamp. The American or common eider. *See* eider.

wandering cell. One not fixed; one that moves from place to place, as certain macrophages.

wapiti. The American elk (*Cervus canadensis*).

warble. **1.** A swelling on the back of domestic animals (horses, cattle) which contains the larva of a warble fly. **2.** The larva of the warble fly.

warble fly. One of several flies of the family Oestridae whose larvae develop under the skin of various animals causing warbles, as *Hypoderma bovis*, the ox warble.

warbler. **1.** A bird that warbles or sings in a trilling manner. **2.** An Old World songbird of the family Sylviidae. **3.** One of a number of small, brightly-colored, insectivorous birds of the family Parulidae.

warm-blooded. Having a relatively high and constant body temperature: HOMOIOTHERMOUS.

warmouth. A sunfish (*Chaenobryttus coronarius*); also called warmouth bass.

warning coloration. Conspicuous marks or colors possessed by animals with injurious, poisonous, or offensive protective mechanisms; aposematic coloration.

wart. *zool.* **1.** In mammals, a small, circumscribed growth of the epidermis, usually of a hard, horny texture. **2.** In holothurians, a modified tube foot. **3.** In toads, a compact mass of poison glands.

warthog. A large, grotesque, wild pig of Africa (*Phachochoerus aethiopicus*) with wartlike excrescences on the sides of its head; also called vlakvark.

wasp. Any of a number of hymenopterous insects characterized by a slender body with abdomen attached by a slender stalk or peduncle. Females commonly possess a sting. Typical wasps and paper wasps (yellow jackets, hornets) belong to the family Vespidae; solitary wasps to the Sphecidae. Some wasps are social, constructing nests and living in colonies.

waste. Material eliminated from the body consisting of undigested food, unusable products of metabolism, glandular secretions, and excess substances not utilized by the body.

water. H₂O, the principal constituent of protoplasm, comprising about 75% of most organisms. It is of primary importance because of its role as a solvent, an ionizing medium, and for its heat-regulating properties. Its high chemical stability and high surface tension are of significance in the physical properties of protoplasm, and its role in hydrolytic reactions is of importance in body metabolism.

water bear. A tardigrade.

water beetle. Any of a number of beetles which frequent water or whose larvae live in water.

waterbuck. A large antelope (*Kobus ellipsiprymnus*) of South Africa.

water buffalo. An often domesticated Indian buffalo (*Bubalus bubalis*) of Southeast Asia. Also called arna and carabao.

waterbug. A croton bug or cockroach.

water bug. Any of a number of hemipterous insects which frequent ponds and streams, as the giant water bug (*Bonacus grisseus*).

water caltrop. The water chestnut; also called red caltrop or red ling.

water cells. Pouchlike diverticula of the stomach (rumen and reticulum) of camels.

water chestnut. An aquatic annual (*Trapa natans*) of the family Hydrocaryaceae, introduced from the Orient and now common along the East Coast.

water dog. One of a number of aquatic salamanders, as **a.** *Taricha torosa*, a Pacific Coast newt; **b.** *Cryptobranchus*, the hellbender; **c.** *Necturus*, the mud puppy.

water fern. An aquatic fern of the families Marsileaceae and Salviniaceae.

water flea. One of a number of small, aquatic crustaceans of the order Cladocera, as *Daphnia, Alona, Leptodora*.

613

water fowl. 1. Any bird which frequents water, especially a swimming bird. **2.** Swimming game birds (ducks and geese), as distinguished from shorebirds and upland game.

water hen. The Florida gallinule (*Gallinula chloropus*), the moorhen of Europe.

water hyacinth. An aquatic plant (*Eichornia crassipes*) often forming dense masses blocking waterways.

water lily. Any plant of the water lily family, Nymphaeaceae, especially those of the genera *Nuphar* and *Nymphaea*.

watermelon. A large, oval or spherical fruit of a vine (*Citrullus vulgaris*) or the plant which produces it.

water mite. An aquatic mite of the family Hydrachnidae; adults red colored and predaceous; larvae parasitic on insects and clams, as *Hydrachna*.

water moccasin. *See* moccasin.

water mold. An aquatic mold of the order Saprolegniales, as *Saprolegnia*, sometimes parasitic on fish or fish eggs.

water ring. The circular or ring canal of a water-vascular system.

water scorpion. A predaceous water bug of the family Nepidae, as *Nepa* and *Ranatra*, with raptorial front legs and caudal cerci modified to form a breathing tube.

water snake. A harmless snake of the genus *Natrix*.

water strider. Any of several hemipterous insects of the families Gerridae and Veliidae, as those of the genera *Gerris* and *Rhagovelia*, with long, slender legs used in skating over the surface of the water.

water tiger. The predaceous larva of various diving beetles, as *Dytiscus*.

water tube. One of the numerous tubes in the gill of a bivalve mollusk which open dorsally into the suprabranchial chamber.

water turkey. The darter or snakebird (*Anhinga anhinga*) with a long snakelike neck.

water-vascular system. In echinoderms, a system of fluid-filled canals and structures which function in locomotion and food-getting. In a starfish, it comprises the madreporite, stone canal, ring canal (water ring), and radial canals with lateral canals connected to tube feet (podia), the inner ends of which bear ampullae.

water witch. The horned or pied-billed grebe.

wattle. 1. One of a pair of fleshy appendages which hang from the throat of various birds, especially domestic fowl and most galliform birds. **2.** In reptiles, a median, ventral fold of skin hanging from the neck.

wattlebird. A honey eater of New Zealand, the huia (*Neomorpha acutirostris*) with large orange and blue wattles.

wave. 1. A disturbance propagated in a medium such as air or water which advances through the medium at a rate dependent upon the physical qualities of the medium, as a sound *wave*. **2.** A moving ridge or swell of water. **3.** An undulation traced by a recording stylus.

wax. A substance, consisting of a mixture of esters of higher fatty acids and higher monohydroxy alcohols, which can be of plant origin (carnauba wax, myrtle wax), animal origin (beeswax, sperm oil, lanolin), or mineral origin (paraffin, ceresin).

waxbill. An Old World weaver finch, especially one of the genus *Estrilda*.

wax gland. A gland that produces wax, as those of a honeybee or wax insect.

wax insect. A homopterous insect which produces wax used commercially. *See* wax scale.

wax moth. A moth (*Galleria mellonella*) whose larvae (wax worms) live in the hives of bees feeding upon wax. Also called bee moth.

wax myrtle. A tree or shrub (*Myrica cerifera*) which, with the bayberry (*M. carolinensis*), is the source of myrtle wax used in the manufacture of candles.

wax scale. One of several homopterous insects which produce wax used commercially, as the Chinese wax scale (*Ericerus pe-la*).

waxwing. A passerine bird of the family Bombycillidae, as the cedar waxwing (*Bombycilla cedrorum*).

wax worm. The larva of the wax moth.

weakfish. A marine food fish of the genus *Cynoscion*, especially *C. regalis* of the East and Gulf Coasts.

wean. To deprive of mother's milk; to cause the young to cease to suckle or nurse at the breast.

weasel. A small, carnivorous, fur-bearing mammal of the genus *Mustela. See* ermine.

weaverbird. One of a number of finchlike birds of Asia and Africa of the family Ploceidae which construct elaborate nests by interlacing grass and other materials. Some are parasitic.

weaver finch. One of a number of Old World passerine birds of the family Ploceidae which includes the English sparrow (*Passer domesticus*).

web. A thin, membranelike structure, as **a.** that uniting the digits of various water animals; **b.** that formed of interlacing threads woven by spiders; **c.** that forming the nest or abode of various webworms; **d.** the vane of a feather. *See* patagium.

Weberian ossicles. A group of four small bones (scaphium, claustrum, intercalarium, tripus) which, in certain teleost fishes, lie between the swim bladder and perilymphatic space of the ear.

webfoot. A foot with toes connected by a web, as in water birds.

web spinner. An insect of the order Embioptera which constructs and lives in a silken gallery or tunnel, as *Anisembia texana*, of southern United States.

webworm. One of a number of caterpillars, the larvae of moths of the family Arctiidae, which construct and live in large webs, as *Hyphantria cunea*, the fall webworm.

weed. A plant that has little value or grows to the disadvantage of other more desirable plants.

weel. A structure consisting of hairlike processes arranged so as to prevent undesirable insect visitors from entering flowers.

weep. **1.** *physiol.* To shed tears. **2.** *bot.* To exude water; to bleed.

weever. An edible marine fish of the genus *Trachinus* but bearing poisonous dorsal spines.

weevil. **1.** Any of a number of snout beetles, as the cotton boll weevil. **2.** One of a number of beetles whose larvae infest pods and seeds, as bean and pea weevils. *See* snout beetle, curculio.

wether. A castrated ram.

whale. An air-breathing aquatic mammal of the order Cetacea, the largest of living mammals. They comprise two suborders, the Mysticeti or whalebone whales, and Odontoceti or toothed whales.

whalebone. *See* baleen.

whale louse. A crustacean (*Cyamus ceti*), a highly modified amphipod.

whale shark. A large shark (*Rhincodon typus*) of tropical waters, sometimes reaching a length of 50 feet.

wheat. 1. One of a number of grasses of the genus *Triticum*, especially *T. aestivum*, cultivated wheat. **2.** The cereal grain produced by these plants.

wheatear. A small thrush (*Oenanthe oenanthe*) inhabiting barren northern regions.

wheat rust. 1. A basidiomycete (*Puccinia graminis*) whose life cycle includes stages on wheat and stages on the European or common barberry. **2.** The disease caused by the wheat rust fungus.

wheatworm. A nematode (*Anguina tritici*) which infests wheat.

wheel animalcule. A rotifer.

wheel bug. An assassin bug (*Arilus cristatus*) with a pronotum bearing a wheellike crest. It sometimes inflicts vicious bites.

whelk. A large, marine snail of the order Prosobranchiata, as the giant whelk (*Busycon*) and the edible whelk (*Buccinum*).

¹whelp. The young of a dog, wolf, or other carnivores.

²whelp. To bring forth or give birth to young, applied especially to canine animals.

whimbrel. The Hudsonian curlew (*Numenius phaeopus*).

whip coral. *See* sea whip.

whiplash flagellum. A smooth-surfaced flagellum; one lacking mastigonemes. *See* mastigoneme.

whippoorwill. A night-flying bird (*Caprimulgus vociferus*) of the goatsucker family, Caprimulgidae.

whip scorpion. An arachnid of the order Uropygi. *See* vinegarroon.

whip snake. A slender-bodied, long-tailed, terrestrial snake of the genus *Masticophis*, as *M. flagellum*, the coachwhip.

whiptail. One of several lizards of the genus *Cnemidophorus*, with slender elongated tails.

whipworm. A parasitic nematode (*Trichuris trichiura*) inhabiting the intestine of man or related species, common in the intestines of various mammals.

whirligig beetle. One of a number of beetles which swim on the surface of quiet waters with a continuous, gyrating movement, as *Dineutes americanus*, of the family Gyrinidae.

whisker. 1. A hair of the beard. **2.** One of the long hairs (vibrissae) which grow near the mouth of various animals, as the cat.

whistler. 1. The goldeneye (*Bucephala clangula*), a diving duck, so called because of the high-pitched whirr of its wings. **2.** The hoary marmot (*Marmota caligata*) of western North America.

white ant. A termite.

white blood cell. A leukocyte.

white corpuscle. A leukocyte.

whitefish. One of a number of fishes of the family Coregonidae, especially those of the genera *Prosopium* and *Coregonus*, as *C. clupeiformis*, the lake whitefish of commercial importance.

whitefly. A homopterous insect of the family Aleyrodidae, minute insects covered with a white dust or powder, as those of the genus *Aleurocanthus*, a serious citrus pest.

white matter. *anat.* Nervous tissue of the brain and spinal cord consisting principally of myelinated axons.

white potato. *See* potato.

whitewood. Basswood, a tree of the genus *Tilia*.

whiting. 1. A European marine fish (*Merlangus merlangus*). **2.** The New England hake (*Merluccius bilinearis*). **3.** A North Atlantic food fish (*Menticirrhus*).

whorl. 1. *bot.* A group of three or more structures, as leaves, arising at one point and arranged in the form of a circle; a verticil. **2.** *anat.* **a.** A complete turn in the line of a finger print. **b.** A spiral arrangement of the hairs of the head; a vortex. **3.** *zool.* A turn in the spire of a gastropod mollusk; a volution.

widgeon. One of several species of freshwater, dabbling ducks of the genus *Mareca* (*Anas*). *See* baldpate.

wiggler. A wriggler; an animal that wiggles, as the larva or pupa of a mosquito or the nymph of a mayfly.

wild. In a natural state; not domesticated; savage; uncivilized; not under the supervision or care of man.

wildcat. 1. The bobcat or bay lynx (*Lynx rufus*). **2.** The European wildcat (*Felis sylvestris*).

wildebeest. A gnu.

willet. A large shorebird (*Catoptrophorus semipalmatus*) of North America.

willow. One of a number of trees or shrubs of the genus *Salix*, as the white willow (*S. alba*).

618

¹**wilt.** To lose freshness and turgidity; to become limp and droopy.

²**wilt.** A disease of plants. *See* wilt disease.

wilt disease. Any of a number of diseases of plants characterized by wilting, as that caused by fungi of the genus *Fusarium* or by bacteria.

wind dispersal. The scattering or spreading of spores, pollen, seeds, or fruits by the agency of the wind. *See* anemochorous.

windpipe. The trachea.

wind pollination. Pollination by air-borne pollen. *See* anemophilous.

wind timber. Trees deformed by the strong winds, as those seen on exposed slopes and mountain tops.

wing. 1. *zool.* An organ of flight, as **a.** one of a pair of movable appendages possessed by flying animals (insects, birds, bats); **b.** a winglike structure used for flight, as the pectoral fin of a flying fish or the gliding membranes of flying squirrels or lemurs. **2.** *anat.* A winglike process of an organ or structure; an ala. **3.** *bot.* A thin, flattened expansion or appendage, as that on various stems, seeds, or fruits. *See* samara.

wing cover. In an insect, an elytron.

wing covert. A tectrix.

winged. 1. Provided with wings or winglike structures. **2.** Wounded or killed when in flight.

wingless. 1. Lacking wings or winglike processes. **2.** Possessing only rudimentary wings.

wing pad. A compact, undeveloped wing of an insect nymph or pupa.

wing shell. 1. A boring clam (*Pholas campechiensis*). **2.** A bivalve of the genus *Pteria*.

winkle. 1. A periwinkle. **2.** Any of a number of marine snails, especially those of the genus *Busycon*.

winter egg. 1. A fertilized egg produced in late summer or fall by parthenogenetic organisms, as rotifers, aphids. **2.** A thick-shelled egg with abundant yolk produced by certain flatworms.

wintergreen. Checkerberry, a flavoring obtained from a low, creeping, evergreen plant (*Gaultheria procumbens*). Its leaves are the source of oil of wintergreen (*methyl salicylate*).

winter kill. The death of large numbers of organisms, as fishes, as occurs in shallow lakes and ponds containing much organic matter when ice covers the surface for prolonged periods.

wireworm. 1. The shiny, slender, hard-bodied larva of the click beetle which feeds on the roots of corn, wheat, cotton, and many other plants. **2.** A slender-bodied millipede.

wisent. *See* aurochs.

wishbone. A furcula.

witches'-broom. An abnormal, brushlike growth that is a symptom of a number of fungus or virus diseases in plants; also called hexenbesen.

witches' milk. Milk secreted by the mammary glands of a new-born infant.

withers. The ridge between the scapulae or shoulder bones of a horse formed by spinous processes of the vertebrae.

withy. A slender, flexible twig.

wolf pl. **wolves.** A doglike, carnivorous mammal (*Canis lupus*), the common gray or timber wolf, or the coyote (*C. latrans*), the prairie wolf. *See* Tasmanian wolf.

wolffian body. The mesonephros.

wolffian duct. The mesonephric duct which, in anamniotes, serves as a common reproductive and excretory duct. In amniotes, it gives rise to certain male reproductive structures, as the epididymis, ductus deferens, and seminal vesicles.

wolf spider. A large, active, hunting spider (*Lycosa*).

wolverine. An American carnivorous mammal (*Gulo luscus*) of the family Mustelidae, noted for its cunning; also called carcajou.

womb. The uterus.

wombat. A burrowing, badgerlike marsupial of Australia, as the common wombat (*Vombatus hirsutus*).

wood. The hard, fibrous substance within the bark which comprises the major portion of the stems and branches of trees and shrubs; xylem. *See* hardwood, softwood, summerwood, springwood, lignin, annual ring.

wood alcohol. Methanol or methyl alcohol, CH_3OH, obtained by destructive distillation of wood. It is a violent poison.

woodbine. 1. The Virginia creeper *Parthenocissus quinquefolia*). **2.** The honeysuckle of Europe (*Lonicera periclymenum*).

woodchuck A stout-bodied burrowing rodent (*Marmota monax*). Also called groundhog.

woodcock. 1. An American game bird (*Philohela minor*), a shorebird of the family Scolopacidae. **2.** In the Old World, *Scolopax rusticola.*

wood duck. A surface-feeding duck (*Aix sponsa*).

wood frog. A small frog (*Rana sylvatica*) of damp woodlands.

wood louse. Any of a number of terrestrial crustaceans of the order Isopoda, commonly called pill bugs, as *Armadillidium, Oniscus. See* sow bug.

woodpecker. One of a large number of climbing birds of the family Picidae with a chisellike bill used for drilling into trees, as the redheaded woodpecker (*Melanerpes erythrocephalus*). *See* flicker.

wood rat. The Florida pack rat or trade rat (*Neotoma floridans*) of the southeastern United States. Also *N. fuscipes* and *N. cinerea of* western United States.

woody. Of or pertaining to wood; composed of wood or woody fibers.

wool. 1. *zool.* **a.** The soft, curly, hairy covering of domesticated sheep and other animals. **b.** Short, crisp, curled hair. **c.** The thick, hairy covering of certain caterpillars. **2.** *bot.* Long, soft, matted hairs, resembling animal wool.

wool fat. Lanolin, a fatlike wax which covers the fibers of sheep's wool.

woolly. 1. *zool.* Of, pertaining to, resembling, or consisting of wool. **2.** *bot.* Lanate; bearing long, soft, and more or less matted hairs; tomentose.

woolly apple aphid. A homopterous insect (*Eriosoma lanigerum*) which secretes white, waxen threads forming woollike masses on twigs; a serious pest.

woolly bear. A woolly worm.

woolly worm. 1. The woolly larva of several moths, especially tiger moths of the family Arctiidae. **2.** A woolly bear or woolly bear caterpillar.

wool maggot. The larva of various flies that infest sheep, as those of the genera *Phaenicia* in Australia, and *Phormia* and *Sarcophaga* in the United States.

621

wool sponge. The soft-fibered sponge of commerce (*Hippospongia lachne*) of the Mediterranean and Carribean Seas and the Gulf of Mexico.

worker. In social insects (ants, bees, wasps, termites), a member of a caste adapted for the performance of nonreproductive functions. They are imperfectly developed females.

worm. 1. Any small, elongated, creeping or crawling animal, especially those of the phyla Annelida, Platyhelminthes, Aschelminthes, and Acanthocephala; a helminth. **2.** Any of a number of other phyla whose members are elongated, creeping animals, as a rhynchocoelan, sipunculid, echiurid, onycophoran, pentastomidan, or hemichordate. **3.** The larva of various insects, especially caterpillars and grubs.

worm disease. In oysters, infestation by a polychaete worm (*Polydora*).

worm lizard. A legless lizard of the family Amphisbaenidae, as *Rhineura floridans*, the florida worm lizard or thunderworm.

worm salamander. A salamander of the genus *Batrachoseps*, with a long, slender body and small limbs with a reduced number of digits.

wormseed. Chenopodium, the fruit of *Chenopodium ambrosioides*, a goosefoot, used as an anthelmintic.

wormseed oil. Oil of chenopodium.

worm shell. A gastropod mollusk with an elongated, coiled shell resembling a worm in appearance, as *Vermicularia*.

worm snake. *See* blind snake.

wormwood. A European herb (*Artemisia*), the source of a bitter oil used in making absinthe.

wrasse. One of several brilliantly-colored food fishes of the family Labridae.

wren. A small, insectivorous, passerine bird of the family Troglodytidae, as *Troglodytes aëdon*, the house wren.

wriggler. A wiggler.

wrist. The region between the hand and the forearm; the carpus.

wryneck. An Old World bird of the genus *Jynx* of the woodpecker family, which holds its head in a peculiar manner when disturbed.

X

xanthophore. A melanophore with yellow pigment.

X-A ratio. The ratio of sex chromosomes to autosomes.

X chromosome. A sex chromosome.

xenia. The direct, visible effects of pollen upon the development of the endosperm and related structures in the development of a seed.

xenic. Of or pertaining to foreign or strange organisms. *Compare* axenic.

xenobiotic. A compound foreign or strange to life, as DDT or other pesticides.

xenodiagnosis. Diagnosis of a disease by using a vector such as an arthropod to transmit disease organisms from a patient to a susceptible laboratory animal.

xenology. The study of host-parasite relationships.

xeric. Pertaining to arid or dry conditions. *Compare* hydric.

xeriobole. A plant whose seed dispersal is accomplished by dehiscence resulting from dryness.

xerochase. A fruit which opens under dry conditions and closes when moist.

xerochastic. Pertaining to plants whose seeds or spores are spread as a result of dessication and bursting of fruits or fruiting bodies.

xerophilous. Drought-loving, with reference to plants adapted for and preferring dry habitats, as desert plants.

xerophthalmia. Abnormal dryness of the eyeball.

xerophyte. A desert plant or one adapted for living in a dry environment; one able to withstand drought or a limited water supply, as cactus, sagebrush.

xerophytic. Of, pertaining to, or adapted for dry conditions.

xerosere. A plant succession or sere having its origin in a dry soil.

xiphiplastron. One of a pair of bony plates located posteriorly in the plastron of a turtle.

xiphisternum. The posterior element in the sternum of many vertebrates; in man, the xiphoid process.

xiphoid. Ensiform or sword-shaped.

xiphosuran. Any arthropod of the subclass Xiphosura, class Merostomata, which includes the king crabs.

X-organ. 1. A neurosecretory structure in the eyestalk of certain crustaceans which produces hormones involved in molting, gonad development, and metabolic activities. **2.** A saclike structure in freshwater gastrotrichs through which eggs pass to the outside.

X-rays. ROENTGEN RAYS.

xylan. A water-soluble polysaccharide present in the cell walls of certain algae and in woody tissue.

xylem. Wood or woody tissue; the plant tissue consisting of tracheids, vessel members, fibers, and parenchyma which forms the woody elements of vascular tissue. It functions principally in the conduction of water and mineral salts but also provides mechanical support and functions in the storage of water and food. *Compare* phloem.

xylem ray. *See* medullary ray.

xylene. Xylol, a colorless, inflammable liquid consisting of three isomers obtained from coal tar, that is used in microscopy as a solvent and clearing agent.

xyloid. Resembling wood; of the nature of wood; ligneous.

xylol. XYLENE.

xylophagous. Destructive to wood, with reference to organisms which feed upon wood, as insects, certain mollusks, fungi.

xylophilous. Preferring wood, said of organisms which live in or on wood.

xylose. Wood sugar, $C_5H_{10}O_5$, a pentose sugar.

xylotomous. Capable of cutting into or boring into wood.

xylotomy. The art of cutting thin sections of wood for microscopic examination.

X-zone. The fetal cortex or boundary zone between the definitive cortex and the medulla of the adrenal gland, well-developed in a fetus.

Y

yak. A wild bovine (*Poephagus grunniens*) of the highlands of Tibet, hunted for its flesh and hair and sometimes domesticated.

yam. 1. A climbing vine (*Dioscorea*) or its root, widely used as a food in the tropics. **2.** In the southern United States, the sweet potato.

yard. A winter feeding area for deer.

yaws. An infectious, nonvenereal disease of the tropics, caused by a spirochaete (*Treponema pertenue*); also called frambesia, pian.

Y chromosome. A sex chromosome in the males of certain animals, as *Drosophila. See* sex chromosome.

yearling. An animal a year old or in the second year of its development.

yeast. 1. A semifluid substance consisting principally of unicellular fungi which forms on the surface or as a sediment in fermenting fruit juices, malt wort, and sugar-containing liquids and used to induce alcoholic fermentation, as a leavening agent in bread, in the manufacture of ethyl alcohol, and as a source of vitamins, especially those of the B-complex. **2.** Any of a number of species of unicellular, ascomycete fungi, especially those of the genus *Saccharomyces*, which reproduce by budding, producing chains of cells but no true mycelium.

yellow body. A corpus luteum.

yellow fever. An acute, infectious disease of tropical and subtropical regions caused by a virus transmitted by the bite of a female mosquito (*Aedes aegypti*).

yellow grub. The encysted metacercaria of a fluke (*Clinostomum*) found in the flesh of freshwater fishes, especially perch. The adult is a parasite of water birds. Also called black spot grub.

yellow jacket. A social wasp of the genus *Vespula* which usually constructs an underground nest of paper.

yellowlegs. An American shorebird of the family Scolopacidae, as the greater yellowlegs (*Totanus melanoleuca*) and lesser yellowlegs (*T. flavipes*).

yellow spot. The macula lutea of the retina of the eye.

yew. A coniferous tree or shrub of the yew family, Taxaceae, especially those of the genus *Taxus*, as *T. canadensis*, the American yew or ground hemlock.

yield. Productivity; the rate at which a breeding stock produces an additional crop or additional stock.

yogurt also **yoghurt** or **yoghourt.** A fermented milk of the consistency of custard made from concentrated milk and milk solids, fermentation being induced by addition of cultures of bacteria, usually *Lactobacillus bulgaricus*.

yolk. 1. Nutritive material in the form of granules or droplets in the cytoplasm of an ovum, usually concentrated at the vegetal pole. **2.** The large, yellow, spheroidal mass of food material in the center of an egg of a reptile or bird. **3.** A greasy material present in sheep wool.

yolk gland. *See* vitellarium.

yolk plug. A mass of yolk-filled cells lying between the lips of the blastopore of a developing amphibian embryo.

yolk sac. 1. A large, yolk-containing sac present in the embryos of fishes, birds, and reptiles formed by growth of the splanchnopleure downward over the massive yolk. **2.** In placental mammals, a vestigial structure consisting of an endoderm-lined sac connected to the gut by a slender yolk stalk. It contains no yolk.

yolk-sac placenta. That present in egg-laying monotremes in which the yolk sac unites with the chorion to form a temporary placenta. Term sometimes applied to a comparable structure in elasmobranchs.

yolk stalk. The connection between the yolk sac and the gut.

Y-organ. An endocrine structure in crustaceans which produces a hormone essential for molting, located in the maxilla or base of the antenna.

ypsiloid. The epipubic bone of a salamander.

yucca. A xerophytic plant of the genus *Yucca* of the lily family, as *Yucca brevifolia*, the Joshua tree.

yucca moth. A moth of the genus *Tegeticula*, family Prodoxidae, essential for the pollination of yucca flowers.

Z

Z-chromosome. One of the two sex chromosomes in the males of birds and some insects.

zeaxanthin. A xanthophyll present in the seeds of yellow corn; also present in various algae, bryophytes, and some vascular plants.

zebra. 1. A striped, African, horselike mammal of the family Equidae, as *Equus burchelli*, Burchell's zebra or bonteauagga. **2.** A striped butterfly (*Heliconius charitonius*) of the family Nymphalidae.

zebu. A domesticated, Asiatic bovine (*Bos indicus*) with a pronounced shoulder hump and a prominent dewlap.

zein. An incomplete protein present in maize, deficient in the amino acids tryptophan and lysine.

Z-line. A thin, dark line or disc which bisects an I band (isotropic disc) in a striated muscle fiber.

zeran. A Mongolian gazelle (*Procapra gutturosa*).

zero. 1. A character, 0, designating the absence of a quantity (naught or nothing) or the lowest point on a scale. **2.** The point on a thermometer from which a graduated scale is established. On the centigrade scale, it is the temperature at which water freezes.

zeuglodont. An extinct, whalelike cetacean of the suborder Zeuglodontia with mammallike teeth.

zinc. A mineral element, symbol Zn, at. no. 30, essential for plant growth. Zinc deficiency results in yellowing and malformation of leaves. In animals, it forms a part of certain enzyme systems and is present in crystallized insulin.

zingiber. Ginger.

zinnia. A widely grown ornamental flower of the genus *Zinnia*, a composite.

zoantharian. Any coelenterate of the subclass Zoantharia, class Anthozoa, which includes the sea anemones and true corals.

zoarium. A colony of colonial bryozoans. *See* ectoproct.

zoea larva. A stage in the development of certain crustaceans

627

ZONA

between the protozoea and mysis stages characterized by a
distinct cephalothorax, abdomen, and eight pairs of appendages, as in the shrimp.

zona. An encircling belt, band, or girdle; a zone.

zonal. Of or pertaining to a zone.

zona pellucida. **1.** A thin, transparent, noncellular layer surrounding a mammalian oocyte or ovum. **2.** The zona radiata of
certain vertebrate oocytes, as in reptiles.

zona radiata. **1.** The vitelline membrane of certain vertebrate
ova, as those of amphibians and birds. **2.** A thin membrane
with radial striations surrounding the oocyte of reptiles; also
called zona pellucida.

zonation. The location and distribution of organisms in definite
zones.

zone. A delimited area or region.

zonite. A body division or segment of a kinorynch.

zonula ciliaris. The ciliary zonule (suspensory ligament) which
attaches the lens of the eye to the ciliary processes; the zonule
of Zinn.

zoo. A collection of animals, especially for exhibition purposes;
a zoological garden.

zoobenthos. The animals in the bottom regions of a body of
water. *See* benthos.

zoochlorella. Unicellular green algae of the genus *Chlorella*
which occur as symbionts in freshwater protozoans, the
gastrodermis of *Hydra*, and in the amebocytes of sponges.

zooecium. In an ectoproct, the body wall which encloses the soft
parts (polypide).

zooflagellate. An animallike flagellate of the class Zoomastigo-
phorea; one lacking chlorophyll.

zoogeography. The study of the geographical distribution of
animals.

zooglea. A mass of bacteria embedded in a gelatinous matrix.

zoogleal film. A thin, gelatinous film containing bacteria, algae,
fungi, and protozoans which coats the filtering medium in a
sewage treatment plant.

zooid. A member of a closely associated group of organisms, as
a. individuals in a colony of protozoans, as *Volvox;* **b.** one of
the individuals in a polymorphic hydroid colony; **c.** an indi-

628

vidual of an ectoproct colony; **d.** one of a chain of individuals produced asexually, as in certain oligochaetes (*Chaetogaster*) or in turbellarians (*Stenostomum*); **e.** a member of a tunicate colony. *See* polyp, polypide.

zoological. 1. Of or pertaining to the science of zoology. **2.** Of or pertaining to animals.

zoology. 1. The science which deals with animals and animal life. **2.** A book or treatise on animals. **3.** The animal life of a region.

zoonosis pl. **zoonoses.** A disease of animals which is transmissible to man, as rabies.

zooparasite. An animal parasite.

zoophagous. Flesh-eating; feeding upon animals.

zoophilous. 1. Animal-loving, or having an affinity for animals. **2.** Pollinated by animals other than insects, as bats, birds.

zoophyte. An animal which has a plantlike form, as a hydroid or bryozoan.

zooplankton. The animals of plankton.

zoosporangium. A sporangium which produces zoospores.

zoospore. A motile or swimming spore; a spore with flagella or cilia, as those produced by various algae and fungi; also called planospore.

zootomy. The anatomy of animals, especially those other than man; comparative anatomy.

zootoxin. A toxin of animal origin, as snake venom.

zooxanthellae. Dinoflagellate protozoans containing yellow chromatophores which live as symbionts in foraminiferans, radiolarians, and some metazoans. *Compare* zoochlorella.

zorapteran. An insect of the order Zoraptera comprising the single genus *Zoortypus*, a colonial, termitelike insect of southern United States.

zorille. An African carnivore (*Ictonyx striatus*) of the family Mustelidae, about the size of a ferret; also called ferret-badger, zoril, zorilla.

zubr. European name for Indian humped cattle.

zygantra. Posterior articular facets on the neural arches of certain reptiles. They articulate with zygosphenes.

zygapophysis. One of two pairs of articular processes on the neural arch of a vertebra.

zygodactyl. Having two toes in front and two behind, as in woodpeckers. *Compare* heterodactyl.

zygogenesis. Reproduction in which male and female gametes fuse, forming a zygote.

zygoid. 1. DIPLOID. **2.** Like a zygote, as in *zygoid* parthenogenesis.

zygoma. The zygomatic arch.

zygomatic arch. An arch on the lateral surface of the skull formed by processes of the temporal and zygomatic bones.

zygomatic bone. The malar bone forming the prominence of the cheek.

zygomorphic. Irregular; bilaterally symmetrical, with reference to flowers.

zygophore. A specialized hyphal branch in certain fungi, the tip which forms a gametangium which fuses with its mate to form a coenozygote, as in the Mucorales.

zygopteran. A damselfly, an odonatan of the suborder Zygoptera.

zygosphene. One of two articular processes on the neural arch of certain reptiles. *See* zygantra.

zygospore. A zygote resulting from the union of isogametes, especially one which forms a thick-walled resting spore, as in various phycomycetes.

zygote. 1. A fertilized egg or ovum. **2.** A diploid cell resulting from the fusion of two haploid gametes.

zygotene. A stage in mitosis following the leptotene stage in which the pairing of chromosomes occurs. Also called synapsis or syndesis.

zygotic. Of, pertaining to, or derived from a zygote.

zymase. An enzyme complex present in plants, especially yeasts, which catalyzes the decomposition of sugars with resultant production of alcohol and carbon dioxide, as occurs in fermentation.

zymogen. The precursor or inactive form of an enzyme, as pepsinogen which, upon activation, is converted into pepsin.

zymogenic. 1. Of, pertaining to, or causing fermentation. **2.** Of or pertaining to a zymogen.

zymology. The science dealing with fermentation; the study of enzymes and their actions; enzymology.

"A man of knowledge is free ... he has no honor, no dignity, no family, no home, no country, but only life to be lived!"

— don Juan